Animal Diversity, Natural History and Conservation
— Vol. 5 —

The Editors

Dr. Vijay Kumar Gupta, Ph.D., FLS, London (born 1953-) Chief Scientist, CSIR- Indian Institute of Integrative Medicine, Jammu, India. He did his M.Sc. (1975) and Ph.D. (1979) in Zoology both from University of Jammu, Jammu-India. His research capabilities are substantiated by his excellent work on histopathology, ecology and reproductive biology of fishes, turtles, birds and mammals, which has already got recognition in India and abroad.

Dr. Gupta has to his credit more than 100 scientific publications and review articles which have appeared in internationally recognized Indian and foreign journals. Founder fellow, life member and office bearer of many national societies, academies and associations. He has successfully completed a number of research/consultancy projects funded by government, private and multinational agencies. His current areas of interest are histopathology, toxicology, pre-clinical safety pharmacology, reproductive efficacy studies of laboratory animals and biodiversity.

He is the Series Editor of the recently published multi-volume set of books, "**Comprehensive Bioactive Natural Products (Vols. 1-8)"**, published by M/S Studium Press, LLC, USA. He is also Editor-in-Chief of the books, "**Utilisation and Management of Medicinal Plants (Vols. 1-3)"**, "**Medicinal Plants: Phytochemistry, Pharmacology and Therapeutics (Vols.1-4)"**, "**Traditional and Folk Herbal Medicine (Vols. 1-3)"**, "**Natural Products: Research Reviews (Vols. 1-4)"**, "**Bioactive Phytochemicals: Perspectives for Modern Medicine (Vols. 1-3)"**, "**Perspectives in Animal Ecology and Reproduction (Vols. 1-10)**" and **"Animal Diversity, Natural History & Conservation (Vols. 1-5)"**. The Editor-in-chief of the American Biographical Institute, USA, has appointed him as *Consulting Editor* of *The Contemporary Who's Who*. Dr. Gupta also appointed as Nominee for the *Committee for the Purpose of Control and Supervision of Experiments on Animals* (CPCSEA, Govt. of India). The *Linnaean Society of London, U.K.* has awarded fellowship to him in November 2009 in recognition of his contribution towards the cultivation of knowledge in Science of Natural History. Recently, Modern Scientific Press, USA has nominated Dr. Gupta as the Editor of the *International Journal of Traditional and Natural Medicine*.

Dr. Anil K. Verma, Ph.D., M.N.A.Sc., FLS, London (born 1963-) Prof. & Head, Department of Zoology, Govt. (P.G.) Degree College Rajouri, J&K State, did his M.Sc. in Zoology (1986) from University of Jammu, Jammu. He has undergone his M.Phil. (1988) and awarded first rank and Ph.D.(1993) in the field of animal reproduction at the same University. Dr. Verma has published about 60 research papers and review articles in reputed journals and books. He is also a member Editorial Board of the book series "Advances in Fish and Wildlife: Ecology and Biology" a Daya Publishing House, New Delhi. In recognition of his standing in greater scientific community, the Board of Directors of the American Association for the advancement of science (AAAS) New York, Washington, has awarded membership to him. The *Linnaean Society of London, U.K.* has awarded fellowship to him in October 2006 in recognition of his contribution towards the cultivation of knowledge of Science of Natural History.

Dr. Verma was also nominated as Nominee, CPCSEA by Ministry of Environment & Forests, Govt. of India, and also awarded "The Rashtriya Gourav Award" and "The Best Citizen of India" award by the India International Friendship Society in 2009-2010. He is also a nominated member of the Global Academy of Science, India.

Animal Diversity, Natural History and Conservation
— Vol. 5 —

— Editor-in-Chief —
Dr. V.K. Gupta
Formerly Chief Scientist
CSIR-Indian Institute of Integrative Medicine,
Canal Road, Jammu – 180 001, India

Editor
Dr. Anil K. Verma
Professor and Head
Department of Zoology, Government (P.G.) Degree College,
Rajouri, Jammu – 180 001, India

2015
Daya Publishing House®
A Division of
Astral International Pvt. Ltd.
New Delhi – 110 002

Cataloging in Publication Data--DK

Courtesy: D.K. Agencies (P) Ltd. <docinfo@dkagencies.com>

Animal diversity, natural history and conservation / editor-in-chief, Dr. V.K. Gupta ; editor, Dr. Anil K. Verma.
  volume 5 cm
  Includes bibliographical references and index.
  ISBN 9789351306610 (International Edition)

  1. Animal diversity conservation--India. 2. Animal diversity conservation. 3. Biodiversity conservation--India. 4. Biodiversity conservation. 5. Natural history--India. 6. Natural history.   I. Gupta, V. K. (Vijay Kumar), 1953-, editor. II. Verma, Anil K. (Anil Kumar), 1963-, editor.

  DDC 333.954160954   23

Published by   : **Daya Publishing House®**
                         A Division of
                         Astral International Pvt. Ltd.
                         – ISO 9001:2008 Certified Company –
                         4760-61/23, Ansari Road, Darya Ganj
                         New Delhi-110 002
                         Ph. 011-43549197, 23278134
                         E-mail: info@astralint.com
                         Website: www.astralint.com

Laser Typesetting   : **Classic Computer Services**, Delhi - 110 035

Printed at   : **Replika Press Pvt. Ltd.**

PRINTED IN INDIA

Editorial Board

Prof. Cosmas Nathanailides

Department of Aquaculture and Fisheries
TEI of EPIRUS, Igoumenitsa, GREECE
E-mail: cosmasfax@yahoo.com

Prof. A.K. Choubey

Nematology Laboratory,
Department of Zoology, Chaudhary Charan Singh University,
Meerut – 250 005, INDIA
E-mail: kumar_ac2001@rediffmail.com

Prof. D.N. Sahi

Department of Zoology,
University of Jammu ,
Jammu – 180 005, INDIA
E-mail: dnsahi@gmail.com

Prof. B.L. Koul

Ex-Principal
Govt. S.P.M.R.College of Commerce & Management,
Jammu – 180 001, INDIA
E-mail: proban@sancharnet.in

UNIVERSITY OF GHANA
DEPARTMENT OF ANIMAL BIOLOGY & CONSERVATION SCIENCE

Telegrams & Cables: UNIVERSITY, LEGON
Telex: 2556UGL GH
Telephone: 500381 Ext. 3277
  Mobile: 0278582307
E-mail: dabcs@ug.edu.gh

P. O. Box LG 67
Legon, Accra

Our Ref.............................

Your Ref.............................

Foreword

The Indo-Malayan ecozone is one of the 17 most diverse on Earth, and India represents an extraordinary high species concentration; for instance about 12, 8, 14 and 9 per cent of respectively all fish, reptilian, bird and mammal species can be found in India, albeit the country only constitutes little over 2 per cent of the global land area. With a standing human population of in excess of 1.2 b, India also faces unprecedented challenges with respect to biodiversity conservation and protected areas management. As one of the globally fastest growing emerging megaeconomies India has the potential to integrate national development with biodiversity conservation objectives, and a huge research base within the management of wildlife and natural resources shows important trends in that direction. In such a huge biogeographcal region basic knowledge on animal diversity, their natural history and conservation requirements are a prerequisite for safeguarding this outstanding scenario of species, habitats and landscapes, which include the highest elevations on the planet, and ranges from arid deserts to humid tropical forests and alpine biomes.

This Volume 5 of "*Animal Diversity, Natural History and Conservation*" continues the accounts of the previous four volumes, and contains varied and up-to-date basic ecological knowledge on both invertebrates and vertebrates, from both the aquatic and terrestrial biomes, including a thematic focus on the interrelationships between on one hand wildlife and habitats contra human exploitation and anthropogenic disturbances.

DABCS
UNIVERSITY OF GHANA
P. O. BOX 67, LEGON

Dr. Lars H. Holbech
Department of Animal Biology & Conservation Science
University of Ghana
P.O. Box LG 67, Legon, Accra, Ghana
E-mail: lholbech@ug.edu.gh

Preface

Today, one of the major areas of concern is the complete loss of some of the species, many of which are not known to us but are already threatened due to the destruction of their natural habitat. Conservation seeks to prevent complete elimination of a species for which there may be no alternatives. All organisms are linked in the food chain and interact with their abiotic environment in order to make the ecosystems self-sustaining units.

The disappearance of a link in the food chain upsets nature's balance, thereby creating problems and disrupting the overall ecological balance. It is, therefore, in our interest to conserve our resources. There is now a global realization about the urgent need to conserve the world's biological diversity.

Considering the immense values of biodiversity and subsequent rapid loss of the same, during the past couple of decades, enormous conservation efforts are being made for restoring the biodiversity of various habitats, through the on-site and off-site protection. The UNGA has decided to declare 2011-2020 as the "Decade on Biodiversity" which coincides with and also supports the implementation of the strategic plan for bio-diversity and there appears to be a basic need of humans to connect with nature and also with bio-diversity. It is in this back drop that it is felt that there should be a serious introspection for finding ways and means to suggest potential remedial measures to arrest the deterioration of the fragile ecological balance.

The earth has experienced and recovered from many natural distress in the past, but we now seem to be at a critical point, with humans facing responsibility for the future of the planet. All species, while evolving and adapting to the demands of their habitats, climatic and geological conditions, can change dramatically, subjecting environments and their ecologies to a dynamic swirl of physical forces.

Within the last 200 years, the human population has exploded, dominating many habitats at the cost of valuable natural systems, such as forests and wetlands, polluting air, water and land. At such, the resulting disruption of natural systems by human activity has revealed the fragility of the system threatening to destroy entire ecologies and render many species extinct.

In the present volume of the book series, an attempt has been made to encapsulate the scientific information pertaining to the modern day research in the field of animal diversity, biology and conservation. This effort is likely to serve as a catalyst for the development of innovations and approaches for future studies in the fields. The chapters have been contributed by the experts with exhaustive, relevant and update information. We are greatly indebted to the contributors who have whole heartedly support us in this endeavour.

We hope, that the book shall be a landmark and useful for an array of biologists, ecologists, naturalists and conservationists to a considerable extent.

Dr. V.K.Gupta, *FLS*

Contents

2015, **Animal Diversity, Natural History and Conservation, Vol. 5**   *Pages 1–13*
Editors: **V.K. Gupta and Anil K. Verma**
Published by: **DAYA PUBLISHING HOUSE, NEW DELHI**

Chapter 1

Review on Diversity and Distribution of Phytoparasitic Nematodes of Banana in West Bengal

Viswa Venkat Gantait*

Zoological Survey of India,
M-Block, New Alipur, Kolkata – 700 053, West Bengal, India

ABSTRACT

Banana (*Musa paradisiaca* L.) is one of the major fruit crops of tropics and sub-tropics forming a staple diet of millions across the globe. It provides a supplementary nutritious food and is commercially very important one. It is used as a green vegetable and possesses a great medicinal value too. The fruit is also associated with various social and religious activities of the people.

Nematodes constitute one of the major limiting factors in banana production. Phytoparasitic nematodes have been recognized as a major constraint in banana production and are responsible for serious yield losses mainly by causing root damages. They also serve as predisposing agents in the development of various disease complexes with fungi, bacteria and different viruses to this important fruit crop. About 154 species belonging to 51 genera are associated with banana through out the world. Five potentially plant pathogenic nematodes *viz., Radopholus similis, Helicotylenchus multicinctus, Meloidogyne incognita, Pratylenchus coffeae* and *Rotylenchulus reniformis* are globally very important. Out of these, the burrowing nematode, *Radopholus similis* is the most serious one causing heavy economic losses. *Helicotylenchus multicinctus* is the most widespread and abundant one after *Radopholus similis*. In India, as many as 71 species belonging to 33 genera of plant

* E-mail: v.gantait@rediffmail.com

parasitic nematodes have been reported associated with this crop. Forty one phytoparasitic nematode species of banana belonging to 18 genera of 11 families under the orders Dorylaimida and Tylenchida have been described and recorded so far from West Bengal. The diversity and distribution of those species in the state are being presented herein.

Keywords: *Phytoparasitic nematode, Diversity, Distribution, Banana, West Bengal.*

Introduction

Nematodes, the members of the phylum Nematoda represent one of the most abundant groups and probably the second largest one in the animal kingdom, immediately behind the arthropods. They are highly diverse in their habitats ranging from Himalayan peaks to sea floor, from Arctic to Antarctic (Lal, 1998). Based on the habitat, the total nematode population can be categorized into three groups: marine, animal-parasitic, soil and freshwater nematodes (Ayoub, 1980). The marine nematodes constitute about 50 per cent of the total nematode population, where as animal-parasitic nematodes make up only 15 per cent of the known nematode species. The soil and freshwater nematodes can be splitted into two finer divisions: free-living and plant-parasitic. The plant-parasitic nematodes constitute only 10 per cent of the total nematodes (Figure 1.1). Yet a plant, growing in soil would often be attacked by

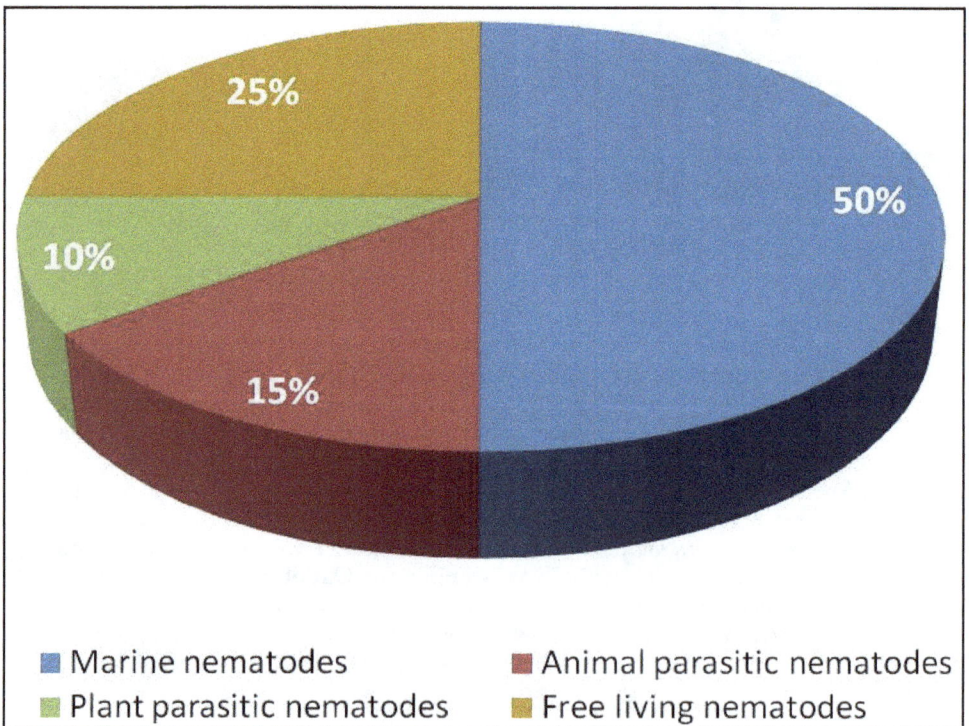

Figure 1.1: Pie Diagram of different Groups of Nematodes.

one or more species of phytoparasitic nematodes. The phytoparasitic nematodes have been known for their virulence causing significant losses to agricultural and horticultural crops.

Banana (*Musa paradisiaca* L.) is an economically important crop which has been extensively planted in tropical and subtropical countries of the world. It is considered as very important staple food globally. It is used as a green vegetable and thought to have great medicinal value too. It's also associated with various social and religious activities of the people. In India, banana is grown in an area of 332.2 thousand hectares with an annual production of 3633 thousand tones (Jonathan and Rajendran, 2003). In West Bengal, banana plantation covers 25.73 thousand hectares under different agro-climate zones; the total production being 502.11 thousand tones per annum. Among fruits, banana demand is only second to mango in this state.

Nematodes constitute one of the major limiting factors in banana production. Phytoparasitic nematodes have been recognized as a major constraint in banana production and are responsible for serious yield losses (Sundararaju, 2006). Beside direct root damages they serve as predisposing agents in the development of various disease complexes with fungi, bacteria and different viruses to the banana plantations (Khan, 2006). More than 150 species under 51 genera of phytoparasitic nematodes are associated with this crop through out the world (Gowen and Queneherve, 1990). Mukherjee *et al.* (1994) stated that five potentially plant pathogenic nematodes of banana *viz.*, *Radopholus similis* (Cobb, 1893) Thorne, 1949, *Helicotylenchus multicinctus* (Cobb, 1893) Golden, 1956, *Meloidogyne incognita* (Kofoid and White, 1919) Chitwood, 1949, *Pratylenchus coffeae* (Zimmermann, 1898) Filipjev and Schuurmans Stekhoven, 1941 and *Rotylenchulus reniformis* Linford and Oliveira, 1940 are globally very important. Out of these, the yield losses caused only by *R. similis* may be as high as 50 per cent (Speijer *et al.,* 1995; Sarah *et al.,* 1996; Speijer and Kajumba, 2000). *H. multicinctus* is the most widespread and abundant species after *R. similis* (Blake, 1972; Gowen and Queneherve, 1990).

In India, as many as 71 species belonging to 33 genera of phytoparasitic nematodes are known to be associated with banana (Devarajan *et al.,* 2003). Till to date 41 species belonging to 18 genera of 11 families under the orders Dorylaimida and Tylenchida have been described and recorded so far from West Bengal. The diversity of those species in the state is being presented in this article. The distribution chart (Table 1.1) of the species in different districts is also cited herein. It would provide a valuable data base to the banana producers as well as researchers regarding these noxious pests of this important fruit crop of the state, which may also in turn serve as a valuable tool for adapting control measures against these hidden enemies. The systematic position of the species is mentioned here following Jairajpuri and Ahmad (1982) for the order Dorylaimida and Siddiqi (2000) for Tylenchida.

SYSTEMATIC POSITION OF PPNs

Order: DORYLAIMIDA Pearse, 1942

Suborder: Dorylaimna Pearse, 1936

Superfamily: Longidoroidea Thorne, 1935

Family: Longidoridae Thorne, 1935

Subfamily: Longidorinae Thorne, 1935

Genus: *Paralongidorus* Siddiqi, Hooper and Khan, 1963

Species 1. *Paralongidorus citri* (Siddiqi, 1959) Siddiqi, Hooper and Khan, 1963

    Syn. *Xiphinema citri* Siddiqi, 1959

Family: Xiphinematidae Dalmasso, 1969

Subfamily: Xiphinematinae Dalmasso, 1969

Genus: *Xiphinema* Cobb, 1913

Species 2. *Xiphinema index* Thorne and Allen, 1950

Species 3. *X. insigne* Loos, 1949

Order: TYLENCHIDA Thorne, 1949

Suborder: Tylenchina Chitwood in Chitwood and Chitwood, 1950

Infraorder: Tylenchata Sidiqi, 2000

Superfamily: Tylenchoidea Örley, 1880

Family: Tylenchidae Örley, 1880

Subfamily: Tylenchinae Örley, 1880

Genus: *Polenchus* Andrássy, 1980

Species 4. *Polenchus shamimi* Baqri, 1991

Genus: *Sakia* Khan, 1964

Species 5. *Sakia indica* (Husain and Khan, 1965) Khan, Mathur, Nand and Prasad, 1968

    Syn. *Basiliophora indica* Husain and Khan, 1965

Subfamily: Pleurotylenchinae Andrassy, 1976

Genus: *Cephalenchus* Goodey, 1962 (Gereart, 1968)

Species 6: *Cephalenchus leptus* Siddiqi, 1963

Infraorder: Anguinata Siddiqi, 2000

Superfamily: Anguinoidea Nicoll, 1935

Family: Anguinidae Nicoll, 1935

Subfamily: Anguininae Nicoll, 1935

Genus: *Nothotylenchus* Thorne, 1941

Species 7. *Nothotylenchus hexaglyphus* Khan and Siddiqi, 1968

Suborder: Hoplolaimina Chizhov and Berezina, 1988

Superfamily: Hoplolaimoidea Filipjev, 1934 (Paramonov, 1967)

Family: Hoplolaimidae Filipjev, 1934 (Wieser, 1953)

Subfamily: Hoplolaiminae Filipjev, 1934

Genus: Hoplolaimus von Daday, 1905

Subgenus: *Basirolaimus* (Shamsi, 1979) Siddiqi, 2000

Species 8. *Hoplolaimus (Basirolaimus) columbus* Sher, 1963

   Syn. *Basirolaimus columbus* (Sher) Shamsi, 1979

Species 9. *H. (B.) indicus* Sher, 1963

   Syn. *Basirolaimus indicus* (Sher) Shamsi, 1979

Species 10. *H. (B.) seinhorsti* (Luc, 1958) Shamsi, 1979

   Syn. *H. seinhorsti* Luc, 1958

   Syn. *Basirolaimus seinhorsti* (Luc) Shamsi, 1979

Genus: *Scutellonema* Andrassy, 1958

Species 11: *Scutellonema siamense* Timm, 1965

Subfamily: Rotylenchoidinae Whitehead, 1958

Genus: *Helicotylenchus* Steiner, 1945

Species 12. *Helicotylenchus abunaamai* Siddiqi, 1972

Species 13. *H. crenacauda* Sher, 1966

Species 14. *H. dihystera* (Cobb, 1893) Sher, 1961

   Syn. *Tylenchus dihystera* Cobb, 1893

Species 15. *H. goodi* Tikyani, Khera and Bhatnagar, 1969

   Syn. *H. gratus* Patil and Khan, 1983

Species 16. *H. hydrophilus* Sher, 1966

Species 17. *H. incisus* Darekar and Khan, 1979

Species 18. *H. indicus* Siddiqi, 1963

   Syn. *H. plumariae* Khan and Basir, 1964

Species 19. *H. medinipurensis* Gantait, Bhattacharya and Chatterjee, 2007

Species 20. *H. microcephalus* Sher, 1966

   Syn. *H. belurensis* Singh and Khera, 1980

Species 21. *H. multicinctus* (Cobb, 1893) Golden, 1956

   Syn. *Tylenchus multicinctus* Cobb, 1893

Species 22. *H. wasimi* Gantait, Bhattacharya and Chatterjee, 2011

Genus: *Varotylus* Siddiqi, 1986

Species 23. *Varotylus siddiqii* Gantait, Bhattacharya and Chatterjee, 2011

Genus: *Rotylenchus* Filipjev, 1936

Subgenus: *Rotylenchus* Filipjev, 1936

Species 24. *Rotylenchus (Rotylenchus) alii* Maqbool and Shahina, 1986

Family: Rotylenchulidae Husain and Khan, 1967 (Husain, 1976)

Subfamily: Rotylenchulinae Husain and Khan, 1967

   Syn. Rotylenchulinae Allen and Sher, 1967

Genus: *Rotylenchulus* Linford and Oliveira, 1940

   Syn. Leiperotylenchus Das, 1960

Species 25. *Rotylenchulus reniformis* Linford and Oliveira, 1940

Family: Pratylenchidae Thorne, 1949 (Siddiqi, 1963)

Subfamily: Pratylenchinae Thorne, 1949

Genus: *Pratylenchus* Filipjev, 1936

Species 26. *Pratylenchus brachyurus* (Godfrey, 1929) Filipjev and Schuurmans Stekhoven, 1941

   Syn. *Tylenchus brachyurus* Godfrey, 1929

Species 27. *P. coffeae* (Zimmermann, 1898) Filipjev and Schuurmans Stekhoven, 1941

   Syn. *Tylenchus coffeae* Zimmermann, 1898

Species 28. *P. neglectus* (Rensch, 1924) Filipjev and Schuurmans Stekhoven, 1941

   Syn. *Aphelenchus neglectus* Rensch, 1924

   Syn. *P. similis* Khan and Singh, 1975

Subfamily: Hirschmanniellinae Fotedar and Handoo, 1978

Genus: *Hirschmanniella* Luc and Goodey, 1964

Species 29. *Hirschmanniella gracilis* (De man, 1880) Luc and Goodey, 1964

   Syn. *Tylenchus gracilis* De man, 1880

Species 30. *H. mannai* Gantait, Bhattacharya and Chatterjee, 2007

Species 31. *H. mucronata* (Das, 1960) Khan, Siddiqi, Khan, Hassan and Saxena, 1964

   Syn. *Radopholus mucronata* Das, 1960

Family: Meloidogynidae Skarbilovich, 1959 (Wouts, 1973)

Subfamily: Meloidogyninae Skarbilovich, 1959

Genus: *Meloidogyne* Goeldi, 1892

Species 32. *Meloidogyne incognita* (Kofoid and White, 1919) Chitwood, 1949

   Syn. *Oxyuris incognita* Kofoid and White, 1919

Species 33. *M. javanica* (Treub, 1885) Chitwood, 1949

   Syn. *Heterodera javanica* Treub, 1885

Superfamily: Dolichodoroidea Chitwood in Chitwood and Chitwood, 1950 (Siddiqi, 1986)

Family: Telotylenchidae Siddiqi, 1960

Subfamily: Telotylenchinae Siddiqi, 1960

Genus: *Tylenchorhynchus* Cobb, 1913

Species 34. *Tylenchorhynchus coffeae* Siddiqi and Basir, 1959

Species 35. *T. crassicaudatus* Williams, 1960

Species 36. *T. leviterminalis* Siddiqi, Mukherjee and Dasgupta, 1982

Species 37. *T. mashhoodi* Siddiqi and Basir, 1959

   Syn. *Macrorhynchus mashhoodi* (Siddiqi and Basir) Sultan, Singh and Sashuja, 1991

Species 38. *T. nudus* Allen, 1955

   Syn. *Macrorhynchus nudus* (Allen) Sultan, Singh and Sashuja, 1991

Species 39. *T. zeae* Sethi and Swarup, 1968

   Syn. *Macrorhynchus zeae* (Sethi and Swarup) Sultan, Singh and Sashuja, 1991

Suborder: Criconematina Siddiqi, 1980

Superfamily: Criconematoidea Taylor, 1936 (1914) (Geraert, 1966)

Family: Criconematidae Taylor, 1936 (1914) Thorne, 1949

Subfamily: Hemicriconemoidinae Andrassy, 1979

Genus: *Hemicriconemoides* Chitwood and Birchfield, 1957

Species 40. *Hemicriconemoides mangiferae* Siddiqi, 1961

   Syn. *Hemicriconemoides litchi* Edward and Mishra, 1964

Hemicriconemoides birchfieldi Edward, Mishra and Singh, 1964

Superfamily: Hemicycliophoroidea Skarbilovitch, 1959 (Siddiqi, 1980)

Family: Caloosiidae Siddiqi, 1980

Subfamily: Caloosiinae Siddiqi, 1980

Genus: *Caloosia* Siddiqi and Goodey, 1964

Species 41. *Caloosiia parlona* Khan, Chawla and Saha, 1979

Discussion

   The review of nematology literature reveals that few surveys have been carried out by different workers in the banana growing districts of West Bengal like Hooghly, Nadia, Bardhaman, Birbhum, Murshidabad, Paschim Medinipur, even in Darjeeling, Jalpaiguri and Coochbehar; as a result 5 species belonging to 4 genera have been described and 36 species under 14 genera have been reported so far from the state. Three species *viz. Tylenchorhynchus zeae, Xiphinema index* and *Helicotylenchus indicus* were reported by Mukhopadhyaya and Haque in 1974. A survey was carried out in

Table 1.1: District-wise Distribution Chart of the Species

Sl.No.	Species	Family	District	Reference
1.	*Paralongidorus citri*	**Longidoridae**	Jalpaiguri	Jana and Baqri, 1984
2.	*Xiphinema index*	**Xiphinematidae**	Hooghly, Bardhaman	Mukhopadhyaya and Haque, 1974; Khan and Hassan, 2010
3.	*X. insigne*		Hooghly, Bardhaman	Mukherjee and Dasgupta, 1983; Khan and Hassan, 2010
4.	*Polenchus shamimi*	**Tylenchidae**	Paschim Medinipur	Gantait, Bhattacharya and Chatterjee, 2010
5.	*Sakia indica*		Hooghly,	Mukherjee and Dasgupta, 1983; Khan and Hassan, 2010
6.	*Cephalenchus leptus*		Hooghly	Mukherjee and Dasgupta, 1983
7.	*Nothotylenchus hexaglyphus*	**Anguinidae**	Paschim Medinipur	Gantait, Bhattacharya and Chatterjee, 2010
8.	*Hoplolaimus (Basirolaimus) columbus*	**Hoplolaimidae**	Hooghly	Mukherjee and Dasgupta, 1983
9.	*H. (B.) indicus*		Hooghly, Paschim Medinipur	Mukherjee and Dasgupta, 1983; Gantait, Bhattacharya and Chatterjee, 2010
10.	*H. (B.) seinhorsti*		Nadia, Hooghly	Mukherjee and Dasgupta, 1981; Khan and Hassan, 2010
11.	*Scutellonema siamense*		Hooghly, Jalpaiguri	Mukherjee and Dasgupta, 1983; Khan and Hassan, 2010
12.	*Helicotylenchus abunaamai*		Nadia, Hooghly, 24 Parganas North, Purba Medinipur, Paschim Medinipur	Khan and Hassan, 2010
13.	*H. crenacauda*		Paschim Medinipur	Gantait, Bhattacharya and Chatterjee, 2010
14.	*H. dihystera*		Darjeeling, Jalpaigui, Coochbehar, Birbhum, Paschim Medinipur, Nadia, Hooghly	Baqri and Ahmad, 1983; Chatterjee, 1998; Khan and Hassan, 2010
15.	*H. goodi*		Nadia	Khan and Hassan, 2010
16.	*H. hydrophilus*		Paschim Medinipur	Gantait, Bhattacharya and Chatterjee, 2010
17.	*H. incisus*		Nadia	Khan and Hassan, 2010

Contd...

Table 1.1–*Contd...*

Sl.No.	Species	Family	District	Reference
18.	*H. indicus*		Hooghly, Nadia, Bardhaman, Birbhum, 24 Parganas North, Purba Medinipur, Paschim Medinipur, Murshidabad, Coochbehar	Mukhopadhyaya and Haque, 1974; Mukherjee and Dasgupta, 1983; Khan and Hassan, 2010
19.	**H. medinipurensis*		Paschim Medinipur	Gantait, Bhattacharya and Chatterjee, 2007
20.	*H. microcephalus*		Nadia, Hooghly, 24 Parganas North	Khan and Hassan, 2010
21.	*H. multicinctus*		Hooghly, Nadia, Purba Medinipur, Paschim Medinipur, Jalpaiguri	Mukherjee and Dasgupta, 1983; Khan and Hassan, 2010
22.	**H. wasimi*		Paschim Medinipur	Gantait, Bhattacharya and Chatterjee, 2010
23.	**Varotylus siddiqii*		Paschim Medinipur	Gantait, Bhattacharya and Chatterjee, 2011
24.	*Rotylenchus (Rotylenchus) alii*		Paschim Medinipur	Gantait, Bhattacharya and Chatterjee, 2010
25.	*Rotylenchulus reniformis*	**Rotylenchulidae**	Hooghly, Bardhaman, Birbhum, Purba Medinipur, Paschim Medinipur, Murshidabad, Jalpaiguri	Mukherjee and Dasgupta, 1981; Mukherjee and Dasgupta, 1983; Khan and Hassan, 2010; Gantait, Bhattacharya and Chatterjee, 2010
26.	*Pratylenchus brachyurus*	**Pratylenchidae**	Hooghly	Mukherjee and Dasgupta, 1983; Khan and Hassan, 2010
27.	*P. coffeae*		Hooghly, Paschim Medinipur, Bardhaman, 24 Parganas North	Mukherjee and Dasgupta, 1983; Khan and Hassan, 2010; Gantait, Bhattacharya and Chatterjee, 2010
28.	*P. neglectus*		Nadia	Khan and Hassan, 2010
29.	*Hirschmanniella gracilis*		Paschim Medinipur	Gantait, Bhattacharya and Chatterjee, 2010
30.	**H. mannai*		Paschim Medinipur	Gantait, Bhattacharya and Chatterjee, 2007
31.	*H. mucronata*		Nadia	Khan and Hassan, 2010

Contd...

Table **1.1**–*Contd...*

Sl.No.	Species	Family	District	Reference
32.	*Meloidogyne incognita*	**Meloidogynidae**	Hooghly, Paschim Medinipur	Mukherjee and Dasgupta, 1983; Gantait, Bhattacharya and Chatterjee, 2010
33.	*M. javanica*		Nadia, Hooghly	Khan and Hassan, 2010
34.	*Tylenchorhynchus coffeae*	**Telotylenchidae**	Hooghly, Paschim Medinipur, Bardhaman, Murshidabad	Mukherjee and Dasgupta, 1983; Khan and Hassan, 2010; Gantait, Bhattacharya and Chatterjee, 2010
35.	**T. crassicaudatus*		Hooghly	Siddiqi, Mukherjee and Dasgupta, 1982; Mukherjee and Dasgupta, 1983
36.	*T. leviterminalis*		Hooghly	Khan and Hassan, 2010
37.	*T. mashhoodi*		Paschim Medinipur, Nadia, 24 Parganas North, Birbhum	Khan and Hassan, 2010; Gantait, Bhattacharya and Chatterjee, 2010
38.	*T. nudus*		Nadia, Bardhaman	Khan and Hassan, 2010
39.	*T. zeae*		Hooghly, Bardhaman	Mukhopadhyaya and Haque, 1974; Khan and Hassan, 2010
40.	*Hemicriconemoides mangifrae*	**Criconematidae**	Hooghly	Mukherjee and Dasgupta, 1983; Chatterjee, 1998
41.	*Caloosia parlona*	**Calooslidae**		Mukherjee and Dasgupta, 1981

* Described from West Bengal.

different agro-ecological tracts of the state from July, 1978 to December, 1979 by Mukherjee and Dasgupta in 1981. They reported 3 species *viz. Hoplolaimus* (*Basirolaimus*) *seinhorsti* (most abundant species), *Rotylenchulus reniformis* and *Caloosia parlona* associated with banana. Siddiqi, Mukherjee and Dasgupta (1982) described *Tylenchorhynchus leviterminalis* from Chandan Nagar of the district Hooghly. Mukherjee and Dasgupta (1983) collected soil and root samples from banana plantations of different age groups in five locations (Chandan Nagar, Chinsurah, Hooghly, Mankundu and Adisaptagram) in the Hooghly district during October-November, 1980. They reported 17 species belonging to 13 genera of which *Tylenchorhynchus coffeae* was the most abundant in most of the plantations. A survey was undertaken by Mukherjee and Dasgupta (1983) in four different agro-climatic zones of the state during 1978-1981 and reported 21 species under 13 genera of plant parasitic nematodes. Baqri and Ahmad (1983) reported *Helicotylenchus dihystera* from Phoolsering, Shaktigarh, Matigarh (Darjeeling district), Collegepara (Jalpaiguri) and Dakshingwrihati (Coochbehar). *Paralongidorus citri* was reported by Jana and Baqri (1984) from New Jalpaiguri of Jalpaiguri district. Dasgupta, Sen, Mukherjee and Rama (1985) reported 21 species belonging to 10 genera from different localities of the state. Chatterjee (1998) reported *Hemicriconemoides mangiferae* from Coochbehar district and *Helicotylenchus dihystera* from Darjeeling and Birbhum districts. Gantait, Bhattacharya and Chatterjee (2010) had done an extensive survey in all the 29 blocks of Paschim Medinipur district during March 2004 to February 2006 and reported 17 species of phytoparasitic nematodes belonging to 11 genera under the order Tylenchida, out of which 4 species are new to science. Khan and Hassan (2010) presented the distribution of 32 species in the important banana growing districts of the state like Nadia, Hooghly, Bardhman, North 24-Parganas, Birbhum, Bankura, Purba and Paschim Medinipur, Murshidabad, Coochbehar and Jalpaiguri and concluded that in West Bengal the banana roots are prone to attack from the most serious nematode species *viz. Pratylenchus coffeae, P. brachyurus, Helicotylenchus multicinctus* and *Meloidogynae incognita*. Only few works had been done to study the diversity and distribution of soil and plant parasitic nematodes of banana in West Bengal. Extensive surveys have to be needed regarding this matter which will ultimately be helpful to identify more species associated with this valuable fruit crop.

References

Ayoub, S.M. 1980. *Plant Nematology: An Agricultural Training Aid*. Nema Aid Publication, 195 p.

Baqri, Q.H., and Ahmad, N. 1983. Nematodes from West Bengal (India) XVI. On the species of the genus *Helicotylenchus* Steiner, 1945 (Hoplolaimidae:Tylenchida). *Journal of Zoological Society of India*, **35** (1 and 2): 29-48.

Blake, C.D. 1972. Nematode diseases of banana plantations. In: *Economic Nematology,* (Ed.) J. M. Webster. Academic Press, New York, pp. 245-267.

Chatterjee, A. 1998. Plant parasitic nematodes (Tylenchida: Nematoda). *State Fauna Series* **3**, *Fauna of West Bengal, Part II*, Zoological Survey of India: 297-339.

Dasgupta, M.K., Sen, K., Mukherjee, B., and Rama, K. 1985. Plant pathogenic nematodes associated with horticultural crops in West Bengal. In: *Horticulture in West Bengal,* (Ed.) D. Mukherjee, pp. 83-93.

Devarajan, K., Rajendran, G., and Seenivasan, N. 2003. Nutrient status and photosynthetic efficiency of banana (*Musa* sp.) influenced by *Meloidogyne incognita* infected with *Pasteuria penetrans. NematologiaMediterranea,* **31**: 197-200.

Gantait, V.V., Bhattacharya, T., and Chatterjee, A. 2010. *Nematodes of Banana.* Lambert Academic Publishing, Germany. 300 p.

Gantait, V.V., Bhattacharya, T., and Chatterjee, A. 2007. Two new species of plant parasitic nematodes associated with banana plantations from West Bengal, India. *Journal of Environment and Sociobiology,* **4** (2): 139-147.

Gantait, V.V., Bhattacharya, T., and Chatterjee, A. 2010. One new and three known species of the genus *Helicotylenchus* Steiner, 1945 associated with banana from West Bengal, India. *Records of Zoological survey of India,* **110** (part-3): 7-12.

Gantait, V.V., Bhattacharya, T., and Chatterjee, A. 2011. Description of *Varotylus jairajpurii,* new species (Tylenchida: Hoplolaimidae: Hoplolaiminae) from Paschim Medinipur, West Bengal, India with key to its world species. *Pakistan Journal of Zoology,* **43** (4): 1- 4.

Gowen, S.R., and Queneherve, P. 1990. Nematode parasites of bananas, plantains and abaca. In: *Plant Parasitic Nematodes in Subtropical and Tropical Agriculture,* (Eds.) M. Luc, R. A. Sikora and J. Bridge. CAB International, Wallinford, England, p. 431-460.

Jairajpuri, M.S., and Ahmad, W. 1992. *Dorylaimida: Free-living, Predaceous and Plant Parasitic Nematodes.* Oxford and IBH Publishing Co. Pvt. Ltd., New Delhi, India, 458 pp.

Jana, A., and Baqri, Q. H. 1984. Nematodes from West Bengal (India) XIV. On the occurrence of ectoparasitic nematodes of Longidoroidea and Trichodoroidea (Dorylaimida), with remarks on the validity of genus *Siddiqia* (Longidoridae). *Bulletin of Zoological Survey of India,* **6** (1-3): 75-79.

Jonathan, E.I., and Rajendran, G. 2003. Spatial distribution of root-knot nematode, *Meloidogune incognita* in Banana, *Musa* sp. *Indian Journal of Nematology,* **33** (1): 47-51.

Khan, M.R., and Hassan, M.A. 2010. Nematode diversity in banana rhizosphere from West Bengal, India. *Journal of Plant Protection Research,* **50** (3): 263-268.

Khan, M.R., 2006. Current options for manging nematode pest of crops in India. In: *Plant Nematology in India,* (Eds.) N. Mohilal and R. K. Gambhir. Parasitology Laboratory, Department of Life Sciences, Manipur University, Manipur, India, pp. 16-50.

Lal, A. 1998. Application of computers in nematode identification. In: *Recent Advances in Plant Nematology* (Ed.) P. C. Trivedi. CBS Publishers and Distributors, Daryaganj, New Delhi, pp. 107-114.

Mukherjee, B., and Dasgupta, M.K. 1981. Soil and plant nematodes of West Bengal, India. *Indian Journal of Nematology*, **11** (1): 72.

Mukherjee, B., and Dasgupta, M.K. 1983. Community analysis of nematodes associated with banana plantations in the Hooghly District, West Bengal, India. *Nematologia Mediterranea*, **11**: 43-48.

Mukherjee, B., and Dasgupta, M.K. 1983. Occurrence and distribution of plant parasitic nematodes in West Bengal. *Madras Agricultural Journal*, **70** (10): 650-655.

Mukherjee, B., Nath, R.C., and Dasgupta, M.K. 1994. Abundance and distribution of plant parasitic nematodes associated with banana plantations in Tripura. *Current Nematology*, **5** (2): 117-126.

Mukhopadhyay, M.C., and Haque, M.S. 1974. Nematodes associated with field crops, fruits and fodder crops in West Bengal. *Indian Journal of Nematology*, **4**: 104-107.

Sarah, J.L., Pinochet, J., and Stanton, J. 1996. The burrowing nematode of bananas, *Radopholus similis* Cobb. *Musa pest fact sheet no. 1.* Montpellier, France, INIBAP, p. 1.

Siddiqi, M.R. 2000. *Tylenchida: Parasites of Plants and Insects, 2nd Edition*. CABI Publishing, CAB International, Wallingford, UK, 833 p.

Siddiqi, M.R., Mukherjee, B., and Dasgupta, M.K. 1982. *Tylenchorhynchus microconus* n. sp., *T. crassicaudatus leviterminalis* n. sub. sp., and *T. coffeae* Siddiqi and Basir, 1959 (Nematoda: Tylenchida). *Systematic Parasitology*, **4**: 257-262.

Speijer, P.R., and Kajumba, C. 2000. Yield loss from plant parasitic nematodes in East African highland banana (*Musa* spp. AAA). *Acta Horticulturae*, **540**: 453-459.

Speijer, P.R., Gold, C. S., Kajumba, C., and Karamura, E.B. 1995. Nematode infestation of 'clean' banana planting materials in farmer's fields in Uganda. *Nematologica*, **41**: 344.

Sundararaju, P. 2006. Community structure of plant parasitic nematodes in banana plantations of Andhra Pradesh, India. *Indian Journal of Nematology*, **36** (2): 226-229.

2015, Animal Diversity, Natural History and Conservation, Vol. 5   *Pages 15–40*
Editors: V.K. Gupta and Anil K. Verma
Published by: DAYA PUBLISHING HOUSE, NEW DELHI

Chapter 2

Community Ecology and Taxonomy of Polychaetes from a Tropical Estuary, South West Coast of India

P.N. Geetha, S. Bijoy Nandan* and T.A. Thasneem

*Department of Marine Biology, Microbiology and Biochemistry,
School of Marine Sciences, Cochin University of Science and Technology,
Cochin – 682 016, India*

ABSTRACT

Polychaetes belonging to the phylum Annelida, commonly called "bristle worms" or "rag worms" are cosmopolitan in distribution with high bioturbation efficiency contributing significantly to the benthic food chain. They are mostly raptorial feeders but some are scavengers, deposit feeders or selective or filter feeders and are grouped into errant or sedentary burrowing in mud or living in tubes made by them. This contribution presents the community ecology and taxonomy of polychaetes from Cochin estuary in Kerala. Polychaetes contributed 38 per cent of macrobenthos in the present study, representing 39 species belonging to 13 families *Pulliella armata* belonging to the family Capitellidae were the dominant species among polychaetes followed by *Capitella capitata, Prionospio polybranchiata* belonging to family Spionidae, *Dendroneries estuarina* southern to Nereidae and *Cossura coasta* to Cossuridae. The capitellids dominated in the middle and northern zone of the estuary. The polychaete assemblages were characteristically different in the three zones of the estuary. The diversity increased in the middle region of the estuary. The numerical abundance of deposit feeders (capitellids) was highest in this region of the estuary with higher salinity and percentage of silt and clay. Dominance of deposit feeders with presence of opportunistic species indicated the negative impacts of anthropogenic activity in the estuary.

Keywords: Benthos, Polychaetes, Taxonomy, Cochin estuary.

---

* *Corresponding author.* E-mail: bijoynandan@yahoo.co.in

Introduction

Polychaetes are multi-segmented worms that live in all environments in the world's oceans from abyssal depths to shallow estuaries, rocky shores and in open water (Read and Fauchald, 2013). Polychaetes exhibit a wide range of feeding habits and play an important role in marine and estuarine food chains, because they exhibit a key role in a various trophic levels as a food for small invertebrates to large wading birds. Polychaetes show many different reproductive modes (Wilson, 1991; Giangrande, 1997) and they belong to 20 different trophic categories (Fauchald and Jumars, 1979). They can also be used as markers of different ecological conditions (Gambi and Giangrande, 1986). Recent classification of polychaetes based on 124 characters of 80 accepted families does not follow the Linnaean categories. Based on this classification, polychaetes are divided into two clades, Scolecida and Palpata (Rouse and Fauchald, 1997; Rouse and Pleijel, 2001).

Worldwide, polychaetes contributes a large proportion of the total macrofauna in soft bottoms (Hutchison, 1998), with more than 16,000 species known so far, that are placed fourth in ranking marine invertebrate species richness (Blake, 1995; Bouchet, 2000). Globally, approximately 12620 species of polychaetes have been reported and in India, around 1093 species representing 8.66 per cent of the total number of polychaetes species that are known (Kumaraswamy *et al.*, 2005). The study of polychaetes have been conducted by different workers in India, initially by Willy (1908), Southern (1921), Gravely (1927), Fauvel (1932 and 1953), Ganapathy and Radhakrishnan (1958), Thampi and Rangarajan (1964), Cheriyan (1966), Achari (1969 and 1972), Misra *et al.* (1984) and Misra (1995). Fauvel (1953), recorded 283 species of marine and estuarine polychaetes from different parts of India, among these 47 species were estuarine. Hartman (1974) prepared a bibliography of polychaetes from India that included 59 families, 315 genera and 860 species. Misra (1998) reported 167 polychaete species belonging to 38 families from different brackish water bodies in India. A total of 30 species belonging to 25 genera and 17 families of polychaetes from Cochin estuary were described by Pillai (2001). Polychaete diversity in Indian estuaries was studied by Ajmal Khan and Murugesan (2005) and they recorded 153 species of polychaetes from India representing about 37.46 per cent of the total polychaetes present in Indian estuaries.

Loss or reduction of native species diversity, coupled with changing environmental conditions can push ecosystems beyond critical thresholds and drastically alter community structure, ecological functioning, and provisioning of services as has been seen in coastal –soft-bottom ecosystems (Myers, 2007). Maintaining of species diversity, composition and functional redundancy is essential for sustaining productive and resilient ecosystems and provision of several ecosystem services (Worm *et al.*, 2006; Stachowicz *et al.*, 2007; Palumbi *et al.*, 2009). Species diversity- the variety and abundance of species within an area or ecosystem – tends to be positively correlated with ecosystem health by increasing the functioning of marine ecosystems (Duffy, 2009). More diverse assemblages support greater ecosystem function (Naeem, 2006). Functional diversity measures the variety of types of organisms that serve different functional roles within a community irrespective of their taxonomic grouping (Steneck and Dethier, 1994; Petchey and Gaston, 2006).

The first procedure in ecological research is exercise in systematics (Hedgpeth, 1957). To this we need a better understanding of the origin, reasons, diversity of organisms and its role in community level in an estuarine ecosystem functioning. Estuarine biodiversity is changing rapidly due to the natural and anthropogenic interventions. This change alters the estuarine ecosystem, especially the trophic status. Habitat diversity is often used as a proxy for species diversity (Mumby *et al.,* 1999) for management and conservation planning purposes (Mumby *et al.,* 2008). The main objective of this paper understands the community ecology and taxonomy of polychaetes in Cochin estuary.

The ecological health of Cochin estuarine system has degraded due to various developmental projects, and anthropogenic interventions. The pollution increased due to the impact of dredging by Rasheed and Balchand (1995, 1997, and 1999), waste disposal (Menon *et al.,* 2000; Martin *et al.,* 2008 and 2011) and metal pollution (Anu *et al.,* 2014). The reduction of polychaete diversity and increase of deposit feeders denotes the deterioration of the estuarine health (Martin *et al.,* 2008 and 2011; Akhil *et al.,* 2014; Bijoy Nandan *et al.,* 2014). This may affect the fishery potential and food web of the estuary. Therefore, this contribution discusses the composition, abundance and community structure of macrobenthic polychaetes from Cochin estuary in relation to the changing environmental characteristics based on a time scale.

Materials and Methods

Study Area

The Cochin estuary is a part of a long chain of lakes and canals extending from 9°40'12" and 10°10'46" N and 76°09' 52" and 76°23' 57" E with its northern boundary at Azheekode and the southern boundary at Thanneermukkam Bund. Its length varies from 60–80 km, width from 500 and 4000 meters and depth range of 2-7 meters. The estuary is connected to the Arabian Sea through a wide channel of about 450m width at Cochin and another at Azhikode. This ecosystem has two seasonal openings into the Arabian Sea- one at Paravur (North) and the second at Andhakaranazhi and a permanent opening at Cochin. Seven major rivers (Chalakkudy, Periyar, Muvattupuzha, Meenachil, Manimala, Pamba and Achencoil) discharge freshwater into the Cochin estuary. The map of the study stations in Cochin estuary is given in Figure 2.1.

The nine sampling stations were, Station 1 (Barmouth), Station 2 (Fishing Harbour), Station 3 (Thevara), Station 4 (Marine Science Jetty), Station 5 (Vallarpadam – Bolgatty), Station 6 (Chitoor), Station 7 (Moolampilly- Pazhala), Station 8 (Tatatpilly) and Station 9 (FACT Eloor), to collect and analyze different parameters in bottom water, sediment and macrobenthos. The selected study stations in the estuary could be classified into three zones, corresponding to the location from the mouth of the estuary. The northern zone was represented by Stations 6, 7, 8 and 9; southern zone was represented by Stations 2 and 3; middle zone was represented by Stations 1, 4 and 5.

Sampling

Field sampling was undertaken for twenty four months from June 2009 to May 2011 from nine selected stations of Cochin estuary to collect and analyze water,

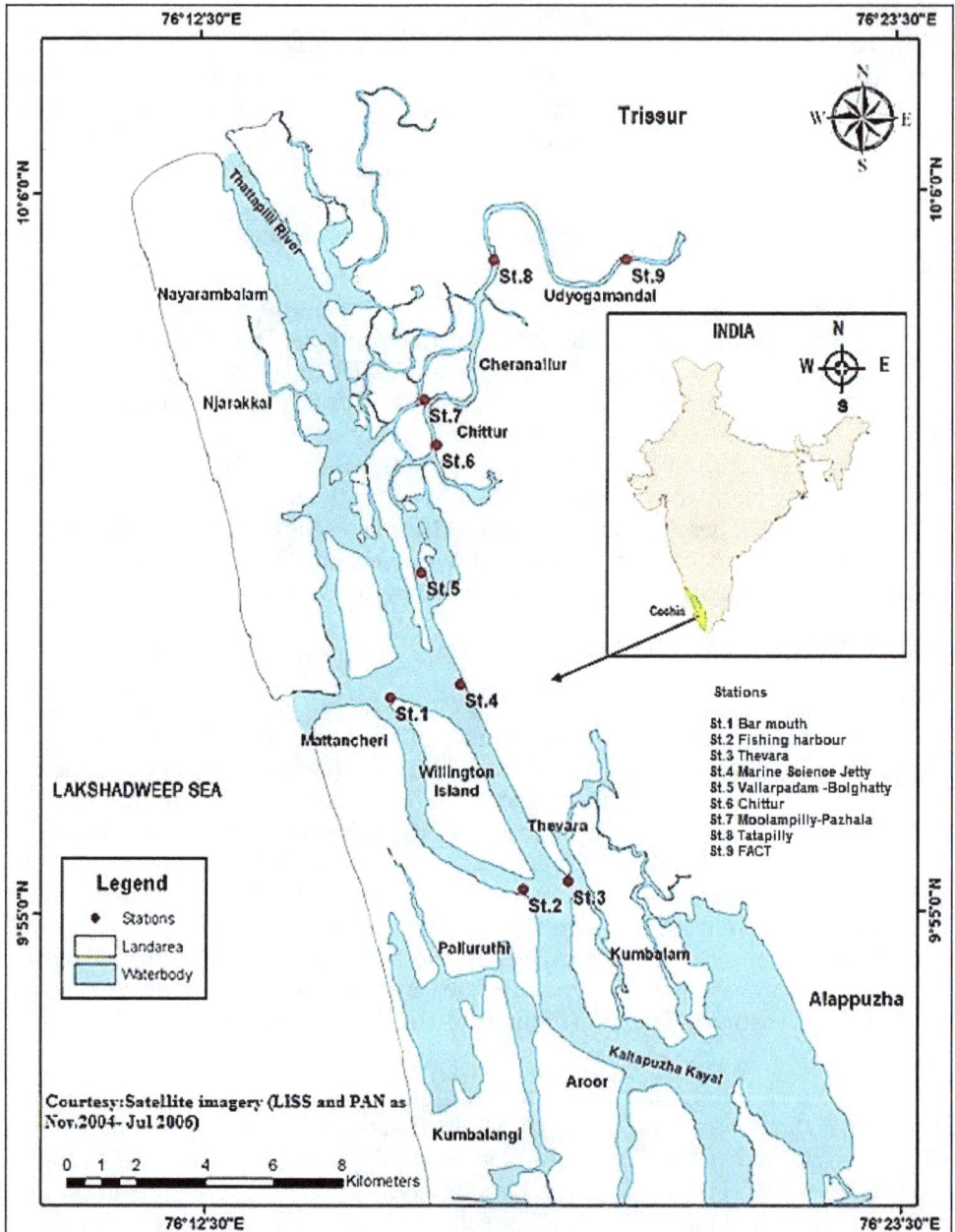

Figure 2.1: Map of Cochin Estuary Indicating the Study Stations.

sediment and macrobenthos. All sampling was invariably carried out during the early morning hours using a motorized boat and bottom water samples were collected using a 2L Niskin sampler. The temperature of the bottom water and sediment was determined in field by a standard degree centigrade thermometer, pH with pH meter

(Systronics Model no: 132; APHA, 2005) and salinity by Mohr-Knudsen method (Strickland and Parsons, 1972; Grasshoff *et al.*, 1983). Sediment organic carbon was estimated by Walkey and Black (1934) method (Jackson, 1973) and the percentage of sand, silt and clay was estimated by Sucell particle size analyzer - SYMPATEC H70010 (Germany, 2000).

Macro fauna samples were collected using a standard Van Veen grab having an area of 0.04 m² (Anastasios Eleftheriou and Alasdair McIntyre 2005; Holme and McIntyre, 1971). The samples were preserved in 4 per cent formalin in plastic bottles. The preserved samples were washed through suitable sieves of mesh size 500 µm for macro fauna, those that are retained in the sieve were collected in a labeled plastic container and preserved in 4 per cent formalin; stained in Rose Bengal for identification (Holme and McIntyre, 1971; Merrit and Cummins 1996; Anastasios Eleftheriou and Alasdair McIntyre, 2005). The Rose Bengal dye at strength of 0.1 per cent selectivity colored all the living organisms in the sample (Claudiu *et al.*, 1979; Zabbey, 2002; Idowu and Ugwumba, 2005).

Taxonomy of Polychaetes

The washed and preserved benthic invertebrates were poured into a white enamel tray and sorted. The sorting was made effective by adding moderate volume of water into the container to improve visibility. The benthos was picked using forceps and organisms were sorted into their different groups into plastic vials and preserved in 5 per cent formalin. The preserved animals were later identified to their lowest taxonomic group under light and stereo dissecting microscope and counted. The dissected portions were analyzed using the compound microscope- Leica-DM 500. The standard as well as published references were employed for identification of different fauna. Taxonomic and morphological studies were conducted using standard literature, monographs and keys (Fauvel, 1953; Day 1967 a, b; Gosner, 1971; Pennak, 1978; Glasby and Fauchald, 2003; Pettibone, 1982; Böggemann, 2002; Sol, 2007).

Data Analysis

Community structure analysis was done by PRIMER v6 software package developed at the Plymouth Marine Laboratory (Clarke and Warwick, 1994, 1998; Clarke and Gorley, 2001). The Ecological indices, the Shannon-Weiner index (H′) (Shannon and Weaver, 1963) diversity index and Pielou evenness index (J) (Pielou, 1969) were calculated. The species were ranked in terms of abundance done by Geoplot (x^2geometric abundance class plot) and Dominance plot (Gray and Pearson, 1982). Cluster analysis was done to find out the similarities between groups. The most commonly used clustering technique is the hierarchical agglomerative method. Bray – Curtis coefficient (Bray and Curtis, 1957) was used to produce the dendrogram.

Results and Discussion

Environmental Parameters

The annual depth ranged from 0.10m to 8.50m during the period 2009-10 and 0.80m to 7m during 2010-11. Bottom water temperature ranged from 25°C to 33.1°C and sediment from 23°C to 33°C at all stations during the study period. Mean station wise sediment temperature was minimum in St.1 (27.8°C) and maximum in St.9 (28.9°C). The annual mean temperature varied from 27°C to 28.2°C during 2009-10 and 28.5°C to 29.9°C during 2010-11. Salinity in the bottom water depicted wide variations among the stations of the estuary and ranged from 0.01 to 34.70ppt during the study period. Mean seasonal salinity varied from a minimum 3.05 ppt during monsoon 2010 to a maximum of 11.4 ppt during post monsoon in 2009-10. Sediment pH varied from 3.03 to 10.48 in all stations. The mean seasonal pH varied between from 6.68 during pre-monsoon 2011 to 7.7 during monsoon 2009-10. The sediment pH in the northern limb of the estuary was slightly acidic nature in all months and at the same time in the southern limb it was slightly alkaline in nature. The mean station wise organic carbon was found to be minimum (1.07 per cent) in St.5 and maximum (2.81 per cent) in St.8. Mean monthly variation of organic carbon ranged from 1.30 per cent in May 2011 to 2.67 per cent in February 2010 (Figures 2.2 to 2.7).

Figures 2.2–2.7: Bottom Water and Sediment Characteristics of Cochin Estuary during 2009-11 Period.

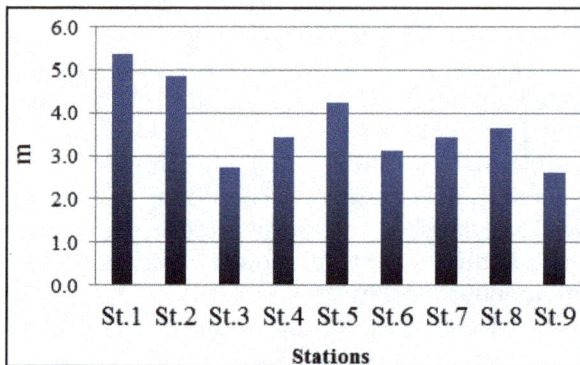

Figure 2.2: Mean Station-wise Depth Variation in Cochin Estuary during June 2009–May 2011.

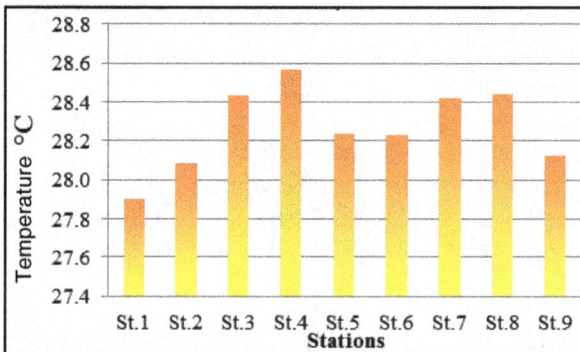

Figure 2.3: Mean Station-wise Bottom Water Temperature in Cochin Estuary during June 2009–May 2011.

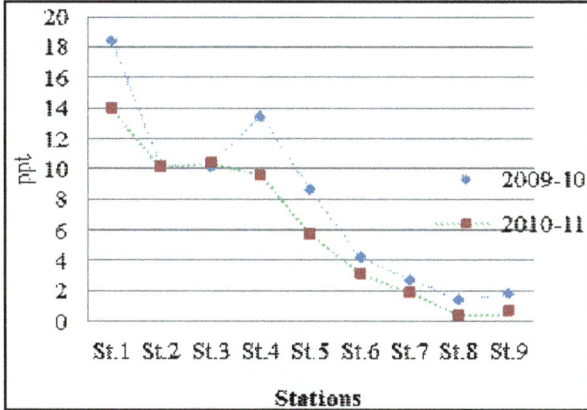

Figure 2.4: Mean Annual Variation of Salinity in Cochin Estuary during 2009–2011.

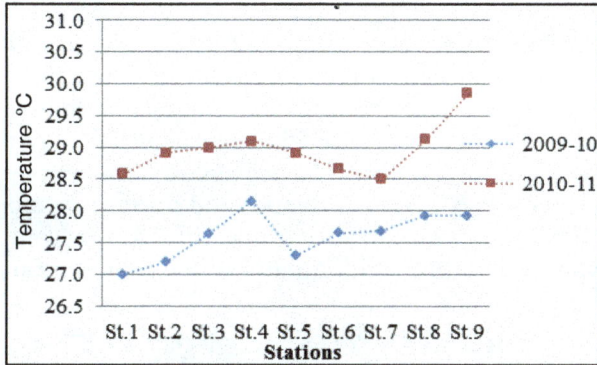

Figure 2.5: Mean Annual Variation of Sediment Temperature in Cochin Estuary during June 2009– May 2011.

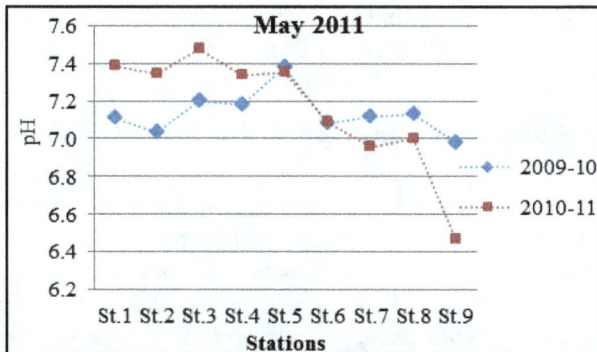

Figure 2.6: Mean Annual Variation of Sediment pH in Cochin Estuary during June 2009– May 2011.

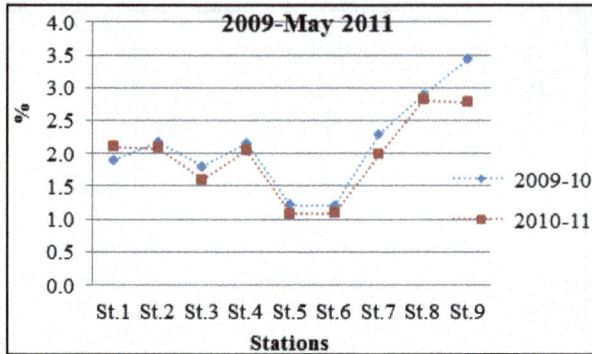

Figure 2.7: Mean Annual Variation of Organic Carbon in Cochin Estuary during June 2009– May 2011.

The sediment in Cochin estuary showed a remarkable variation in composition. The mean annual percentage of clay, silt and sand observed in the estuary was 28.22 per cent, 35.67 per cent and 15.65 per cent during 2009-10 and 27.8 per cent, 35.6 per cent and 15.0 per cent respectively during 2010-11 periods. The silt was the major fraction in all seasons in the estuary followed by clay and sand. Generally an increase of silt content may be due to the continuous dredging and mining activities in the estuary. The seasonal sediment composition revealed that clay was high during monsoon and post monsoon season (92.08 per cent to 92.16 per cent) and silt during monsoon (70.67 per cent to 85.25 per cent) and sand during pre- monsoon (69.25 per cent to 78.25 per cent). The annual mean variation of sediment composition showed no remarkable variation in distribution pattern in Cochin estuary (Figures 2.8 to 2.13).

Sediment was characterized by 22.68 per cent clay, 43.96 per cent silt and 33.36 per cent sand in the northern zone followed by 64 per cent clay, 33.83 per cent silt and 1.39 per cent sand in the middle zone and 79.12 per cent clay, 20.65 per cent silt and 0.23 per cent sand in the southern zone. This indicated that clay was predominant in the southern zone, clayey silt in the middle zone and silty sand in northern zones. The riverine regions of the northern stations were sandy or silty-clay. Martin *et al.* (2011); Sheeba (2000) reported that clay was dominant in the southern zone and middle zone of the Cochin estuary. But in the present study it was found that the sediment in the middle zone was silty clayey.

Composition, Abundance and Distribution of Polychaetes

Thirty nine species of polychaetes belonging to 29 genera and 16 families were recorded from nine stations in Cochin estuary during the study period. The distribution of different polychaete species varied considerably from station to station. In St.1, Bar mouth, the dominant polychaete species was *Cossura coasta* followed by *Prionospio pinnata, Prionospio polybranchiata, Prionospio cirrifera, Ancistrosyllis constricta* Southern, *Diopatra neapolitana, Boccardia polybranchia* and *Nephthys oligobranchia* Southern whereas *Notomastus favueli* was the subdominant species. In St.2, Fishing harbour, *Prionospio* sp., *Prionospio polybranchiata, Capitella capitata* and *Pulliella armata* were the dominant species and *Parahetromastus tenuis* Monro, *Notomastus favueli* were the

Figures 2.8–2.13: Mean Seasonal Variation of Sediment Grain Size in Cochin Estuary during 2009-11 Period.

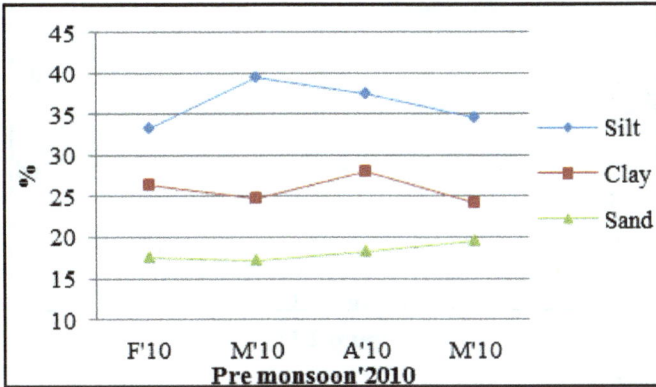

Figure 2.8

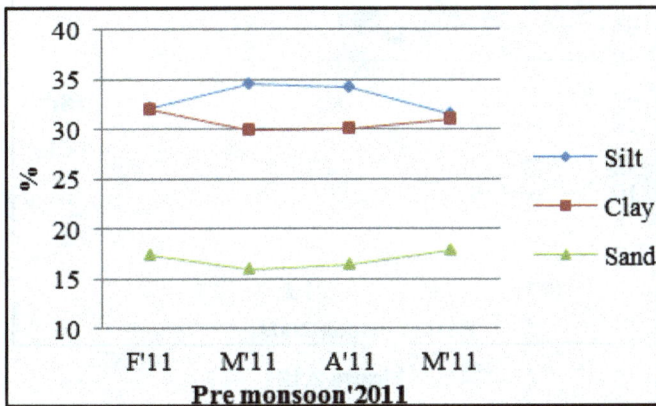

Figure 2.9

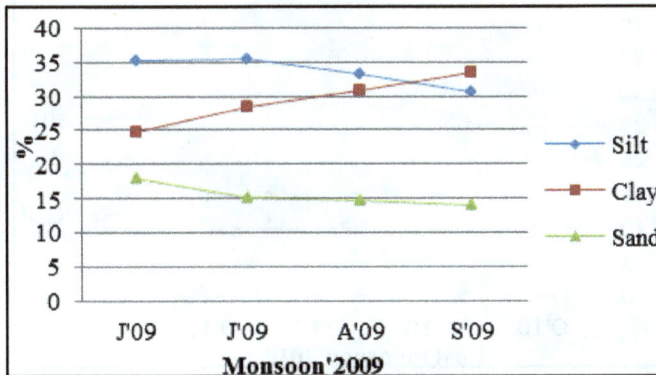

Figure 2.10

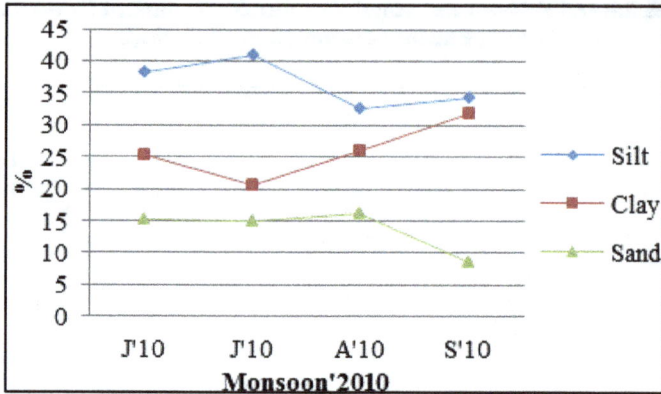

Figure 2.11

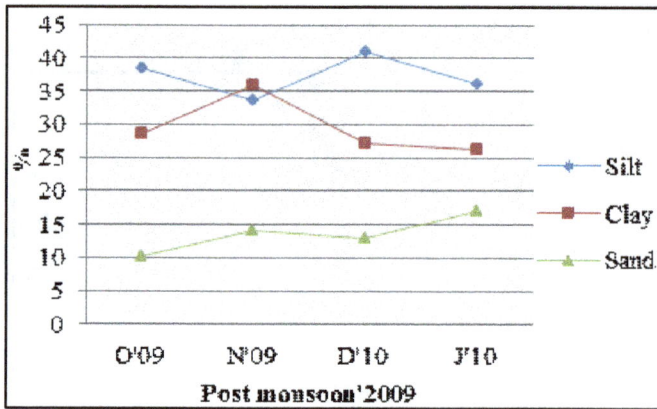

Figure 2.12

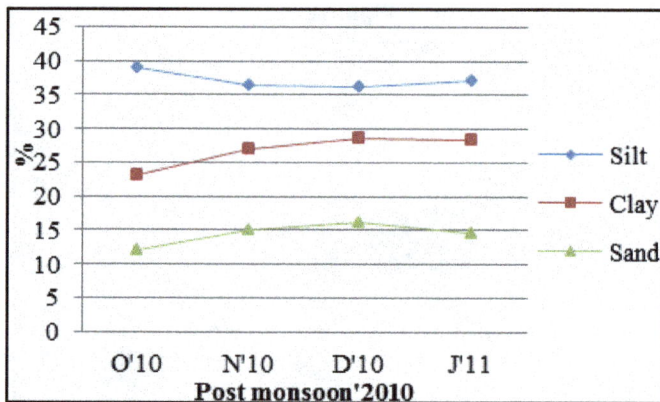

Figure 2.13

subdominant species. In St.3, Thevara, *Parahetromastus tenuis* Monro, *Capitella capitata*, *Lycastis indica* Southern, *Nephthys polybranchia* Southern, *Dendroneries estuarina* Southern, *Cossura coasta, Nephthys dibranchiata* were the dominant species followed by *Spio filicornis, Notomastus favueli*, and *Ancistrosylllis parva* Day were the subdominant species. In St.4, *Cossura coasta, Pulliella armata, Prionospio polybranchiata, Ancistrosylllis parva* Day, *Nephthys oligobracnhia* Southern, *Nephthys polybranchia* Southern the dominant species and *Nephthys dibranchiata* were the subdominant species. Among polychaetes, the abundance of *Capitellids* were increased in the middle zone of estuary (Stations 4 and 5) and in northern zone (Station 6) during the study period and the representative species were *Capitella capitata* and *Pulliella armata* were dominant in these stations. Besides *Notomastus favueli, Lumbriconeries simplex* Southern and *Spio filicornis* were the subdominant species. In Stations 6 and 7, *Pulliella armata* was the dominant species followed by *Capitella capitata* and *Dendronereis estuarina* Southern, that were the subdominant species. In Station 8, *Capitella capitata* was the dominant species and *Lycastis indica* Southern and *Dendronereis* sp. was the dominant species in Station 9. Fifteen top ten abundant species of polychaetes in the estuary are given in Table 2.1.

Table 2.1: Top Fifteen Polychaetes Species in the Order of Abundance in Cochin estuary during 2009-11 Period

Sl.No.	Species	Mean (no/m²)	Rank
1.	*Spio filicornis*	117	XV
2.	*Prionospio cirrifera*	115	XIV
3.	*Prionospio pinnata*	249	XI
4.	*Prionospio polybranchiata*	653	IV
5.	*Cossura coasta*	520	VI
6.	*Notomastus favueli*	300	IX
7.	*Parahetromastus tenuis* Monro	234	XII
8.	*Capitella capitata*	1259	II
9.	*Pulliella armata*	1363	I
10.	*Ancistrosyllis parva* Day	321	VIII
11.	*Dendronereis* sp.	602	V
12.	*Dendronereis estuarina* Southern	693	III
13.	*Lycastis indica* Southern	161	XIII
14.	*Nephthys oligobracnhia* Southern	403	VII
15.	*Nephthys dibranchiata*	275	X

Kurian *et al.* (1975) and Pillai (1977) worked on macrobenthos of Cochin estuary and observed that the bottom fauna were mainly composed of polychaetes, crustaceans and molluscs. From their study polychaetes were the dominant community and major species identified were, A. *constricta, Nephthys ologobranchia* Southern, *Prionospio pinnata, Prionospio polybranchiata* and *Parahetromastus tenuis* (Capitellid) that were abundant during the study period. Sarala Devi (1986) reported that *P. polybranchiata* was a selective deposit feeder and an indicator of organic pollution and resilient to industrial pollution. Pillai (1977) studied the macrobenthos of Cochin estuary and reported that the *Prionospio polybranchiata* was the dominant polychaete species in Cochin estuary.

The numerical abundance of polychaetes varied in different stations (Figure 2.14). The abundance of polychaetes was minimum (190 no/m²) in St.8 and maximum (1312 nom²) in St.6. Of these, polychaete species, *Pulliella armata* (1363 no/m²) followed by *Capitalla capitata* (1259no/m²), *Dendronereis aestuarina* Southern (693no/m²), *Prionospio polybranchiata* (653no/m²), *Dendronereis* sp. (602no/m²) and *Cossura coasta* (520no/m²) were common and most abundant in the estuary. *Sabellaria cementorium* (45no/m²·) was the least abundant species in the study. The mean percentage abundance of polychaete family showed that the family Capitellidae was dominant followed by Spionidae, Nereidae and Nephthyidae in Cochin estuary during the study period (Figure 2.15). *Pulliella armata, Capitella capitata, Notomastus favueli* and *Parahetromastus tenuis* Monro were the representatives of family Capitellidae.

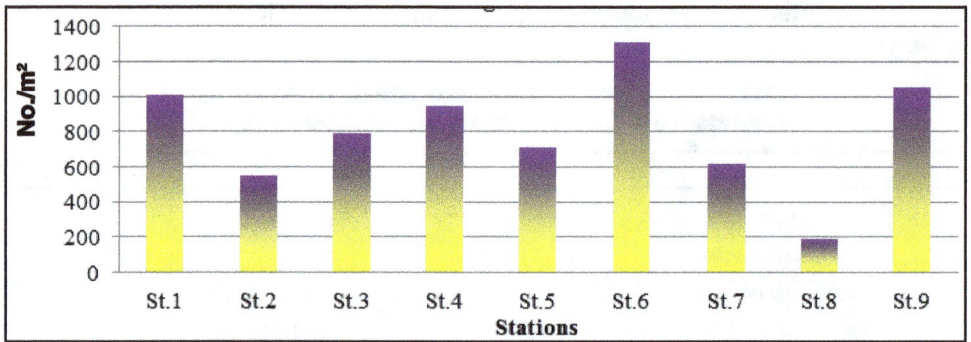

Figure 2.14: Mean Numerical Abundance of Polychaetes in Cochin Estuary during 2009–2011.

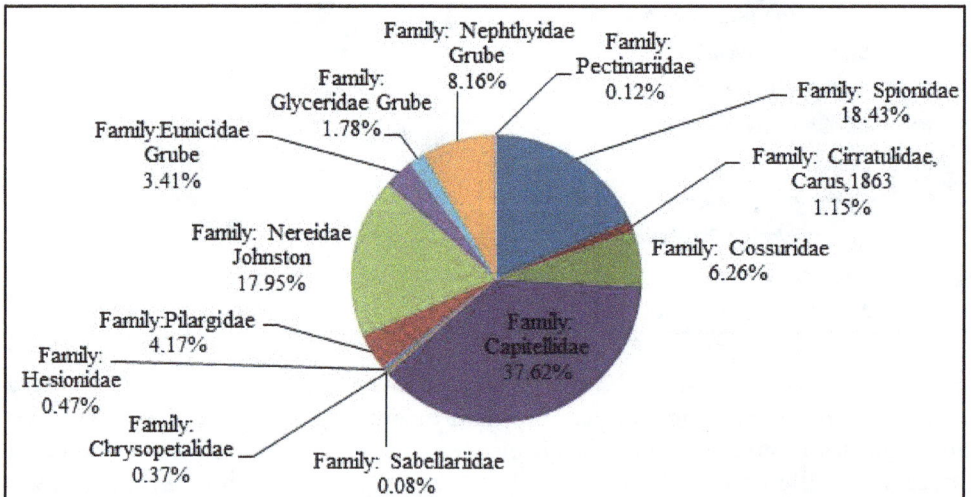

Figure 2.15: Mean Percentange Abundance of Major Polychaetes Families in Cochin Estuary during 2009–2011.

In the present study, the polychaetes showed a remarkable variation in occurrence and distribution in different stations and zones of the Cochin estuary. The occurrence of different polychaetes varied from St. 1 to St. 9 in Cochin estuary. The distributions of polychaetes are given in (Table 2.2).

Table 2.2: Distribution of Polychaetes in Cochin Estuary during 2009- 2011 Period

Polychaetes	St.1	St.2	St.3	St.4	St.5	St.6	St.7	St.8	St.9
Scololepis squamata	+	–	++	–	–	–	–	–	–
Spio filicornis	+	–	++	–	++	++	–	–	–
Paraprionospio sp.	+	+	–	–	–	–	–	–	–
Boccardia polybranchia	++	+	+	–	–	–	–	–	–
Malacoceros indicus	–	–	–	–	–	+	–	–	–
Polydora antennata Claparedae	–	–	–	–	–	–	–	–	–
Polydora armata	+	–	–	–	–	+	–	–	–
Prionospio cirrifera	++	+	+	++	+	+	–	–	–
Prionospio multibranchiata	+	–	–	–	+	–	–	–	–
Prionospio pinnata	+++	++	+	+	+	+	+	+	–
Prionospio polybranchiata	+++	+++	+++	+++	+	++	+	+	–
Prionospio saldhana Day	+	+	–	–	–	++	–	–	–
Cirratulus chrysoderma Claparede	+	+	–	++	+	–	–	–	–
Cossura coasta	+++	++	+	+++	+	++	+	+	+
Notomastus favueli	+	++	+	+	++	++	+	+	–
Parahetromastus tenuis Monro	–	–	+++	+	–	–	+	–	–
Capitella capitata	+	++	+++	++	+++	+++	+++	++	–
Pulliella armata	+	++	++	+++	+++	+++	+++	+	+
Sabellaria cementarium Moore	–	–	+	–	–	–	–	–	–
Paleanotus debilis	+	+	+	+	–	–	–	–	–
Ophiodromus angustifrons	+	+	+	–	+	+	–	–	–
Ancistrosyllis constricta Southern	++	+	–	–	+	–	–	–	–
Ancistrosylllis parva Day	++	++	++	+ +	+	+	–	+	+
Neries glandicincta Southern	+	+	+	+	+	+	+	–	–
Creatonereis sp.	++	+	+	+	–	–	–	–	–
Dendronereis sp.	++	+	+	+	++	+++	++	++	++
Dendroneriana estuarina Southern	–	+	++	+	++	+++	++	++	+
Lycastis indica Southern	+	+	+	+	++	++	+	++	++
Marphysa sanguinnea Montagu	–	+	+	–	–	+	–	–	–
Epidiopatara gilchristi Day	+	–	–	+	–	–	–	–	–
Diopatra Neapoliatana Delle Chiage	++	–	–	+	++	–	–	–	–
Lumbriconeries simplex Southern	+	+	+	++	+	+	–	–	–

Contd...

Table 2.2–*Contd...*

Polychaetes	St.1	St.2	St.3	St.4	St.5	St.6	St.7	St.8	St.9
Glycera alba var cochinensis	+	+	+	+	+	+	+	–	–
Glycera longipinnis Grube	+	+	+	–	++	–	–	–	–
Goniada agnesiae Fauvel.	+	+	+	+	+	+	–	–	–
Nephthys oligobracnhia Southern	++	+	++	++	+	++	+	+	–
Nephthys polybranchia Southern	+	+	–	++	–	–	–	–	–
Nephthys dibranchiata	+	++	++	++	+	+	+	+	–
Pectinaria sp.	+	–	+	+	–	–	–	–	–

+++: Highly abundant; ++: Moderately abundant; +: Occasional; –: Not found

Seasonal Abundance

In St.1, numerical abundance were minimum (23 no/m^2) in November 2009 (post monsoon 2009) and maximum (3632no/m^2) in September (monsoon 2010). Polychaete community showed a seasonal reduction in numerical abundance during post monsoon (653no/m^2) and increased in number during pre-monsoon (950no/m^2) in 2009-10 period. Whereas, during 2010-11 period, polychaetes observed were minimum during post monsoon (425no/m^2) and maximum during monsoon (1094no/m^2) (Figure 2.16).

The monsoonal rainfall has an important role in regulating benthic fauna in tropical estuaries. The faunal abundance decreased during the monsoon followed by substantial increase in post- monsoon indicating faster recovery of bottom fauna with stabilized estuarine conditions (Panickar, 1969; Paruleker *et al.*, 1980). Seasonally polychaete fauna showed a minimum abundance during monsoon and a maximum abundance during the pre-monsoon period (Pillai, 1977). In the present study also the numerical abundance increased during monsoon. This may be due to the increase of detritus feeders in the estuary due to the availability of large deposition of organic

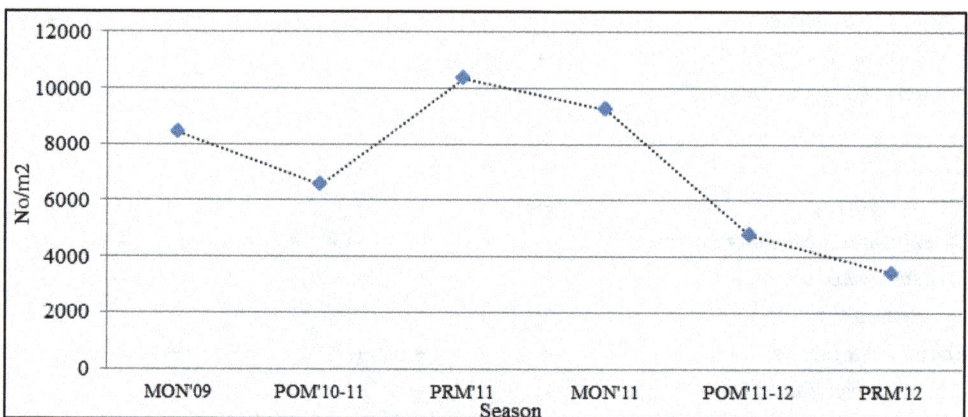

Figure 2.16: Mean Seasonal Abundance of Polychaetes in Cochin Estuary during 2009-11 Period.

content from river runoff during monsoon. Other possible reason is that the reproduction in polychaetes increased by the absence or decline in monsoon rainfall. The high salinity and food availability from organic detritus deposited in the estuarine bottom by rainfall may help to increase the reproduction in polychaetes.

Several studies reported that the rainfall pattern shift and climate change may affect the abundance of organisms living in estuaries and coastal waters. The reduction in abundance of benthos directly or indirectly could be related to the present climate change scenario. These shifts are likely to bring about new assemblages of species (Williams and Jackson, 2007), sometimes leading to interspecific interactions or abundance that increased in particular species or family and in worst case scenarios results in some extinctions (Barnosky *et al.,* 2011) and sometimes leads to invasion of new members.

Community Structure

The diversity indices of polychaetes were moderate in all stations during the study period. Compared to three zones of the estuary the diversity indices were highest in the middle zone. During 2009-11 period, the highest mean value of richness index (d) was 5.6774, evenness index (J') was 0.9487, diversity (H') index was 3.4511 and dominance (Lambda'1) was (0.9644), noted in the estuary (Figure 2.17). The more diversified polychaete fauna occurred in bar mouth of the estuary. Similar observations were reported by Desai and Krishnankutty (1967) and Kurian (1973), where they observed the higher abundance of bottom fauna in stations near bar mouth than in the interior part of the estuary due to higher salinity and a favorable substratum with rich supply of nutrients. The geometric class plot is useful to measure the effects of pollution on biota, which plot the number of species that fall into a set of geometric (x^2) abundance classes as a means of detecting the effects of pollution stress to the biota. Geometric class plot of mean abundance of polychaetes displayed a tremendous irregularity (jagged) with the curve becoming zigzag during June 2009-May 2011 periods signifying that the stations were relatively polluted (Figure 2.18). The dominance plot of polychaetes showed that the *Pulliella armata* was the most dominant species among polychaetes. Second abundant species was *Capitella capitata* followed by *Dendroneries estuarina* Southern, *Prionospio polybranchiata* and *Cossura*

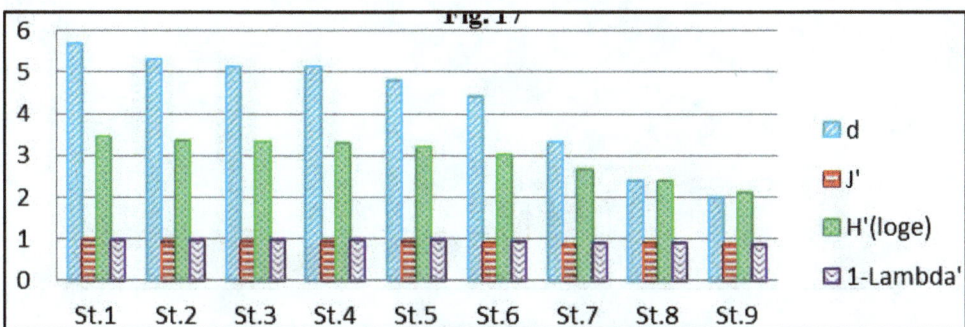

Figure 2.17: Diversity Indices of Polychaetes in Cochin Estuary during 2009-2011 Period.

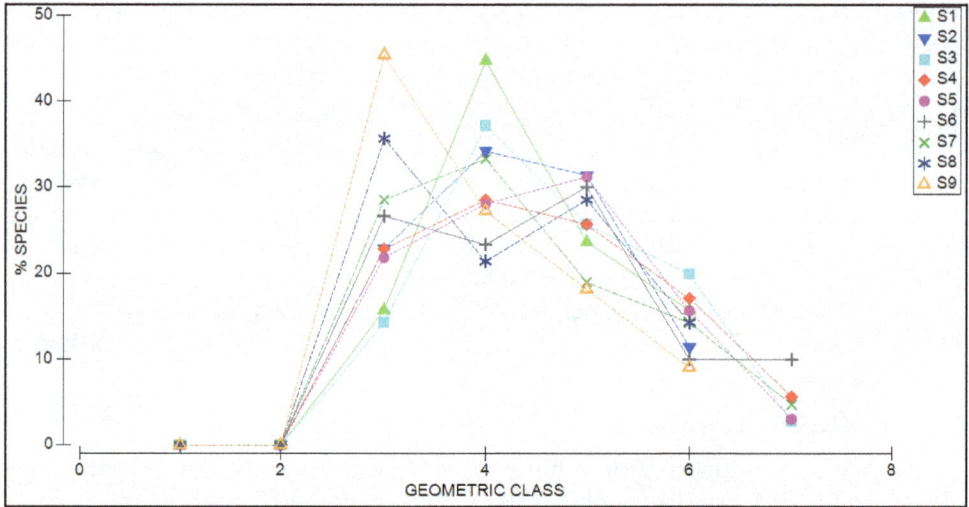

Figure 2.18: Geometric Class Plot of Mean Polychaetes in Cochin Estuary during 2009-2011 Period.

coasta. Higher abundance of capitellids indicates the dominance of deposit feeders among the polychaetes in Cochin estuary due to organic pollution and industrial activity leading to decreased diversity in most of the stations (Figure 2.19). Sarala Devi, (1986) reported that *D. estuarina* Southern was a pollution resistant species observed from the northern limb of the Cochin estuary. Tube dwelling polychaete, *Diopatra neapolitana* was abundant in the bar mouth area of Cochin estuary. A similar observation was made by Desai and Krishnankutty, (1967); Pillai, (1977); Batcha, (1984) from Cochin estuary. But in the present study, it was found to be *Cossura coasta*

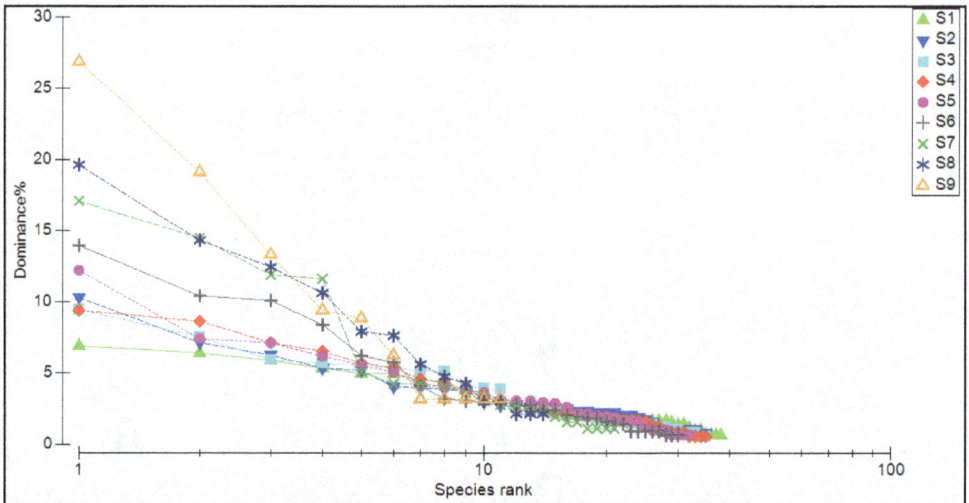

Figure 2.19: Dominance Plot of Mean Polychaetes in Cochin Estuary during 2009-11 Period.

that was the dominant species in bar mouth area. It may due to the alteration in environmental condition in the area. At the time of sand dredging the harmful effects include changes in grain size, bathymetry and lethal stress that may lead to alteration of community in a disturbed area and possibly lead to preventing the decolonization of the original community (Drucker *et al.,* 1996 and Hobbs, 2002).

Dendrogram of similarity showed that 39 species of polychaetes were formed in 7 major clusters in Cochin estuary during June 2009 - May 2011. However, these species were arranged as 4 groups. They were different from all other species. The polychaetes, *Pulliella armata, Capitella capitata, Paraheteromastus tenuis* and *Dendroneries* sp. formed in a separate cluster. *Sabellaria cementarium* Moore was the species that was not similar to other species. Twenty six species formed another cluster (Figure 2.20).

Trophic Structure

Feeding guilds and functional groups play an important role in understanding benthic communities (Gallagher, 2008). Feeding guild of polychaetes were first applied by Jumars and Fauchald (1977). The feeding guild of any organism may be defined as the set of relations among food particle size and composition, the mechanism involved in food-intake, and the motility patterns associated with feeding (Fauchald and Jumars, 1979). They explained that polychaetes have two modes of feeding (macrophagy and microphagy), five submodes (herbivores, carnivores, filter-feeders, surface deposit-feeders, and burrowers), and a total of a dozen morphological subgroups.

Based on the present study details, the feeding guild of polychaetes in Cochin estuary belonged to five groups that were composed of carnivores and surface deposit feeders and suspension feeders. Terborgh and Robinson (1986) defined guilds as species deriving sustenance from shared resources and argued that guilds could be used to compare the functional similarity of communities that shared no species.

Among the polychaetes, sub- surface deposit feeders (SSDF), surface deposit feeders (SDF) and carnivores (CVR) were the major groups observed in the present study in Cochin estuary. A minor group of suspension feeders with scavengers were also identified. Polychaetes belonging to Capitellidae are sub surface-deposit feeders (SSDF) and their higher abundance denotes the deterioration in estuarine health. They increased in the stations having higher silt and sand fractions. The Spionidae are sub surface deposit feeders; Nereidae, Glyceridae and Pilargidae that are major carnivore feeders and Sabellariidae were suspension feeders. The feeding guild of polychaetes in Cochin estuary is given in Table 2.3.

The present study confirmed the increase of capitellids in the middle and northern zone of the estuary. Capitellids can tolerate a broad range of salinity and their higher abundance in such environment was reported by Guelorget and Perthuisot, (1992) and Gray *et al.* (2002). The capitellids were abundant in all stations of the estuary. In stations (Sts. 1 and 4) near bar mouth area, *Cossura coasta* was the dominant species. Ajao and Fagade, (1990) reported that, *Capitella capitata, Nereis* sp. and *Polydora* sp. were found associated with sites that were grossly polluted with

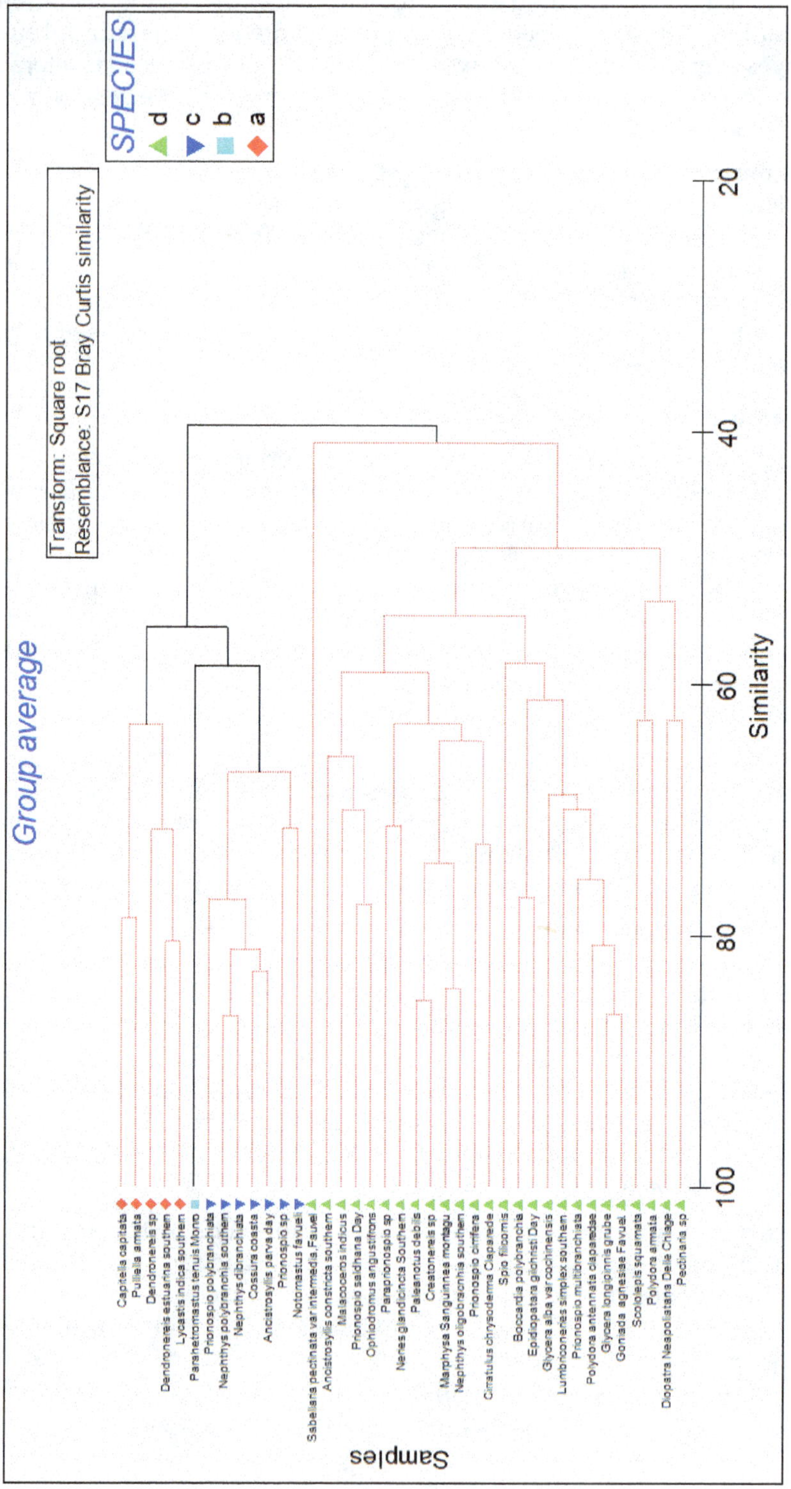

Figure 2.20: Dendrogram Showing the Bray-Curtis Similarity Index of Polychaetes in Cochin Estuary during 2009–11.

**Table 2.3: Feeding Guild of Polychaete Families in
Cochin estuary during 2009-11 Period**

Family	Feeding Guild
Pilargidae	Carnivore (CVR)
Hesionidae	Carnivore (CVR)
Nereidae	Carnivore (CVR)
Nephtyidae	Carnivore (CVR)
Glyceridae	Carnivore (CVR)
Goniadidae	Carnivore (CVR)
Eunicidae	Carnivore (CVR)
Lumbrinereidae	Carnivore (CVR)
Spionidae	Surface deposit feeder (SDF)
Cirratulidae	Surface deposit feeder (SDF)
Cossuridae	Sub-surface deposit feeder (SSDF)
Chrysopetalidae	Scavangers
Capitellidae	Sub-surface deposit feeder (SSDF)
Pectinariidae	Surface deposit feeder (SDF)
Sabellariidae	Suspension feeder (SF)

organic matter, heavy metals and petroleum hydrocarbons. *Capitella capitata* and *Pulliella armata* were recognized as indicators of estuarine pollution in the tropical and temperate regions (Gangaev, 1996; Wlodarska- Kowalczuk *et al.*, 1998). Sarala Devi (1986) reported that, capitellids and *Dendroneries estuarina* Southern were pollution resistant species and capitellids constituted about 98.8 per cent of the benthic population in the northern limb of Cochin estuary. Capitellids were also recorded from all stations, but was abundant in middle zone of the estuary that may be due to the high amount of organic pollutants in the area. Similar observations were reported by Feeba Rani (2009), that small deposit feeders dominate the Cochin estuary, which reflects a more dynamic and less stable environment. In the present study it was observed that decrease in polychaete diversity and increase in abundance was observed from lower to upper regions of the estuary, similar observations of increase in abundance of some polychaetes was reported by Varshney (1982) and Mirza (1981).

Feeding guild of polychaetes in three zones of the estuary that reflects the pattern of particle size and nature of the substratum. The *Pulliella armata* (SSDF) were increased in northern zone, *Cossura coasta* (SSDF) in the middle zone and *Parahetromastus* in the southern zone. It denotes the status of the polychaete distribution in three zones of the estuary. The result confirmed that the subsurface deposit feeders were dominated in three zones of the estuary.

The patterns of dominance by feeding type are good for detecting the estuarine environmental health. Feeding guild of polychaete families in Cochin estuary showed that the sub-surface feeders are the most common and abundant group in northern stations and some in middle stations. It indicates that the increased organic pollution in the estuary and also succession of some feeding groups. Functional classification from species or family data is useful to detect the trophic structure of benthos. Some studies reported the reduction of deposit feeders after the dredging (Brooks, 2004). Dominance of deposit feeders and suspension feeders in Gulf of Mexico (Fauchald and Jumars, 1979) is an example of this type of studies. The dominant species of polychaetes and their feeding habit in Cochin estuary is given in Table 2.4.

Table 2.4: The Dominant Polychaetes and their Feeding Habit in Three Zones of Cochin Estuary during 2009-11 Period

Station	Polychaete	No/m²	Feeding Habit	Zone
St. 1	*Cossura coasta*	2225	SSDF	M
St. 2	*Pulliella armata*	1861	SSDF	S
St. 3	*Parahetromastus tenuis* Monro	5198	SSDF	S
St. 4	*Cossura coasta*	5198	SSDF	M
St. 5	*Capitella capitata*	6447	SSDF	M
St. 6	*Pulliella armata*	10510	SSDF	N
St. 7	*Pulliella armata*	5062	SSDF	N
St. 8	*Capitella capitata*	1930	SSDF	N
St. 9	*Lycastis indica* Southern	1657	CVR	N

SSDF: Sub-surface deposit feeder; CVR: Carnivore; M: Middle zone; S: South zone; N: North zone.

The study has thus proved that the higher abundance of capitellids may affect the species diversity in the estuary due to their high adaptability to organic pollution. The other species may decrease due to competition, predation, invasion and high mortality due to stress from different types of pollution and anthropogenic activities. The trophic level analysis of polychaetes revealed the current status on the dominance pattern of polychaetes in Cochin estuary. The increase of sub surface deposit feeders in all stations indicates the deterioration of estuarine health. In short, the succession of a particular species may lead to the alteration in trophic structure and community ecology of polychaetes in the estuary and may affect the biodiversity in Cochin estuary.

Acknowledgements

The authors are thankful to the Head, Department of Marine Biology, Microbiology and Biochemistry, School of Marine Sciences, Cochin University of Science and Technology for the facilities provided and also to the Directorate of Environment and Climate Change, Govt. of Kerala for the funding a research project to conduct this study.

References

Achary, G.P.K. 1969. Catalogue of polychaetes, reference collection of the Central Marine Fisheries research Institute. *Bull. Cent. Mar. Fish. Res. Inst.*, **7**:31-40.

Achary, G.P.K. 1972. Polychaetes of the family Sabellariidae with special reference to their inter-tidal habitat. Proc. *Indian. National Science Academy 38 Part B*, **(5 and 6)**: 442-455.

Ajao, E.A., and S.O. Fagade, 1990. A study of sediment communities in Lagos Lagoon, Nigeria. *J. Oil Chem. Pollut.*, **7**: 85-105.

Ajmal Khan, S. and Murugesan, P., 2005. Polychaete diversity in Indian estuaries. *Indian J. Mar. Sci.*, **34**: 114-119.

Akhil, P.S., and Sujatha, C. H. 2014. Spatial budgetary evaluation of organo chlorine contaminants in the sediments of Cochin Estuary. *India Marine Pollution Bulletin* vol. **78** issue 1-2 January 15, p. 246-251

Anastasios Eleftherioo and Alasdair McIntyre, 2005. Methods for the study of Marine Benthos, 3 editions, Blackwell Science publishers (Oxford).

APHA, 2005. American Public Health Association Standard methods for the examination of analysis of water and wastewater, 21st edition.

Balachandran, K. K., C.M. Laluraj, M. Nair, T. Joseph, P. Sheeba and P. Venugopal, 2005. Heavy metal accumulation in a flow restricted, tropical estuary Estuarine. *Coastal and Shelf Science*, **65**: 361-370.

Barnosky AD, Matzke N, Tomiya S, Wogan GOU, Swartz B, Quental TB, Marshall C, McGuire JL, Lindsey EL, Maguire KC, Mersey B, and Ferrer EA. 2011. Has the Earth's sixth mass extinction already arrived? *Nature*, **471**: 51-57.

Batcha, A. S. M. 1984. Studies on the bottom fauna of north Vembanad Lake. *PhD Thesis. Cochin University of Science and Technology.* Cochin, India.

Bijoy Nandan, S, P. N. Geetha and T. A.Thasneem (2014). Current status on the biotic potential in relation to environmental quality of the Cochin estuary, India. Project completion report submitted to *Directorate of Environment and Climate Change, Govt. of Kerala.*1-200 pp.

Blake, J. 1995. Introduction to the Polychaeta. Chapter 4. In: J. Blake, C. Erseus, B. Milbig (eds.), Taxonomic Atlas of the benthic fauna of the Santa Maria Basin and Western Santa Barbara Channel, pp. 39-51 Santa Barbara Museum of Natural History, California.

Böggemann, M. 2002. Revision of the Goniadidae. Abundlungendes Naturwissenschaftlichen Vereins in Hamburg NF 39.Goecke and Evers, KelternWeiler.1-354 pp.

Böggemann, M. 2002. Revision of the Glyceridae Grube 1850 (Annelida: Polychaeta). Abh. Senckenb. *Naturforsch. Ges.*, **555**: 1- 249

Bouchet, P. 2000. L'insaisissable inventaire des especes. *La recherché*, **333**: 40-44.

Bray, J. R., and J. T. Curtis. 1957. An introduction of the upland forest communities of southern Wisconsin. *Ecol. Monogr.*, **27**:325-349.

Brooks, R. A., S. S. Bell, C. N. Purdy, and K. J. Sulak. 2004. The benthic community of offshore sand banks: a literature synopsis of the benthic fauna resources in potential MMS OCS sand mining areas. USGS Outer Continental Shelf Ecosystem Program Report USGS SIR-2004-5198 (CEC NEGOM Program Investigation Report No. 2004-01, February 2004); Minerals Management Service, OCS Study MMS-2004.

Cherian, P.V.1966. Polychaetes from the Cochin Harbor Area. Bull. Dept. Mar. Boil., and Oceanography. *Univ. Kerala*, **2**: 41-50.

Cladiu, T., H.G.Rogers and H. Judit, 1979. Structure dynamics and production of benthic fauna in Lake Maniotba.*Hydrob.*, **64**:59-95.

Clarke, K. R., and R. M. Warwick. 1994. Similarity-based testing for community pattern: the two-way layout with no replication. *Mar. Biol.*, **118**: 167- 176.

Clarke, K. R., and R. M. Warwick. 1998. A taxonomic distinctness index and its statistical properties. *Journal of Applied Ecology*, **35**: 523-531.

Clarke, K. R., and Gorley, R. N. 2006. PRIMERv6: User manual/Tutorial PRIMER-E, Plymouth.

Day, J.H. 1967. Polychaetes of South Africa. Pt.I Errantia; Pt.II Sedentaria published by the British Museum of Natural History, pp. 1-878.

Day, J. H. 1967a. A monograph on the Polychaeta of Southern Africa. Part 1 Errantia, Trustees of the British Museum London, 458 p.

Day, J. H. 1967b. A monograph on the Polychaeta of Southern Africa. Part II Sedentaria, Trustees of the British Museum London, 877 p.

Desai, B. N and M. Krishnan Kutty, 1967. A comparison of the marine and estuarine benthic fauna of the near shore regions of the Arabian Sea. *Symp.on Indian Ocean. Bull. NISI.*, **38**: 677-683.

Drucker, B. S., N. J. Blake, L. J. Doyle, and J. K. Cutler. 1996. Evaluation of dredging impact on benthic organisms: The Minerals Management Service's West Florida shelf benthic repopulation study: Coastal Zone Meeting, Tampa, FL. 1995:545.

Duffy, J.E.2009. Why biodiversity is important to the functioning of ecosystems. *Frontiers in Ecology and the Environment*, **7(8)**: 437–44.

Fauchald, K., Jumars, P. 1979. The diet of worms: a study of polychaete feeding guilds. *Oceanogr Mar Biol Ann Rev.*, **1711**: 93-284

Fauvel, P. 1932. Annelidae polychaeta of Indian Museum, Calcutta. Mem. Indian Mus. XII, No. I PP 1-262 PII-IX.

Fauvel, P. 1953. The fauna of India including Pakistan, Ceylon, Burma and Malaya. The Indian press Allahabad, 507 pp.

Feeba Rani John, 2009. Meiobenthos of Cochin backwaters in relation to macrobenthos and environmental parameters. *Ph.D. Thesis,* Cochin University of Science and Technology, Cochin, India.

Gallagher, E.D., 2008. Chapter 1, Benthic feeding guilds and functional groups. http://ocw.umb.edu/environmental-earth-and-ocean-sciences/eeos-630-biological-oceanography.

Gampi, M.C and Giangrande, A. 1986. Distribution of soft-bottom polychaetes in two coastal areas of the Tyrrehenian Sea (Italy): Structural Analysis, Estuarine, Coast. *Shelf Sci.,* **23**: 847-862.

Ganapathy, P.N., and Y. Radhakrishna. 1958. Studies on the Polychaete Larvae in the plankton of Waltair Coast. Andra Univ. *Mem. Oceanography*, **2**: 210-237.

Gangaev, S. Y. 1996. Ecology and bioenergetics of some polychaete populations in Chaun Bay of the East Siberian Sea. *Okeanologiya*, **36(5):** 756–758.

Giangrande, A. 1997. Polychaete reproductive patterns, life cycles and life histories: an overview. *Oceanography and Marine Biology. Ann. Rev.*, **35**: 323-38

Glasby Christopher, J., and Kristian Fauchald. 2003. An information system for polychaete families and higher taxa, Version 2: 5 June 2003

Gosner, R.L.1971. Guide to identification of marine invertebrates (Jhon wiley and Sons) 693

Grasshoff, K., 1983. *Methods of Sea Water Analysis.* Edited by Grasshoff, K., Ehrhardt, M. and Kremling (Second edn.). Verlag Chemie, Weinheim, 419.

Gray, J.S., and T.H. Pearson 1982. Objective selection of sensitive species indicative of pollution- induced changes in the benthic communities. I. Comparative methodology. *Mar.Ecol.Prog.Ser*, **9**: 111-119.

Gravely, F.H. 1927. Chaetopoda. The littoral fauna of Krusadi Island in the Gulf of Mannar. *Bull Madras Govt. Mus. Nat. Hist. (N.S.)*, **1**: 1-32.

Gray, J.S., R.S.S. Wu, and Y.Y. Or. 2002. Effects of hypoxia and organic enrichment on the coastal marine environment. *Marine Ecology Progress Series* 238: 249–279. Doi: 10.3354/meps238249.

Guelorget, O., and J.P. Perthuisot. 1992. Paralic ecosystems. Biological organization and functioning. *Vie et Milieu.*, **42**: 215–251.

Hartman, O. 1974. Polychaetous Annelids from the Indian Ocean including an account of species collected by members of the International Indian Ocean expendition, 1963-64 and A Catalogue and Bibilography of Species from India. Part I and Part II. *J. Mar. Biol. Ass., India*. **16(1)**: 191-252 and **16(2)**: 609-644

Hedgepeth, J.W.1957. Estuaries and Lagoons. II Biological aspects. In Treatise on Marine ecology and paleoecology, I, 693-729.

Hobbs, C. H. 2002. An investigation of potential consequences of marine mining in shallow water: An example from the mid-Atlantic coast of the United States. *Journal of Coastal Research*, **18**: 94.

Holme N. A and Mc Intyre, A. D. 1971. Methods for study of Marine Benthos, IBP Hand book No.6, Blackwell Scientific Publications.

Hutchinson, P. 1998. Biodiversity and functioning of polychaetes in benthic sediments. *Biodiversity Conserv.*, **7**: 133-145.

Idowu, E.O and A. A. A. Ugwumba, 2005. Physical chemical and faunal characteristics of a southern Nigeria reservoir. *Zoologist*, **3**:15-25.

Jackson, M.L., 1973. Soil Chemical Analysis, Prentice-hall of India Pvt.Ltd. New Delhi.

Jumars, P.A. and Fauchald, K., 1977. In: *Ecology of Marine Benthos*. Edited by B.C. Coull, University of South Carolina Press, Columbia, S.C., 1-20.

Kumaraswamy Achary, G.P. Gurudas Chakravarty, S. K. Chakraborty, P. K. Jaya Surya and K. Sarala Devi. 2005. Benthos- polychaetes. MANGROVE ECOSYSTEMS: A Manual for the Assessment of Biodiversity. Central Marine Fisheries Research Institute, Cochin. Niseema Printers and Publishers, Cochin - 18. Kerala, India. pp. 125-132

Kurian, C.V., 1973. Ecology of benthos in a tropical estuary. *Proc. Indian Natl. Sci. Acad.*, **38(3-4)**: 156-163.

Kurian C. V., R. Damoradan and A, Antony, 1975. Bottom fauna of the Vembanad Lake. *Bull. Dept. Mar. Sci. Univ. Cochin,* **7(4)**: 987-994.

Linnaeus, C.1767. Systema naturae per regna tria naturae, secundum classes, ordines, genera, species, characteribus, differentiis, synonymis, locis, 12th edn. Copenhagen.

Martin, G. D., P. A. Nisha, K. K. Balachandran, N. V. Madhu, M. Nair, P. Shaiju, T. Joseph. K. Srinivas, G. V. M. Gupta. 2011. Eutrophication induced changes in benthic community structure of a flow-restricted tropical (Cochin backwaters), India. *Environ Monit. Assess,* **176 (1-4)**:427-38.

Martin, G. D., Vijay, J. G., Laluraj, C. M., Madhu, N. V., Joseph, T., Nair, M., Gupta, G. V. M., Balachandran, K. K. 2008. Freshwater influence on nutrient stoichiometry in a tropical estuary, southwest coast of India. *Applied Ecology and Environmental Research,* **6(1)**: 57-64.

Menon, N. N., Balachand, A. N., and Menon, N. R. 2000. Hydrobiology of the Cochin backwater system-a review. T. J. Pandian (ed.), Advances in Indian Aquatic Research. *Hydrobiologia,* **430**: 149-183.

Merritt, R.W., and K.W. Cummins, 1996. An Introduction to the Aquatic Insects of North America, 3rd Edn. Kendal/Hunt Publishing Company.

Mirza, F. B and Gray, J.S.1981. *J.exp. Mar. Biol. Eco.*, **54**:181.

Misra, A., 1995. Polychaetes. Estuarine Ecosystem Series, Part 2: Hugli Matla Estuary. *Zool. Surv., India,* p. 93-155.

Misra, A., Soota T. D., and Choudhury, A. 1984. On some polychaetes from Gangetic delta, West Bengal, India. *Rec. Zool. Survey. India,* **8**:41-54.

Misra. A., 1998. Polychaeta of West Bengal. State Fauna Series 3: Fauna of West Bengal, Part 10 Zoological Survey of India: 125-225.

Mumby, P.J, Broad, K, Brumbaugh, D.R, Dahlgren, C.P, Harborne, A.R, Hastings, A. 2008.Coral reef habitats as surrogates of species, ecological functions, and ecosystem services. *Conservation Biolog.,* **22(4)**:941–51.

Mumby, P.J, Green, E.P, Edwards, A.J, Clark, C.D, 1999. The cost-effectiveness of remote sensing for tropical coastal resources assessment and management. *Journal of Environmental Management,* **55(3)**:157–66

Myers RA, Baum JK, Shepherd TD, Powers SP, Peterson CH. 2007. Cascading effects of the loss of apex predatory sharks from a coastal ocean. *Science, 315(5820)*:1846–50.

Naeem S.2006. Expanding scales in biodiversity-based research: challenges and solutions for marine systems. *Marine Ecology-Progress Series*, **311**: 273–283

Palumbi S.R, Sandifer P.A, Allan J.D, Beck M.W, Fautin D.G, Fogarty M.J. 2009. Managing for ocean biodiversity to sustain marine ecosystem services. *Frontiers in Ecology and the Environment*, **7(4)**:204–11.

Panickar N. K. (1969). New perspectives in estuarine biology; *Proc. all India Symp.in Estuarine biology,* Madras 27- 30 December, 8 pp.

Parulekar, A.H.V.K. Dhugalkar and S.Y.S. Singbal, 1980. Benthic studies in Goa estuaries: Part ill - Annual cycle of Macrofaunal Distribution, Production and Trophic Relations. *I11dian J. Mar. Sci.,* **9**: 189-200.

Pennak, R.W., 1978. Freshwater invertebrates of the United States, 2nd Ed. John Wiley and Sons, New York. pp: 810.

Pettibone, M. H., 1982 In: Parker, S.P. (Ed.) Synopsis and classification of living organisms. Volume 2. McGraw-Hill Book Company, New York, 1-1232. pp. 3-43

Petchey, O.L, Gaston, K.J. 2006. Functional diversity: back to basics and looking forward. *Ecology Letters*, **9(6)**:741–58.

Pielou, E. C. 1969. The measurement of diversity in different types of biological collections. *J. Theor. Biol.*, **13**:131-144

Pillai, N. G. K. 1977. Distribution and seasonal abundance of macrobenthos of the Cochin backwaters. *Indian J. Mar. Sci.,* **6**: 1-5.

P.R. Anu, P.R. Jayachandran, P.K. Sreekumar, S. Bijoy Nandan 2014). A Review on Heavy Metal Pollution in Cochin Backwaters, Southwest Coast of India. *International Journal of Marine Science,* **4(10)**: 92-98 (doi: 10.5376/ijms.2014.04.0010).

Rasheed, K., and A. N. Balchand, 1995. Dredging impact assessment (DIA). Sedimentation VS desilting. *Proc. Nat. Sem. Environ. Aspects of Water Resour. Develop and Manage.*, Trivandrum, p. 59-62.

Rasheed, K., and A. N. Balchand, 1997. Dredging impact assessment (DIA) at Cochin Port. *Proc. of Second Indian National Conference on Harbour and Ocean Engineering*, Trivandrum, p. 586-594.

Rasheed, K., and A. N. Balchand, 1999. A documentation of case on harbour development viz –a viz dredging and sedimentation. *Proc. of Symposia COPEDEC'99,* Cape Town (in press).

Read, G, Fauchald, K. 2013. World Polychaeta database.http://www. Marine species.org/polychaeta.

Rouse, G. W., Fauchald, and K. 1997. Cladistics and polychaetes. *Zoologica Scripta*, **26 (2)**: 139-204

Rouse, G., Pleijel, F. 2001. Polychaetes. Oxford University Press, Oxford.

Sarala Devi. K. 1986. Effect on industrial pollution on the benthos communities of a tropical estuary. *Ph.D. Thesis,* Cochin University of Science and Technology, Cochin 380.

Shannon, C. E and W. Weaver. 1963. The mathematical theory of communication. Urbana press, Illinoise, 117p.

Sheeba, P., 2000. Distribution of benthic infauna in the Cochin backwaters in relation to environmental parameter. *Ph.D. Thesis,* Cochin University of Science and Technology, Cochin. 241.

Sol Felty Light. 2007. The Light and Smith Manual: Intertidal Invertebrates from Central California to Oregon (4th Ed.), University of California Press, 1001 pp.Southern, R. 1921. Polychaetes of Chilka Lake and also of fresh and brackish waters of other parts of India. Mem. Indian Mus. V PP 563-569 Pts XIX – XXXI.

Southern, R., 1921. Polychaeta of the Chilka Lake and also of fresh and brackish waters in other parts of India. *Memoirs of the Indian Museum, Calcutta,* **5**: 563-659.

Stachowicz, J. J, Bruno, J.F, Duffy, J.E. 2007. Understanding the effects of marine biodiversity on communities and ecosystems. *Annual Review of Ecology Evolution and Systematics,* **38**:739–66.

Steneck, R.S, Dethier, M.N.1994. A functional-group approach to the structure of algal-dominated communities. *Oikos,* **69(3)**:476–98.

Strickland, J.D.H. and Parsons, T.R., 1972. *A Practical Handbook of Sea Water Analysis.* Fisheries Research Board of Canada, Bull. No. 167.

Terborgh, J. and S. Robinson, 1986. Guilds and their utility in ecology. Pp. 65-90 in J. Kikkawa and D.J. Anderson, (eds.) *Community Ecology: Patterns and Process.* Blackwell Scientific Publications, Melbourne.

Thampi, P.R.S., and Rangarajan. 1964. Some polychaetous annelids from the Indian waters. J. *Mar. Biol. Ass. India,* **6 (1)**: 98-121.

Varhney, P.K, 1982.PhD Thesis. University of Bombay.

Walkley, A. and I.A. Black, 1934. An examination of the Degtjareff method for determining organic carbon in soils: Effect of variations in digestion conditions and of inorganic soil constituents. *Soil Sci.,* **63**: 251-263.

Willey, A. 1908. The Fauna of brackish water ponds of Port Canning, Lower Bengal. XII Description of new species of a Polychaete worm. Rec. *Indian Mus.,* **2 (4)**: 389-390.

Williams, J. W and Jackson S. T. 2007. Novel climates, no-analog communities, and ecological surprises. *Frontiers in Ecology and the Environment,* **5**: 475-482.

Wilson, W.H. 1991. Competition and predation in marine soft-sediment communities. *Ann Rev Ecol. Syst.,* **21**:221-241.

Wlodarska-Kowalczuk M, Weslawski J. M, Kotwicki L. 1998. Spitsbergen glacial bays macrobenthos—a comparative study. *Polar Biol.,* **20**: 66–73.

Worm B, Barbier E.B, Beaumont, N, Duffy, J.E, Folke, C, Halpern, B.S, 2006. Impacts of biodiversity loss on ocean ecosystem services. *Science,* **314(5800)**:787–90.

Zabby, N., 2002. An ecological survey of benthic macro-invertebrates of woji week, off the bonny river system rivers state. M.Sc.Thesis, University of Port Harcourt, pp: 102.

2015, Animal Diversity, Natural History and Conservation, Vol. 5   *Pages 41–48*
Editors: V.K. Gupta and Anil K. Verma
Published by: DAYA PUBLISHING HOUSE, NEW DELHI

Chapter 3

Distribution of Nymphalid Butterflies (Lepidoptera: Nymphalidae) at Different Altitudinal Ranges in Godavari-Phulchoki Mountain Forest, Central Nepal

Bhaiya Khanal[1]*, Mukesh Kumar Chalise[2] and
Ghan Shyam Solanki[3]
[1]Natural History Museum, Swayambhu, Kathmandu, Nepal
[2]Central Department of Botany, Tribhuvan University, Kirtipur,
Kathmandu, Nepal
[3]Department of Zoology, Mizoram University, Aizawal, Mizoram, India

ABSTRACT

Godavari-Phulchoki, the southern part of Kathmandu city still retains pristine forest specially at its vertical range to the Phulchoki Mountain top. Within this range at 1500-2734 m, a changing pattern of ecosystems has been displayed offering potential habitats for butterflies of different status categories. The members of the family Nymphalidae represented here were mostly rare and colorful. Endangered species of this family like *Phaedyma aspasia kathmandia* is an endemic subspecies which so far has not been reported elsewhere except a narrow range (1600-2100m) of this part. *Teinopalpus imperialis* and *Troides aeacus* (Papilionidae) which now are enlisted in CITES Appendix II also occur here at the upper Mountain range. Similarly, *Neptis manasa, Achillides krishna* (Papilionidae*), Chraxes* sps., *Polyura* sps. were other notable species observed in this study.

---

* *Corresponding author.* E-mail: baya2000@live.com

The study of butterflies at 1500-2730 m of elevation of Godavari base to Phulchoki Mountain forest revealed a total record of 34 species of Nymphalids of different status. This paper presents the list of the recorded species of Nymphalids in its Appendix Section.

Keywords*: Biodiversity, Ecosystem, Endemic, Habitat, Pristine, Status.*

Introduction

Godavari- Phulchoki is the southern mountain range of Kathmandu Valley. This place is well known for diverse butterfly species where butterflies of different status categories are found. It is situated at 27°58′ N to 85°39′ E and 27°83′ N to 85°385′E, in the south-eastern corner of the valley.

Study Area

Ecology

Two types of forests are predominant in this district which include coniferous evergreen forest and deciduous mixed forest. The former forest can be seen above 2134 m where vegetations like *Pinus wallichiana, Rhododendron arboreum, Alnus nepalensis, Quercus semicarpifolia, Arundinaria, Berberis* etc. are found. *Quercus lamellosa*–Laurus (laurels) occurs in the middle of Phulchowki Mountain (Government of Nepal, 1969). The deciduous mixed forest which lies below 2134 m displays *Schima wallichii, Castanopsis indica, Sambucus* sps., *Dalbergia sisoo,* etc.

The faunal components of this region include *Panthera pardus, Muntiacus muntijak, Melursus ursinus, Felis bengalensis, Hystris indica, Herpestes edwardsii, Canis aureus,*

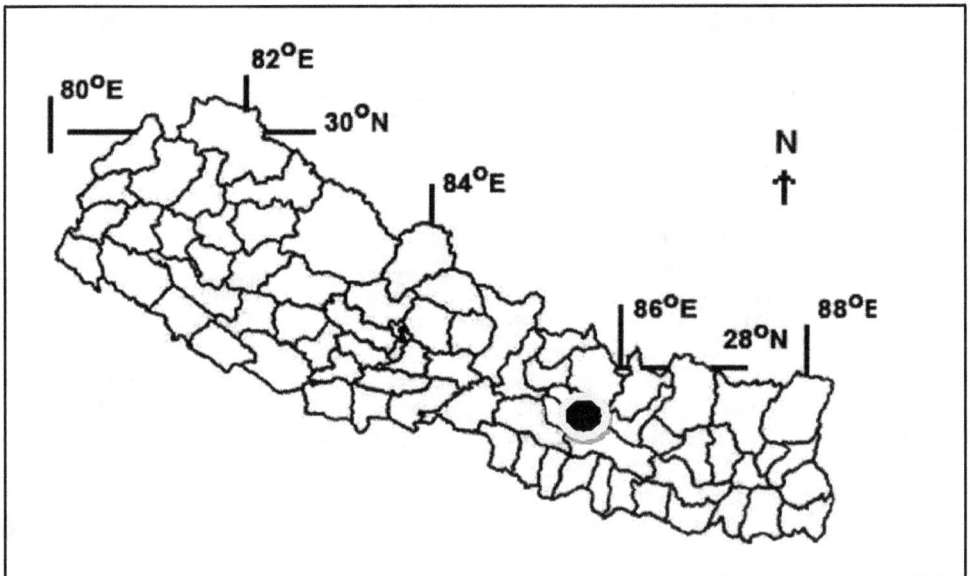

Map 3.1: ● **Location Map of Lalitpur District.**

Martens etc. Bird life is very diverse in this part which includes common birds like *Turdoides nepalensis, Milvus migrans, Apus apus, Spilornis cheela, Columbus livia, Columba leuconata, Streptopelia chinensis,Gallus gallus* etc.

Climate

The temperature ranges from 10° to 14 °C during winter and 15° to 30 °C in summer (Figure 3.1). Snow fall occurs in the hills in winter and heavy rainfall occurs during summer. Phulchoki Mountain and Lankuri Bhanjyang of this district are the notable places for snowfall during extreme winter condition (Anonymous, 1971).

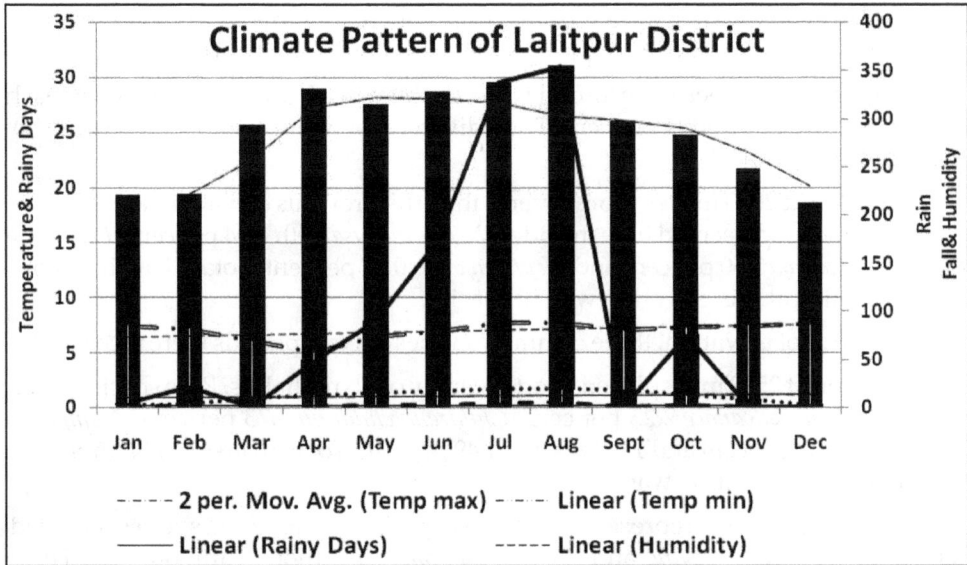

Figure 3.1: Climatic Pattern of Lalitpur.

Methodology

Following methods were adopted for studying biological parameters of butterfly namely; butterfly documentation, distribution and analysis.

Observation and identification of butterflies in the field were undertaken with sufficient accuracy mostly through direct observation and some confusing species with Capture and Release method (Sutherland, 1996). Every specimen, noted was recorded with accurate collateral field data. This study was done inside the established quadrates and on linear walk at the vertical range of 1500 m to 2734 m.

Identification and Status Determination

Identification of butterfly was done by standard procedures available for identification. Talbot (1975), Smith (1989) and Haribal (1992) were consulted for identification besides confirming with reference specimens deposited at the Natural History Museum of Tribhuvan University, Nepal. The status included here is the local status assessed at various studied spots of the Godavari-Phulchoki range.

Seasonal Pattern of Observations

Studies were scheduled at different seasons of 2010/2011 covering months of March/April (spring), June (summer), August (mid-monsoon), September (post monsoon), October (autumn) and November end (pre-winter).

Results

Elevation at 1500 m displayed rare species of Nymphalids like *Stibochiona nicea* with relative density of 2 per cent and *Dilipa morgiana* with 1.26 per cent. Nymphalids recorded at this elevation were of six species.

At 1700 m of elevation, rare *Euthalia patala* with relative density of 8 per cent was recorded. Total diversity of Nymphalids recorded here was five species.

At 1800 m, rare species recorded were *Euthalia patala* and *Dilipa morgiana* each with relative densities of 4 per cent. Total diversity of Nymphalids noted here was eight species.

Elevation at 2000 m was more diverse than the previous elevation. Rare species of Nymphalids represented here included *Neptis manasa* with 1.94 per cent, *Phaedyma aspasia kathmandia* 3.50 per cent and *Neptis nycteus* 0.77 per cent. Total diversity record of Nymphalids at this elevation was 10 species.

At 2400 m of elevation, three common species of Nymphalids were recorded.

Elevation at 2500 m displayed good diversity of Nymphalids. Rare species noted here were *Kaniska canace* 2.23 per cent, *Childrena childreni* 3.73 per cent, *Nymphalis xanthomelas* 1.49 per cent and *Euthalia nara* 1.49 per cent. Total diversity of Nymphalids recorded at this elevation was 8 species.

Elevation at 2700 m represented very few species. Endangered species recorded here were *Achillides krishna* and *Teinopalpus imperialis* (both Papilionidae). Three common species of Nymphalids were recorded at this elevation.

Total species record of all the families in Godavari/Phulchoki included 43 species at 1500 m, 62 species at 2000 m, 21 species at 2500 m and 7 species at 2700m. This also included 34 species of Nymphalids at the entire range.

Discussion

Altitude can be taken as an important gradient to control the temperature which in turn affects distribution of vegetations and butterflies at different elevations. With elevation rise the temperature starts dropping down and above 3500 m of elevation the vegetation diversity is scarce. The mid-mountain range at 2000 m displayed rich and diverse species of butterflies. The uniformity in distribution for butterflies was determined using Equitability Index which indicated less equality in species distribution at Godavari-Phulchoki range.

Nymphalid butterflies like *Phalanta phalantha*, *Vanessa indica*, *Issoria issaea*, and *Aglais cashmirensis* were the popular species at 3500 – 4300 m of elevation. These species also shared their habitats at different elevations of the subtropical to temperate zones.

At 2700 m, rare species like *Teinopalpus imperialis,* and *Achillides krishna* were recorded. These species appear only for a month at the end of May to June end every year (Smith, 1989). Next species *Phaedyma aspasia kathmandia,* an endemic subspecies (Khanal and Smith, 1997) was reported at 1800 – 2100 m of elevation. This species also appeared only for a month from June to July end each year (Smith, 1989).

Godavari-Phulchoki (1500-2734 m) is influenced mainly with subtropical to temperate bio-climatic zones as this district showed a stepwise change in elevation followed with change in ecosystem where flora and fauna of different status categories are sheltered.

Godavari- Phulchoki represented altogether 123 species of butterflies which included 22 species in the family Papilionidae, 8 species in Pieridae, 33 in Lycaenidae, 4 in Nemeobiidae, 11 species in Satyridae, 34 in Nymphalidae, 1 in Libytheidae, 2 in Acraeidae, 3 in Danaidae, and 5 in Hesperiidae. No species record was made from the family Amathusiidae. Nymphalidae showed the highest diversity and least diversity was displayed by the family Libytheidae.

According to Smith (1980) as being a geographically complex country, Nepal does not have the same season for the emergence of butterflies. May to October is the best time for the mountain butterflies for their emergence and March to November for lowland species. Species diversity reaches to its maximum in July-August period. With this regard, Khanal (2006) made an extensive study on late season butterflies of Koshi Tappu Wild Life Reserve, a lowland area of eastern Nepal. Though November is the late season for butterflies he still made a record of 54 species mostly of common status.

Godavari-Phulchoki range at 1400 to 2734 m represented varied habitats to offer. Many rare species of Nymphalids including rare and endangered Papilionid butterflies like *Achillides krishna* and *Teinopalpus imperialis* are found at 2100-2734 m in Phulchoki Mountain. The later species and *Troides aeacus* which also occur here are listed under the Appendix II of CITES.

According to Khanal *et al.* (2012), habitat preferences for diverse species were found at the elevation of 1500 – 2000 m in the Rasuwa district of central Nepal where 85 species were recorded. This elevation is influenced basically with the warm temperate climate sheltering butterflies of different habitat types. Cold climatic condition, scarce vegetation and less preferred habitats are the main causes for decreased diversity in higher elevation. This elevation at Godavari- Phulchoki Mountain displays less diverse butterflies due to differing bio-physical gradients mainly the temperature, precipitation and vegetations types.

The increasing rate of human encroachment has led direct impact on the habitats of many rare and endangered butterflies in this area. The marble quarry which now has stopped its operation and construction of roads at different directions of this mountain also caused destruction to the significant habitats of many butterfly species.

Acknowledgements

I would like to express my sincere thanks to Associate Professor Dr. Ramesh Shrestha, Chief of Natural History Museum for the facility to prepare this paper.

Associate Professor Dr. Nirmala Pradhan is acknowledged for her kind help to identify some of the plant specimens brought from the field. Mr. Madan Krishna Shrestha, Zoologist, is highly appreciated for his help and cooperation in the field.

References

Anonymous, 1971. *Mechi Dekhi Mahakali, Department of Information*. Nepal Government, Part II: 714-717.

Government of Nepal, 1969. *Flora of Phulchoki and Godawari*. Ministry of Forests, Department of Medicinal Plants, pp.144.

Haribal, M. 1992. *The butterflies of Sikkim Himalaya, Sikkim Nature*. Conservation Foundation, Gangtok Sikkim, India, pp. 12- 207.

Khanal, B., Chalise, M.K., and Solanki, G.S. 2012. Diversity of butterflies with respect to altitudinal rise at various pockets of the Langtang National Park, central Nepal. *International Multidisciplinary Research Journal*, ISSN: 2231-6302, **2(2)**:41-48.

Khanal, B. 2006. The Late Season Butterflies of Koshi Tappu Wildlife Reserve, Eastern Nepal. *Our Nature*, **4**: 42-47.

Khanal, B., and Smith, C. 1997. *Butterflies of Kathmandu Valley*. TecPress Press, Bangkok, Thailand, pp. 3-70.

Smith, C. 1989. *Butterflies of Nepal*, TecPress Service, Bangkok, Thailand, pp. 350.

Sutherland, W.J., 1996. *Ecological Census Techniques: A Handbook*, Cambridge University Press.

Talbot, G. 1975. *Butterflies*. The fauna of British India including Ceylon and Burma, Today and Tomorrow's Printers and publishers, New Delhi, India, pp.40-41.

APPENDIX I

Sl.No.	Altitude (m)	Genus and Species	Count	Relative Density (per cent)	Local Status
1.	1500	*Ariadne merione* Cramer	19	4.76	Common
2.		*Aglais cashmirensis* Kollar	51	12.78	Common
3.		*Athyma opalina* Kollar	28	7.01	Common
4.		*Neptis ananta* Moore	20	5.01	Common
5.		*Symbrenthia hypselis*	12	3.00	Uncommon
6.		*Stibochiona nicea*	8	2.00	Common
7.	1700	*Euthalia patala* Kollar	8	3.22	Rare
8.		*Symbrenthia hypselis* Godart	8	3.22	Common
9.		*Aglais cashmirensis* Kollar	29	11.70	Common
10.		*Athyma jina* Moore	6	2.41	Rare
11.		*Precis atlites* Linnaus	7	2.82	Uncommon
12.	1800	*Neptis hylas* Linnaeus	38	12.06	Common
13.		*Symbrenthia hypselis* Godart	9	2.85	Common
14.		*Euthalia sahadeva* Moore	10	3.17	Common
15.		*Euthalia patala* Kollar	4	1.26	Rare
16.		*Athyma perius* Linnaeus	12	3.80	Common
17.		*Cyrestis thyodamus* Boisduval	8	2.53	Common
18.		*Precis iphita* Cramer	45	14.28	Common
19.		*Dilipa morgiana* Westwood	4	1.26	Rare
20.	2000	*Sumalia dudu* Westwood and Doubleday	2	1.85	Rare
21.		*Abrota ganga* Moore	8	7.40	Common
22.		*Athyma perius* Linnaeus	11	10.18	Common
23.		*Athyma jina* Moore	12	11.11	Common
24.		*Neptis manasa* Moore	5	1.94	Rare
25.		*Neptis nycteus* DeNiceville	2	0.77	Rare
26.		*Phaedyma aspasia* Leech	9	3.50	Rare
27.		*Euthalia duda* Staudinger	4	1.55	Rare
28.		*Polyura athamus* Drury	2	0.77	Rare
29.		*Hestina nama* Doubleday	5	1.94	Rare
30.	2400	*Hypolymnas bolina* Linnaeus	11	5.16	Common
31.		*Athyma opalina* Kollar	26	12.20	Common
32.		*Athyma selenophora*	11	5.16	Common
33.	2500	*Childrena children* Gray	5	3.73	Rare
34.		*Aglais cashmirensis* Kollar	22	16.41	Common

Contd...

Appendix I—*Contd...*

Sl.No.	Altitude (m)	Genus and Species	Count	Relative Density (per cent)	Local Status
35.		*Precis orithya* Linnaeus	8	5.97	Common
36.		*Precis hierta* Fabricius	12	8.95	Common
37.		*Vanessa cardui* Linnaeus	20	14.92	Common
38.		*Kaniska canace* Linnaeus	3	2.23	Rare
39.		*Nymphalis xanthomelas* Denise and Schieff	2	1.49	Rare
40.		*Euthalia nara* Moore	2	1.49	Rare
41.	2700	*Vanessa indica* Herbst	11	9.16	Common
42.		*Aglais cashmirensis* Kollar	18	20.22	Common
43.		*Kaniska canace* Linnaeus	3	3.37	Rare

2015, **Animal Diversity, Natural History and Conservation, Vol. 5** *Pages 49–55*
Editors: **V.K. Gupta and Anil K. Verma**
Published by: **DAYA PUBLISHING HOUSE, NEW DELHI**

Chapter 4

A Taxonomic Review of the Genus *Cyrtolabulus* van der Vecht, 1969 (Hymenoptera: Vespidae: Eumeninae) from India

P. Girish Kumar*[1], J. M. Carpenter[2] and P.M. Sureshan[3]
[1]*Zoological Survey of India,*
M-Block, New Alipore, Kolkata – 700 053, West Bengal, India
[2]*Division of Invertebrate Zoology, American Museum of Natural History,*
Central Park West at 79th Street, New York, NY 10024, U.S.A.
[3]*Western Ghat Regional Centre, Zoological Survey of India,*
Kozhikode – 673 006, Kerala, India

ABSTRACT

The potter wasp genus *Cyrtolabulus* van der Vecht is reviewed from India. The species *Cyrtolabulus suavis* (van der Vecht, 1963) is newly recorded from Kerala and Uttarakhand. A key to species of India is provided.

Keywords: *Cyrtolabulus, Eumeninae, India, Key, Review, Vespidae.*

Introduction

Van der Vecht (1969) introduced the name *Cyrtolabulus* as a replacement name for *Cyrtolabus* van der Vecht, 1963, a genus that he had separated from *Labus* de Saussure, 1867. *Cyrtolabulus* is distributed in the Ethiopian, Palaearctic and Oriental

* *Corresponding author.* E-mail: kpgiris@gmail.com

Regions. Fifty four species are included in this genus worldwide of which five species are recorded from the Indian subcontinent and three species from India (Cameron, 1902; Meade-Waldo, 1910; van der Vecht, 1963; Gusenleitner, 2006). The Indian species are reviewed here and a key to species provided. The species *Cyrtolabulus suavis* (van der Vecht, 1963) is recorded for the first time from Kerala and Uttarakhand.

Materials and Methods

The Indian specimens were studied and photographed by using a Leica Stereo microscope with LAS software version 3.6.0. These specimens were properly preserved and added to the 'National Zoological Collections' of the Hymenoptera Section of the Zoological Survey of India, Kolkata (NZC).

Abbreviations Used for the Museums

AMNH = American Museum of Natural History, New York, USA; BMNH = Natural History Museum, London, UK; ML = Rijksmuseum van Natuurlijke Historie, Leiden, Netherlands; MR = Natuurhistorisch Museum, Rotterdam, Netherlands; NHMB = Naturhistorisches Museum, Basel, Switzerland; NZC = 'National Zoological Collections' of Zoological Survey of India, Kolkata, India.

Abbreviations Used for the Terms

H = Head; M = Mesosoma; S2 = Metasomal sterna 2; T1-T2 = Metasomal terga 1 and 2.

Results and Discussion

Genus *Cyrtolabulus* van der Vecht, 1969

Cyrtolabulus van der Vecht, 1969: 1, replacement name for *Cyrtolabus* van der Vecht, 1963, *non* Voss, 1925. Type species: *Cyrtolabus suavis* van der Vecht, 1963, by original designation as type species of *Cyrtolabus* van der Vecht.

Cyrtolabus van der Vecht, 1963: 11, genus. Type species: *Cyrtolabus suavis* van der Vecht, 1963, by original designation.

Diagnosis

Female without fovea anterior to mid-ocellus; tegula exceeding parategula posteriorly; metanotum not dentate; propodeum with extensive horizontal portion, somewhat narrowed apically, abruptly sloping posteriorly, as seen in profile the two surfaces forming almost a right angle; submarginal carina produced in to a pointed lamella; second submarginal cell forming obtuse angle basally; midtibia with one spur; metasomal petiole not conspicuously swollen in apical half.

Distribution

Ethiopian, Palaearctic and Oriental Regions.

Key to species of the Genus
Cyrtolabulus van der Vecht from India
(Modified from van der Vecht, 1935; Giordani Soika, 1941)

1. Tegulae ferruginous with yellow markings at base
   and apex (Figure 4.5); Length (H+M+T1+T2)
   5-7.5 mm .... ***C. suavis*** (van der Vecht, 1963)

– Tegulae entirely ferruginous without yellow markings 2

2. Propodeum very long and dorsal surface at the same
   level as metanotum; two curved yellow marks on
   scutellum; large species, length (H+M+T1+T2)
   10 mm .... ***C. interstitialis*** (Cameron, 1902)

– Propodeum visibly shorter, dorsal surface located
   at a level below metanotum; scutellum without yellow
   marks; small species as *C. suavis* .... ***C. punctatus*** (Meade-Waldo, 1910)

1. *Cyrtolabulus interstitialis* (Cameron, 1902)

Zethus interstitialis Cameron, 1902: 291, female, male, "**Matheran" (BMNH).**

Calligaster interstitialis: Dalla Torre, 1904: 17 (cat.).

Labus interstitialis: Meade-Waldo *In* Meade-Waldo and Morley, 1914: 404; Bequaert, 1928: 150 (lectotype: female); van der Vecht, 1935: 159 (note); Giordani Soika, 1941: 218 (notes on types).

Cyrtolabus interstitialis: van der Vecht, 1963: 14.

Diagnosis

Female. Antennae black, scape with yellow mark at apex, flagellum rufous beneath at apex; clypeus sparsely punctured, with a transverse yellow mark at base, apex of clypeus narrowed in the middle and broadly but not deeply incised; front and vertex strongly and closely punctured; mesosoma rather strongly punctured; pronotal angle project sharply; tegulae ferruginous; two curved yellow marks on scutellum; propodeum very long and its dorsal surface lies at the same level of metanotum; legs black with yellow marks on fore femora and fore tibiae; Wings hyaline with greater part of costal and radial cells smoky; petiole irregularly punctured at dorsal side, less curved in lateral view, sides rufous except at base and apex; apex of petiole and second gastral segment with a yellow band. Length (H+M+T1+T2) = 10 mm. **Male**. Clypeus almost entirely yellow, anterior apex sharply bidentate.

Material Examined

No specimens examined.

Distribution

India: Maharashtra.

Figure 4.1: *Cyrtolabulus suavis* van der Vecht ♀.
1: Body Profile; 2: Head Frontal View; 3: Clypeus; 4: Mandibles;
5: Head and Mesosoma Dorsal View; 6: Gastral Petiole and T2 Dorsal View.

Remarks

No material was available for our studies; hence the diagnostic characters were taken from Cameron (1902), Giordani Soika (1941) and van der Vecht (1963).

2. *Cyrtolabulus punctatus* (Meade-Waldo, 1910)

Labus punctatus Meade-Waldo, 1910: 37, male, "Kangra Valley, 4500 feet" [presently in Himachal Pradesh] (BMNH); Bequaert, 1928: 150 (note on type); Dover, 1931: 252 (compared to *L. exigua* (de Saussure)); van der Vecht, 1935: 159 (notes on type); Giordani Soika, 1941: 217 (notes on holotype).

Cyrtolabulus punctatus: Gusenleitner, 1988: 188.

Diagnosis

Female. Propodeum visibly shorter, its dorsal face located at a level below metanotum; metanotum with a transverse carinate projection, which emarginate medially and look like bidentate; petiole coarsely punctate, more sharply gibbous to the base, then swollen towards apical half; scutellum and tegula without yellow markings; legs partly ferruginous. Smaller species like *L. suavis*.

Material Examined

No specimens examined.

Distribution

India: Himachal Pradesh (Kangra Valley).

Remarks

No material was available for our studies; hence the diagnostic characters were taken from Dover (1931), van der Vecht (1935) and Giordani Soika (1941).

3. *Cyrtolabulus suavis* (van der Vecht, 1963)

Cyrtolabus suavis van der Vecht, 1963: 11, fig.2a, female, male, holotype female "Ceylon: Central Province, Dambulla" (NHMB); N. W. Province, Kalpitiya, allotype male (NHMB); Uva, Kataragama, male paratype (ML); India: Coimbatore (1 ♀ 1♂ MR, 1♂ ML); 1969: 1 (transferred to *Cyrtolabulus*).

Diagnosis

Female. Clypeus slightly convex, wider than long, anterior margin shallowly emarginate; mandible with three somewhat irregular and rather short teeth; ocelli in a flat triangle; anterior lateral angle of pronotum distinct but hardly projecting; metanotum with low and somewhat irregular transverse crest anteriorly; propodeum forming a rectangular hump, horizontal area as long as scutellum, gradually narrowed towards the end, where it is truncate and slightly emarginate, posterior surface concave, concavity narrow and shallow at top, wider and deeper below; petiole slightly longer than second gastral segment; petiole moderately curved in profile view; apical margin of T2 and S2 depressed and translucent, its base with a row of coarse, but superficial punctures. Punctures generally very dense on head and mesosoma; anterior face of pronotum coarsely punctate on each side of the median impression; sides of propodeum almost impunctate anteriorly, very coarsely, but superficially and reticulately, punctate posteriorly and on the sloping areas on each side of the median horizontal part; tegulae finely punctate; petiole coarsely and densely punctate, except at extreme base and on apical rim; T2 densely but superficially punctate; S2 more distinctly but less densely punctate with distinct interspaces. Body black; mandibles, ventral side of flagellomeres and part of tegulae ferruginous; legs and petiole partly reddish; yellow parts as follows: a narrow transverse band at base of clypeus, ventral side of scape, a narrow band at anterior margin of horizontal part of pronotum (narrowly interrupted in the middle), a spot at base and apex of tegulae, parategulae, two small transverse spots on scutellum, propodeal spines, apical rim of petiole, a narrow subapical band on T2 and S2. Petiole red with dark blotch on basal two thirds of dorsal side (in one specimen from Uttarakhand, dorsal side of petiole almost entirely black except apical yellow rim). Legs ferruginous, tips of claws black; mid tarsi and hind legs beyond trochanters more or less infuscated; yellow markings: a mark on outer side of fore and mid femora at apex (sometimes absent at mid femora), outer side of fore and mid tibiae, a line on dorsal side of hind tibiae. Wings hyaline. Length (H+M+T1+T2) = 6.5-7 mm. *Male.* Structurally similar to female aside from the usual sexual dimorphism; colouration with more yellow, including clypeus entirely, ventral side of scape and flagellomeres, most of tibiae, outer sides of fore and mid femora. Length 5 mm.

Material Examined

INDIA: Kerala, Idukki district, Chinnar Wildlife Sanctuary, Chambakad, 1♀, 18.xi.1996, Coll. P.M. Sureshan and Party, NZC No. 14588/H3; Tamil Nadu, Coimbatore district, Mathuramalai Hills, 1♀, 11.xii.1978, Coll. R.C. Basu and Party, NZC No. 14589/H3; Uttarakhand, Dehradun district, Rishikesh, 1♀, 5.xi.1978, Coll. R.K. Varshney and Party, NZC No. 14590/H3. SRI LANKA: Pol. Dist., Mannampitiya, 1♀, 24-26.xii.1969, Coll. P.B. Karunaratne (AMNH); Ham. Dist., Palatupana, 1♂, 3-6.ii.1975, Coll. K.V. Krombein, P.B. Karunaratne, P. Fernando, *E.G.* Dabrera (AMNH).

Distribution

India: Kerala (new record), Tamil Nadu, Uttarakhand (new record). Elsewhere: Sri Lanka.

Acknowledgements

The authors are grateful to Dr. K. Venkataraman, Director, Zoological Survey of India, Kolkata for providing facilities and encouragements. PGK is also grateful to Dr. Kailash Chandra, Additional Director and Officer-in-Charge of Entomology Division (A), Zoological Survey of India, Kolkata and Dr. Gaurav Sharma, Officer-in-Charge, Hymenoptera Section, Zoological Survey of India, Kolkata for providing facilities and encouragements.

References

Bequaert, J. 1928. A study of certain types of diplopterous wasps in the collection of the British Museum. *Annals and Magazine of Natural History*, **(10)2**: 138-176.

Cameron, P. 1902. Descriptions of new genera and new species of *Hymenoptera* collected by Mayor C.S. Nurse at Deesa, Simla and Ferozepore, Part I. *Journal of Bombay Natural History Society,* **14:** 267-293.

Dalla Torre, K.W. 1904. Vespidae, *Genera Insectorum,* **19**: 1-108.

Dover, C. 1931. The Vespidae in the F.M.S. Museums. *Journal of the Federated Malay States Museum,* **16**: 251-260.

Giordani Soika, A. 1941. Studi sui Vespidi Solitari. *Bollettino della Societa Veneziana Storia Naturale,* **2(3):** 130-279.

Gusenleitner, J. 1988. Uber Eumenidae aus Thailand, mit einer Bestimmungstabelle fur orientalischer *Labus-* Arten (Hymenoptera Vespoidea). *Linzer Biologische Beitrage,* **20(1):** 173-198.

Gusenleitner, J. 2006. Über Eumeninae, aufgesammelt in Pakistan (Hymenoptera: Vespidae). *Linzer Biologische Beitrage,* **38(2):** 1295-1305.

Meade-Waldo, G. 1910. New species of Diploptera in the collection of the British Museum. *Annals and Magazine of Natural History,* (8)5: 30-51. http://dx.doi.org/10.1080/00222931008692723

Meade-Waldo, G., and Morley, C. 1914. Notes and synonymy of Hymenoptera in the collection of the British Museum. *Annals and Magazine of Natural History,* **(8)14:** 402-408.

Vecht van der, J. 1935. Notes on Oriental *Labus,* with descriptions of three new species from Java (Hym., Vespidae). *Treubia,* **15(2):** 157-167.

Van Der Vecht, J. 1963. Studies on Indo-Australian and East Asiatic Eumenidae (Hymenoptera: Vespoidea). *Zoologische Verhandelingen Leiden,* **60:** 1-116.

Vecht, J. van der. 1969. A new name for the genus *Cyrtolabus* van der Vecht (Hymenoptera, Eumenidae). *Entomologische Berichten, Amsterdam* **29:** 1-2.

2015, Animal Diversity, Natural History and Conservation, Vol. 5   Pages *57–64*
Editors: **V.K. Gupta and Anil K. Verma**
Published by: **DAYA PUBLISHING HOUSE, NEW DELHI**

Chapter 5

Community Composition of Termites (Isoptera) in Different Habitats and Seasons in Kurukshetra, Haryana, India

Sanjeev K. Gupta and Nidhi Kakkar*
University College Kurukshetra University,
Kurukshetra – 136 119, Haryana, India

ABSTRACT

To know the community composition of termites in different habitats and seasons in Kurukshetra the samples were collected randomly for two consecutive years by hand picking method. It showed 11 species of termites belonging to 4 genera, 3 subfamilies and 2 families which is 4.35 per cent of the total species found in India. Community as a whole is dominated by subfamily Termitinae and genus *Odontotermes*. Four species such as *O.feae, O.gurdaspurensis, O.obesus* and *M.obesi* were found in both the years and in most of the habitats. But *O.feae* and *O.obesus* were available in most of the time period of the year. Thus it was found all the species showed variability due to food preference and environmental conditions.

Keywords: *Termitinae, Odontotermes, Coprotermes, O. feae, Temporal variation.*

Introduction

Termites are social insects, having three casts. Termites are generally considered notorious insects but they are highly significant from the ecological point of view.

---

* *Corresponding author.* E-mail: nidhikakkar12@yahoo.com

Termites are one of the most abundant terrestrial animals on earth. Their biological diversity is not high but population is extremal. They are reported from all zoo geographical regions of the world. These are mainly tropical and sub-tropical insects, categorized into four distinct type *viz.* Damp wood, Subterranean, arboreal and mound builders. The extraordinary abundance of termites is the result of their highly developed social organization (Myles, 1988; Noirot, 1990; Wilson, 1990; Nelepa, 1994) and symbiosis with microorganisms (Martin, 1987; Wood and Thomas, 1989; Brezank and Brunce, 1994). Termites play important role of super decomposer (Matsmoto and Abe, 1979) and carbon nitrogen balancer thus forming the basis for a large food web (Deligne *et al.,* 1981).

Their role in recycling wood and plant material is highly significant. Their tunneling efforts also help to ensure that soils are porous, contain nutrients and are healthy enough to support plant growth. Termites are the ecosystem engineers (Jones *et al.,* 1994; Lawton, 1994) that modify the soil structure by constructing mounds and subterranean nests (Lee and Wood, 1971; Wood, 1998) providing many species of animals and plants with diverse habitats (Glover *et al.,* 1964).

The Indian subcontinent consists of various types of physical and climatic features that are suitable for termite growth and development. Overall, termites comprise nearly 2858 species distributed in about 286 genera. About 253 species under 54 genera of seven families have been documented from the Indian region (Wild Life of India News Letter, 2005). And about 35 species have been reported to damage agricultural crops and timber in buildings (Sociobiology, 2002).They are also major source of methane, one of the greenhouse gases. Most are soil inhabiting, either as mound builders or as subterranean nest builders. However, only limited information is available on the termite fauna of Haryana. Therefore the present attempt was made to document the termite fauna of Kurukshetra, Haryana.

Survey Methods

Field observations have been carried in and around Kurukshetra which is situated at 29°58 N and 76°51 E longitudes. Ecological data was randomly surveyed for termites from Oct.2008 to December 2010. Intensive observations were made on ground, trees, branches, twigs and dead stumps for presence of termites. Information relevant to sample collections was recorded. Specimens were preserved in 70 per cent ethyl alcohol containing a few drops of glycerol. The relevant labels depicting the source, locality, date and time of collection were added to the tubes containing the collection sample. All castes were collected but priority was given to find soldiers and workers as these are easy to identify.

The hand picking method provides a semi quantitative measure of the relative abundance of termites based on the number of encounters with each species in each catch. An encounter is defined as the recorded presence of a species in a single catch. The identities of the species were confirmed using keys, descriptions and measurements given in Roonwal and Chhotani (1989) and Chhotani (1997).

Results

Species Richness and Composition

Overall Eleven termite species belonging to, 4 genera, 3 subfamilies and 2 families were recorded from different habitats and seasons during the study period (List 1).

List 1: Termite Species Collected

1 Family	Rhinotermitidae			
	1 Subfamily	Coptotermitinae		
		1 Genus	*Coptotermes*	
			Species 1	*heimi* (Wasmann)
2 Family	Termitidae			
	2 Subfamily	Amitermitinae		
		2 Genus	*Microcerotermes*	
			Species 2	*beesoni* Snyder
	3 Subfamily	Tremitinae		
		3 Genus	*Odontotermes*	
			Species 3	*bhagwatii* Chatterjee and Thakur
			Species 4	*guptai* Roonwal and Bose
			Species 5	*gurdaspurensis* Holmgren and Holmgren
			Species 6	*feae* (Wasmann)
			Species 7	*microdentatus* Roonwal and Sen-Sharma
			Species 8	*obesus* (Ranbur)
		4 Genus	*Microtermes*	
			Species 9	*imphalensis* Roonwal and Chhotani
			Species 10	*mycophagus* Desneux
			Species 11	*obesi* Holmgren

The community as a whole (as recorded by the random collection) is dominated by subfamily Termitinae (90.9 per cent) of the total species with the maximum share of genus *Odontotermes* (55 per cent of total species) (Figure 5.1). Temporal variation in terms of number of species can also be seen in termite community in Kurukshetra. It was found that maximum numbers of species were avaialable in 2009 (Table 5.1, Figure 5.2).

Habitat and Seasonality of Termite Species in and around Kurukshetra

Out of 11 species, 8 species were recorded in 2008 and 2010 and 10 species in 2009. *C.heimi* was found only in November 2008 but in 2009 in the month of July, August and November and in March in 2010 from Popular and Bottle brush tree. *M.beesoni* was found only in December 2008 in soil and bark of tree. *O.bhagwati* was

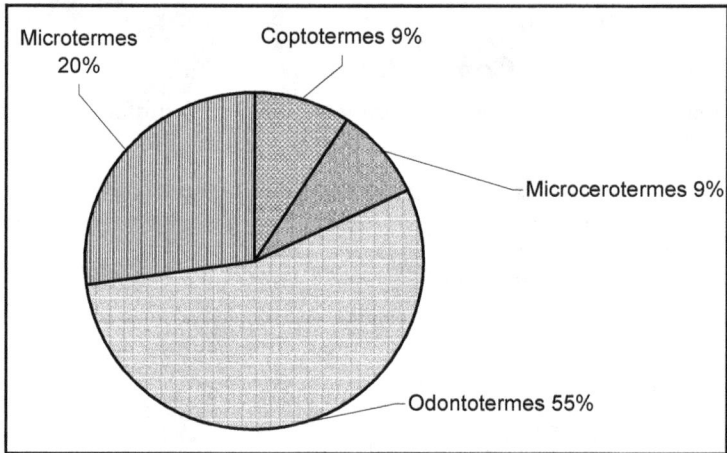

Figure 5.1: Percentage Sharing of different Genera in Termite Community.

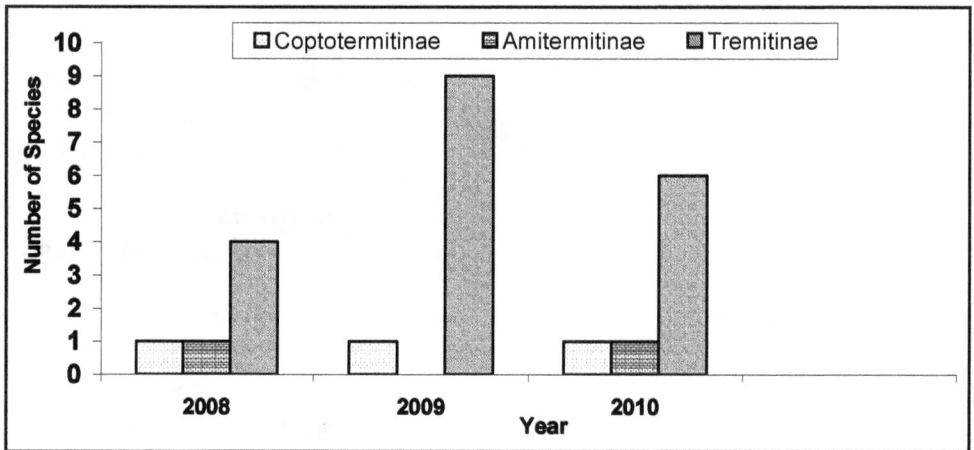

Figure 5.2: Maximum Share in Terms of Number of Species.

found only in July 2009 in the University Campus. *O.feae* was available in most of the occasion throughout the period in November and December in 2008, in March, April, July to September and December in 2009 and in October and November in 2010 in Chameli, Neem tree and soil. *O.guptai* was found in July 2008 and March 2010 in soil and Seesam tree. *O.gurdaspurensis* was found in January and December in 2008 and 2009, only in November in 2010 in mango tree, soil and drain. *O.microdentatus* was absent in 2008, found only in December 2009 and October in 2010 in bark of tree. *O.obesus* was found in most of the occasions and in all possible habitats *i.e.* in May, July, November and December in 2008, January, May to December in 2009 and in October and November in 2010 in maize and Mulberry plant, dung, soil, in Neem and Popular trees. Only the imago of *M. imphalensis* was found in July 2009 in Camphor tree. *M. mycophagus* was found only in January and September in 2009 in

Seesam tree. *M. obesi* was found only in December in 2008, in July and December in 2009 and only in October in 2010 in soil, Saal and Bamboo trees (Tables 5.2 and 5.3).

Table 5.1: Temporal Variation in Termite Species of Kurukshetra

Species	2008	2009	2010
C.heimi	+	+	–
M.beesoni	–	+	–
O.bhagwatii	–	+	–
O.feae	+	+	+
O.guptai	+	–	+
O.gurdaspurensis	+	+	+
O.microdentatus	–	+	+
O.obesus	+	+	+
M.imphalensis	–	+	–
M.mycophagus	+	+	–
M.obesi	+	+	+

Table 5.2: Seasonality of Termite Species in and around Kurukshetra (2008-2010)

Species	Jan	Feb	Mar	Apr	May	Jun	Jul	Aug	Sep	Oct	Nov	Dec
C.heimi	–	–	+	–	–	–	+	+	–	–	+	–
M.beesoni	–	–	–	–	–	–	–	–	–	–	–	+
O.bhagwatii	–	–	–	–	–	–	+	–	–	–	–	–
O.feae	–	–	+	+	–	–	+	+	+	+	+	+
O.guptai	–	–	+	–	–	–	+	–	–	–	–	–
O.gurdaspurensis	+	–	–	–	–	–	–	–	–	–	+	+
O.microdentatus	–	–	–	–	–	–	–	–	–	+	–	+
O.obesus	+	–	+		+	+	+	+	+	+	+	+
M.imphalensis	–	–	–	–	–	–	+	–	–	–	–	–
M.mycophagus	+	–	–	–	–	–	–	–	+	–	–	–
M.obesi	–	–	–	–	–	–	+	–	–	+	–	+

It was found that termite species were available throughout of the collection period but most of the species (7 species) were recorded in July. Each species prefer a suitable habitat to grow in its colony. It was found that *O.feae, O.gurdaspurensis, O.obesus* and *M.obesi* were widely distributed in all possible habitats. In the recorded data always the number of workers was greater as compared to soldiers; Imago of *M.imphalensis* and *M.obesi* were recorded only once.

Discussion

The subfamily Termitinae was found to be most dominant in Kurukshetra. There were 11 species (4.35 per cent of species available in India) of 4 genus, 3 subfamilies

Table 5.3: Occurrence of Termite Species

Sl.No.	Name of the Species	Habitat
1.	*Coptotermes heimi*	Popular tree and Bottle brush tree from Kurukshetra
2.	*Microcerotermes beesoni*	Soil, bark of tree Seonthi Forest
3.	*Odontotermes bhagwatii*	Kurukshetra University campus
4.	*Odontotermes feae*	Chameli, Neem tree, soil from NIT, University campus and Seonthi forest
5.	*Odontotermes guptai*	Soil, Seesam tree from Kirmich Village
6.	*Odontotermes gurdaspurensis*	Mango tree, drain and soil from University campus
7.	*Odontotermes microdentatus*	Bark of tree
8.	*Odontotermes obesus*	Maize plant, Neem tree, popular tree, Mulberry, soil and dung, NIT and Dayalpur, Kurukshetra
9.	*Microtermes imphalensis*	Camphor tree
10.	*Microtermes mycophagus*	Seesam tree
11.	*Microtermes obesi*	Soil, Bamboo tree, Saal tree from Dayalpur, University Campus

and 2 families. Overall, termites comprise nearly 2858 species distributed in about 286 genera. About 253 species under 54 genera of seven families have been documented from the Indian region (Wild Life of India News Letter, 2005). Remaining one species *i.e., C. heimi* from subfamily Coptotermitinae has been reported from study area which is widely distributed and occurs throughout India and Pakistan (Thakur, 1991).

C. heimi was reported by Thakur (2000) as the most common wood destroying termite in India. According to Chhotani (1980), this is the common species found in the subcontinent. In Gujarat it is recorded from I and III ecological zones (Thakur, 1991). *C. heimi* is one of the wood destroying subterranean termite species and feed the inner portion of the attacked material leaving the outer sheath intact (Sen-Sarma, 1989).

Results of the present study clearly indicate that, maximum numbers of species were found in 2009 and in subfamily Termitinae genus *Odontotermes* was the most abundant and available in whole the period except in the month of February. Out of all the 3 subfamilies Termitinae was showing its maximum share in all the three years in terms of number of species. Being a common and having a wide niche breadth they (pest species) can be referred as generalists. Based on the ecological specialization hypothesis, species with the widest niche breadth should be more widespread and locally abundant than specialized species (Gaston and Lawton, 1990; Pomeroy and Ssekabiira, 1990). A subterranean termite species, *O. obesus* is distributed throughout India (Sen-Sarma, 1989). During present study *O.feae, O.gurdaspurensis, O. obesus* and *M.obesi* have widest range in habitat and seasonally among all species. However, Termites remain active throughout the year (Imms, 1919).

It was observed that the food availability promotes the activity of termites since wooden structures, tree bark, leaf litter, crop residue and weed heaps are good food and habitat for termites (Kumar and Pardeshi, 2011). More plant cover of the crop increases the ground shade and reduces the water evaporation rate thus increasing the humidity and moisture of the soil which is the main factor that provide the appropriate habitat to promote termite activity (Pearce, 1997). The major and common problem in India was due to many species of termites being subterranean necessitating ground connection for moisture requirement (Thakur, 2000). Termite are most sensitive to their environment especially temperature, heat, humidity and CO_2 concentration (Srivastava *et al.*, 2012). The food preference and environmental conditions affect the presence or absence of termites in different season and habitat these factors showed the presence of maximum number of species in July then in December and *O.obesus* was found in most of the period of the year.

Acknowledgement

We are grateful to University Grant Commission, Delhi for giving us financial support to run the research work in the University College, Kurukshetra University, Kurukshetra.

References

Breznak, J.A., and Brune, A. 1994. Role of microorganisms in the digestion of Lignocellulose by Termites. *Annual Review of Entomology*, **39**: 453-487.

Chhotani, O.B., 1980. Termite pests of Agriculture in Indian region and their control. *Tech. Monograph*, **4**: 1-84.

Chhotani, O.B., 1997. *The fauna of India and the adjacent countries Isoptera* (Termites) Vol II (Famili- Termitidae) 794 pp.

Deligne, J.A., Quennedey, A.C., and Blum, M.S., 1981. *Hermann, H.R. (ed.) Social Insects. Vol.2 Academic Press*, New York. pp. 1-76.

Gaston, K.J., and Lawton, J.H., 1990. Effects of scale and habitat on the relationship between regional distribution and local abundance. *Oikos* **58**: 329-335.

Glover, P.E., E.C. Trump and L.E.D. Wateridge, 1964. Termitaria and vegetation patterns on the. Loita plains of Kenya. *The Journal of Ecology*, **52**: 367-377.

Imms, A.D., 1919. On the structure and biology of Archotermopsis together with discussion of new species of intestinal protozoa and general observations on the Isoptera. *Phils. Trans. Roy. Soc. B*, **209**: 75-180.

Jones, C.G., Lawton, J.H., and Shachak. M., 1994. Organisms as ecosystem engineers. *Oikos*, **69**, 373-386.

Kumar, D., and Pardeshi, M., 2011. Biodiversity of termites in Agro-ecosystem and relation between their nitch breadth and pest status. *Journal of Entomology*, 250-258.

Lawton, J.H., 1994. What do species do in ecosystems? *Oikos*, **71**, 367-374.

Lee, K.E., and Wood, T.G., 1971. *Termites and soils. Academic Press* London. 251 pp.

Martin, M.M., 1987. Invertebrate Microbial Interactions, Ingested Fungal Enzymes in *Arthopod Biology*. Cornell University Press, Itaca.

Matsumoto, T., and Abe, T., 1979. The role of termites in an Equatorial Rain Forest System of West Malaysia. *Oecologia*,**38**: 261-274.

Myles, T.G., 1988. in: Slobodchikoff, C.N. (ed.) *The Ecology of Social behavior, Academic Press*, San Diego, pp. 379-423.

Nalepa, C.A., 1994.: Hunt, J.H., and Nalepa, C.A. (eds.) *Nourishment and Evolution in Insect Societies.* Westview Press Boulder, Co. pp. 57-104.

Pearce, M.J., 1997. *Termites, Biology and Pest Management. CAB International*, Willing Ford UK., pp: 172.

Pomeroy, D., and Ssekabiira, D., 1990. An analysis of terrestrial distributions in Africa. *Afri. J. Ecol.*, **28**: 1-13

Roonwal, M.L., and Chhotani, O.B., 1989. *The fauna of India and the adjacent countries Isoptera* (Termite) Vol I 665 pp.

Sen-Sarma, P.K., 1989. Protection of cultural heritage against insect borers and termites. *Indian J. Ent.,* **51**: 256-260.

Srivastava, B., Dwivedi, A.K., and Pandey, V.N., 2012. Sociobiology and Natural Adaptation of Termite and Termitomyces in Different Forest Division of Gorakhpur Region. *Bull. Env. Pharmacol. Life Sci.*, 32-36.

Thakur, M.L., 2000. *Forest Entomology (Ecology and Management). Sai Publishers*, Dheradun, India, p. 609.

Thakur, R.K., 1991. Field ecology, eco-biogeography and economic importance of Gujarat termites. *Ind. For. Rec.*, **16**: 1-42.

Wilson, E.O. 1990. *Success and dominance in ecosystems : the case of the social insects, EcologyInstitute.* Oldendorf Luhe, Germany.

Wood, T.G., and Thomas, R.J. 1989. in: *Wilding, N., Collins, N.M., Hammond, P.M., and Webber, J.F. (eds.) Insect-Fungus Interactions. Academic Press*, London. pp. 69-92.

2015, Animal Diversity, Natural History and Conservation, Vol. 5   *Pages 65–73*
Editors: V.K. Gupta and Anil K. Verma
Published by: DAYA PUBLISHING HOUSE, NEW DELHI

Chapter 6

Web Dynamics and Estimation of Capture Area of the Common Signature Spider, *Argiope pulchella* Thorell (Araneae: Araneidae)

Rahi Soren[1,2]*

[1]Department of Zoology,
University of Calcutta, Kolkata – 700 019, West Bengal, India
[2]Centre for Biodiversity and Ecological Studies,
Kolkata – 700084, West Bengal, India

ABSTRACT

Webs primarily serve as catchment areas for trapping prey for the spiders. The fundamental unit of behavior in orbweb spiders is the construction and design of the web. A variety of orb-weaving spider species construct stabilimenta, typically near the hub of the orb. *Argiope* spiders display considerable inter- and intra-species, inter- and intra-individual and ontogenetic variation in stabilimentum-building. The distance between the capture spirals (mesh height) may affect the visibility of the web. The webs of 15 adult and 15 juvenile *Argiope pulchella* Thorell (Araneidae) were studied and were analyzed for capture area estimation which directly influences the number and types of prey entangled. Student's t test was done and a significant (< 0.05) difference of web decorations was found and hence difference in prey capture between adult and juvenile *Argiope pulchella*. The most striking result was that sub adult spiders built dramatically oversized decorations.

* *Corresponding author.* E-mail: rahisoren@gmail.com

These tests help validate surrogate variables and provide ecologists with appropriate tools for estimating orb web parameters in the field. These studies can be implemented as a tool in biological pest management in regard to the agro economy.

Keywords: *Argiope pulchella, Capture area, Orb web, Stabilimenta.*

Introduction

Spiders have often been thought of as a mere annoyance, and have been universally shunned and feared. But few deserve it. They are extremely shy creatures, fearing man, more than they are feared. Spiders can easily be distinguished from other arachnids by their lack of visible segmentation and the marked constriction between the prosoma and the opisthosoma, dividing the body into cephalothorax and abdomen, respectively. Spiders occupy nearly all terrestrial environments and can be found wherever there are other terrestrial animals to prey upon. Research on spider biology, particularly the diversity of their silks, webs and venoms; together with the associated ecology and behavior has increased greatly in the recent decades. Moreover, phylogenetic advances are beginning to provide the context for comparisons between spider taxa and between spiders, other arachnids, and other terrestrial arthropods.

Spiders generally build webs made out of proteinaceous spider silk extruded from their spinnerets. Also known as cobweb (from the obsolete word "coppe", meaning "spider"), the structure of a spider web is a composite material with stiff, stress-resistant crystals interspersed in an extensible, energy-absorbing matrix. Webs primarily serve as catchment areas for trapping several faunal elements as prey for the spiders. Webs have also been found to be utilized for the purpose of dispersal of spiderlings, an event called *Ballooning*. However, not all spiders build webs to catch prey. Some species do not build webs at all. Some even, have been found to build communal webs that are found to be inhabited by several individuals.

The silk producing organs of the spiders are the most complicated silk organs known. A single species of spider spins several distinct kinds of silk. Seven different kinds of silk glands have been recognized. No spider possesses all of the seven kinds; but three of the kinds have been found in all species studied; others are present in a group specific manner (Foelix, 2010).

Web decorations are conspicuous silk structures included in the webs of many species of diurnal orb-web spiders, belonging to the families Araneidae, Uloboridae and Tetragnathidae. A variety of orb-weaving spider species construct stabilimenta – 'patterned areas of dense silk', typically near the hub of the orb (Schoener, 1992) Although web decorations are common in a number of spider species they are probably best known from spiders of the genus *Argiope*. However, spiders in this genus also construct web decorations as a vertical line and juveniles commonly construct disc-shaped decorations. Other spiders also construct round structures covering the entire hub of the web. There is much controversy surrounding the function of these structures, and it is likely that different species use it for different purposes.

The genus *Argiope* (Family: Araneidae) is cosmopolitan in distribution, primarily distributed in tropical and subtropical areas of the world. About 75 species have been recorded on a global scale (Platnick, 2003). In India, nine species of *Argiope* are found (Tikader, 1987). Interpretations of phylogenetic relationships between different web types and their evolution have been made for many web builders by several authors on a speculative basis (Kaston, 1964; Kullman, 1958; Peters, 1939; Szlep, 1966; Wiehle, 1929). Web data support the likelihood of a monophyletic evolution (Herberstein *et al.*, 2000). *Argiope* spiders display considerable inter- and intra-species, inter- and intra-individual and ontogenetic variation in stabilimentum-building behaviour (Herberstein *et al.*, 2000). For example, in many species, juveniles decorate their webs with disc-like (*discoid*) stabilimenta, while adults spin linear or cross-like (*cruciform*) stabilimenta (Herberstein *et al.*, 2000) (Figure 6.1). The typical pattern of cruciform stabilimenta in adult *Argiope* has given its common name as 'Signature Spider'. Many spiders build daily orb webs with a characteristic geometric pattern. This geometric arrangement, an expression of movements in time and space, is a relatively permanent record of a significant part of the behavior of the thread-laying animal; it contains an enormous amount of information, which can be defined, measured, and evaluated. Certain web features are correlated with age, size, weight, sex, and species (Witt *et al.*, 1972).

Following characters have been observed in terms of *Argiope* web:

a. The decorating silk is produced by the aciniform and piriform glands (Peters, 1993; Foelix, 1996) and arranged in various patterns on the web.

b. The expression of web decorations is characterized by very high levels of ontogenetic and interspecific variability. In the genus *Argiope*, juveniles commonly include discoid decorations in their webs, whilst adults include cruciate or linear decorations (Herberstein *et al.*, 2000a).

c. There is significant interspecific variation with at least six forms of decorations found in adult female spiders (Herberstein *et al.*, 2000a). Within just one genus, *Argiope* adults are known to construct three different decoration forms; linear, cruciate, and discoid (Herberstein *et al.*, 2000a; Bruce *et al.*, 2005).

d. The interspecific variation is not limited to the form of decorations but also the frequency of spiders including decorations in their webs (Bruce, 2005).

The present study focuses on the web dynamics of adult and juvenile *Argiope pulchella* Thorell and comparing the results of their potency in prey capture, depending upon stabilimenta and whether it corresponds to the body size of the spider.

Study Site

Lying along the fertile banks of the Adiganga Drainage Scheme Canal at Narendrapur (South 24 Parganas, West Bengal), Chintamoni Kar Bird Sanctuary (Lat. 22°27′N; Long. 88°22′E) is a small patch of primarily mixed deciduous forest (Soren and Chowdhury, 2010). The present study deals with the *Argiope pulchella*, Thorell of CKBS surveyed for one year (March, 2010–February, 2011). Specimens were identified with the help of Tikader (1987), Sebastian and Peter (2009) and

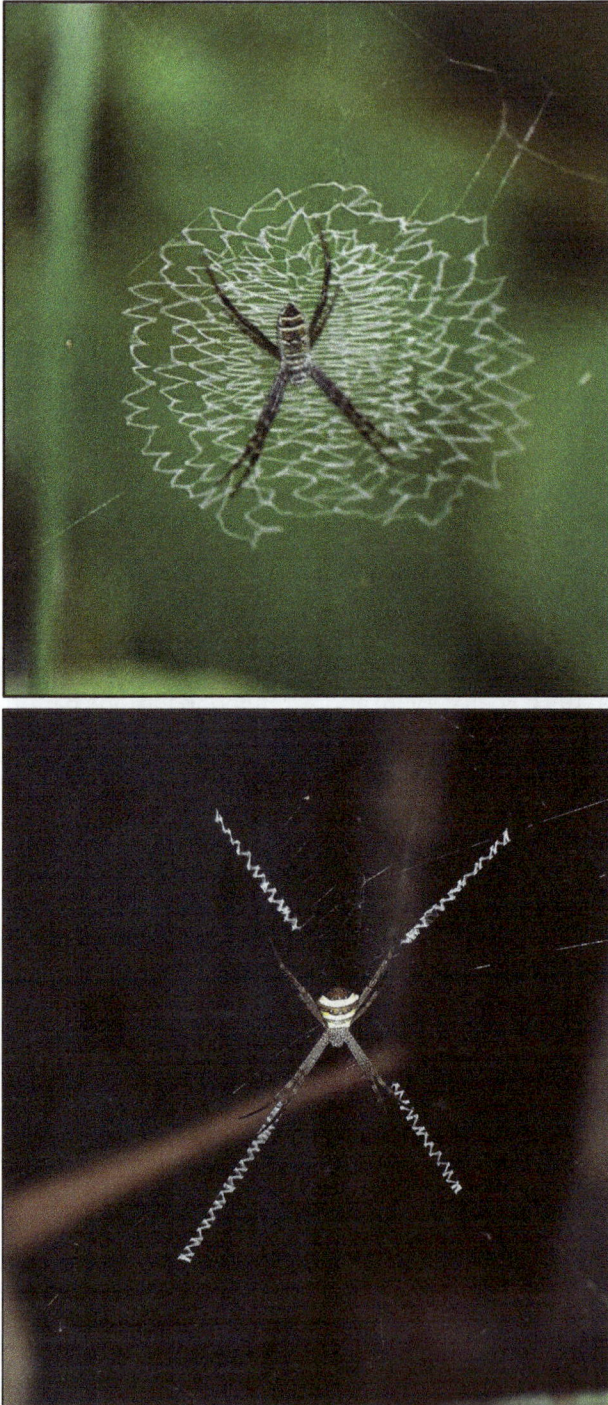

Figure 6.1: Web Decorations in the Signature Spider of the Genus *Argiope*:
Discoid in Juvenile (Left) and Crusiate in Adult (Right).

Chakrabarti (2009). The overall health of such a fast-degrading urban sanctuary can be well represented by the complex interaction of spider and associated fauna among the diverse vegetation and habitat forms therein.

Methodology

The webs of fifteen female and fifteen juvenile *Argiope pulchella* were studied in the field. The exact capture areas of the webs of the individuals studied were obtained by summing up the area covered by spirals in each web sector. The exact capture area excluded the area of the hub, which is not covered by sticky spirals and therefore does not function in capturing prey. The capture areas of the webs were estimated by applying the equation 1 given by Herberstine and Tso (2000). The distance between the capture spirals (mesh height) may affect the visibility of the web and the size of prey entangled. So, Average mesh height has been calculated. Web parameters considered in the equations 1 and 2 are explained in the Table 1 as well as in Figure 2. Measurement of each parameter was taken by the help of metal tape and transparent 'mm' scale in natural conditions of the webs.

$$\textbf{Capture Area} = \left[\frac{1}{2}\pi r_{au}{}^2 - \frac{1}{2}\pi(Hr_u)^2\right] + \left[\frac{1}{2}\pi r_{al}{}^2 - \frac{1}{2}\pi(Hr_l)^2\right] \quad Eq.\ 1$$

Average mesh height for each web was calculated from the equation 2, given by Tso (1996).

$$\textbf{Average Mesh Height} = \frac{1}{2}\left(\frac{r_u - Hr_u}{(S_u - 1)} + \frac{r_l - Hr_l}{(S_l - 1)}\right) \quad Eq.\ 2$$

Table 6.1: Web Parameters have been taken into Consideration While Calculating the Capture Area (Equation 1) and Average Mesh Height (Equation 2) of *Argiope pulchella* Webs

Abbreviations Used in Equations	Web Parameters
H_{ru}	upper hub radii
H_{rl}	lower hub radii
H	vertical hub diameter = (Hru + Hrl)
d_h	radius from the horizontal web diameter
d_v	radius from the vertical web
r_u	upper vertical radius
r_l	lower vertical radius
r_{au}	adjusted upper vertical radius
r_{al}	adjusted lower vertical radius
$(r_u + r_l)$	diameter
S_u	Sticky spirals in upper web half
S_l	Sticky spirals in lower web half
S	Sticky spirals = (S u + S l)

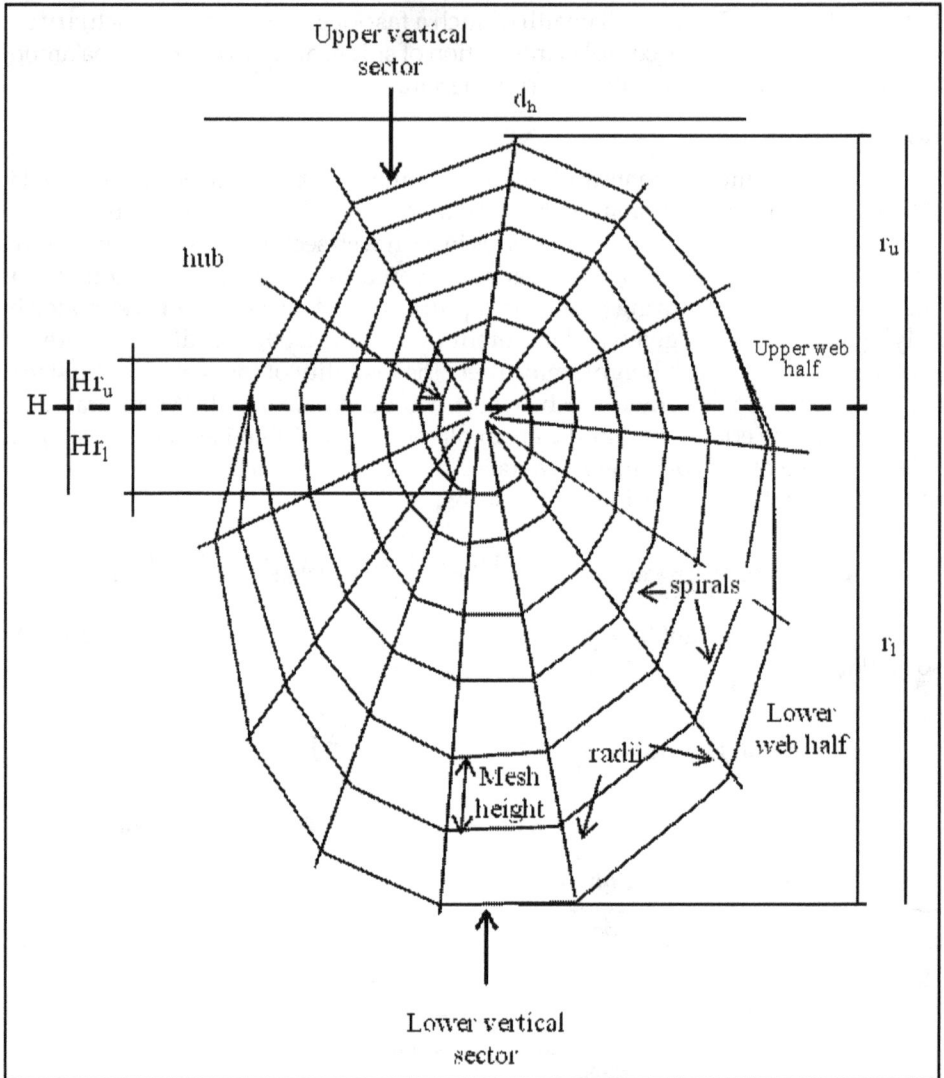

Figure 6.2: A Schematic Representation of an Asymmetric Orb-Web (Modified from Heiling and Herberstein, 1998), Defining the Parameters Used in the Equations 1 and 2 in the Study (See Table 6.1 for explanations of the abbreviations used in the figure).

While the geometric nature of orb webs aids the measurement and consequent comparison of web elements such as web size and mesh height, these are sometimes difficult to obtain, particularly in the field. Therefore, some studies have used the length of the web radius (Higgins and Buskirk, 1992) or web diameter (McReynolds and Polis, 1987) as a very rough approximation of web size.

Results

In all the sampled specimens, the stabilimenta in juvenile *Argiope pulchella* were found to be discoid in shape forming dense rings around the spirals. Here the hub was found to be absent as the juvenile tend to make a dense silk deposition at the middle of the web. In contrast to the juveniles, the adults formed cruciform stabilimenta distinguished by dense silk deposition in a zigzag pattern.

The most accurate estimates are generated by the 'Adjusted Radii - Hub' formula, because vertical asymmetry has been considered by incorporating the horizontal radii as well as calculating the upper and lower hub region separately. Additionally, this formula generates separate values for the upper and lower web regions, which can be used for further analyses. The average value of different web parameters necessary to calculate the capture areas as well as the mesh height are summarized in Table 6.2.

Table 6.2: Average Web Parameters taken into Consideration while Calculating the Capture Area (Equation 1) and Average Mesh Height (Equation 2) of *Argiope pulchella* Webs

Web Parameters (in cm.)	Argiope pulchella	
	Adjult	Junvenile
Hr_u	0.54	0
Hr_l	0.4866667	0
H	1.0333333	0
d_h	29.233333	11.766667
d_v	28	4.3133333
r_u	11.506667	6.2733333
r_l	16.573333	5.2533333
r_{au}	13.061667	6.0783333
r_{al}	15.595	5.5683333
S_u	59.4	17.84
S_l	81.8	27.88
S	141.66667	45.72
Capture Area	727.62657	107.45433
Mesh Height	0.188876	0.283573

From the calculated values, it was evident that the juveniles had no hub. This was indicated by null values of H, *i.e.* the vertical hub diameter. The adults, however, were found to have a distinct hub with a mean diameter of 1.03 cm. This was also reflected in case of the horizontal as well as vertical web radius (d_h and d_v respectively). The mean d_h in adults and juveniles were found to be 29.23 cm and 11.76 cm, while that of d_v were found to be 28.0 cm and 4.31 cm respectively. Further, average Capture Area for the adults and juveniles were found to be 727.62 cm and 107.45 cm respectively.

Student's t test revealed significant difference of capture area (< 0.05, 0.00032) between the adult and juvenile *A. pulchella*.

Furthermore, average radii of the lower and upper web for the juveniles and adults were found to vary accordingly with the body and leg size. Mesh Height was calculated and remarkable difference was found among the adult and juvenile specimens. Mesh height was found to be greater (0.28 cm) in the juvenile *A. pulchella* than that of adult specimens (0.18 cm). Student's t test revealed significant difference of Average Mesh Height (< 0.05, 0.0000058) as well, revealing the difference in prey capture between the adult and juvenile specimens. However, the most striking result was the dramatically oversized decorations built by the sub adults of *A. pulchella*.

Discussion

Stabilimentum is an integral part of the spider web and its functions are much debated. The primary function of orb web is prey capture (Eberhard, 1990); but webs can also be a valuable defense (Edmunds and Edmunds, 1986; Jackson *et al.*, 1993; Cloudsley-Thompson, 1995). The sticky capture silk of spider web is capable of entangling predators such as jumping spiders and wasps, and it is also a noxious stimulus avoided by vertebrate predators such as the birds (Horton, 1980; Eisner and Nowicki, 1983). Yet there is little advertisement of these noxious and sometimes lethal aspects of webs to the predators of spiders. Instead, orb webs are usually cryptic, indicating the great importance of low visibility of webs for effective prey capture (Rypstra, 1982; Uetz, 1990; Craig and Freeman, 1991). The fundamental unit of behavior in orbweb spiders is the construction and design of the web. These web variations directly influence the number and types of prey entangled. Similarly, the distance between the capture spirals (mesh height) affects the visibility of the web and the size of prey entangled.

In the present study it is evident that the web dynamics of the adult and the juvenile *Argiope pulchella* are different, indicating the diversity of the prey captured. This shows a significant change in the ecological role it plays during its various life stages. The difference in the number of sticky spirals and mesh height also indicates the potentiality of resource partitioning in their ecosystem, which in turn indicates towards a change in their feeding guild. While the geometric nature of orb webs aids the measurement and consequent comparison of web elements such as web size and mesh height, these are sometimes difficult to obtain, particularly in the field. These studies can be implemented as a tool in biological pest management of tropical agroecosystems of India, where spiders can act as effective biocontrol agents.

Acknowledgements

I intend to express my gratitude and thanks Dr. Soumyajit Chowdhury for his guidance, constructive criticism and encouragement during the course of preparation of the present article. I extend my thanks to Centre for Biodiversity and Ecological Studies for its logistic support both during the field studies and preparation of the manuscript.

References

Bruce, M.J., and Herberstein, M. E. 2005. Web decoration polymorphism in Argiope Audouin, 1826 (Araneidae) spiders: ontogenetic and interspecific variation. *Journal of Natural History*, **39(44)**: 3833–3845.

Eberhard, W. G. 1990. Function and phylogeny of spider webs. *Annu. Rev. Ecol. Syst.* **21**: 341–372.

Foelix, R. 2010. *Biology of spiders.* Oxford University Press. UK, 376 pp.

Herberstein, M.E., and Tso, I.M. 2000. Evaluation of Formulae to estimate the Capture Area and Mesh Height of Orb Webs (Araneoidea, Araneae). *Journal of Arachnology*, **28 (2)**: 180-184.

Higgins, L. E., and Buskirk, R. E. 1992. A trap-building predator exhibits different tactics for different aspects of foraging behaviour. *Animal Behaviour*, **44**: 485-499.

Horton, C. C. 1980. A defensive function for the stabilimenta of two orb weaving spiders (Araneae, Araneidae). *Psyche*, **87.1-2**: 13-20.

McReynolds, C. N. 1987. Ecomorphological factors influencing prey use by two sympatric species of orb-web spiders, *Argiope aurantia* and *Argiope trifasciata* (Araneidae). *The Journal of arachnology.*

Platnick, N. I. 2000-2014. The World Spider Catalog, Version 3.5. American Museum of Natural History. Available from: http://research. amnh. org/entomology/ spiders/catalog81-87/index. html. Accessed on: 22 February, 2014.

Schoener, T. W., and Spiller, D. A.1992. Stabilimenta characteristics of the spider *Argiope argentata* on small islands: support of the predator-defense hypothesis. *Behav. Ecol. Sociobiol.*, **31**: 309-318.

Sebastian, P. A., and Peter, K. V. 2009. *Spiders of India.* Universities Press Ltd, India, 734 pp.

Soren, R., and Chowdhury, S. 2010. Spider Fauna of Chintamoni Kar Bird Sanctuary, West Bengal, India. *Bionotes,* **12** (3): 103.

Tikader, B.K.1987. *Handbook: Indian Spiders*, Zoological Survey of India, Calcutta, 251 pp.

Witt, P. N., Rawlings, J. O., and Reed C. F. 1972. Ontogeny of web-building behavior in two orb-weaving spiders. *American Zoologist*, **12.3**: 445-454.

2015, Animal Diversity, Natural History and Conservation, Vol. 5   *Pages 75–83*
Editors: V.K. Gupta and Anil K. Verma
Published by: DAYA PUBLISHING HOUSE, NEW DELHI

Chapter 7

Throwing More Light on Morphology of *Cyrtophora bidenta* Tikader, 1970 (Araneidae): An Endemic Spider Species of India, with Checklist of the Genus from Southern Western Ghats

Elizabeth V. Mathew* and P.A. Sebastian
[1]Division of Arachnology,
Department of Zoology, Sacred Heart College,
Cochin– 682013, Kerala, India

ABSTRACT

The spider genus *Cyrtophora* Simon, 1864 is unique among the whole of the family Araneidae as they construct tightly woven, tent-like, horizontal orb webs with a network of supporting threads above and below. *Cyrtophora bidenta* is endemic to mainland India and was first described by Tikader in 1970 from the state of Sikkim. This is the first report since the original description and the maiden report from the Western Ghats. We collected *C. bidenta* from Southern Western Ghats in May 2012. This paper attempts a detailed description of the species with more taxonomic details than published in the original paper. Total length reaches 18mm as against 5.90mm long holotype. Cephalothorax and legs dirty brown, abdomen pale brown with white and dark brown patches. Ocular quad longer than wide. AMEs projected outwards. Chelicerae long, stout and blackish brown, with teeth on retromargin and promargin. Sternum heart shaped, clothed with hairs. Legs long and clothed with hairs and spines. Dorsal abdomen with a pair of

* *Corresponding author.* E-mail: lizshz@gmail.com

prominent black shoulder humps which give the species its name. Mid longitudinal broad brown patch observed in the posterior half. Ventrum with mid longitudinal dark brown patch guarded by chalk white stripes. Epigastric furrow distinct, slightly concave and lip like. Epigyne is unique in having a prominent ring-like structure projecting outwards.

We have observed 5 species of this genus in Southern Western Ghats, of which *C. cicatrosa* and *C. citricola* are the most prevalent. *C. unicolor* is reported from India after more than a century. The checklist is provided.

Keywords: *Cyrtophora, Araneidae, Endemic, Holotype, Checklist.*

Introduction

The araneid spider genus *Cyrtophora* Simon 1864 is unique among the whole of the family Araneidae. Although technically orb-web spiders (family Araneidae), these spiders do not build orb webs. Instead, they construct tightly woven, tent-like, horizontal orb-webs with a network of supporting threads above. This highly complex non-sticky web is sometimes considered a precursor of the simplified orb-web (Scharff and Coddington, 1997). These spiders often live in colonies unlike other spiders, which prefer to lead a solitary life. The genus currently includes 43 species and 9 subspecies (Platnick 2013). *Cyrtophora* mainly occurs in the Indo-Australasian region, with some species described from Africa. In India this genus is represented by seven species (Mathew *et al.*, 2008) *viz. C. cicatrosa, C. citricola, C. feae, C. moluccensis, C. ksudra, C. jabalpurensis* and *C. bidenta* of which the last three are endemic. *C. moluccensis* is a semi-social spider (Berry, 1987). Males are small and not commonly observed in webs. A recent taxonomic revision has revealed that three colour morphs of the same species *i.e., Cyrtophora mollucensis* (Doleschall, 1857) distributed in Japan and Taiwan were actually three independent nominal species based on molecular phylogeny and morphological characteristics (Tanikawa *et al.,* 2010). Considering the fact that this species has a much wider distribution from India to Japan to Australia, it is highly probable that country-specific individuals hitherto assigned to *C. moluccensis* might not be conspecific. This could be true for other species of *Cyrtophora* as well since a comparative study of the specimens lodged in different museums/ collections of the genus has not been undertaken so far. This signifies a much needed taxonomic revision.

Cyrtophora bidenta is endemic to mainland India and was first described by Tikader in 1970 from the state of Sikkim in North India. This is the first report from a site other than that of holotype. Description of holotype lacks mention of many important features. This paper attempts further description of the species.

Materials and Methods

Specimen was collected from its web and preserved in 70 per cent alcohol. Morphology and genitalia were examined and measured using Leica M205 C equipped with Automontage software. Epigyne was examined after clearing in 10 per cent KOH. Species status is confirmed using description and illustration by Tikader 1970 and 1980.

Type Material

Female holotype, Sikkim, India

Material Examined

1 female, Kalakkad, Tamil Nadu. Coll. By Balasubramanyam, 03.05.2012

Habitat

Shrub

Abbreviations Used in the Text

AE = Anterior Eyes

ALE = Anterior Lateral Eyes

AME = Anterior Median Eyes

MOQ = Median Ocular Quadrangle

PE = Posterior Eyes

PLE = Posterior Lateral Eyes

PME = Posterior Median Eyes

All measurements are in mm

Distribution

Sikkim, India (Tikader 1970); Tamil Nadu, India (New Record): Kalakkad, Tirunelveli district.

Results

General

Total length 18 mm, Cephalothorax 7.5 mm, Abdomen 10.2 mm. Cephalothorax yellowish brown bordered by white lateral stripes. Legs yellowish green, long and clothed with hairs and spines. Leg measurements as in Table 7.1.

Cephalothorax

6.0 mm long, 6 mm wide near middle; yellowish brown, longer than wide, clothed with hairs, narrowing in front with dark patches laterally and behind the ocular quad. Ocular quad longer than wide, as wide in front as behind. Eyes encircled with black margins in two recurved rows arranged in the pattern typical of Araneidae – laterals placed close to each other and away from the medians. Laterals situated on prominent tubercles. AME projected anteriorly. Eye diameter: PLE = ALE = 0.220, PME = 0.419, AME = 0.367. Eye interdistances: AME – AME = 0.435, AME – ALE = 0.632, AME – PME = 0.687, PME –PME = 0.447, PME – PLE = 0.653. Distinct fovea. Sternum pale brown, heart shaped, clothed with intermittent hairs; a broken pale stripe extend across its length. Maxillae broad and brownish with pale white margins on inner tips; provided with distinct scopulae. Labium small, convex, pale with scopulae. Chelicerae long and stout with 3 teeth each on retro- and promargins. Fangs blackish brown. Legs long, hairy and spinous; tarsus with three claws. Leg

Plate 7.1: *C. bidenta.*

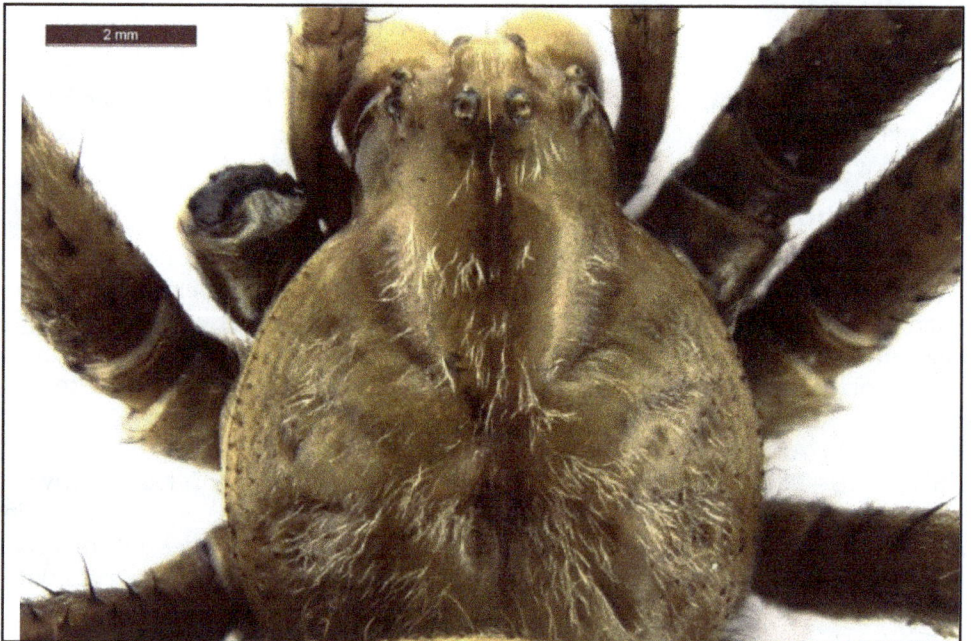

Plate 7.2: Cephalothorax.

Plate 7.3: Sternum.

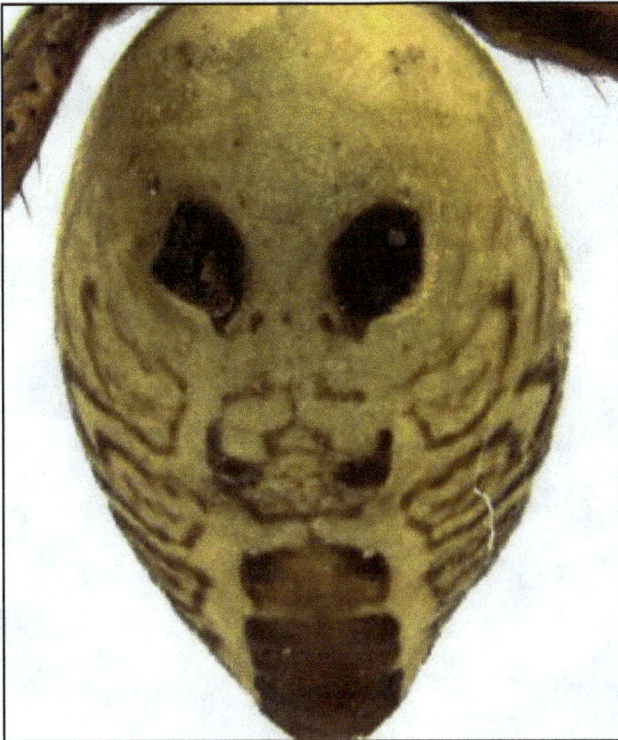

Plate 7.4: Abdomen (Dorsum).

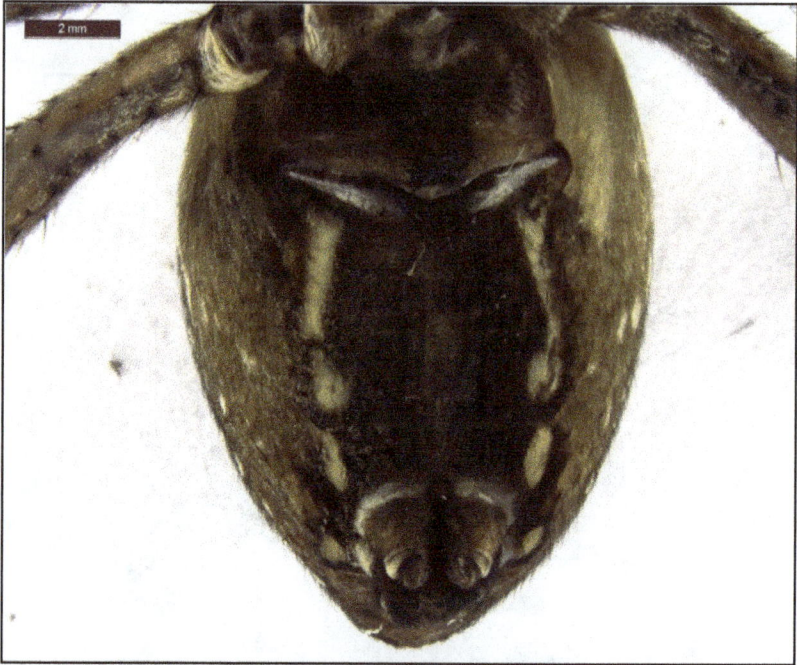

Plate 7.5: Abdomen (Ventrum).

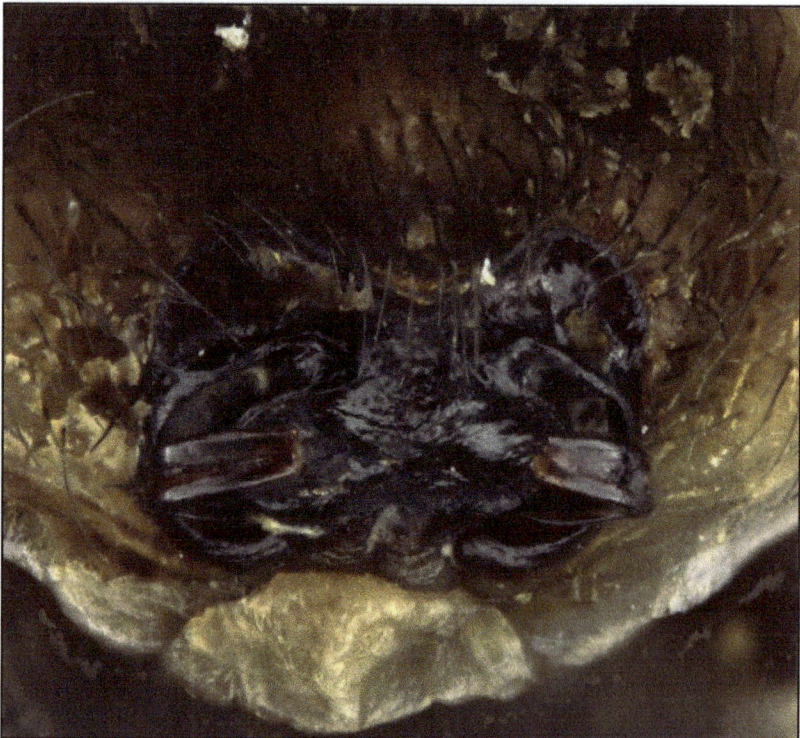

Plate 7.6: Epigyne.

Plate 7.7: Ring of Epigyne.

formula 1243; segment joints pale except tarsus-metatarsus junction. Long, slender palps similar in colour to legs.

Table 7.1: Leg and Palp Measurements of *C. bidenta*

Leg	Femur	Patella	Tibia	Metatarsus	Tarus	Total
I	9.50	3.25	4.63	4.60	2.80	30.95
II	9.5	3.5	7.75	7.5	2.25	30.5
III	6.7	2.75	3.0	4.3	2.4	19.15
IV	9.5	2.75	5.9	7.7	2.6	28.45
Pedipalp	2.94	1.23	1.83	–	2.06	8.06

Abdomen

10.2 mm long, 8.5 mm wide, broadest near the middle, high up anteriorly and slightly overlapping the carapace. Dorsum pale brown with one pair of prominent, black shoulder humps. Posterior half distinguished by a broad brown mid-longitudinal patch, either side of which is adorned by mirror images of patterns in brown and white. Ventrum dirty brown, clothed with hairs; mid longitudinal brown patch guarded on either sides by chalk white stripes in between epigastric furrow and spinnerets. Two white spots posterior to the spinnerets and four white spots around it. Colulus prominent, epigastric furrow surrounded by black hairs and outer rim bordered by brown curves. Epigyne prominent due to the presence of a ring like structure that projects outwards. Internal genitalia as follows: fertilization tubes short; spermathecae well distinguished.

Table 7.2: Checklist of *Cyrtophora* species from Southern Western Ghats

Sl.No.	Species
1.	*C. cicatrosa* (Stoliczka, 1869)
2.	*C. citricola* (Forsskål, 1775)
3.	*C. moluccensis* (Doleschall, 1857)
4.	*C. bidenta* (Tikader, 1970)
5.	*C. unicolor* (Doleschall, 1857)

Conclusion

The described specimen is a fully mature female reaching 18mm in length; the holotype measures only 5.8 mm. *C. bidenta* is very large in size compared to *C. cicatrosa* and *C. citricola* and is similar to *C. moluccensis* in having a pair of shoulder humps but colour and body pattern are spectacularly different, hence can be easily distinguished.

C. bicauda Saito, 1933 and C. exanthematica too have two humps but in the caudal region. Majority of *Cyrtophora* species including the described one have a pair of chalk white broken stripes between epigyne and spinnerets. The region of AME bulges out unlike other species which have only a slight projection. Epigyne of this species is unique in having the ring.

This is the first report for *C. unicolor* from India since 1900 and for *C. bidenta* from the Western Ghats. Pocock reported *C. unicolor* from mainland India in 1900 after which there have been no confirmed reports. B. K. Tikader (1980) has not listed *C. unicolor* in his monograph on Indian Araneae. This paper confirms Pocock's report and raises the number of *Cyrtophora* species in India to eight. The presence of *C. unicolor* doesn't come as a surprise since this species is confirmed for Sri Lanka and a few other countries on the immediate east of India *viz.* Myanmar, Burma Thailand, Taiwan and Philippines, with which India share much similarities in geographic and climatic features.

C. bidenta, endemic to India was not known from the Western Ghats until this study. Sikkim, the type locality of this species nestles in the Himalayan Mountains and has a temperate climate. The geography and climate of the Southern Western Ghats is indisputably different. Hence the present finding may possibly throw light on the ability of this species to adapt to various topographic and climatic conditions. Further study on seasonality and microhabitat has to be undertaken before coming to any conclusion. The origin of this species also remains to be determined. Whether it moved from South to North or vice versa is a valid question. The other three species are invasive among which *C. cicatrosa* and *C. citricola* are the most prevalent.

References

Berry, J. W. 1987. Notes on the life history and behavior of the communal spider *Cyrtophora moluccensis* (Doleschall) (Araneae, Araneidae) in Yap, Caroline Islands. *Journal of Arachnology*, **15**:309-319.

Mathew, M.J., P. A. Sebastian and E. Sunish (2008). Updated checklist of Indian Spiders (With references). In P.A. Sebastian and K.V. Peter (eds). *Spiders of India.* Orient Blackswan Pvt. Ltd., Hyderabad, India, 734 pp.

Platnick, N. I. 2013. The world spider catalog, version 13.5. American Museum of Natural History, online at http://research.amnh.org/iz/spiders/catalog/index.html.

Pocock, R.I. 1900. The Fauna of British India, including Ceylon and Burma. Arachnida. Taylor and Francis, London. 279 pp.

Scharff, N., and J. A. Coddington. 1997. A phylogenetic analysis of the orb-weaving spider family Araneidae (Arachnida, Araneae). *Zoological Journal of the Linnean Society,* **120**: 355-434.

Saito, S. 1933. Notes on the spiders from Formosa., *Trans. Sapporo nat. Hist. Soc.,* **13**: 32-61.

Tanikawa A., Y. Chang and I. Tso (2010). Taxonomic revision of Taiwanese and Japanese *Cyrtophora* spiders hitherto identified with *C. moluccensis* (Arachnida: Araneae), using molecular and morphological data. *Acta Arachnologica,* **59(1)**: 31-38.

Tikader, B.K. 1970. Spiders of Sikkim. Zool. Surv. India, Kolkata. 32pp.

Tikader, B.K. 1980. Fauna of India-Spiders-Araneae (Araneidae and Gnaphosidae). *Zool. Surv. India,* Kolkata, 174-176.

2015, Animal Diversity, Natural History and Conservation, Vol. 5   *Pages 85–97*
Editors: V.K. Gupta and Anil K. Verma
Published by: DAYA PUBLISHING HOUSE, NEW DELHI

Chapter 8

Effect of Insecticides on Spiders of Kuttanad Rice Agroecosystem, Kerala

A.V. Sudhikumar*

Centre for Animal Taxonomy and Ecology,
Department of Zoology, Christ College, Irinjalakuda, Kerala, India

ABSTRACT

Spiders are obligate carnivores and hold the unique position of being the only large class of arthropods which are entirely predatory in nature. To form a basis for research into the role of spiders to determine the economic importance of them in the rice agroecosystem of Kuttanad region of Kerala, a toxicity study of 3 commonly used insecticides on dominant spiders were conducted in the laboratory. Topical application (spraying) and dipping method were used for study. Of the three insecticides tested, Methyl parathion recorded the lowest lethal concentration values indicating its comparatively high toxicity in both methods. This is followed by Quinalphos and Monocrotophos. The exposure to Methyl parathion resulted in 80 per cent mortality of experimental spiders compared to 65 per cent and 40 per cent of mortality with Quinalphos and Monocrotophos respectively. This is suggestive of the usefulness of Monocrotophos as a component of integrated pest management strategy for sustainable paddy cultivation. Of the three dominant species tested, *Pardosa pseudoannulata* was the least susceptible to application of insecticides both by topical application method and the dipping method under laboratory conditions. *Tetragnatha mandibulata* was the most susceptible to the insecticides tested under laboratory conditions. Among the two methods used, dipping method was found to be more fatal compared to topical application. Spiders are very important biological control agents in agroecosystems

* E-mail: avsudhi@rediffmail.com

and play a major role as potential defenders by suppressing the pest population to a safe level which emphasizes the concept of Integrated Pest Management (IPM) in modern agriculture. Faced with the need to reduce pesticide usage on crops and optimize natural biological control, full investigation on the means by which spiders influence pest abundance and effect of insecticides on spiders is long overdue.

Keywords: *Spiders, Rice agroecosystem, Insecticide toxicity, Spraying, Dipping.*

Introduction

Integrated pest management is a part of the broader overall phenomenon of natural control. Natural control may be defined as the regulation of populations within certain more or less regular upper and lower limits over a period of time by any one or any combination of natural factors. Natural enemy populations have the unique ability of being able to interact with their prey or host populations and to regulate them at lower levels. In biological control, natural enemies are referred to as parasites, predators and pathogens. In order to appreciate the biological workings and ecological basis of biological control, it is first desirable to have an idea of different pest groups and their major characteristic natural enemies. If the natural enemies of a potential pest under complete biological control are killed off by insecticides, a great upset will occur. If the potential pest is under partial biological control only a mild upset may occur (De Bach, 1974).

The past experience in the pest control has indicated that no single method is successful. Viewing this multifactor problem associated with ecology, behaviour and biology of the pest and inherent limitations of each method of control, it is now almost universally accepted that the final solution lies with Integrated Pest Management (IPM) and biological control. The utilization of naturally occurring predatory arthropods is a major component of IPM. The compatibility of chemical and biological control may be achieved by determining the relative susceptibility of parasites and predators to the insecticides and to use only those which are relatively harmless to them. The integration of chemical and biological control should be brought about in such a manner that both of them should be able to exert their optimum beneficial effects.

Effect of persistent insecticides to non-target and useful organisms such as predators, parasites and pollinators is causing a chain of reactions leading to the appearance of several minor pests in epidemic form. When insecticides are used to control pests, care should be taken to ensure that the most important beneficial organisms are not seriously affected. When predators are allowed to operate undisturbed in the fields, less insecticidal treatments may be needed to control pests. Evaluation of the effects of insecticides on beneficial arthropods has attracted increasing attention of scientists in many parts of the world. The increasing use of insecticides for the pest control has come to the need for evaluating their effects on the beneficial arthropods. Even though insecticide application is the most commonly used method for controlling the insect pests in the rice fields; its usage has inevitably been followed by pest resistance and outbreaks of secondary pests. In the present study, the effect of certain insecticides used in the field on some dominant spiders was studied.

Materials and Methods

Sensitivity of spiders to widely used insecticides in the fields was evaluated in the laboratory condition. Adult females of three dominant species, *Araneus ellipticus* (Araneidae), *Pardosa pseudoannulata* (Lycosidae) and *Tetragnatha mandibulata* (Tetragnathidae) from the Kuttanad rice agroecosystem were selected for the toxicological studies. The insecticides tested were Ekalux EC 25 (*Quinalphos* 25 per cent EC), Hilcron 36 SL (*Monocrotophos* 36 SL) and Metacid (*Methayl parathion* 50 per cent EC). All these three insecticides are commonly used in the study area for the control of rice bug, brown plant hopper, green leaf hopper and other insect pests of paddy. The insecticides were diluted to four different concentrations (0.02 per cent, 0.04 per cent, 0.06 per cent and 0.08 per cent) by adding water. Before the tests, the field collected and reared spiders were kept at 27°C to 30°C at 70 per cent R.H. They were fed on *Drosophila melanogaster* Meiger and *Corcyra cephalonica* Stainton in the laboratory.

Susceptibility of three insecticides was evaluated in the laboratory by two methods *viz.*, dipping method and topical application as described by Tanaka *et al.* (2000). In the topical application, the various concentrations of insecticides were applied with a hand compression sprayer to the dorsal side of each adult female spider keeping them in the plastic petri dishes. Distilled water was sprayed as control. A total of 10 spiders were examined for each concentration. The treated spiders were transferred to individual plastic tubes (rearing chambers) containing moist cotton. Mortality counts were made after 24 and 48 hours of the treatment.

In the dipping method, the bottom end of a glass tube was covered with nylon gauze, which was fixed with a rubber band. Then 10 pre-anesthetized test spiders were put into the glass tube. The bottom end of this tube was dipped in the insecticide solution in a petri dish for 20 seconds. The tube was then placed on several sheets of filter paper to remove the insecticide solution. The treated spiders were transferred to individual plastic tubes (rearing chambers) containing moist cotton. The process was repeated for all the concentration of these insecticides. In the control treatment, the test individuals were dipped in distilled water. Mortality counts were recorded for 24 and 48 hours after treatment. The observed mortality was corrected using Abbot's equation (Abbot, 1925) and the LC_{50} (median lethal concentration) and LC_{90} values were calculated by probit analysis (Finney, 1971).

Results

Effect of Insecticides on Dominant Spiders

Araneus ellipticus (Family: Araneidae)

Of the three insecticides tested, Methyl parathion recorded the lowest lethal concentration values (LC_{50} of 0.042 and LC_{90} of 0.134) indicating its comparatively high toxicity in the topical application method (Table 8.1). This is followed by Quinalphos in the order of toxicity with an LC_{50} value of 0.067 and LC_{90} of 0.124.

Monocrotophos showed the least toxicity to *A. ellipticus* (LC_{50} of 0.085 and LC_{90} of 0.282). Mortality responses were high to all the concentrations of Methyl parathion, the highest being 77.80 per cent mortality at a dose of 0.08 per cent, 48 hours after treatment. As it would be expected, Monocrotophos proved to the safest among the three formulations, effecting only 44.40 per cent mortality to the test spiders at its highest concentration of 0.08 at 48 hours after treatment.

Table 8.1: Susceptibility Assays using Adult Females of *Araneus ellipticus* Exposed to different Concentrations of Insecticides by Topical Application

Insecticide	Dose (a.i. ml/L)	Per cent Mortality* (48 HAT**)	Lethal Concentration (ml/L)	Slope±SE
Quinalphos 25 per cent EC	0.02	00.00	LC_{50} = 0.067	5.02±2.14
	0.04	00.00	LC_{90} = 0.124	
	0.06	44.40	X^2 = 0.566	
	0.08	55.60		
Monocrotophos 36 per cent SL	0.02	00.00	LC_{50} = 0.085	2.45±1.15
	0.04	00.00	LC_{90} = 0.282	
	0.06	33.30	X^2 = 1.114	
	0.08	44.40		
Methyl parathion 50 per cent EC	0.02	11.10	LC_{50} = 0.042	2.61±0.97
	0.04	44.40	LC_{90} = 0.134	
	0.06	55.60	X^2 = 0.223	
	0.08	77.80		

* Corrected using Abbot's formula.

** Hours after treatment.

In the dipping method, Methyl parathion again showed the highest toxicity recording the least lethal concentration values as shown in Table 8.2 (LC_{50} of 0.038 and LC_{90} of 0.122). Quinalphos showed an LC_{50} value of 0.048 and LC_{90} of 0.112. Monocrotophos recorded the highest LC_{50} of 0.058 and LC_{90} of 0.137, indicating its least toxicity to *A. ellipticus* as was the case in the topical application method.

As far as mortality responses are concerned, Methyl parathion caused the highest response of 77.80 per cent, 55.50 per cent, 44.40 per cent and 11.10 per cent at 0.08 per cent, 0.06 per cent, 0.04 per cent and 0.02 per cent concentrations respectively in topical application and 77.80 per cent, 77.80 per cent, 44.40 per cent and 11.10 per cent at 0.08 per cent, 0.06 per cent, 0.04 per cent and 0.02 per cent concentrations respectively in dipping method, 48 hours after the treatment. Monocrotophos affected the lowest mortality responses, causing only 44.40 per cent and 33.30 per cent at 0.08 per cent and 0.06 per cent concentrations respectively in topical application and 77.80 per cent, 55.60, 22.20 and 11.10 per cent at 0.08 per cent, 0.06 per cent, 0.04 per cent and 0.02 per cent concentrations respectively in dipping method, 48 hours after treatment. The lowest two concentrations of topical application did not cause any response to *A. ellipticus*. This is suggestive of the usefulness of Monocrotophos as a component of integrated pest management strategy for sustainable paddy cultivation. This study revealed that Monocrotophos and Quinalphos was the least toxic to this spider in spraying and dipping method respectively.

Table 8.2: Susceptibility Assays using Adult Females of *Araneus ellipticus* Exposed to different Concentrations of Insecticides by Dipping Method

Insecticide	Dose (a.i. ml/L)	Per cent Mortality* (48 HAT**)	Lethal Concentration (ml/L)	Slope±SE
Quinalphos 25 per cent EC	0.02	00.00	LC_{50} = 0.048	4.61±2.02
	0.04	11.10	LC_{90} = 0.112	
	0.06	55.60	X^2 = 0.344	
	0.08	66.70		
Monocrotophos 36 per cent SL	0.02	11.10	LC_{50} = 0.058	2.78±1.01
	0.04	22.20	LC_{90} = 0.137	
	0.06	55.60	X^2 = 1.007	
	0.08	77.80		
Methyl parathion 50 per cent EC	0.02	11.10	LC_{50} = 0.038	3.03±1.00
	0.04	44.40	LC_{90} = 0.122	
	0.06	77.80	X^2 = 0.402	
	0.08	77.80		

* Corrected using Abbot's formula.

** Hours after treatment.

Pardosa pseudoannulata (Family: Lycosidae)

Of the three dominant species tested, *P. pseudoannulata* was the least susceptible to the application of insecticides both by topical application method and the dipping method under laboratory conditions. Among the three insecticides tested, Monocrotophos showed the highest toxicity to the spiders recording the lowest LC_{50} value of 0.072 and LC_{90} value of 0.125 (Table 8.3) in the topical application method.

Table 8.3: Susceptibility Assays using Adult Females of *Pardosa pseudoannulata* Exposed to different Concentrations of Insecticides by Topical Application

Insecticide	Dose (a.i. ml/L)	Per cent Mortality* (48 HAT**)	Lethal Concentration (ml/L)	Slope±SE
Quinalphos 25 per cent EC	0.02	00.00	LC_{50} = 0.105	2.16±1.98
	0.04	00.00	LC_{90} = 0.408	
	0.06	30.00	X^2 = 0.254	
	0.08	40.00		
Monocrotophos 36 per cent SL	0.02	00.00	LC_{50} = 0.072	5.05±2.19
	0.04	10.00	LC_{90} = 0.125	
	0.06	40.00	X^2 = 0.064	
	0.08	60.00		
Methyl parathion 50 per cent EC	0.02	00.00	LC_{50} = 0.084	2.78±1.01
	0.04	20.00	LC_{90} = 0.243	
	0.06	30.00	X^2 = 0.116	
	0.08	50.00		

* Corrected using Abbot's formula.

** Hours after treatment.

Methyl parathion recorded an LC_{50} of 0.084 and LC_{90} of 0.243. Quinalphos was the least toxic recording the highest lethal concentration values (LC_{50} 0.105 and LC_{90} 0.408). Mortality responses were the highest to Monocrotophos recording per cent mortalities of 60 per cent, 40 per cent and 10 per cent at concentrations of 0.08, 0.06 and 0.04 respectively, 48 hours after treatment. The lowest concentration, 0.02 did not have any effect on *P. pseudoannulata* at 48 hours after treatment.

In the dipping method, the three insecticides did not show much variation in the toxicity and mortality responses. However, Monocrotophos recorded the lowest lethal concentration values (LC_{50} of 0.064 and LC_{90} of 0.012) as shown in Table 8.4 indicating its comparatively higher toxicity to *P. pseudoannulata*. Methyl parathion was the least toxic among the three insecticides with an LC_{50} of 0.075 and LC_{90} of 0.271. Monocrotophos affected the highest mortality of 70 per cent at 0.08 concentration, 48 hours after treatment. Quinalphos and Methyl parathion affected 60 per cent mortality each at the same concentration. Monocrotophos and Quinalphos did not cause any mortality response to *P. pseudoannulata* at 0.02, 48 hours after treatment. This study revealed that Quinalphos and Methyl parathion was the least toxic to this spider in spraying and dipping method respectively.

Table 8.4: Susceptibility Assays using Adult Females of *Pardosa pseudoannulata* Exposed to different Concentrations of Insecticides by Dipping Method

Insecticide	Dose (a.i. ml/L)	Per cent Mortality* (48 HAT**)	Lethal Concentration (ml/L)	Slope±SE
Quinalphos 25 per cent EC	0.02	00.00	$LC_{50} = 0.071$	5.05±2.19
	0.04	10.00	$LC_{90} = 0.125$	
	0.06	40.00	$X^2 = 0.064$	
	0.08	60.00		
Monocrotophos 36 per cent SL	0.02	00.00	$LC_{50} = 0.064$	4.49±2.05
	0.04	20.00	$LC_{90} = 0.012$	
	0.06	40.00	$X^2 = 0.174$	
	0.08	70.00		
Methyl parathion 50 per cent EC	0.02	10.00	$LC_{50} = 0.075$	2.28±1.06
	0.04	30.00	$LC_{90} = 0.271$	
	0.06	30.00	$X^2 = 0.805$	
	0.08	60.00		

* Corrected using Abbot's formula.

** Hours after treatment.

Tetragnatha mandibulata (Family: Tetragnathidae)

T. mandibulata was the most susceptible to the insecticides tested under laboratory conditions. Among the three insecticides tested, Methyl parathion was the most toxic in the topical application method recording an LC_{50} value of 0.028 and LC_{90} of 0.062 (Table 8.5). Monocrotophos was the least toxic to *T. mandibulata* having lethal concentrations of 0.053 and 0.115 respectively. Complete mortality was resulted by Methyl parathion at 0.06 and 0.08, as well as Quinalphos 0.08, at 48 hours after

Table 8.5: Susceptibility Assays using Adult Female of *Tetragnatha mandibulata* Exposed to different Concentrations of Insecticides by Topical Application

Insecticide	Dose (a.i. ml/L)	Per cent Mortality* (48 HAT**)	Lethal Concentration (ml/L)	Slope±SE
Quinalphos 25 per cent EC	0.02	00.00	$LC_{50} = 0.051$	7.49±2.29
	0.04	11.10	$LC_{90} = 0.074$	
	0.06	77.70	$X^2 = 0.538$	
	0.08	100.0		
Monocrotophos 36 per cent SL	0.02	00.00	$LC_{50} = 0.053$	3.78±1.19
	0.04	11.10	$LC_{90} = 0.115$	
	0.06	55.50	$X^2 = 1.218$	
	0.08	77.80		
Methyl parathion 50 per cent EC	0.02	22.20	$LC_{50} = 0.028$	3.71±1.34
	0.04	66.70	$LC_{90} = 0.062$	
	0.06	100.0	$X^2 = 0.026$	
	0.08	100.0		

* Corrected using Abbot's formula.

** Hours after treatment.

Table 8.6: Susceptibility Assays using Adult Females *Tetragnatha mandibulata* Exposed to different Concentrations of Insecticides by Dipping Method

Insecticide	Dose (a.i. ml/L)	Per cent Mortality* (48 HAT**)	Lethal Concentration (ml/L)	Slope±SE
Quinalphos 25 per cent EC	0.02	00.00	$LC_{50} = 0.041$	5.15±1.66
	0.04	12.50	$LC_{90} = 0.073$	
	0.06	87.50	$X^2 = 2.285$	
	0.08	100.0		
Monocrotophos 36 per cent SL	0.02	00.00	$LC_{50} = 0.041$	3.36±1.04
	0.04	25.00	$LC_{90} = 0.097$	
	0.06	62.50	$X^2 = 0.812$	
	0.08	87.50		
Methyl parathion 50 per cent EC	0.02	62.50	$LC_{50} = 0.009$	1.58±1.29
	0.04	100.01	$LC_{90} = 0.060$	
	0.06	00.01	$X^2 = 0.905$	
	0.08	00.0		

* Corrected using Abbot's formula.

** Hours after treatment.

treatment. Monocrotophos effected only 77.8 per cent mortality at its highest concentration revealing its superiority over other insecticides as far as safety to the spiders is concerned.

The same trend was observed in the dipping method. However, dipping method proved rather more toxic to *T. mandibulata* in comparison with topical application. In the dipping method, Methyl parathion recorded the lowest LC_{50} value of 0.009 and

LC_{90} of 0.06, revealing its high toxicity to the test arthropods (Table 8.6). Again, Monocrotophos was the least toxic among the three insecticides tested with values 0.041 (LC_{50}) and 0.097 (LC_{90}) and effecting maximum mortality of 87.5 per cent at its highest concentration tested. Methyl parathion at all concentrations except at 0.02, as well as Quinalphos at 0.08 effected 100 per cent mortality to *T. mandibulata* 48 hours after the treatment. The data clearly suggest that among the three insecticides tested, Methyl parathion was the most toxic to *T. mandibulata*.

The overall LC_{50} values of the three chemicals to the three spiders by two different methods is given in Table 8.7. This study revealed that *P. pseudoannulata* was most resistant spider to these chemicals followed by *A. ellipticus* and *T. mandibulata*.

Table 8.7: LC_{50} Value of Three different Insecticides on Three Dominant Spiders by Two Methods of Aapplication

Spider Species	LC_{50} (ml/L)	
	Spraying	Dipping
Quinalphos 25 per cent EC		
A. ellipticus	0.067	0.058
P. pseudoannulata	0.105	0.071
T. mandibulata	0.051	0.041
Monocrotophos 36 per cent SL		
A. ellipticus	0.085	0.048
P. pseudoannulata	0.072	0.064
T. mandibulata	0.053	0.041
Methyl parathion 50 per cent EC		
A. ellipticus	0.042	0.038
P. pseudoannulata	0.084	0.075
T. mandibulata	0.028	0.009

Discussion

Laboratory experiments to test the insecticidal effect on spiders revealed that the spiders are quite sensitive to insecticides which kill them by contact. Kiritani (1979) reported the reduction of spiders through contact toxicity by ingesting pests contaminated with insecticides. However, their behaviour to feed only on moving insects perhaps saves them from greater reductions in population as a result of insecticidal usage. The susceptibility to insecticides varied between species of spiders. In the present study, the tetragnathid spider (*Tetrgantha mandibulata*) was more susceptible to insecticides than the other spiders, *Araneus ellipticus* and *Pardosa pseudoannulata*. The similar result was obtained in experiments by Tanaka *et al.* (2000) with *T. maxillosa*.

Mortality of predators usually occurs not only by direct contact but also by ingesting prey that had taken up the insecticides (Kiritani and Kakiya, 1975). Since the predator has to search out its prey, it is expected to pick up greater amounts of toxicant and thus suffer greater mortality than the more sedentary pests occupying the same habitat. In the present study, the effects of commonly used insecticides were tested for their toxicity to three common spiders in rice fields which are potential predators of hoppers. Among them, *P. pseudoannulata* is a hunter and their chances of coming across pesticide residues are greater while the other two, *A. ellipticus* and *T. mandibulata* are orb weavers which have a greater chance of being directly exposed to pesticide spray.

Thang *et al.*(1988) reported that *P. pseudoannulata* could tolerate common chemicals used in agroecosystems for pest eradication. The results of present study also agree with this. The safety of different chemicals to different spiders has been reported earlier by Fabellar and Heinrichs (1986). This study also confirms the earlier report of Feber *et al.* (1998) that *P. pseudoannulata* could tolerate the common chemicals. One of the theories involved in the selectivity of certain insecticides is the involvement of the mixed function oxidases. In general, organophosphates have been reported to be highly selective to *P. pseudoannulata* compared to carbamate compounds (Thang *et al.*, 1988). Kiritani and Kakiya (1975) and Thang *et al.* (1988) have observed significant mortality of predators not only from direct toxicity but also from ingestion of prey that had taken up the insecticides. They have theorized that the coupling of low mixed function oxidase activity with slow penetration, weak binding to the integument and tissue and speedy excretion in *P. pseudoannulata* could be the explanation for its lesser susceptibility to insecticides.

Spider communities did change in many ways over the season, but this was not related to an incremental increase in the rate of pesticide application. One major factor which influenced the spider community composition over the season was the profound effect of differences in spider phenology. Apart from the fact that insecticide use is rarely necessary, it also poses a risk to farmer health and the environment (Heong *et al.*, 1995). Continued insecticide use stresses the need to bridge the gap between research and farmers. FAO has supported Farmers' Field Schools in many countries and provided farmers with a practical understanding of integrated pest and nutrient management (Matteson, 2000). The expectation is that the farmers who receive training will pass their new knowledge on to other farmers. Another approach was developed by Heong *et al.* (1998), where farmers were motivated to 'test' a simple rule of thumb (no spray necessary in the first 40 days after sowing) by the use of communication media, including the radio. The practice of no early spray is now adopted by many farmers in southern Vietnam, and recommended by the National Agricultural Research and Extension Agencies in Malaysia, the Philippines, and Thailand (Matteson, 2000).

Many farmers use chemical pesticides to help control pests. An ideal biological control agent, therefore, would be one that is tolerant to synthetic insecticides. Although spiders may be more sensitive to insecticides than insects due in part to their relatively long life spans, some spiders show tolerance, perhaps even resistance, to some pesticides. Spiders are less affected by fungicides and herbicides than by insecticides (Yardim and Edwards, 1998). Spiders such as the wolf spider *P. pseudoannulata* are highly tolerant of botanical insecticides such as neem-based chemicals (Theiling and Croft, 1988; Markandeya and Divakar, 1999). They are also generally more tolerant of organophosphates and carbamates than of pyrethroids, organochlorines, and various acaricides, although this tolerance may be due to genetic resistance bred over a period of continuous exposure (Theiling and Croft, 1988; Wisniewska and Prokopy, 1997; Yardim and Edwards, 1998; Marc *et al.,* 1999; Tanaka *et al.,* 2000). For example, *P. pseudoannulata* (Lycosidae), *Tetragnatha maxillosa* (Tetragnathidae), *Ummeliata insecticeps* (Linyphiidae) and *Gnathonarium exsiccatum* (Linyphiidae) were highly sensitive to the pyrethroid deltamethrin, but very tolerant of the organophosphate diazinon and the carbamate carbaryl (Tanaka *et al.,* 2000).

Some broad-spectrum organophosphates are highly toxic to spiders. For example, dimethoate sprays resulted in 100 per cent mortality to the lycosid *Trochosa ruricola* at concentrations below recommended field application rates (Birnie *et al.,* 1998). The organophosphate methyl parathion and the pyrethroid cypermethrin are highly toxic to spiders in the genus *Erigone* (Linyphiidae), while the carbamate pirimicarb is almost harmless (Brown, 1981; Huusela-Veistola, 1998). Toft and Jensen (1998) found that sublethal doses of dimethoate and cypermethrin had no effect on development and predation rates of the wolf spider *Pardosa amentata.* In fact, with very low doses of cypermethrin, killing rates of the adult and penultimate females increased. However, the insecticides did have knockdown effects that, although not influencing survival in the laboratory, would likely result in death in the field due to desiccation or predation (Toft and Jensen, 1998).

Other factors that influence effects of pesticides on spiders are type of solvent, soil type, moisture, percent organic matter, temperature, and time of day of spraying. Further, the microhabitat, hunting style, prey preference, and behaviour of the spider also influence their response to pesticide application (Marc *et al.,* 1999). Wisniewska and Prokopy (1997) reported that if pesticides were only used early in the growing season, spider populations increased. Spatial limitation of pesticides (such as only applying the pesticides to certain plants or certain plots) also results in higher spider numbers, since they can move out of the treated areas and return when the chemicals dissipate (Riechert and Lockley, 1984; Balanca and de Visscher, 1997).

Spider density and diversity are significantly higher in orchards and fields where no pesticides have been sprayed (Feber *et al.,* 1998; Huusela-Veistola, 1998; Yardim and Edwards, Bogya and Marko, 1999; Marc *et al.,* 1999; Holland *et al.,* 2000; Amalin *et al.,* 2001). Restricting insecticide treatment to crucial periods in the pest life cycle or limiting spraying to midday when many wandering spiders are inactive and in sheltered locations can help conserve spider numbers (Riechert and Lockley, 1984). Spiders can recolonize if the interval between chemical applications is long enough, but several applications per season can destroy spider communities. Some pesticides are also retained in the webs of spiders and can be detrimental to those spiders that ingest their webs daily (Marc *et al.,* 1999).

Spiders are the dominant predators in rice fields. Among them *Pardosa* sp., *Araneus* sp. and *Tetragnatha* sp. are the most common. Evidences are coming out of different countries that these spiders, because of their high population densities in rice fields have a damping influence on the populations of pest insects especially the plant and leaf hoppers. Regular application of insecticides is found to almost totally suppress the spider population. The present study was taken up to assess the safety of a few insecticides to these spiders. Application of insecticides is the most commonly practiced methods for controlling the rice hoppers. However, extensive use of insecticides has exposed limitations of providing temporary control and posing some adverse toxicological problems. Furthermore, the use of broad spectrum insecticides has almost inevitably been followed by pest resistance, resurgence and secondary pest outbreaks. In all probability, the hoppers have gained importance as pests due to the destruction of their natural enemies. Selectivity of insecticides is important for pest management. Careful choice of insecticides might not only restrict the adverse effects of chemical application on the spider fauna but also on the predator community as a whole. Restriction of the application of chemicals to local areas or only to randomly selected hills (Ressig *et al.*, 1982) could also permit recolonisation of predators. However, unsprayed, irrigated rice fields have relatively few insect pest problems. This is largely attributed to natural biological control, which keep planthoppers and other potential pests in check (Kenmore, 1991; Way and Heong, 1994).

Acknowledgements

Author is grateful to Principal, Christ College, Irinjalakuda, Kerala for providing laboratory facilities. He also acknowledges Idea Wild Organization, USA and Encyclopaedia of Life Programme, National Museum of Natural History, Smithsonian Institution, USA for financial assistance.

References

Abbot, W. S., 1925. A method of computing the effectiveness of an insecticide. *J. Econ. Entomol.*, **18**: 265-267.

Amalin, D. M., Pena, J. E., Duncan, R. E., Browning, H. W., and McSorley, R., 2001. Natural mortality factors acting on citrus leafminer, *Phyllocnistis citrella*, in lime orchards in south Florida. *Biocontrol*, **47**: 327-347.

Balanca, G., and de Visscher, M. N., 1997. Impacts on nontarget insects of a new insecticide compound used against the Desert Locust (*Schistocerca gregaria*). *Arch. Envi. Cont. Toxicol.*, **32**:58-62.

Birnie, L., Shaw, K., Pye, B., and Denholm, I., 1998. Considerations with the use of multiple dose bioassays for assessing pesticide effects on non-target arthropods. In *Proc. Brighton Crop Protection Conference. Pests and Diseases.*UK. 296pp.

Bogya, S., and Marko, V., 1999. Effect of pest management systems on ground-dwelling spider assemblages in an apple orchard in Hungary. *Agri. Ecosys. Environ.*, **73**: 7-18.

Brown, K., 1981. Foraging ecology and niche partitioning in orb-weaving spiders.*Oecologia*, **50**: 380-385.

De Bach, P., 1974. *Biological control by natural enemies.*Cambridge Univ. Press.London. 323pp.

Fabellar, L. T., and Heinrichs, E. A., 1986. Relative toxicity of insecticides to rice planthoppers and leafhoppers and their predators. *Crop Protection*, **5**: 254-258.

Feber, R. E., J. Bell, Johnson, P. J., Firbank, L. G., and Macdonald, D. W.,. 1998. The effects of organic farming on surface-active spider (Araneae) assemblages in wheat in southern England, UK. *J. Arachnol.*, **26**: 190-202.

Finney, D. J., 1971. *Probit analysis.* Cambridge university press, London. 333pp.

Heong, K. L., Escalada, M. M., and Lazaro, A. A., 1995. Misuse of pesticides among rice farmers in Leyte, Philippines. *Bull. Entomol. Res.*, **81**: 97-108.

Heong, K. L., Escalada, M. M., Huan, N. H., and Mai, V., 1998. Use of communication media in changing rice farmers' pest management in the Mekong Delta, Vietnam. *Crop Prot.*, **17**: 413-25.

Holland, J. M., Winder, L., and Perry, J. N., 2000. The impact of dimethoate on the spatial distribution of beneficial arthropods in winter wheat. *Ann. Appl. Biol.*, **136**: 93-105.

Huusela-Veistola, E., 1998.Effects of perennial grass strips on spiders (Araneae) in cereal fields and impact on pesticide side-effects. *J. Appl. Ent.*, **122**: 575-583.

Kenmore, P. E., 1991. Indonesia's integrated pest management program: A model for Asia. Manila, Philippines: FAO.

Kiritani, K., and Kakiya, N. 1975. An analysis of the predator-prey system in the paddy field. *Res. Popul. Ecol.*, **17**: 2938.

Kiritani, K., 1979. Pest management in rice. *Ann. Rev. Entomol.*, **24**: 279-312.

Marc, P., Canard, A., and Ysnel, F., 1999. Spiders (Araneae) useful for pest limitation and bioindication. *Agric. Ecosyst. Environ.*, **74**: 229-273.

Markandeya, V., and Divakar, B. J., 1999. Effect of a neem formulation on four bioagents. *Plant Prot. Bull.*, **51**: 28-29.

Matteson, P. C., 2000. Insect pest management in tropical Asian irrigated rice. *Ann. Rev. Entomol.*, **45**: 549-574.

Reissig, W. H., Heinrichs, E. A., and Valencia, S. L., 1982. Effects of insecticides on *Nilaparvata lugens* and its predators: Spiders. *Microvelia atrolineata* and *Cyrtohinus lividipennis*. *Environ. Entomol.*, **11**: 193-99.

Riechert, S. E., and Lockley, T., 1984. Spiders as biological control agents. *Annu. Rev. Entomol.*, **29**: 299-320.

Tanaka, K., Endo, S., and Kazano, H., 2000. Toxicity of insecticides to predators of rice planthoppers: spiders, the mired bug, and the dryinid wasp. *Appl. Entomol. Zool.*, **35**: 177-187.

Thang, M. H., Mochida, O., and Rejesus, B. M., 1988. Mass rearing of the wolf spider *Lycosa pseudoannulata* (Araneae :Lycosidae). *Phillipp. Ent.*, **1**: 51-66.

Theiling, K. M., and Croft, B. A., 1988. Pesticide side-effects on arthropod natural enemies: a database summary. *Agric., Ecosyst. Environ.*, **21**: 191-218.

Toft, S., and Jensen, A. P., 1998. No negative sublethal effects of two insecticides on prey capture and development of a spider. *Pestic. Sci.*, **52**: 223-228.

Way, M. J., and Heong, K. L., 1994. The role of biodiversity in the dynamics and management of insect pests of tropical irrigated rice - a review. *Bull. Entomol. Res.*, **84**: 567-587.

Wisniewska, J., and Prokopy, R. J., 1997. Pesticide effect on faunal composition, abundance, and body length of spiders (Araneae) in apple orchards. *Environ. Entomol.*, **26**: 763-776.

Yardim, E. N., and Edwards, C. A., 1998. The influence of chemical management of pests, diseases and weeds on pest and predatory arthropods associated with tomatoes. *Agric. Ecosyst. Environ.*, **70**: 31-48.

2015, Animal Diversity, Natural History and Conservation, Vol. 5   *Pages 99–111*
Editors: **V.K. Gupta and Anil K. Verma**
Published by: **DAYA PUBLISHING HOUSE, NEW DELHI**

Chapter 9

Prediction of Number of Generations of American Serpentine Leaf Miner, *Liriomyza trifolii* (Burgess) based on Thermal Degree-Days Requirement under Climate Change Scenario

V. Sridhar*, L.S. Vinesh and M. Jayashankar

Division of Entomology and Nematology,
Indian Institute of Horticultural Research,
Hessaraghatta Lake Post, Bengaluru – 560 089, Karnataka, India

ABSTRACT

Number of generations of American serpentine leaf miner, *Liriomyza trifolii* (Burgess) was assessed under current and expected future climate change scenarios based on its Thermal Degree-Days (TDD) requirement by using both horizontal and vertical cut off methods (HCO and VCO, respectively). The assessment was done for A1B climate change scenario for three different locations in India *viz.*, Ludhiana (Northern India), Bengaluru (Southern India) and Kalyani (Eastern India). Number of generations of *L. trifolii* estimated by Horizontal cut-off method was 15, 17, 18 in Ludhiana; 16, 18, 20 in Bengaluru and 17, 19, 21 in Kalyani under the base line, and future time frames of 2030 and 2075, respectively. It is therefore expected that number of generations of *L. trifolii* would increase with the rising temperatures under climate change situations. However, with vertical cut-off

* *Corresponding author.* E-mail: vsridhar@iihr.ernet.in

method the number of generations is expected to decrease in Ludhiana and Kalyani. *L. trifolii* would have had 12 generations, in base line and attain same number of generations in 2035 and only 11 generations in 2085 time frame under Ludhiana conditions. Under Bengaluru conditions, it would have had 16 generations in base line and is expected to complete the same number of generations in 2035, while 17 in 2085 scenarios. Under Kalyani conditions 14, 17, 15 number of generations were assessed for above mentioned time frames. Horizontal and vertical cut-off methods predict different number of generations in the selected locations. Relevance of HCO and VCO methods in assessing the number of generations of insect pests in the diverse climatic conditions of the country are discussed.

Keywords: *Liriomyza trifolii, Climate change, TDD, A1B, India.*

Introduction

Climate exerts powerful effects on the distribution and abundance of the insect species, and climate warming is expected to generate changes for many insect populations and the ecosystems they inhabit (Stange and Ayres, 2010). Several phenology models based on temperature are used to predict exotic pest establishment (Herms, 2004; Nietschke *et al.*, 2007; Shi, 2010) as it plays a major role in the growth and development of insects, compared to other abiotic factors. The amount of heat required to complete a generation in *poikilothermic* organisms like insects is fairly constant, with a species specific optimum temperature range restricted with lower and upper threshold temperatures. This optimum range can be used to calculate the degree-day units for the target insect and subsequently predict insect development under present and climate change scenarios (Roltsch *et al.*, 1999). After obtaining the effective degree-days accumulation, the number of generations is calculated manually or with the aid of software programmes. Different authors have reported the utility of TDD for predicting number of generations in *Linepithema humile* (Mayr) (Hartley and Lester, 2003); *Heliothrips haemorrhoidalis* Bouche (Chhagan and Stevens, 2007); *Bactrocera zonata* (Saunders) (Khalil *et al.*, 2010); *Tuta absoluta* (Abolmaaty *et al.*, 2010).

Leaf miner, *Liriomyza trifolii* (Burgess) (Diptera: Agromyzidae) is an invasive polyphagous pest on variety of vegetables and ornamental foliages causing significant economic loss in India (Viraktamath *et al.*, 1993; Srinivasan *et al.*, 1995; Kumar *et al.*, 2004; Kaur *et al.*, 2010; Chakraborty, 2011). Though the insect is reported in India in early 1990's (Anonymous, 1991) it has spread and established across India with peak infestation during summer months in different parts of the country (Saradhi and Patnaik, 2004; Durairaj, 2007; Singh *et al.*, 2005). Basic developmental biology of *L. trifolii* including Thermal Degree-Days requirement was studied by Liebee (1984). Estimation of number of generations of *L. trifolii* for present and future climate change time frames in India was keeping in view the climate change aspect into consideration. Possible number of generations of *L. trifolii* was estimated also for the time frame prior to the introduction of the insect into India, for the sake of comparing it with successive future time frames. Three locations are selected for the study representing different agro-climatic locations where leaf miner incidence was reported *i.e.*, Ludhiana (Sohi *et al.*, 1994; Kaur *et al.*, 2010), Bengaluru (Kumar *et al.*, 2004), and Kalyani (Chakraborty, 2011).

Prediction of number of generations of *L. trifolii* were estimated for IPCC (Inter-governmental Panel on Climate Change) A1B scenario for two future time frames *i.e.,* 2035 (2021-2050) and 2085 (2071-2098) compared to base line of 1975 (1961-1990). The IPCC Scenario A1 assumes a world of very rapid economic growth, a global population that peaks in mid-century and rapid introduction of new and more efficient technologies. A1 is divided into three groups that describe alternative directions of technological change: fossil intensive (A1FI), non-fossil energy resources (A1T) and a balance across all sources (A1B) (IPCC Special Report on Emissions Scenarios. 2001).

Materials and Methods

Representative Regions and Climate Data

The three locations selected in India for the present study are Ludhiana (N 30.70°E 75.79°- Northern India), Bengaluru (N 13.00°E 77.56°-Southern India) and Kalyani (N 23.18°E 88.62°- Eastern India). Data on minimum and maximum daily temperature for three time frames each comprising thirty year duration *viz.,* baseline (1961-1990), (2021-2050) and (2071-2098) under A1B scenario was obtained from IITM, Pune which has used Met office Hadley Center (U.K) Regional Climate Model PRECIS (Providing Regional Climates for Impact Studies). Average minimum and maximum temperatures for the above said time frames were calculated starting from 1st January to 31st December to arrive at the mean year's data for each duration *viz.,* 1975, 2035 and 2085. Henceforth in this paper time frame would be used referring to the mean year. Average temperatures for each day were used to calculate accumulated degree-days and for further predicting leaf miner generations based on the TDD requirements.

Thermal Requirements

Threshold temperatures for development and Thermal degree-days requirements for *L. trifolii* were followed after Leibee (1984) *i.e.,* 9.70°C (lower threshold), 35°C (upper threshold) and 313.70°C (TDD requirement for egg-adult development). Expected number of generations were estimated using UC IPM online software (www.ipm.ucdavis.edu/WEATHER/index.html), wherein accumulated TDD was calculated by horizontal cut-off (assumes that development continues at a constant rate at temperatures in excess of the upper threshold) and vertical cut-off (assumes that no development occurs when a temperature is above the upper threshold) methods for selected locations. The acronyms for the cut-off methods *viz.,* Horizontal and Vertical Cut-off methods used in the text and tables are HCO and VCO, respectively.

Results and Discussion

Number of Generations in Representative Regions

In the present analysis-number of generations of *L. trifolii* estimated with HCO was 15, 17, 18 in Ludhiana; 16, 18, 20 in Bengaluru and 17, 19, 21 in Kalyani in different time frames *i.e.,* base line, 2035 and 2085, respectively (Tables 9.1–9.3). Thus, a gradual increase in the number of generations is expected in the three selected locations by 2100 (Figure 9.1). The highest additional generations possible were

Table 9.1: Number of Generations of *L. trifolii* under Present and Climate Change Scenarios in Ludhiana

No. of Generations	Current Climate 1975				Future Climate 2035				Future Climate 2085			
	HCO		VCO		HCO		VCO		HCO		VCO	
	Days	TDD	Days	TDD	Days	TDD	Days	TDD	Days	TDD	Days	TDD
1	64	318	64	318	60	314	60	314	48	317	48	317
2	23	315	23	312	22	317	24	324	23	319	24	324
3	18	319	25	312	17	326	28	318	17	321	27	317
4	16	326	33	313	15	316	45	316	15	312	74	316
5	14	321	35	326	14	319	34	324	14	330	28	327
6	13	313	16	328	13	320	16	338	13	321	20	325
7	13	321	16	314	13	326	15	312	13	328	22	320
8	14	328	16	317	13	325	15	317	13	329	25	322
9	15	320	16	328	13	333	15	317	13	327	20	316
10	16	316	17	323	15	331	17	322	20	334	20	311
11	16	316	20	321	15	313	21	322	13	326	42	315
12	16	326	29	318	15	315	29	312	15	323		
13	16	313			15	320			14	326		
14	19	323			15	313			14	329		
15	25	319			18	314			15	323		
16					23	314			15	320		
17					50	313			16	318		
18									24	318		
Mean	20	320	26	319	20	320	27	320	18	323	32	320

HCO: Horizontal cut-off method; VCO: Vertical cut-off method; TDD: Thermal degree days.

Table 9.2: Number of Generations of *L. trifolii* under Present and Climate Change Scenarios in Bengaluru

No. of Generations	Current Climate				Future Climate							
	1975				2035				2085			
	HCO		VCO		HCO		VCO		HCO		VCO	
	Days	TDD	Days	TDD	Days	TDD	Days	TDD	Days	TDD	Days	TDD
1	25	319	13	319	22	323	22	323	19	314	19	314
2	23	322	23	322	20	320	20	320	18	315	19	326
3	20	325	20	325	18	317	19	312	17	326	21	315
4	18	324	21	326	17	325	23	324	16	326	24	316
5	17	317	20	317	16	318	24	324	15	320	27	315
6	16	312	24	325	16	330	26	321	15	333	28	322
7	17	329	19	324	15	318	20	320	14	314	21	325
8	18	318	20	312	16	330	18	318	15	333	17	318
9	21	320	22	320	17	326	19	314	15	321	17	319
10	22	321	22	319	19	327	20	325	16	328	16	318
11	22	317	22	318	20	326	20	318	17	330	17	315
12	22	320	21	314	20	323	20	328	17	320	17	319
13	21	317	21	317	20	319	19	326	17	317	17	322
14	22	323	23	319	19	318	20	322	17	315	17	324
15	23	316	26	324	19	321	20	314	17	320	18	318
16	25	321	24	315	20	320	21	317	17	328	19	326
17					21	327			18	338	19	313
18					22	327			18	315		
19									19	323		
20									19	313		
Mean	21	321	21	320	19	323	21	321	20	319	17	322

HCO: Horizontal cut-off method; VCO: Vertical cut-off method; TDD: Thermal degree days.

Table 9.3: Number of Generations of *L. trifolii* under Present and Climate Change Scenarios in Kalyani

No. of Generations	Current Climate 1975				Future Climate 2035				Future Climate 2085			
	HCO		VCO		HCO		VCO		HCO		VCO	
	Days	TDD	Days	TDD	Days	TDD	Days	TDD	Days	TDD	Days	TDD
1	47	317	47	317	28	321	28	321	31	325	30	313
2	23	327	23	323	22	315	22	315	21	329	21	320
3	16	319	25	324	15	327	21	317	16	322	24	322
4	15	328	28	313	14	315	22	321	14	324	38	314
5	14	321	27	322	14	314	25	320	13	314	41	316
6	14	320	20	322	14	330	26	328	13	322	24	314
7	14	326	17	324	13	314	15	314	13	323	15	313
8	14	325	17	327	14	332	16	326	13	322	15	326
9	15	314	17	326	14	325	17	326	13	321	15	312
10	17	324	17	323	15	315	17	323	13	324	15	314
11	17	326	17	317	17	330	17	320	14	322	15	328
12	17	326	18	326	17	322	17	323	14	314	18	326
13	17	322	20	313	17	322	16	320	15	334	19	329
14	17	316	31	315	17	322	17	324	15	335	17	325
15	18	316			16	315	19	324	14	312	22	319
16	22	319			16	316	21	317	15	328		
17	38	317			18	318	26	321	15	332		
18					20	316			15	330		
19					24	326			16	327		
20									19	326		
21									27	322		
Mean	20	321	23	321	17	321	20	321	16	324	22	320

HCO: Horizontal cut-off method; VCO: Vertical cut-off method; TDD: Thermal degree days.

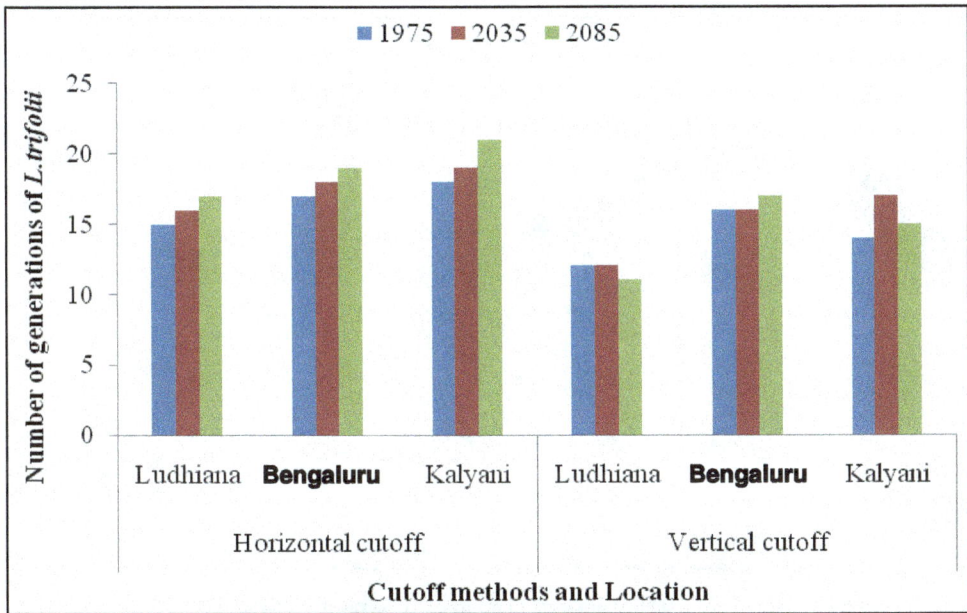

Figure 9.1: Linear Increase in Predicted Number of Generations for Three Time Frames.

recorded in Kalyani and Bengaluru (4 generations each), where available accumulated thermal units were more when compared to Ludhiana (3 generations) with extreme winter and summer by 2085. Lowest number of generations was observed in Ludhiana in 1975 time frame, which can be attributed to extreme lower temperatures in winter and fatal temperatures during summer (http://www.imd.gov.in/doc/climateimp.pdf). With VCO method (Tables 9.1–9.3), the insect would have had 12 generations, in base line and would attain same number of generations by 2035 and only 11 generations in 2075 time frame under Ludhiana conditions. Under Bengaluru conditions, the insect would have had 16 generations in base line and is expected to complete the same number of generations by 2035, while 17 generations by 2085 time frame. Under Kalyani conditions, 14, 17, 15 numbers of generations were calculated for above mentioned time frames, respectively. Abolmaaty *et al.* (2010) estimated increase in number of generations of Tomato Leaf miner moth, *Tuta absoluta* (Meyrick) for 2050 and 2100 future climates for A1 scenario in Egypt based on Degree Day's Units. The results are comparable with similar predictions for peach tree borer, *Synanthedon exitiosa* (Say) (Johnson and Mayes, 1984) and clover root weevil, *Sitona lepidus* Gyll. (Arbab and Mcneill, 2011). The variation in the number of generations estimated by HCO and VCO methods is attributed to the mode of calculating accumulating TDD as mentioned in material and methods. Also, under VCO method, any average temperature above upper threshold is not considered resulting in less number of generations when compared to HCO. In the present findings under Ludhiana (Figure 9.2) and Kalyani (Figure 9.4) conditions, lesser number of generations is predicted due to the expected higher temperatures during summer months with above upper threshold ranges.

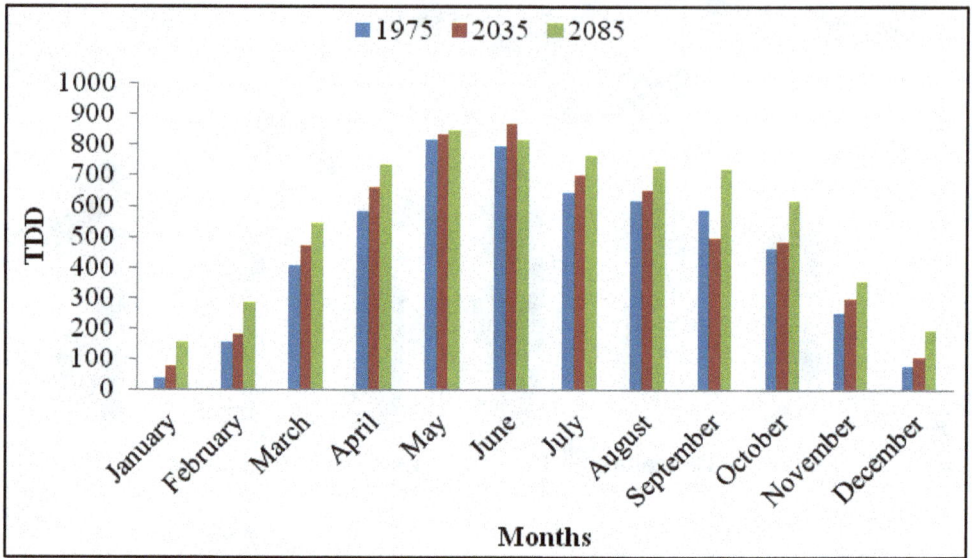

Figure 9.2: Monthly Availability of TDD in Ludhiana for Three Time Frames.

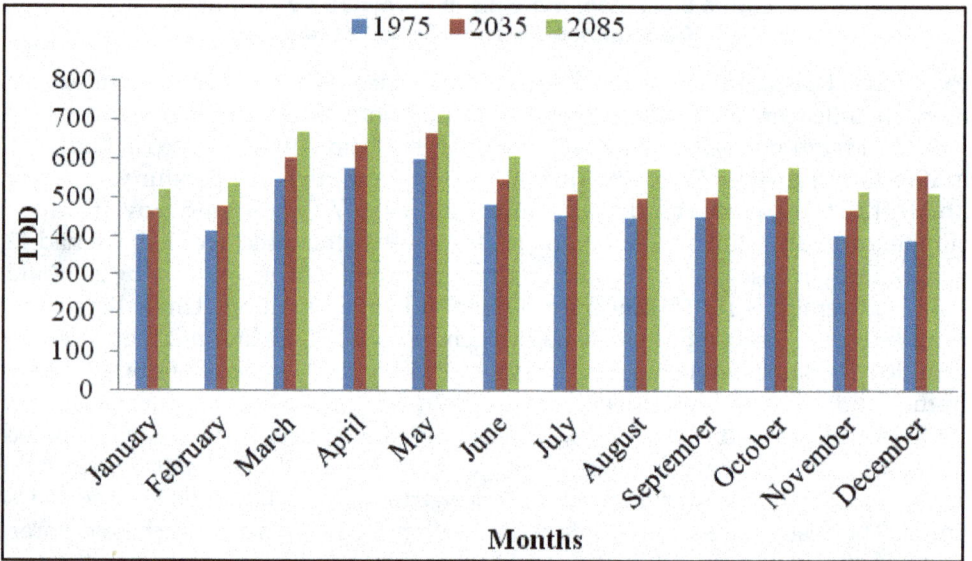

Figure 9.3: Monthly Availability of TDD in Bengaluru for Three Time Frames.

Days for Development and TDD

The mean values of the number of days and thermal degree days (Days-TDD) required for complete development of *L. trifolii* (Tables 9.1–9.3) for Ludhiana, Bengaluru and Kalyani during baseline were 19.87-319.64, 20.75-320.64 and 19.71-321.42, respectively. The number of days required for development ranged between

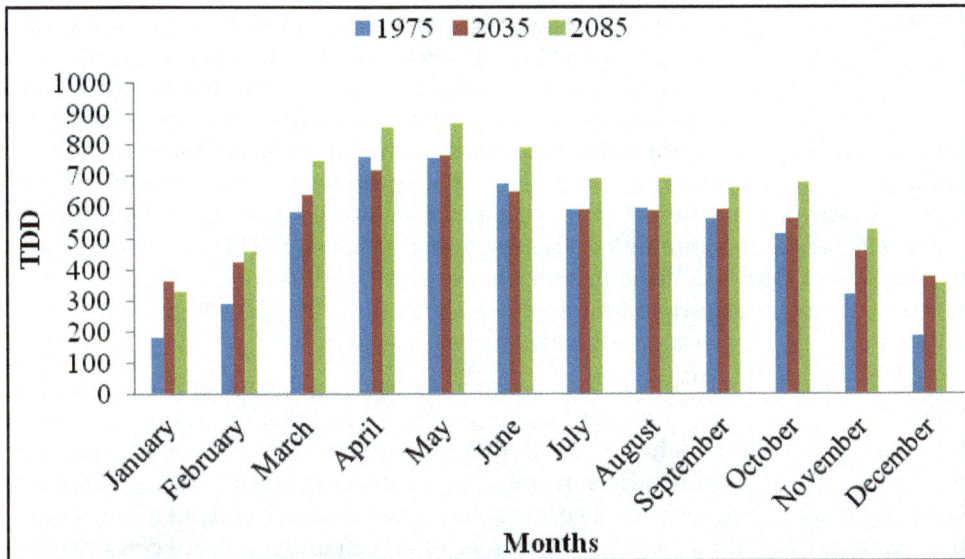

Figure 9.4: Monthly Availability of TDD in Kalyani for Three Time Frames.

13-64, 16-25 and 14-47 for the three selected locations, respectively. For 2035 time frame, estimated Days -TDD was 20.35-320.18, 18.72-323.05 and 17.11-320.78, respectively for Ludhiana, Bengaluru and Kalyani and the number of days required for development ranged between 13-60, 15-22 and 13-28 for the three selected regions, respectively. For 2085 time frame estimated Days-TDD was 17.50-323.31, 19.59-319.16 and 16.14-324.13, respectively for Ludhiana, Bengaluru and Kalyani and the number of days required for development ranged between 13-48, 14-19 and 13-31 for the three selected places, respectively.

In the vertical cut-off method (Tables 9.1–9.3) the number of days and thermal degree days (Days-TDD) required for complete development of *L. trifolii* for Ludhiana, Bengaluru and Kalyani during baseline are 25.83-319.07, 21.31-320.21 and 23.14-321.03, respectively. The number of days required for development ranged between 16-64, 13-26 and 17-47 for three selected locations, respectively. For 2035 it was 26.58-319.66, 20.69-320.78 and 20.12-321.10, respectively for Ludhiana, Bengaluru and Kalyani. The number of days required for development ranged between 15-60, 18-26 and 15-28 for the three selected locations respectively. For 2085 it was found to be 31.82-319.65, 16.80-322.48 and 21.93-319.57 respectively for Ludhiana, Bengaluru and Kalyani. The number of days required for development ranged between 20-74, 16-28 and 15-41 for three selected locations respectively.

Although both the cut-off methods are used to calculate TDD, single sine method with a HCO is usually recommended to use at the upper threshold, if the method used in developing the organism's biological growth relationships is unknown (Roltsch *et al.*, 1999; UC IPM, 2012). Additionally, VCO was used keeping in view the possible extreme temperatures predicted in places like Ludhiana and Kalyani because under those conditions, temperatures would be detrimental for the development of the pest. In spite of variations in the cut –off methods, our analysis indicated that the distribution of *L.trifolii* across India may well be limited by microclimate and region specific climates as in Ludhiana and Kalyani. However, locations like Benguluru are potentially suitable for this species. The present results emphasize the usage of degree-day models like mathematical models to describe the relationship between temperature and development rate for future prediction of pests. However, a multi-factorial approach incorporating other biotic and abiotic factors determining the development of *poikilotherms* is recommended (Sharpe and DeMichele, 1977; Schmaedick and Nyrop, 1993; Arbab and Mcneill, 2011; Rao *et al.,* 2012). Insecticide efficacy may be enhanced through the proper timing of applications at critical developmental stages of the insect pest, assessed through thermal degree-days.

Several studies indicate an increase in global average temperatures leading to regional climate changes that may have immediate effect mainly on the population dynamics of insects which are poikilothermic, affecting their development and distribution. The present results show rapid development of *L. trifolii* during summer months with increased number of generations because of availability of more thermal units, compared to other months (Figures 9.2–9.4) both under present conditions as well as under climate change situations. Various authors have predicted 15 generations per year for *L. trifolii* with life-cycle completion in 23.6 – 25.6 days at 20 °C; 24 generations at 25 °C (15.0 – 16.5 days) (Lanzoni *et al.,* 2002; Sakamaki *et al.,* 2003; Tokumaru and Abe, 2003; Andersen and Hofsvang, 2010). Degree day calculations are impounded by unavoidable circumstances in terms of variation in season and climate of different regions, accuracy of functioning of the weather instruments used and most of all different methods available for calculating degree days. However, phenological models based on degree-day accumulation have been developed to support the integrated pest management of many insects (Johnson and Mayes, 1984; Fettig *et al.,* 2004; Arbab and Mcneill, 2011). In conclusion, TDD accumulation based on degree day units can be a reliable tool to assess insect activity and as an indicator for the estimation of number of generations.

Acknowledgements

The authors are grateful to the Director, Indian Institute of Horticultural Research (IIHR) for his support and encouragement and Indian Council of Agricultural Research (ICAR), for providing financial aid under the project 'National Initiative on Climate Resilient Agriculture' (NICRA).

References

Abolmaaty, S.M., Hassanein, M.K., Khalil, A.A., and Abou-Hadid, A.F., 2010. Impact of Climatic Changes in Egypt on Degree Day's Units and Generation Number for Tomato Leaf miner Moth *Tuta absoluta*, (Meyrick) (*Lepidoptera : Gelechiidae. Nat. Sci.,* **8**: 122-129.

Andersen, A., and Hofsvang, T., 2010. Pest risk assessment of the American Serpentine Leafminer, *Liriomyza trifolii* in Norway. Opinion of the Panel on Plant Health of the Norwegian Scientific Committee for Food Safety, 09/904-5 final, ISBN 978-82-8082-401-1 (Electronic edition). pp. 35.VKM, Oslo, Norway.

Anonymous, 1991. Castor: Annual Report. Directorate of Oilseeds Research, Hyderabad. p.137.

Arbab, A. and Mcneill, R. M., 2011. Determining suitability of thermal development models to estimate temperature parameters for embryonic development of *Sitona lepidus* Gyll. (Coleoptera: Curculionidae). *J. Pest Sci.,* **84**: 303–311.

Chakraborty, K., 2011. Incidence and Abundance of Tomato Leaf Miner, *Liriomyza trifolii* (Burgess) in Relation to the Climatic Conditions of Alipurduar, Jalpaiguri, West Bengal, India. *Asian J. Exp. Biol. Sci.,* **2**: 467-473.

Chhagan, A., and Stevens, P.S., 2007. Effect of temperature on the development, longevity and oviposition of Greenhouse thrips (*Heliothrips haemorrhoidalis*) on Lemon fruit. *New Zealand Pl. Protec.,* **60**: 50-55.

Durairaj, C., 2007. Influence of abiotic factors on the incidence of serpentine leafminer, *Liriomyza trifolii. Indian J. Plant Prot.,* **35**: 232-234.

Fettig, C.J., Dalusky, J.M., and Berisford, W.C., 2004. Controlling Nantucket Pine Tip Moth Infestations in the Southeastern U.S. www.forestpests.org version 2.0, XHTML 1.1, CSS, 508.

Hartley, S., and Lester, P.J., 2003. Temperature-dependent development of the Argentine ant, *Linepithema humile* (Mayr) (Hymenoptera: Formicidae): a degree-day model with implications for range limits in New Zealand. *New Zealand Entomol.,* **26**: 91-100.

Herms, D.A., 2004. Using degree-days and plant phenology to predict pest activity, p. 59-59. In Krischik V, Davidson J (eds) IPM of midwest landscapes. St Paul, University of Minnesota, 316p.

IPCC, 2001. Special Report on Emissions Scenarios. (Eds) Nebojsa Nakicenovic prepared and published to web by GRID-Arendal on web (www.grida.no/climate/ipcc/emission/index.htm).

Johnson, D.T., and Mayes, R.L., 1984. Studies of larval development and adult flight of the peach tree borer, *Synanthedon exitiosa* (Say), in Arkansas. *J. Ga. Entomol. Soc.,* **19:** 216-223.

Kaur, S., Sukhjeet Kaur, Srinivasan, R., Cheema, D.S., Lal, Tarsem, Ghai, T.R. and Chadha, M.L., 2010. Monitoring of major Pests on Cucumber, Sweet Pepper and Tomato under Net-House Conditions in Punjab, India. *Pest Manag. Hort. Ecosyst.,* **16**: 148-155.

Khalil, A.A., Shaker, M.A., Mousad, K.H., Ostafa, M.M., El-Mtewally and Ameha, Moustafa S., 2010. Degree-days units and expected generation numbers of peach fruit fly *Bactrocera zonata* (Saunders) (Diptera: Tephritidae) under climate change in Egypt. *Egypt. Acad. J. Biolog. Sci.,* **3:** 11-19.

Kumar, N.K.K. Venugopalan, R. and Krishnamoorthy, P.N., 2004. A Statistical modelling approach to study the influence of weather parameters on the leafminer, *Liriomyza trifolii* (Burgess) in South India. *Pest Manag. Hort. Ecosyst.,* **10**: 55-59.

Lanzoni, A., Bazzocchi, G.G., Burgio, G. and Fiacconi, M.R., 2002. Comparative life history of *Liriomyza trifolii* and *Liriomyza huidobrensis* (Diptera: Agromyzidae) on beans: Effect of temperature on development. *Environ. Entomol.*, **31**: 797-803.

Leibee, G.L., 1984. Influence of temperature on development and fecundity of *Liriomyza trifolii* (Burgess) (Diptera: Agromyzidae) on celery. *Environ. Entomol.*, **13**: 497-501.

Nietschke, B.S., Magarey, R.D., Borchert, D.M, Calvin, D.D., and Jones, E. A., 2007. Developmental database to support insect phenology models. *Crop Prot.*, **26**: 1444–1448.

Rao, S. M., Rama Rao, C.A., Vennila, S., Raju, B.M.K., Srinivas, K., Padmaja, P.C.M., Rao A.V.M.S., Maheswari, M., Rao, V.U.M., and Venkateswarlu, B., 2012. Meta analysis of Impact of elevated CO_2 on host-insect herbivore interactions. Bulletin No. 2/2012. Central Research Institute for Dryland Agriculture, Santoshnagar, Saidabad, Hyderabad, Andhra Pradesh, India.48 p.

Roltsch, W.J., Zalom, F.G., Strawn, A.J., Strand,J.F., and Piteairn, M.J., 1999. Evaluation of several degree-day estimation methods in California climates. *Int. J. Biometeorol.*, **42**: 169-176.

Sakamaki, Y.S., Chi, Y., and Kushigemachi, K., 2003. Lower threshold temperature and total effective temperature for the development of *Liriomyza sativae* Blanchard on kidney beans. *Bull. Fac. Agric. Kagoshima Univ.*, **53**: 21-28.

Saradhi, P.M.P., and Patnaik, N.C., 2004. Seasonal population fluctuations of serpentine leafminer, *Liriomyza trifolii* (Burgess) in different host plants. *J. Appl. Zool. Res.*, **15**: 60-63.

Schmaedick, M.A. and Nyrop, J.P., 1993. Sampling second generation spotted tentiform leaf miner: A means to reduce overall control costs and facilitate biological control of mites in aople Orchards. *Food Life Sci. Bull.*, NY.

Sharpe, P.J., and DeMichele, D.W., 1977. Reaction kinetics of poikilotherm development. *J. Theor. Biol.*, **64**: 649–670.

Shi, P.A., 2010. Comparison of different thermal performance functions describing temperature-dependent development rates. *J. Thermal. Biol.*, **35**: 225–231.

Singh, R.K., Nath, P., and Singh, P.K., 2005. Effect of sowing time of bottle gourd on the population of serpentine leaf miner, *Liriomyza trifolii* Burgess. *J. Exp. Zool.*, **8**: 145-149.

Sohi, A.S., Mann, H.S., Singh, J., Singh, B., and Dhaliwal, C.S., 1994. Attack of serpentine leaf miner, *Liriomyza trifolii* Burgess on cotton in Punjab - a new record. *J. Insect Sci.*, **7**: 114.

Srinivasan, K. Viraktamath, C. A. Gupta, M., and Tewari, G.C., 1995. Geographical distribution, host range and parasitoids of serpentine leaf miner, *Liriomyza trifolii* (Burgess) in South India. *Pest Manag. Hort. Ecosyst.*, **1**: 93-100.

Stange, E.E. and Ayres, M.P., 2010. Climate Change Impacts: Insects. In: eLS. John Wiley and Sons Ltd, Chichester. http://www.els.net [doi: 10.1002/9780470015902.a0022555]

Tokumaru, S., and Abe, Y., 2003. Effects of temperature and photoperiod on development and reproductive potential of *Liriomyza sativae, L. trifolii*, and *L. bryoniae* (Diptera: Agromyzidae. *Jpn. J. Appl. Entomol. Zool.*, **47**: 143-152.

UCIPM, 2012. University of California Statewide Integrated Pest Management Program, Division of Agriculture and Natural Resources, University of California on web (www.ipm.ucdavis.edu/WEATHER/index.html).

Viraktamath, C.A., and Tiwari, G.C., Srinivasan, K., and Gupta, M., 1993. American serpentine leaf miner is a new threat to crops. *Indian Farming,* **10**: 12.

2015, **Animal Diversity, Natural History and Conservation, Vol. 5**   *Pages 113–132*
Editors: **V.K. Gupta and Anil K. Verma**
Published by: **DAYA PUBLISHING HOUSE, NEW DELHI**

Chapter 10

Breeding Biology and Mating Strategies of Coral Reef Shrimp *Lysmata wurdemanni* Gibbes (Decapoda: Caridea: Hippolytidae) from Southeast Coast of India

S. Prakash[1,2]*, T.T. Ajith Kumar[1,3] and T. Subramoniam[2]

[1]Center of Advanced Study in Marine Biology,
Faculty of Marine Sciences, Annamalai University,
Porto Novo – 608 502, Tamil Nadu, India
[2]Centre for Climate Change Studies,
Sathyabama University, Jeppiaar Nagar, Rajiv Gandhi Road,
Chennai – 600 119, Tamil Nadu, India
[3]National Bureau of Fish Genetic Resources (ICAR),
Canal Ring Road, Dilkusha Post, Lucknow – 126 002, Uttar Pradesh, India

ABSTRACT

Sexual systems present in the crustaceans are diverse, unlike insects. They range from the commonly found gonochorism to different types of hermaphroditism. Although sexes are separate in many caridean shrimps, sequential hermaphroditism is exhibited in diverse species. Caridean shrimps, belonging to Hippolytidae are particularly important that they represent many types of hermaphroditism. Protandric simultaneous hermaphroditism is a newly described sexual system in which the female phase shrimp retain the male

---

* *Corresponding author.* E-mail: prakash.s1311@gmail.com

functionality by mating as a male as well as female. The caridean shrimps exhibit equally complex mating systems, in combination with bizarre life history patterns. Cleaner shrimp *Lysmata* sp. are turning out to be an attractive aquarium ornamental shrimp and hence a study on their life cycle and breeding biology is necessary in order that suitable culture protocols could be developed for commercial purposes. In addition, many *Lysmata* species inhabit the coral reef niches and a few of them are known as fish cleaners that remove ectoparasites from their host. In this study, we have investigated the sexual system of the shrimp *Lysmata wurdemanni* Gibbes from Gulf of Mannar, Tamilnadu, India to determine if the species is protandric or it reaches the status of simultaneous hermaphroditism. Size frequency distributions and morphological identifications indicated that the population comprised of small or immature males, and large functional females. All smaller individuals possessed characters of typical caridean male's including male gonopores, appendix masculina, ejaculatory ducts and cincinnuli on the first pair of pereiopods. However, the gonad of these male-phase (MP) individuals was an ovotestis with an undeveloped ovarian portion. Female phase (FP), which spawn eggs and incubate embryos, also had male gonopores and an ovotestis terminating in ejaculatory ducts containing spermatophores. In FPs, male pleopod characters were absent or reduced, and a brooding chamber, produced by expanded pleopod flanges is present. Transitional individuals with MP characters and an ovotestis containing vitellogenic oocytes, were found to be rare, but when present overlapped completely in size with FP and larger MP individuals.

Further, the simultaneous hermaphroditism was also confirmed using "time-lapse" video observations on the mating pairs of the female phase (FP) individuals maintained in the laboratory. Copulations between FPs resulted in successful spawning and development of embryos. However, FP hermaphrodites maintained in isolation were unable to self-fertilize spawned eggs. It is known that social interactions may mediate MP change into FP hermaphrodites in *L. wurdemanni*.

Keywords: *Caridea, Hippolytidae, Lysmata wurdemanni, Protandry, Sexual biology, Sex change, Simultaneous hermaphroditism.*

Introduction

Many marine invertebrates and fishes are sequential hermaphrodites, a sexual condition that is considered to be an adaptive response of considerable interests in hermaphrodites (Charnov, 1982; Policansky, 1982). Evidence from size-frequency distributions of fish populations suggested that labile sex change was presumably mediated by social interactions in different demographic situations ("environmental sex determination" ESD) (Charnov *et al.,* 1978; Charnov, 1981, 1982; Charnov and Anderson, 1989). By far, protandrous hermaphroditism is the most prevalent form among other sexual patterns in as much as the adaptive significance of this system could be easily explainable. In most species, the size and age of the individuals are positively correlated, as the crustaceans tend to grow continuously during their lifetime (Subramoniam, 2013). These studies are of major importance to researchers dealing with the evolutionary ecology, but also for those willing to successfully keep breeding pairs of decapods in captivity and wishing to maximize their reproductive output.

Different species of the genus *Lysmata* displays a unique and puzzling sexual system: protandrous simultaneous hermaphrodite (PSH) in which shrimps first reproduce as males and then transform to reproductive females (Bauer, 2000). Bundy (1983) and Crompton (1992) have also confirmed this by providing with some histological evidences for the presence of mature oocytes and spermatocytes in the same gonad of *Lysmata wurdemanni*. Further, reports from the aquarium hobbyists describing the viable embryo production by both the members of *Lysmata* pairs have led researchers to re-examine the complex sexuality in these shrimps. Occurrences of protandric simultaneous hermaphroditism were extended to other species of *Lysmata* genus in several other investigations (Fiedler, 1998; Bauer and Holt, 1998; Baeza *et al.*, 2007, 2010, 2014). This sexual condition benefitted shrimp breeders by virtue of the easy way in which these shrimps could find their mating partners, compared with the other pair living shrimps (Lin and Zhang, 2001). The present study deals with the description and characterization of both primary and secondary sexual characters male and female phases in the ontogeny of these sequential hermaphroditic caridean shrimp *Lysmata wurdemanni*. Laboratory observations on mating between two secondary females have also been recorded to substantiate the contention that these shrimps are simultaneous hermaphrodites, capable of mating with each other.

Materials and Methods

Populations of coral reef shrimp species *Lysmata wurdemanni* Gibbes (Caridea: Hippolytidae) were collected from Tuticorin waters (8°50'06.94"N 78°12'48.77"E), Tamilnadu, India on 5th May 2013 and 30th June 2013. The areas sampled were gently sloped sandy bottoms with large boulders and coral patches, slightly covered with coralline algae. All the shrimps were collected by scoop nets and dip nets by adopting snorkeling and SCUBA diving methods at water depths ranging from 2-15 m. Samples collected from the field were immediately preserved in 5-10 per cent sea water formalin initially and later washed with freshwater and transferred to 70 per cent ethanol for permanent storage.

Data Collection

Species identification from preserved specimens was based on standard literature with verification of special key characters for *L. wurdemanni* (Rhyne and Lin, 2006). The body size was measured by calculating the carapace length (CL), (chordal distance from the posterior edge of the eye orbit to the mid-dorsal posterior edge of the carapace). Specimens were sexed, using the presence (male) or absence (female) of an appendix masculina on the endopod of the second pleopod. Individuals within the male size range lacked an appendix masculina were considered as juveniles. Presence or absence of openings (male gonopores) on the coxa of the fifth pereiopods was recorded. An ejaculatory duct was dissected from the body wall/coxa juncture of the fifth pereiopod. The number of appendix masculina spines was counted and width of basipod flange of second pleopod was also measured, which was narrow in males, but expanded in females (Hoglund, 1943).

For comparison with other species of *Lysmata* sp. the measurements were made using a zoom stereomicroscope (Olympus SZ51, Japan) with ocular meter of the individuals collected. The length of the distal segment of third maxilliped and the length of appendix masculina was considered. Measures on the character of basipod flange width of pleopod 2 (breeding dress-Bauer, 2004) was also made. The presence or absence of a vitellogenic ovary, visible through the dorsal carapace in preserved specimens, was determined and scored from 1 (no vitellogenic oocyte observed) to 4 (maximum filling of ovarian portion with vitellogenic oocytes) as in Bauer (1986). Measures of carapace length (mm), ovarian condition, presence or absence of appendix masculina and cincinnuli of pleopod 1 were done on all the specimens collected during the study for construction of size-frequency diagrams of population structure. "Male-phase" (MP) individuals were characterized by the occurrence of cincinnuli on the first pleopods (P1), a male character in most carideans (Bauer 1976; Butler 1980), along with a non-vitellogenic gonad. "Female-phase" (FP) individuals were those lacking pleopod 1 (P1) cincinnuli and transitional individuals were those with presence of P1 cincinnuli. Ejaculatory duct was dissected from the coxa of fifth pereiopods and its structure was observed under the microscope. Spermatophore material and sperm cell was also removed and mounted on a cover slip to observe under the microscope (Olympus Cxl 21i, Japan). In addition, embryos incubated by brooding females were classified with a system suggested by Allen (1966) and Subramoniam (1979 for *Emerita asiatica* Milne Edwards) (Table 10.1).

Mating Behavior

Certain number of living individuals of *Lysmata wurdemanni* collected from the study site without causing any damage to the body parts and pereiopods. Shrimps were sampled by gently driving the individuals in polyethylene bags, packed with sufficient oxygen and transferred to Marine Ornamental Fish Hatchery at Centre of Advanced Study in Marine Biology, Annamalai University, Tamilnadu. Live shrimps were then acclimatized for 2-4 hrs and thereafter transferred to flowing seawater tanks in the hatchery where they were maintained with *ad libitum* food supply, *e.g.* boiled oysters, mussels, shrimps and commercial pellets. Initially, observations of pre-copulatory and mating behaviors of these species were conducted in glass tanks with a size of 60 cm length x 30 cm width x 30 cm height with a water holding capacity of 54 liter fitted with a biological filter. The water quality parameters were maintained at 27°C, 30 ppt, DO 4-6 ml/l and 12:12 light dark cycle to imitate the natural photoperiod. Activities were recorded by Sony DSC-H90 digital camera connected to Octopus Tripod and day illumination was provided by overhead fluorescent lights.

In mating experiments of *Lysmata* shrimps pairs of female phase individuals (carapace length 6-7.5 mm) carrying embryos during mating or just hatched a brood were used. Activities were recorded for 24h daily until the spawning and molting of pre-spawning individual. After molting, observations were made on the presence of molt as well as recently spawned embryos of the individuals. The spawned individuals were monitored daily to confirm the successful embryonic development.

Data Analysis

The variables were tested using statistical analysis (ONE WAY ANOVA and Multiple Regression) using standard statistical tool SPSS v16.0, USA.

Results

Reproductive Morphology and Population Structure

A total of 29 individuals (9 males, 7 transitional and 13 reproductive females) collected during the study period were sexed and measured. Male-phase morphs (MP's) are easily distinguished by the presence of appendix masculina (am) with several spines on the endopod of second pleopod (Plate 10.1D), presence of cincinnuli (coupling hooks) on the endopod of first pleopod (Plate 10.1E) and non-vitellogenic ovotestes. Appendix masculina, a characteristic of male caridean shrimps could be

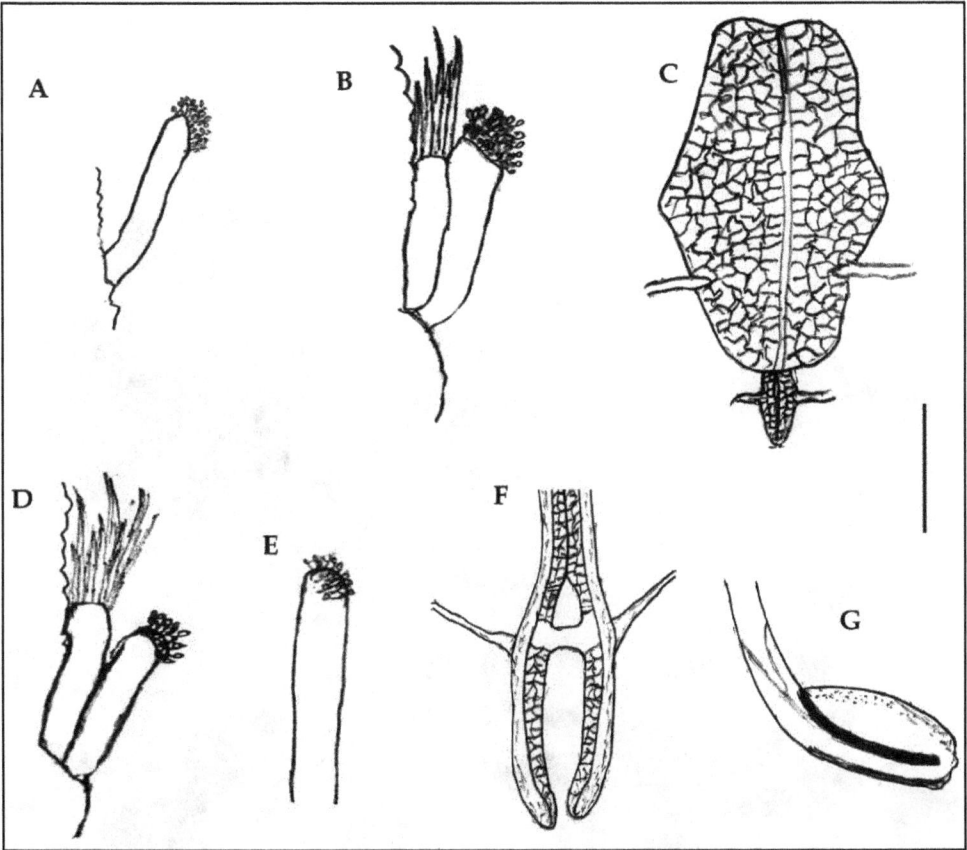

Plate 10.1: Reproductive Morphology of Sexual Morphs of *Lysmata wurdemanni*: Pleopods 1 and 2, ejaculatory ducts.

A: Female phase (FP) individual 8.0 mm CL shows only appendix interna of pleopod 2; B: Transitional phase individual of 5.5 mm CL shows both a.m and a.i; C: FP 8.0 mm CL that carried embryos near hatching; D: Male phase (MP) individual, 4.0 mm carapace length (CL) shows well developed appendix masculina (a.m) and appendix interna (a.i); E: Presence of cincinnuli on pleopod 1 of MP (4.0 mm CL); F: Testicular portion of gonad; G: Ejaculatory duct of FP individual (8.0 mm CL). Scale bars A-D = 1 mm, E = 0.5 mm, F-G = 0.25 mm.

observed in a reduced condition (transitional phase) or absent in individuals of increased CL >7mm (particularly functional females) (Plate 10.1A).

Many FP's have large vitellogenic oocytes in the gonad (Plates 10.1C, 10.2A), which can be observed from the dorsal carapace region. Interestingly, these vitellogenic females constituted that part of the population, which incubated embryos on the abdomen (Plate 10.2B) indicating repetitive nature of the reproductive cycle.

Plate 10.2: Reproductive Morphology of *Lysmata wurdemanni* Showing Primary Sexual Characters: Visible Gonads and Embryos.

A: Presence of fully ripened ovaries (black arrow) on the dorsal carapace and less developed ovaries (red arrows) in reproductive females; B: Presence of embryos on the newly spawned individual (black arrow indicates egg clutch on the abdomen); C: Development of embryo-stage 1 (absence of visible blastoderm); D: Stage 2-presence of distinct blastoderm; E: Stage 4-fully developed larvae with little or no yolk appearance (modified from Allen, 1966 and Subramoniam, 1979). Scale bars A-B = 10 mm, C-E = 1 mm.

Transitionals are individuals with cincinnuli on the endopod of the first pleopod and appendix masculine on the second pleopod (Plate 10.1B); also possess vitellogenic oocytes in the ovary (green coloration) as seen through the dorsal region. Female phase individuals incubating embryos were abundant and recently spawned embryos showed little or no ovarian development. Embryonic stages were evenly distributed from the collected FP's individuals.

Size-Frequency Distributions of Sexual Morphs

The size-frequency distributions of *L. wurdemanni* show that the body sizes (carapace length in mm) of males are smaller than the transitional and reproductive females. There is an overlap in size between the MP, transitional and FP respectively (Figure 10.1a). All transitional and FP's were reproductive, incubating embryos when collected. The percent size-frequency distributions showed that the individuals with carapace length between 5-7 mm are more (58.62 per cent) during the sampling, which indicates that there is some overlap in size occurred among MP's, transitional and FP's (Figure 10.2a). MP's size ranged from 3.2 to 5.4 mm CL (mean 4.5 mm ± 0.76 SD; n=9) followed by transitional 4.9 to 6.2 mm CL (5.6 mm ± 0.50 SD; n=7) and female-phase morphs (FP's) 5.6 to 8.4 mm CL (6.8 mm ± 0.85 SD; n=13) respectively of all the individuals collected and the sex ratio of the samples indicates that it is highly female biased.

Variation of Sexual Characters with Sexual Phase and Size

Male gonopores of flap-like opercula were observed on the coxa juncture of fifth pereiopods of all males by stereo zoom microscope. Other male characters like cincinnuli on endopod of first pleopod and appendix masculina were also developed in these small MP's. The FP's incubating embryos lack cincinnuli on the endopods of first pleopods as well as appendix masculina on the second pleopod confirming that all the individuals lacking these characters was considered as FP's. The gonads of FP individuals also bear oviducts, terminated in a testicular portion (Plate 10.1F) with anterior vas deferens and leading to ejaculatory ducts (Plate 10.1G), which is typical of male carideans. Another change in pleopod structure, which accompanies growth in MP hermaphrodites, is the development of postero-lateral flange, a thin muscular plate on the basipod of anterior three pairs of pleopods. In male phase, the basipod flange is completely reduced or smaller in size with increased CL and the transitionals have slightly greater flange than the MP individuals. This flange is always well developed in females with increased body size (carapace length) and is associated with spawning and embryo incubation ("breeding dress") of reproductive females.

All individuals of *Lysmata wurdemanni* (MP's, Transtionals and FP's: brooding and non-brooding) that were dissected have male gonopores on the coxa juncture of the third pereiopod, stored a spermatophore filled with inverted umbrellas like sperm cells. The posterior portion of male and female gonads, containing sperm was relatively small in ovigerous shrimp. In both brooding and non-brooding shrimps, gonads were connected laterally to anterior vas deferens and leading to ejaculatory duct. Sperm cells were squeezed from the ejaculatory ducts of live embryo carrying FP's. This is the strong evidence that all individuals can function as males. Brooding

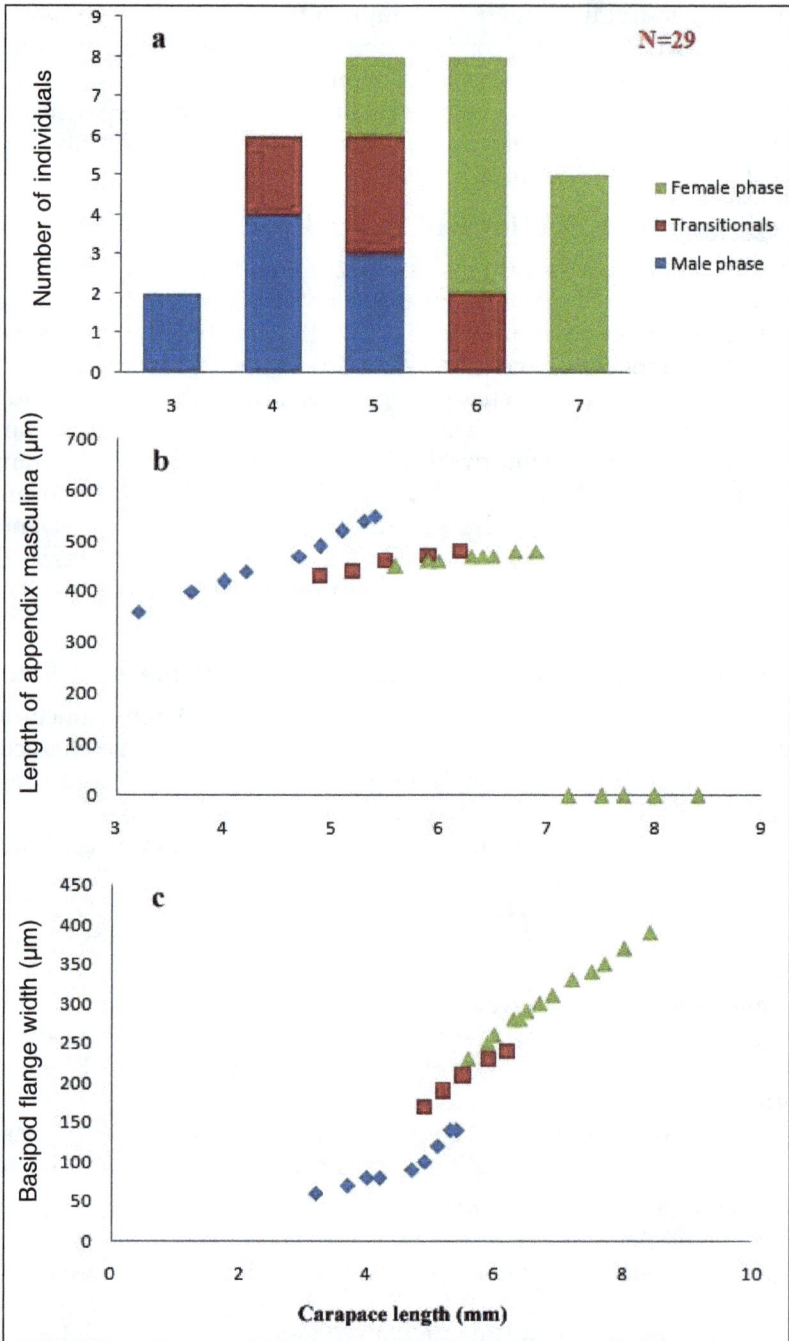

Figure 10.1: *Lysmata wurdemanni* Gibbes.
Population structure and variation as a function of body size (carapace length in mm).
a: Size-frequency distribution, b: Appendix masculina length (μm) and c: Pleopod 2 flange width (μm) (incubatory character-"breeding dress").

shrimps have ovotestes with a relatively large anterior female portion of fully greenish vitellogenic oocytes. In contrast, non-brooding shrimps have non-vitellogenic oocytes lacking green coloration. Embryonic development was observed from the larger individuals with more than 5 mm CL and the frequency of stages from 1-4 was observed (Plates 10.2C-E) (see Table 10.1 for detailed explanation). Embryonic stage 1, 3 and 4 are dominant from the ovigerous shrimps collected during the present study (Figure 10.2b).

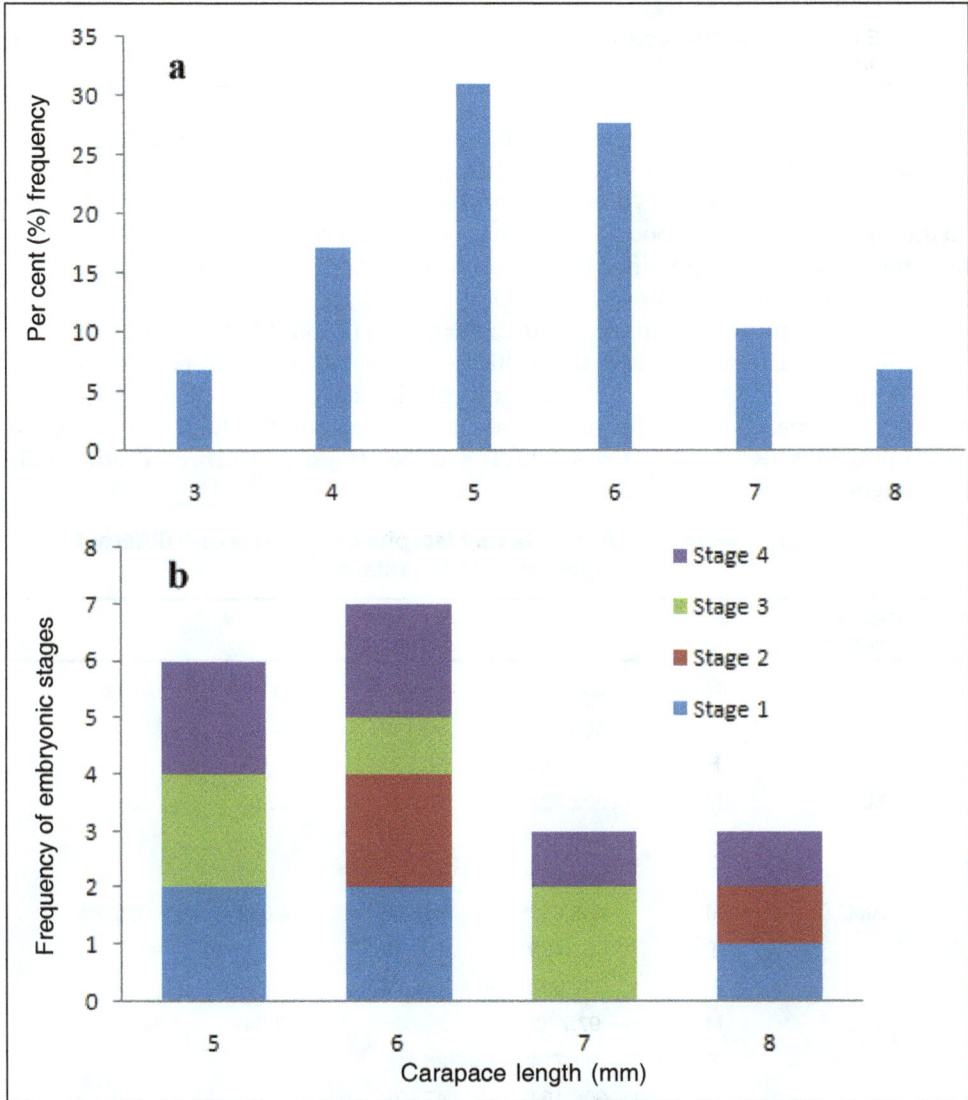

Figure 10.2: *Lysmata wurdemanni* Gibbes.

Population structure and reproductive condition of sexual morphs function of body size (carapace length in mm). a: Per cent (per cent) frequency of size distributions, b: Frequency of embryonic stages.

Animal Diversity, Natural History and Conservation Vol. 5

Table 10.1: Classification of Embryonic Development in Brooding Females of *Lysmata wurdemanni* Gibbes (modified from Allen, 1966 and Subramoniam, 1979)

Stages	Description
1	Recently spawned embryos bright green in color with no visible blastoderm
2	Clevage takes place with bright green color and blastoderm distinct, with no eye development
3	2/3rd of the yolk utilized and presence of eye spots, appendages are visible with small orange color pigments
4	Embryos completely developed and white in color; near to hatch and yolk fully utilized, eyes are prominent and abdomen free from cephalothorax

Variation in sexual characters among individuals of different body sizes were compared among all the individuals sampled during the present study. The individuals identified as FP's used for morphometric observations lacked cincinnuli on the endopod of first pleopod and presence of ventral brood. The MP's individuals from the samples were identified by the absence of brood production and presence of cincinnuli on the endopod of first pleopod. The length of appendix masculina in males varied significantly ($F_{(2, 26)} = 4.072$; $P = 0.029$) (Table 10.2) and are similar in transitional and female phase. In multiple regression analysis, coefficient beta intercept values of AML also showed significant increase in male ($R^2 = 0.997$; $t = 4.7$; $P = 0.005$) compared to significant decrease in transitional ($R^2 = 1.000$; $t = -1.102$; $p = 0.351$) and FP's ($R^2 = 0.998$; $t = -0.320$; $P = 0.756$) (Figures 10.1b, 3) (Table 10.3) respectively.

Table 10.2: One-Way ANOVA of Sexual Morphs of *Lysmata* with different Morphological Characters

Independent Variables	Sex	Mean	Standard Deviation	F Value	P Value
CL	M	450.000a	76.485	25.695	<0.001*
	T	568.571b	50.142		
	F	685.385c	85.791		
MX3	M	362.222a	64.764	22.53	<0.001*
	T	448.571b	19.303		
	F	504.231c	46.630		
AML	M	465.556a	65.595	4.072	0.029**
	T	461.429b	19.518		
	F	287.692b	236.859		
FW	M	97.778a	29.486	77.349	<0.001*
	T	215.714b	26.992		
	F	306.154c	47.878		

*: Significant level at 1 per cent; **: Significant level at 5 per cent; different superscripts indicates significant at 5 per cent based on Duncan-Multiple Range Test (DMRT).

CL: Carapace length; MX3: Third maxilliped; AML: Appendix masculina length; FW: Basipod flange width.

The size of ejaculatory ducts in FP's was comparable with that of MP's and correlation between the ejaculatory duct size and body size (carapace length) was positive and significant for both FP's and MP's. Another change in pleopod structure, which accompanies growth in MP hermaphrodites, is the development of postero-lateral flange, a thin muscular plate on the basipod of anterior three pairs of pleopods and in FP's it is associated with spawning and embryo incubation of reproductive females ("breeding dress"). The FW are significantly (F $_{(2, 26)}$ = 77.349; P = <0.001) (Table 10.2) varied between the male, transition, and female. In multiple regression analysis, coefficient beta intercept values of FW showed a significant decrease in male (t = -3.318; P = 0.021) compared to transitional (R^2 = 1.000; t = 8.660; P = 0.003) and female phase (t = 3.945; 0.003) showed a significant increase (Figures 10.1c, 3) (Table 10.3).

Table 10.3: Multiple Regression Analysis of Sexual Morphs of *Lysmata*. CL as dependent variable and MX3, AML and FW as independent variables (Enter method).

Sexual Morphs	Independent Variables	R^2	t	P
M	MX3	0.997	0.878	0.420
	AML		4.7	0.005**
	FW		−3.318	0.021*
T	MX3	1.000	21.301	0.000**
	AML		−1.102	0.351
	FW		8.660	0.003**
F	MX3	0.998	2.274	0.023*
	AML		−0.320	0.756
	FW		3.945	0.003**

*: Significant values at 5 per cent; **: Significant values at 1 per cent; values without asterisk indicates non-significant at 5 per cent based on Duncan Multiple Range Test (DMRT).

Mating Behavior

All mating events occurred during the darkness (nights) initiated by the female molt. FP's were able to copulate successfully both as male and female during the reproductive period was demonstrated. Both individuals of these pairs were incubating embryos at the beginning of observation period (Plate 10.3A); it was thus certain that they had previously reproduced as females. Recordings were made until the pre-spawning female had molted and spawned.

The mating events were actually a sequence of few stereotyped actions.

1. *Initial contact and climbing*: potential reproductive females (here after mentioned as males) detect receptive females (those with mature ovaries, which have recently molted) after touching the female with their antennal flagellum (Plate 10.3B). After detecting a receptive female, the behaviors of reproductive females change dramatically and try to seize the receptive females with its pereiopods. After contact, the male tries to crawl up to the

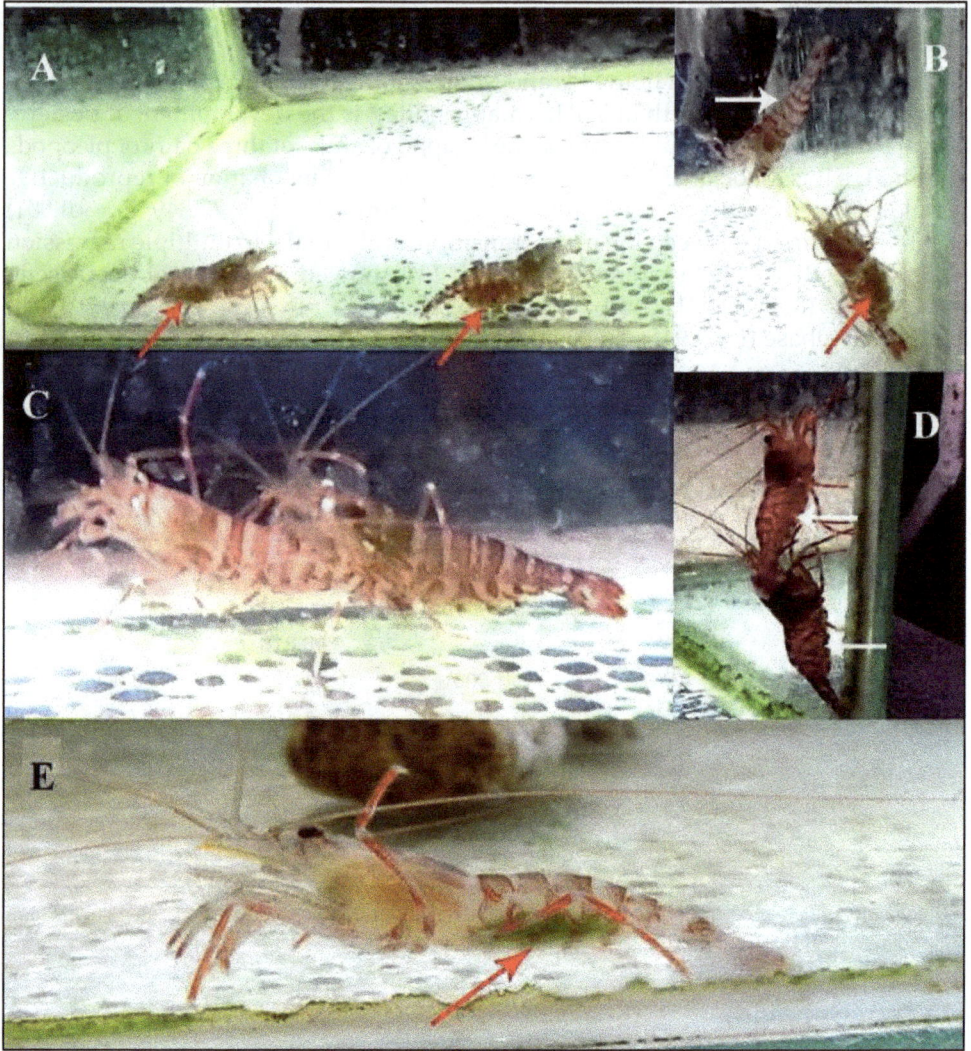

Plate 10.3: Mating Events of *Lysmata wurdemanni*.
A: Both the individuals incubating embryos (red arrows); B: Newly molted female (white arrow) and ovigerous female (red arrows); C: Ovigerous female approaching the newly molted/receptive female; D: Both the individuals molted on the same day (white arrows indicates absence of egg clutch) and presence of ripened ovary on the dorsal carapace region; E: Spawning in isolated individuals. Scale bars A-D = 10 mm, E = 5 mm.

dorsal midsection of the receptive female (Plate 10.3C). If the contact has been made at the anterior of the female, there are usually a brief flurry of rapid blows and strikes by the pereiopods between male and female. It is apparently some sort of recognition or pre-copulatory behavior.

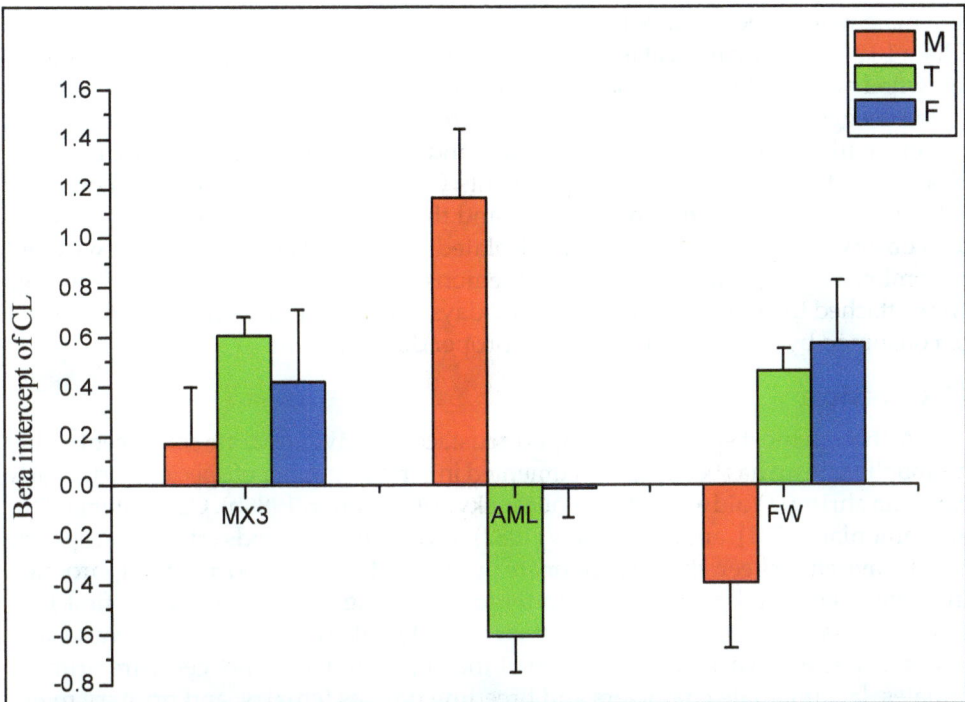

Figure 10.3: Allometry of Growth in Sexual Characters in *Lysmata wurdemanni*/ Gibbes. Beta intercepts (histogram bars) with 95 per cent confidence intervals of regression lines calculated from measures of sexual characters on carapace length (size), with log transformation of both dependent variables and carapace length in MP's (M), transitional (T) and FP's (F). MX3, length of terminal article of maxilliped; AML, length of appendix masculina; FW, basipod flange width of pleopod 2.

2. *Overlapping*: Clutching the female with its walking legs and the male astride to the females mid dorsal line with anterior end of both the shrimps facing same direction. This allows the female to settle down and accept the male. In certain cases, rejection by the female takes place during contact and climb phases. The male swings his body either side of the female, so that the abdomen is down along the females' abdominal region. The male swings under the female, positioning the thoraco-abdominal junction beneath and perpendicular to the females' first abdominal sternite.

3. *Sperm transfer and withdrawal*: Male briefly beat the pleopods in the dip position and the spermatophores are emitted and transferred that lasts for few to several seconds. Separation of male and female occurs after pleopod beat. Females may disengage by jumping backward.

In certain cases, the individuals acting as male seized the molting pair before it gets molted completely. In one case, both FP's of each were near a spawn and molted within few hours of each other on the same night and they are successfully copulated as a male when other member was molted and then it acts as a female when molted (Plate 10.3D). The reciprocal mating was observed successful and both the FP's have spawned with embryos on the next day morning. All the individuals spawned a brood of eggs (embryos are visible with full green color covering the entire abdominal region) within several minutes of mating and fully formed ejaculatory ducts was observed in FP's after the mating experiments. Gonadal condition of FP's had potential role in serving as male mating partners and the degree of ovarian conditions were also observed. Spawning in solitary/isolated individuals was observed twice but the embryos occupying the abdominal regions are very less (Plate 10.3E). The eggs were attached to the abdominal part for 1-2 days after spawning, which was discarded or consumed by the individual without proper development.

Discussion

Although most species of decapod crustaceans have separate sexes; sequential hermaphroditism has been well documented in certain number of species, particularly caridean shrimp (Yaldwyn, 1966; Policansky, 1982; Bauer, 1986b). Carpenter (1978), Subramonian (1981), and Policansky (1982) have listed decapods crustacean species which have either been shown to be or are purported to be protandric hermaphrodites. In some species, all populations undergo sex change as protandry *i.e.* first they reproduce as males and then change their sex to female with increasing age and size. In other species, a variable proportion of the population (<50 per cent) are primary females, lacking male characters and breeding only as females, and primary males are individuals that breed only as males throughout their lifetime (Bauer, 1986).

In protandric caridean shrimps, environmental determination of sex reversal through social interactions has been suggested for *Athanas* sp. (Alpheidae) (Nakashima 1987; Gherardi and Calloni, 1993), and it was represented on some pandalid shrimps (Charnov *et al.,* 1978; Bergstrom, 1997). In several species of Pandalids, some variable proportion (< 50 per cent) of the population either go through a brief, non-functional male phase before maturing into female phase ("early maturing females") or mature directly into females. It is still not clear whether these latter are "primary females" *i.e.*, genetically females as in gonochoristic species; or they are same genotype as protandric individuals, but in which male phase is repressed completely by environmental (social) factors, as indicated in Charnov *et al.* (1978) and Charnov (1981).

A species, in which the protandry was well documented among the carideans is a hippolytid *Lysmata seticaudata* Risso (Dohrn, 1950). A complete protandry *i.e.* all individuals function first as males and the change their sex to reproductive females and the transformational characters are visible by the modifications by loss of appendix masculina and on the endopods of first pleopods has been described by various authors (Dohrn, 1950; Butler, 1964, 1980; Charniaux-Cotton, 1975). Ghiselin (1969) proposed a size advantage model for the evolution of protandric hermaphroditism in which the protandry is favored in a species where the reproductive fitness is positively correlated with the female size but not in males. In *Lysmata*, the sexual system clearly indicates functional simultaneous hermaphroditism and the size-frequency distribution shows that the individuals first develop as males

(MP's) and there are no juvenile females. Later, it changes sex to develop as reproductive females with increasing size and incubate embryos (FP's) (Bauer, 1991), but retain the male gonopores and ejaculatory ducts.

The sexual system described in the present study for *L. wurdemanni* agrees well with the documentation of functional simultaneous hermaphroditism. In this species, smaller individuals can only act as males and larger individuals are hermaphrodites, serving as potential brooders of all the populations. Instead of losing male capabilities when entering the female phase, as in other sex-changing carideans, female-phase individuals retain male ducts, sperm, and the ability to copulate successfully as males. Female-phase individuals (FPs) of the "protandric" genus *Lysmata* retain the testicular portions of the gonads and male ducts from the male phase after changing to the terminal female phase: *Lysmata seticaudata* Risso (Spitschakoff, 1912; Berreur-Bonnenfant and Charniaux-Cotton, 1965), *L. grabhami* (Wirtz, 1997), *L. amboinensis* (Fiedler, 1998), *L. californica* (Bauer and NewMan, 2004), *L. nayaritensis* (Baeza *et al.,* 2007), *L. hochi* (Baeza and Anker, 2008) and *L. boggessi* (Baeza *et al.,* 2014).

Our observations on size-frequency distribution of *L. wurdemanni* suggested that there is an overlap in size between the males (MP's) and transitional phase. The sample size of 5-7 mm carapace length was more from the collected individuals, dominated mainly by the female phase (FP's) compared to males (MP's) and transitional. The "size-advantage hypothesis" elaborated for fishes and shrimps by Warner (1975) and Charnov (1979), attempts to explain the evolution of sequential hermaphroditism. The protandry is fairly well recognized in most of the species, but it does not explain why gonochoristic species with small males and larger females have not evolved protandric behaviors. Nor does it explain why more protandric carideans have not evolved a system similar to *L. wurdemanni,* in which male function is retained with no apparent loss of female function. Evidence from size-frequency distributions of protandric pandalids suggested that labile sex change was presumably mediated by social interactions in different demographic situations ("environmental sex determination" or ESD) (Charnov *et al.,* 1978; Charnov, 1981, 1982; Charnov and Anderson, 1989). Alternately, threshold sex ratio (Shapiro and Lubbock, 1980) may regulate proportions among FPs and MPs in a group, with the type and frequency of interactions of an individual shrimp with other sexual morphs as the proximate factor triggering change in sexual type.

The first published report on simultaneous hermaphroditism in a caridean was that of Kagwade (1982) on *Exhippolysmata ensirostris* Kemp, a member of genus which is taxonomically close to the genus *Lysmata.* The male sexual appendices on the pleopod as well as male ducts and sperm were also found in females and concluded that these shrimps perform simultaneous hermaphroditism. Further studies also confirmed that FP's of certain species under the genus *Lysmata* are simultaneous hermaphrodites (Bauer and Holt, 1998; Fiedler, 1998). The smaller individuals MP's possess appendix masculina on second pleopod, presence of cincinnuli (coupling hooks) on endopod of first pleopod (typical identification character of male carideans) and an ovotestes with both male and female ducts. FP's individuals *i.e.* those carrying broods of developing embryos lacks appendix masculina on the second pleopod and absence of cincinnuli on first pleopod, although some vestiges can be observed from the smaller sized reproductive females.

In our observations, the male characters were determined in the MP's and transitional individuals. However, the size of appendix masculina was completely absent or much reduced in the larger sized individuals of transitional and smaller sized FP's. The gonadal observations shown here for *L. wurdemanni* has a distinct testicular portion with vas deferens in addition to an anterior ovarian portion, distended with vitellogenic oocytes as spawning approached? The vas deferens led to fully developed male ejaculatory ducts, which open via gonopores on the coxa juncture of third pereiopods in FP's and fifth pereiopods in MP's. Material squeezed from the ejaculatory ducts contains typical spermatorphore material as well as sperm cell. The observed sexual characters are similar to other species of the genus *Lysmata* sp. as illustrated by Spitschakoff (1912), Berreur-Bonnenfant and Charniaux-Cotton (1965), Bauer and Holt (1998), Fiedler (1998), Bauer and Newman (2004) and Baeza *et al.* (2007, 2014). After maturation, hermaphrodites resemble females of caridean gonochoric species and brood embryos in a pleonal chamber. They retain testicular tissue, male ducts, and gonopores, and hence can reproduce as both male and female (Bauer and Holt, 1998). After becoming hermaphrodites, individuals do not revert to males and no self-fertilization has been ever observed (Bauer, 2002). In general, studies on the sexual biology of *Lysmata* shrimps can provide new insights on the evolution of hermaphroditism among marine invertebrates.

The sexual system described in the present study for *L. wurdemanni*, documented with mating studies confirmed the synchronous hermaphroditism in decapods. Apparently, in this species, smaller (younger) individuals can only reproduce as males; only larger (older) individuals are functional hermaphrodites, serving as the breeding females. Instead of losing male capabilities when entering the female phase, as in other sex-changing carideans, female-phase individuals retain male ducts, sperm, and the ability to copulate successfully as males. Our mating observations showed that FPs of *L. wurdemanni* are true out-crossing hermaphrodites, with the remarkable capability of copulating and in-seminating as males on the very same night in which they themselves undergo a spawning molt and copulate, as females, with another FP. FPs mated with other FPs produced fertile spawns of embryos which developed in an apparently normal manner, *i.e.* went through the stages of embryonic development resembling those described for other caridean species (Bauer, 1991). In the experiments that brooding shrimps are able to fertilize each other but are not capable of self-fertilization. Both the individuals can also spawn at the same time. Isolated individuals are also able to spawn but unable to self fertilize. The eggs were attached to the abdomen for 1-2 days and it has been consumed or detached.

The testicular tissue, sperm ducts, and sperm mass of the male system of FPs are small in size and presumably in energetic costs compared with the ovarian tissue and large mass of yolky oocytes of the female system prior to spawning. However, the ejaculatory ducts, which store sperm, are nearly as large in relative size in FPs as in MPs (Bauer and Holt, 1998; Bauer and Newman, 2004 for *L. californica*). The emission, depletion, and subsequent replacement of spermatophores (sperm masses) during and after copulation must have some energetic cost. There are also behavioral costs to consider. About an hour before a pre-spawning FP molts and was receptive to mating, other FPs behave as do MPs in approaching and remaining near the pre-moulting FP. Bauer (2005) also demonstrated that when FPs are allowed to act as male as well as female, the usual condition in nature, brood production is reduced relative to that of FPs that cannot reproduce as males. Is this cost of maleness, somehow overcome by the *Lysmata* ancestor, so significant that it is a barrier to attainment of FP simultaneous hermaphroditism in other caridean species? The cost of maleness on brood production in FPs, observed here in *L. wurdemanni* and presumably other *Lysmata* species, might be viewed as an evolutionary 'price' paid by individuals of the *Lysmata* ancestor for the high selective advantage of functional simultaneous hermaphroditism.

Conclusion

It is concluded that collaborative efforts involving evolutionary biologists, physiologists and molecular ecologists will not only help to (re)confirm the sexual system of the species from the genus *Lysmata* but also contribute to a better understanding on the evolution of hermaphroditism in caridean shrimps. Future molecular analyses of all species of carideans will help to determine their phylogenetic relationships and shed light onto the evolutionary origins of their exceptional sexual system.

Acknowledgements

Authors are grateful to their Director and Dean, Faculty of Marine Sciences, Annamalai University for his constant support and encouragement throughout the study. Thanks are also extended to the authorities of Annamalai University for the facilities provided. S.P. also acknowledges Council of Scientific and Industrial Research (CSIR), New Delhi for granting a fellowship.

References

Allen, J. A., 1966. *The dynamics and interrelationships of mixed populations of Caridean found off the north-east coast of England*. In: Some contemporary studies in Marine Science, ed. H. Barnes, 45-66. London: Allen and Unwin.

Baeza, J. A., Reitz, J. M., and Collin. R. 2007. Protandric simultaneous hermaphroditism and sex ratio in *Lysmata nayaritensis* Wicksten, 2000 (Decapoda: Caridea). *J. Nat. Hist.*, **41**: 2843-2850.

Baeza, J. A., and Anker, A. 2008. *Lysmata hochi* n. sp., A New Hermaphroditic Shrimp from the Southwestern Caribbean Sea (Caridea: Hippolytidae). *J. Crust. Biol.*, **28(1)**: 148-155.

Baeza J.A., Braga A.A., Lo´pez-Greco L.S., Perez E., Negreiros-Fransozo M.L., and Fransozo A. 2010. Population dynamics, sex ratio and size at sex change in a protandric–simultaneous hermaphrodite, the spiny shrimp *Exhippolysmata oplophoroides*. *Marine Biology*, **157**: 2643-2653.

Baeza J.A., Behringer, D. C., Hart, R. J., Dickson, M. D., and Anderson, J. A. 2014. Reproductive biology of the marine ornamental shrimp *Lysmata boggessi* in the south eastern Gulf of Mexico. *J. Mar Biol. Ass. UK*, **94(1)**: 141-149.

Bauer, R. T., 1976. Mating behavior and spermatophore transfer in the shrimp *Heptacarpus pictus* (stimpson) (Decapoda: Caridea: Hippolytidae). *J. Nat. Hist.*, **10**: 415-440.

Bauer, R. T., 1986. Sex change and life history pattern in the shrimp *Thor manningi* (Decapoda: Caridea): a novel case of partial protandric hermaphroditism. *Biol. Bull.*, **170**: 11-31.

Bauer, R. T., 1991. *Analysis of embryo production in a Caridean shrimp guild from a tropical sea grass meadow.* In: A. Wenner and A. Kuris (eds.) Crustacean egg production. A. A. Balkema, Rotterdam. The Netherlands, pp. 181-191.

Bauer, R. T., and Holt, G. J. 1998. Simultaneous hermaphroditism in the marine shrimp *Lysmata wurdemanni* (Caidea: hippolytidae): an undescribed sexual system in the decapods Crustacea. *Mar. Biol.*, **132**: 223-235.

Bauer, R. T. 2000. Simultaneous hermaphroditism in caridean shrimps: A unique and puzzling sexual system in the Decapoda. *J. Crust. Biol.*, **20(2)**: 116-128.

Bauer, R. T. 2002. Test of hypotheses on the adaptive value of an extended male phase in the hermaphroditic shrimp *Lysmata wurdemanni* (Caridea: Hippolytidae). *Biol. Bull.*, **203**: 347-357.

Bauer, R. T., 2004. *Remarkable shrimps: Adaptations and Natural History of the Carideans.* University of Oklahoma Press, Norman, 282 pp.

Bauer, R. T., and Newman, W. A. 2004. Protandric simultaneous hermaphroditism in the marine shrimp *Lysmata californica* (Caridea: Hippolytidae). *J. Crust. Biol.*, **24**: 131-139.

Bauer, R. T., 2005. Cost of maleness on brood production in the shrimp *Lysmata wurdemanni* (Decapoda: Caridea: Hippolytidae), a protandric simultaneous hermaphrodite. *J. Mar. Biol. Assoc. London,* **85**: 101-116.

Bergstrom, B. I., 1997. Do protandric pandalid shrimp have environmental sex determination? *Mar. Biol.*, **128**: 397-407.

Berreur-Bonnenfant. J., and Charniaux-Cotton, H. 1965. Hermaphrodisme proterandrique et fonctionnement de la zone germinative chez la crevette *Pandalus borealis* Kroyer. *Bull. de la Soc. Zool. de France,* **90**: 240-259.

Bundy, M. H., 1983. *Simultaneous functional hermaphroditism in the shrimp Hippolysmata wurdemanni* (Gibbes) (Decapoda: Caridea: Hippolytidae). MS Thesis, Old Dominion University, Norfolk, VA.

Butler, T. H., 1964. Growth, reproduction and distribution of pandalid shrimps in British Columbia. *J. Fish. Res., Board Canada,* **21**: 1403-1452.

Butler, T. H., 1980. Shrimps of the Pacific coast of Canada. *Canadian Bull. Fish. Aqua. Sci.,* **202**: 1-280.

Carpenter, A., 1978 Protandry in the freshwater shrimps, *Paratya curvirostris* (Heller, 1862) (Decapoda: Atyidae) with a review of the phenomenon and its significance in the Decapoda. *J. Royal Soc. New Zealand,* **8**: 343-358.

Charniaux-Cotton, H., 1975. *Hermaphroditism and gynaadromorphism in malacostracan Crustacea.* p. 91-105 in R. Reinboth (ed.) Intersexuality in the animal kingdom. Springer Verlag. New York.

Charnov, E. L., Gotshall, D. W., and Robinson, J. G. 1978. Sex ratio: adaptive response to population fluctuations in pandalid shrimps. *Science,* **200**: 204-206.

Charnov, E. L., 1981. Sex reversal in *Pandalus borealis*: Effect of a shrimp fishery? *Marine Biological Letters,* **2**: 53-57.

Charnov, E. L., 1982. *The Theory of Sex Allocation.* Princeton, N. J.: Princeton University Press.

Charnov, E. L., and Anderson, P. J. 1989. Sex change and population fluctuations in pandalid shrimp. *American Naturalist,* **134**: 824-827.

Crompton, W. D., 1992. *Laboroatory culture and larval development of the peppermint shrimp Lysmata wurdemanni* (Gibbes) (Caridea: Hippolytidae). MS thesis, Corpus Christi State.

Dohrn, P. F. R., 1950. Studi sulla *Lysmata seticaudata* Risso (Hippolytidae). I-Le Condizioni normally della sessualita in natura. *Pubblicazione de la Stazione Zoologica di Napoli,* **22**: 257-272.

Fiedler, G. C., 1998. Funcional, simultaneous hermaphroditism in female-phase *Lysmata amboinensis* (Decapoda: Caridea: Hippolytidae). *Pacific Science,* **52**: 161-169.

Gherardi, F., and Calloni, C. 1993. Protandrous hermaphroditism in the tropical shrimp *Athanas indicus* (Decapoda: Caridea) a symbiont of sea urchins. *J. Crust. Biol.,* **13**: 675-689.

Ghiselin, M. T., 1969. The evolution of hermaphroditism among animals. *Quar. Rev. Biol.,* **44**: 189-208.

Hoglund, H., 1943. On the biology and larval development of *Leander squilla* (L.) forma typical De Man. *Svenska hydrogr-biol Kommn Skr,* **6**: 1-44.

Kagwade, P. V., 1982. The hermaphrodite prawn *Hippolysmata ensirositris* Kemp. *Indian J. Fish,* **28**: 189-194.

Lin, J., and Zhang, D. 2001. Reproduction in simultaneous hermaphroditic shrimp *Lysmata wurdemanni*: any two will do? *Mar. Biol.,* **139**: 919-922.

Nakashima, Y., 1987. Reproductive strategies in a partially protandrous shrimp, *Athanas kominatoensis* (Decapoda: Alpheidae): sex change as the best of a bad situation for subordinates. *J. Ethology,* **5**: 145-159.

Policansky, D., 1982. Sex change in plants and animals. *Ann. Rev. Ecol. Sys.,* **13**: 471-495.

Rhyne, A. L., and Lin, J. 2006. A western peppermint shrimp complex: re-description of *Lysmata wurdemanni* (Gibbes) description of four new species and remarks on *L. rathbunae* Chace (Crustacea: Decapoda: Hippolytida). *Bull. Mar. Sci.,* **79**: 165-204.

Shapiro, D. Y., and Lubbock. R. 1980. Group sex ratio and sex reversal in coral reef fish. *J. Theor. Biol.*, **83**: 411-426.

Spitschakoff, T., 1912. *Lysmata seticaudata* Risso, als Beispiel eines echten Hermaphroditismus beiden Decapoden. *Z. wiss. Zool.*, **100**: 190-209.

Subramoniam, T., 1979. Some aspects of reproductive ecology of a mole crab *Emerita asiatica* Milne Edwards. *J. Exp. Mar. Biol. Ecol.*, **36**: 259-268.

Subramoniam, T., 1981. Sexual and reproductive endocrinology of Crustacea. *J. Sci. Indus. Res.*, **40**: 396-403.

Subramoniam, T., 2013. Origin and occurrence of sexual and mating systems in Crustacea: A progression towards communal living and eusociatlity. *Journal of Biosciences*, **38(5):** 951-969.

Warner. R. R., 1975. The adaptive significance of sequential hermaphroditism in animals. *American Naturalist*, **109**: 61-82.

Wirtz, P., 1997. Crustaceans symbionts of the sea anemone *Telmatactis cricoides* at Madeira and the Canary Islands. *J. Zool.*, **242**: 799-811.

Yaldwyn, J. C., 1966. Protandrous harmaphroditism in decapods prawns of the families Hippolytidae and Campylonotidae. *Nature*, **209**: 1366.

2015, Animal Diversity, Natural History and Conservation, Vol. 5   *Pages 133–164*
Editors: V.K. Gupta and Anil K. Verma
Published by: DAYA PUBLISHING HOUSE, NEW DELHI

Chapter 11

Checklist of the Freshwater Crabs of Indian Himalyaya, its Threats and Conservation

Santanu Mitra*
Zoological Survey of India,
Fire Proof Spirit Building,
27, Jawharlal Nehru Road, Kolkata – 700 016, W.B., India

ABSTRACT

Among the total 90 species of Freshwater crab occurs in Indian territory 45 species (50 per cent) were reported from the Himalayas of Indian part. The eastern part of Himalyas or the eastern india are more diversed by harbouring a total 34 species exclusively in its nest, where as Western Himalayas are very poor in species diversity and only 5 species are exclusively reported from there which are not available in Eastern india. Interestingly only 6 species are common in both these two widely sparated region. A check-list of 45 species of crabs has been prepared providing recent generic and species names of the group as far as possible along with State-wise distribution in the Himalyas of Indian territory. Threats and conservation measures of these crabs are also discussed in this communication.

Keywords: *Freshwater Crab, Indian Himalyaya, Threats, Conservation.*

Introduction

Freshwater crabs have a great role in nutrient cycle in freshwater ecosystem, these species has a significant use as food in rural peoples. Recently these species are considered as bio-indicator in environmental monitoring. Some species of freshwater

---

* E-mail: santanuzsi@gmail.com

crab recently found as carrier of paragonimiasis a serious disease caused by the Lung-fluke from Manipur and Arunachal Pradesh. Freshwater crab constitutes only a small fraction of the brachyuran fauna of our country. True freshwater crabs are those which spend their entire lives in freshwater without return to the sea for whatever reason. There are some crabs which occasionally wander or even live in freshwater habitats, especially those occurring near the sea, but they are always common in estuarine areas and their larval development occurs in the open sea. True freshwater crabs belong to two superfamilies *viz.* Potamoidea and gecarcinucoidea. All the members of the Potamoidea and Gecarcinucoidea spend their entire lives in freshwater or surrounding wetland area.

In recent years there has been a drastic change in the taxonomy of freshwater crabs. For example, Alcock (1910) dealt all freshwater species under a single family Potamonidae. But presently these are treated under two families namely, Gecarcinucidae and Potamidae. Many of the genera dealt therin are either splitted or merged and several new genera have been erected (Ng *et al.,* 2007).

The Himalyas are known to be a geologically young and dyanamic mountain range system supporting a highly diverse Fauna and Flora, many of which are endemic. The Indian Himalayas extends over 2500 Km from Jammu and Kashmir in the West to Arunachal Pradesh in the East, covering an area of about 5,33,600 sq.km. Geographically, Esatern Himalayas are characterized by high rainfall, heavy snowfall and conditions more akin to temperate regions. Both, the climate condition as well as geographical variations are play a great role in the distribution of Fauna and Flora in Eastern and western Himalayas. Among the total 90 species of Freshwater crab occurs in Indian territory there are only 11 species are recorded from Western Himalayas and 40 species are recorded from eastern Himalayas in Indian part. This available data suggests that the eastern Himalayas are much more diverse than western Himalayas, total 14 genera are recorded from Esatern Himalayas and 6 genera are reported from Western Himalayas. Only a total 6 species namely *Barytelphusa cunicularis* (Westwood, 1836); *Maydelliatelphusa masoniana* (Henderson, 1893); *Sartoriana spinigera* (Wood-Mason, 1871); *Himalayapotamon atkinsonianum* (Wood-Mason, 1871); *Himalayapotamon koolooense* (Rathbun, 1904) and *Eosamon tumidum* (Wood-Mason, 1871); are common in both this part of Himalayas. Where as 34 species were found exclusively only in eastern Himalayas, there are only 5 species namely *Himalayapotamon ambivium* (Alcock, 1909); *Himalayapotamon babaulti* (Bouvier, 1918); *Himalayapotamon kausalis* (Pretzmann, 1966); *Himalayapotamon emphyseteum* (Alcock,1909) and *Larnaudia larnaudi* (A Milne Edwards, 1869) distributed exclusively in Western Himalayas.

In the present communication, a check-list of 45 species of crabs has been prepared providing recent generic and species names of the group as far as possible. State-wise distribution along the Himalayas, of these crabs has been presented. Threats and conservation measures of these crabs are also discussed in this communication.

Review of Literature

From the perusal of literature, it appears that the first freshwater crab reported from freshwater habitat of India collected by Daldorff was *Cancer senex* (= *Oziotelphusa senex* (Fabricius, 1798). Herbst (1799) and Nobili (1903) recorded the species *Potamon leschenaudii* (Edwards) = *Oziotelphusa senex* (Fabricius, 1798) from Pondichery. Lucas (1850) recorded *Thelphusa indica* from the Coromandel Coast. H. Milne Edwards (1853) reported three species from "Inde" (=India) namely, *Thelphusa indica, T. perlata* and *T. leschenaultia*. In 1869, Hilgendorf also recorded the species *Thelphusa leschenaultia* (Milne Edwards) from Pondichery. In addition, he also reported *Telphusa guerini* which was probably collected from India. Heller (1862) described the crab *Thelphusa wüllerstorfi* (= *Spiralothelphusa wüllerstorfi*) based on collections from Madras, Nicobar, Sri Lanka and Tahiti. In1865, he described another species, *Thelphusa corrugata* on the basis of collections from Madras and Java. Both the species are now merged with the species, *Spiralothelphusa wüllerstorfi*. The crab, *Thelphusa leschenaudii* was also recorded by him from Nicobars and Madras. Wood-Mason (1871a, b; 1875) and Bürger (1894).

Alcock (1909a, b) described several species from India. In 1910, he published catalogue of the Indian decapod crustacean which is still considered invaluable publication in the study of freshwater crab of the Indian subcontinent. Henderson (1893, 1912 and 1913), Rathbun (1904, 1905), Bouvier (1918), Roux (1931), Bott (1964, 1969, 1970), Pretzman (1963, 1966a, b) have also studied the freshwater crabs of India and reported several new species. Dutta (1983), Ghatak and Ghosh (2008), Ghosh and Ghatak (1999, 2000), Ghosh *et al.* (1999), have studied the freshwater crabs of Assam, Meghalaya, Manipur and Tripura. Yeo and Ng (2007) have made significant contributions on the taxonomy of freshwater crabs belongs to family Potamidae.

Checklist and Synonymy

The following is the check-list of species with detailed synonymy which have been arranged chronologically for convenience to indicate the extent of work on the species done in Himayan region of India. The distribution of species is shown in paraenthesis. The check-list is prepared based on recent classification of Ng *et al.* (2008).

Superfamily Gecarcinucoidea Rathbun, 1904

Family Gecarcinucidae Rathbun, 1904

Subfamily Gecarcinucinae Rathbun, 1904

1. *Barytelphusa cunicularis* (Westwood, 1836)

1836. *Thelphusa cunicularis* Westwood, In: Sykes and Westwood, *Trans. Entom. Soc. London*, **1**: 183, pl. 19, fig. 1.

1853. *Thelphusa indica*: H. Milne Edwards, *Ann. Sci. nat. (Zool.)*, (3), **20**: 209.

1871. *Thelphusa indica*: Wood-Mason, *J. Asiat. Soc. Bengal*, **40**(2): 196.

1893. *Thelphusa indica*: Henderson, *Trans. Linn. Soc. Lond. Zool.* (2), **5**: 380 (Very common in hill-streams at Kotagiri and elsewhere in Nilgiri Hills at an elevation of 6,000 ft.).

1970. *Barytelphusa (Barytelphusa) cunicularis*: Bott, *Abhandl. Sencken. Naturfors. Ges*, **526**: 31, pl. 2, fig. 18-20; pl. 26, fig. 13 (Bombay, Belaghat, Karnataka – North Kannada (Castle Rock), Mangalore, Dandheli, Shimoga Jog Falls; Malaiyandipattanams Pollachi, Annamalai, Nilgiris – Gudalur and Masinigudi).

1998. *Barytelphusa (Barytelphusa) cunicularis*: Deb, *Zool. Surv. India State Fauna series* **3**: *Fauna of West Bengal, Part 10*: 387 (West Bengal – Darjeeling district: Kalimpong; Jalpaiguri district: Jalpaiguri; Puruliya district: Manbhum and Barabhum).

2005. *Barytelphusa (Barytelphusa) cunicularis*: Ghosh, Ghatak and Roy, *Zool. Surv. India State Fauna Series* **5**: *Fauna of Andhra Pradesh (Part 5)*: 556 (Andhra Pradesh: Pulicat and Station 3, Matchrela, Dist. Nalgonda; Pedamindu, Kolleru Lake).

2005. *Barytelphusa (Barytelphusa) cunicularis*: Srivastava, *Rec. zool. Surv. India*, **104**: 118, pl. 1, fig. 3

2006. *Barytelphusa (Barytelphusa) cunicularis*: Srivastava and Krishnan, *Zool. Surv. India Fauna of Bilgiri Rangaswamy Temple Wildlife Sanctuary, Conservation Area Series*, **27**: 17, 18 (Karnataka – Kolar Road, Honametti Bedaguli, Basavangodu, Kadakkinagandi, Hanakere, Biligiri Rangaswamy Temple, Doddasempige, all from BRTWLS).

2007. *Barytelphusa (Barytelphusa) cunicularis*: Srivastava, *Zool. Surv. India Fauna of Andhra Pradesh, State Fauna Series*, **5** *(Part 4)*: 246, fig. 1(Chittor district: Talakona waterfalls, Naynargardikey; Nellore district: Kadivedu; Karnool district: Lalalabugga; Anantpur district: Pennadam).

2008. *Barytelphusa cunicularis*: Bahir and Yeo, *Raffles Bulletin of Zoology, Supplement*, **16**: 312, fig. 3. (Maharashtra: Poona, Bombay, south western India; Balaghat, south western India. Karnataka: "Indien, Dandheli, Kalu-fluss, N-Kanara"; Dharwar district, freshwater channels, in deep burrows in muddy areas. Kerala: Chathankodu near Ponmudi, Lat. 08°39´ 45.1´´ N, Long. 77°09´ 03.5´´ E., altitude 100m; near Ponmudi, Lat. 08°43´ 04.5´´ N- 08°44´ 19.0´´ N, Long. 77°07´ 41.4´´ E. - 77°09´ 09.7´´ E., altitude 120 – 339m; Kalikavur on Manjeri-Trissur Road, Lat. 11°10´ 0.6.´´ N, Long. 76°19´ 51.7´´ E., altitude 52m; Palaruvi waterfall, Lat. 08°56´ 30.7´´ N, Long. 77°09´ 52.4´´ E., altitude 600m; Chalakudy.

Type locality: Bombay, Western Ghats.

Habitat: Muddy areas beside the water body

Distribution: Darjeeling (W.B.); West Bengal, Andhra Pradesh, Tamil Nadu, Maharashtra, Kerala.Karnataka, Gujarath, Madhya Pradesh, Chhattisgarh, Bihar, Jharkhand, Himachal Pradesh and Odisha.

Conservation Status: It is categorised as Least Concern in IUCN Red list 2013.

Remarks: mainly occupied in peninsular India, but now invaded in Eastern as well as western part of India. Its an Endemic species to India

Genus *Globitelphusa* Alcock, 1909

1909b. *Globitelphusa* Alcock, *Rec. Indian Mus.*, **3**(4): 250.

Type species: *Paratelphusa (Globitelphusa) bakeri* Alcock, 1909, by original designation, gender: feminine.

2. *Globitelphusa bakeri* (Alcock, 1909)

1909b. *Paratelphusa (Globitelphusa) bakeri* Alcock, *Rec. Indian Mus.*, **3** (4): 378.

1910. *Paratelphusa (Globitelphusa) bakeri*: Alcock, *Cat. Indian Decapod Crust. Indian Mus.*, **1**(2): 114, pl. 8, fig. 30 (Assam: Ganjam, north Cachar).

1970. *Liotelphusa laevis bakeri*: Bott, *Abhandl. Sencken.Naturfors. Ges.*, **526**: 50, pl. 33, figs. 13-16 (Ganjan, North Cachar; Darbund, Cachar).

Type locality: Cachar, Ganjam.

Habitat: Hill stream and River

Distribution: Assam (North Cachar.)

Remarks: Endemic to Assam

Conservation Status: It is categorised as Data Defficient in IUCN Red list 2013.

3. *Globitelphusa cylindra* (Alcock, 1909)

1909b. *Paratelphusa (Globitelphusa) bakeri cylindrus* Alcock, *Rec. Indian Mus.*, **3**(4): 378 (Naga Hills and Assam).

1910. *Paratelphusa (Globitelphusa) bakeri cylindrus*: Alcock, *Cat. Indian Decapod Crust. Indian Mus.*, 1(2): 115 (Assam; Nagaland: Naga Hills).

1970. *Liotelphusa cylindra*: Bott, *Abhandl. Sencken.Naturfors. Ges.*, **526**: 51, pl. 33, figs. 21-24 (Assam; Naga Hills).

Type locality: Naga Hills (Nagaland, Northeast India)

Habitat: No data is available

Distribution: Nagaland

Conservation Status: It is categorised as Data Deficient in IUCN Red list 2013.

Remarks: Endemic to Nagaland of Northeast India

4. *Globitelphusa pistorica* (Alcock, 1909)

1909b. *Paratelphusa (Globitelphusa) pistorica* Alcock, *Rec. Indian Mus.*, **3**(4): 378 (Assam: Cachar).

1910. *Paratelphusa (Globitelphusa) pistorica*: Alcock, *Cat. Indian Decapod Crust. Indian Mus.*, **1**(2): 116, pl. 8, fig. 31 (Assam: Sibsagar; Darband Pass, Cachar).

Type locality: Darbund – Pass, Assam

Habitat: Freshwater River and Stream.

Distribution: Assam (North Cachar: Ganjam)

Remarks: Endemic to Assam.

Conservation Status: It is categorised as Data deficient in IUCN Red list 2013.

5. *Liotelphusa gagei* (Alcock, 1909)

1909a. *Paratelphusa (Phricotelphusa) gageii* Alcock, *Rec. Indian Mus.*, **3**(3): 251(Sureil near Kurseong).

1910. *Paratelphusa (Phricotelphusa) gageii*: Alcock, *Cat. Indian Decapod Crust. Indian Mus.*, **1**(2): 105, pl. **13**, fig. 26 (West Bengal: Kurseong, Darjeeling; Sikkim: S. E. Sikkim).

1970. *Liotelphusa laevis gagei*: Bott, *Abhandl. Sencken.Naturfors. Ges.*, **526**: 48, pl. 35, figs. 39-42 (Sikkim).

1998. *Liotelphusa laevis gagei*: Deb, *Zool. Surv. India State Fauna Series* **3**: *Fauna of West Bengal, Part 10*: 385 (West Bengal – Darjeeling district: Kurseong. Sikkim).

2003. *Liotelphusa laevis gagei*: Roy, Ghosh and Ghatak, *Zool. Surv. India State Fauna Series* **9**: *Fauna of Sikkim, Part 5*: 118 (South East Sikkim, no other data, information based on literature).

Type locality: Kurseong, 5000 ft.

Habitat: River and small stream.

Distribution: West Bengal (Darjeeling, Sureil, Kurseong), Sikkim.

Abroad: Bhutan.

Conservation Status: It is categorised as Near Threatened in IUCN Red list 2013.

Remarks: Restricted distribution in eastern Himalaya and Bhutan.

6. *Liotelphusa laevis* (Wood-mason, 1871)

1871. *Telphusa laevis* Wood-Mason, *J. Asiat. Soc. Bengal*, **40**: 201, pl. 14, figs. 1-6.

1910. *Paratelphusa (Liotelphusa) laevis*: Alcock, *Cat. Indian Decapod Crust. Indian Mus.*, **1**(2): 109, pl. 13, fig. 65 (Assam: Cachar; Meghalaya: Shillong, Cherrapunji).

1970. *Liotelphusa laevis laevis*: Bott, *Abhandl. Sencken.Naturfors. Ges.*, **526**: 49, pl. 6, figs. 63-65, pl. 27, fig. 28 and pl. 33, figs. 17-20 (Sibsagar, Cherrapunji, Darjeeling and Naga Hills).

1999. *Liotelphusa laevis laevis*: Ghosh and Ghatak, *Zool. Surv. India State Fauna Series* **4**: *Fauna of Meghalaya, Part 9*: 570 (Meghalaya – Khasi Hills: Cherrapunji; Mardphalang; Mawblong; Mawphlong; Mawpat; Shillong. Garo Hills: Tura near Debasipara hill stream).

2008. *Liotelphusa laevis laevis*: *Zool. Surv. India Fauna of Kopili Hydro Electric Project Site, Wetland Ecosystem Series*, **8**:36, 37 (Assam - River Kopili).

Type locality: Cherrapunji (laevis). Sibsagar (quadrata).

Habitat: Hill Stream and Rivers

Distribution: Meghalaya: (Cherrapunji;) Assam: (Sibsagar;) Nagaland; West Bengal: (Sureil, Darjeeling)

Conservation Status: It is categorised as Near Threatened in IUCN Red list 2013.

Remarks: Ristricted to Eastern Himalaya.

7. *Liotelphusa quadrata* (Alcock, 1909)

1909. *Paratelphusa (Liotelphusa) laevis* (Wood-Mason) var. *quadrata* Alcock, *Rec. Indian Mus.*, **3**(4): 377 (Assam, Meghalaya and Nagaland).

1910. *Paratelphusa (Liotelphusa) laevis* (Wood-Mason) var. *quadrata*:Alcock, *Cat. Indian Decapod Crust. Indian Mus.*, **1**(2): 110, fig. 28(Assam: Sibsagar, Goalpara; Meghalaya: Khasi Hills; Nagaland: Naga Hills).

Type locality: Sibsagar.

Habitat: Hill stream

Distribution: Meghalaya; Assam; Nagaland.

Conservation Status: It is categorised as Vulnarable in IUCN Red list 2013.

Remarks: Endemic to North-eastern states and as per IUCN it's a very threatened species.

Genus *Maydelliathelphusa* Bott, 1969

1969. *Barytelphusa (Maydelliathelphusa)* Bott, *Senckenbergiana biol.*, **50**: 361.

Type species: *Telphusa masoniana* Henderson, 1893, by original designation, gender: feminine.

8. *Maydelliathelphusa edentula* (Alcock, 1909)

1909. *Potamon lugubre edentula* Alcock, *Rec. Indian Mus.*, **3**(4): 247.

1909. *Paratelphusa (Barytelphusa) edentula*: Alcock, *Rec. Indian Mus.*, **3**: 376.

1910. *Paratelphusa (Barytelphusa) edentula*: Alcock, *Cat. Indian Decapod Crust. Indian Mus.*, **1**(2): 84, pl. 5, fig. 19 Assam: Sibsagar, Darbund Pass; Nagaland: Naga Hills).

1970. *Barytelphusa (Maydelliathelphusa) lugubris edentula*: Bott, *Abhandl. Sencken.Naturfors. Ges.*, **526**: 36, pl. 3, fig. 30-32; pl. 26, fig. 17 (Upper Assam; Naga Hills).

1983. *Paratelphusa (Barytelphusa) edentula*: Dutta, *J. Bombay nat. Hist. Soc.*, **80**(2): 539 (Assam– Kamrup district: Maligaon; Sibsagar district: Panbesa near Sibsagar; Lakhimpur district: Corella beel; Dibrugarh district: Dibrugarh).

Type locality: Assam: Sibsagar.

Habitat: Hill stream and River

Distribution: Upper Assam; Naga Hills, Mizoram (Present record).

Abroad: Bhutan

Conservation Status: It is categorised as Near Threatened in IUCN Red list 2013.

Remarks: Restricted to Esatern Himalaya including Bhutan.

9. *Maydelliathelphusa falcidigitis* (Alcock, 1910)

1909a. *Potamon lugubre* var. *falcidigitus* Alcock, *Rec. Indian Mus.*, **3**(3): 248 (Cachar, Cherrapunji, Khasi Hills, Garo Hills and Naga Hills).

1910. *Paratelphusa (Barytelphusa) falcidigitis* Alcock, *Cat. Indian Decapod Crust. Indian Mus.*, **1**(2): 94, pl. 7, fig. 24 (Meghalaya: Cherrapunji, Khasi Hills).

1924. *Paratelphusa (Barytelphusa) falcidigitis*: Kemp, *Rec. Indian Mus.*, **26**: 41 (Common in the stream leading from cave mouth to the Someswari River).

1970. *Barytelphusa (Maydelliathelphusa) lugubris falcidigitis*: Bott, *Abhandl. Sencken.Naturfors.Ges.*, **526**: 35, pl. 34, fig. 29-32 (Ganjam, North Cachar; Dumpep, Khasi Hills).

1999. *Barytelphusa (Maydelliathelphusa) lugubris falcidigitis*: Ghosh and Ghatak, *Zool. Surv. India State Fauna Series* **4**: *Fauna of Meghalaya, Part 9*: 571 (Meghalaya – New Tasku Village, Lailad, East Khasi Hills; Unsing

2008. *Barytelphusa (Maydelliathelphusa) lugubris falcidigitis*: *Zool. Surv. India Fauna of Kopili Hydro Electric Project Site, Wetland Ecosystem Series*, **8**: 35, 36 (Assam - River Kopili).

Type locality: India - Naga Hills.

Habitat: Hill Rivers and streams.

Distribution: Assam, Meghalaya, Nagaland and Mizoram

Conservation Status: It is categorised as Data Deficient in IUCN Red list 2013.

Remarks: Endemic to Eastern Himalaya.

10. *Maydelliathelphusa harpax* (Alcock, 1909)

1909a. *Potamon lugubre* var. *harpax*: Alcock, *Rec. Indian Mus.*, **3**: 247 (Assam, Cachar, Khasi Hills, Garo Hills, Naga Hills, Sylhet).

1910. *Paratelphusa (Barytelphusa) harpax*: Alcock, *Cat. Indian Decapod Crust. Indian Mus.*, **1**(2): 95, pl. 7, fig. 25 (Assam: Cachar, Barak River near Silchar; Hill stream near Harmutti;.Meghalaya: Khasi Hills, Garo Hills; Nagaland: Naga Hills).

1913. *Paratelphusa (Barytelphusa) harpax*: Kemp, *Rec. Indian Mus.*, **8**: 302 (Assam: near Sadiya).

1983. *Paratelphusa (Barytelphusa) harpax*: Dutta, *J. Bombay nat. Hist. Soc.*, **80**(2): 540, fig. 3(Assam – Sibsagar district: Puronipukhuri beel near Gurisagar; Lakhimpur district: North Lakhimpur; Dibrugarh district: Proper Dibrugarh).

Type locality: Nagaland

Habitat: Hill stream and river, Shallow water River bed.

Distribution: India – Assam, Meghalaya, Nagaland, Mizoram. Abroad: Bangladesh.

Conservation Status: It is categorised as Least Concern in IUCN Red list 2013.

Remarks: Mostly confined to the North eastern india and adjacent areas of Bangladesh.

11. *Maydelliathelphusa lugubris* (Wood-Mason, 1871)

1871. *Telphusa lugubris* Wood-Mason, *J. Asiat. Soc. Bengal*, **40**(2): 197, pl. 12, fig. 5-7 (Sikkim: Pankabaree; Teesta Valley; Meghalaya: Cherrapunji).

1893. *Telphusa lugubris*: Henderson, *Trans. Linn. Soc. Lond. Zool.* (2), **5**: 381(Environs of 'Calcutta'. Labelled as *Telphusa indica* and were collected possibly from Himalayas).

1909a. *Potamon lugubre* var. *nigerrimum*: Alcock, *Rec. Indian Mus.*, **3**: 247 (North Lushai).

1909a. *Potamon lugubre* var. *plautum*: Alcock, *Rec. Indian Mus.*, **3**: 247 (Assam and Khasi Hills).

1910. *Paratelphusa (Barytelphusa) lugubre*: Alcock, *Rec. Indian Mus.*, **3**: 376.

1910. *Paratelphusa (Barytelphusa) lugubris*: Alcock, *Cat. Indian Decapod Crust. Indian Mus.*, **1**(2): 91, pl. 12, fig. 58 (Meghalaya: Cherrapunji, Garo Hills; Dafla Hills; Manipur Hills; West Bengal: Teesta Valley, Punkabari; Sikkim).

1910. *Paratelphusa (Barytelphusa) lugubris nigerrima*: Alcock, *Cat. Indian Decapod Crust. Indian Mus.*, **1**(2): 93 (Changsil, North Ludhai).

1910. *Paratelphusa (Barytelphusa) lugubris plauta*: Alcock, *Cat. Indian Decapod Crust. Indian Mus.*, **1**(2): 93, pl. 6, fig. 23 (Sibsagar, Khasi Hills, Naga Hills).

1970. *Barytelphusa (Maydelliathelphusa) lugubris lugubris*: Bott, *Abhandl. Sencken. Naturfors. Ges.*, **526**: 34, pl. 3, fig. 24-26; pl. 26, fig. 15 (Naga Hills; Garo Hills; North India; Kolkata: Assam, North Lushai, Changil).

1998. *Barytelphusa (Maydelliathelphusa) lugubris lugubris*: Deb, *Zool. Surv. India State Fauna Series* **3**: *Fauna of West Bengal, Part 10*: 387 (West Bengal – Darjeeling district: Kalimpong, Darjeeling, Teesta Valley; Jalpaiguri district: Mahananda river, Siliguri).

1999. *Barytelphusa (Maydelliathelphusa) lugubris lugubris*: Ghosh and Ghatak, *Zool. Surv. India State Fauna Series* **4**: *Fauna of Meghalaya, Part 9*: 572 (Meghalaya – **West** Garo Hills: Phulbari, Rangui River; Mahadeo River, Mahadeo; Bogai River; Siju Cave; North East of Barangapara; Rangranchidekgray near Williamnagore; Balat; Chinabat; Nengkhra Crossing; Bogai River; Tura; Dobasipara near hillstream and Valleysite; Jakrem River near hotspring. Khasi Hills: Norblong Village near Byrnihat; Shillong. Kyrdemkulai stream; Kyrdemkulai Dam, No. 1, Damside; Synrangmmanarati River; Mawmai Cave and Mawluh, Cherrapunjee;. Jayantia Hills: Jowai-Chongpung, Bridle Path; Stream near Mawlyngkneng; Jowai stream, Jowai; Garampani; Kollasiv).

2000. *Barytelphusa (Maydelliathelphusa) lugubris lugubris*: Ghosh and Ghatak, *Zool. Surv. India State Fauna Series* **7**: *Fauna of Tripura, Part 4*: 274 (Tripura – North Tripura: Jumpoi Hills: Fuldungsei, Vangmun, Subal Lake; Salema; Boruma).

2003. *Barytelphusa (Maydelliathelphusa) lugubris lugubris*: Roy, Ghosh and Ghatak, *Zool. Surv. India State Fauna Series* **9**: *Fauna of Sikkim, Part 5*: 117 (North Sikkim – Namak; Singhik; Phodong; Mamul. East Sikkim: Pakyong; Singham; Panyong. South Sikkim: Manpur Village, Mohanpur Village, Mantur Village, Rollu Village, Damthang).

2004. *Barytelphusa (Maydelliathelphusa) lugubris lugubris*: Roy, Ghosh and Ghatak, *Zool. Surv. India State Fauna Series* **10**: *Fauna of Manipur, Part 3*: 122 ((Manipur –

Maram and Karong, Dist. Senapati; Loktak Lake; Bishnupur; Keibul-Lamjao; Thanga, Moirang).

2008. *Barytelphusa (Maydelliathelphusa) lugubris lugubris*: *Zool. Surv. India Fauna of Kopili Hydro Electric Project Site, Wetland Ecosystem Series*, **8**:36 (Assam - River Kopili).

Type locality: Sikkim, Pankabaree, Altitude 200ft.

Habitat: River bed, but this crab are used to live in holes beside the rivers in to dense forest area also.

Distribution: West Bengal, Sikkim, Meghalaya, Manipur, Assam, Nagaland, Bihar and Mizoram

Abroad: Bangladesh, Bhutan and Nepal.

Conservation Status: It is categorised as Least Concern in IUCN Red list 2013.

Remarks: Widely distributed in Eastern Himalaya and adjacent Countries, very possibly it my also distributed in Myanmar. It's the most frequently highly priced edible crab in North-east Indian states.

12. *Maydelliathelphusa masoniana* (Henderson, 1893)

1893. *Telphusa masoniana* Henderson, *Trans. Linn. Soc. Lond. Zool.* (2), **5**: 381(River Jumna; North-West Provinces).

1904. *Potamon (Potamon) masonianus*: Rathbun, Nouv. Arch. Mus., sér. 4, 6: 299, pl. 11, fig. 10 (Himalayas).

1910. *Paratelphusa (Barytelphusa) masoniana*: Alcock, *Cat. Indian Decapod Crust. Indian Mus.*, **1**(2): 96, pl. 12, fig. 59 (Uttar Pradesh: Hardwar, Saharanpur, Dehra Dun, Naini Tal and Rurki; Bihar: Darbhanga; Bijnor).

1970. *Barytelphusa (Maydelliathelphusa) lugubris masoniana*: Bott, *Abhandl. Sencken. Naturfors. Ges.*, **526**: 36, pl. 3, figs. 27-29, pl. 26, fig. 16 (Nishangal, North India; Sikkim; Ringrengiri, Meghalaya).

1995. *Paratelphusa (Barytelphusa) masoniana*: Krishnamurthy, *Zool. Surv. India Himalayan Ecosystem Series, Part 1, Uttar Pradesh*: 23.

Type locality: North India - River Jamuna (Henderson).

Habitat: River and hill stream.

Distribution: Assam, Bihar, Chattisgarh, Sikkim, Meghalaya, Himachal Pradesh, Uttaranchal, Uttar Pradesh and Jammu and Kashmir.

Conservation Status: It is categorised as Least Concern in IUCN Red list 2013.

Remarks: Endemic to India. Common in Both the Esatern and Western part of Himalayas. This species is also distributed in Terai region of Himalayan foot hills.

Genus *Travancoriana* Bott, 1969

1969. *Travancoriana* Bott, *Senckenbergiana. biol.*, **50**(5/6): 361.

Type species: *Travancoriana schirnere* Bott, 1969, by original designation, gender: feminine.

13. *Travancoriana napaea* (Alcock, 1909)

1909a. *Potamon napaeum* Alcock, *Rec. Indian Mus.*, **3**(3): 248 (Ganjam, North Cachar, 4000 ft.)

1909b. *Paratelphusa (Barytelphusa) napea*: Alcock, *Rec. Indian Mus.*, **3**(4): 376.

1910. *Paratelphusa (Barytelphusa) napaea*: Alcock, *Cat. Indian Decapod Crust. Indian Mus.* **1**(2): 85, pl. 5, fig. 20 (Assam: Ganjam, North Cachar).

Type locality: India: Assam - Ganjam, North Cachar.

Habitat: Unknown.

Distribution: Assam.

Conservation Status: It is categorised as Data Deficient in IUCN Red list 2013.

Remarks: Endemic to Assam. Thre is no any collection of this species after Alcock, 1909.

Genus *Sartoriana* Bott, 1969

1969. *Sartoriana* Bott, *Senckenbergiana biol.*, **50**: 361.

Type species: *Paratelphusa (Paratelphusa) spinigera* Wood-Mason, 1871, by original designation, gender: feminine.

14. *Sartoriana spinigera* (Wood-Mason, 1871)

1871. *Paratelphusa spinigera* Wood-Mason, *J. Asiat. Soc. Bengal*, **40**(2): 194, pl. 10, figs. 1-4 (Museum Tank of Calcutta).

1876. *Paratelphusa spinigera*: Wood-Mason, *Ann. Mag. nat. Hist.*, ser. 4, **17**: 121, 122.

1893. *Paratelphusa spinigera*: Henderson, *Trans. Linn. Soc. Lond. Zool.* (2), **5**: 386 (Calcutta, Roorke, Ganjam, North-West Provinces, Sind).

1910. *Paratelphusa (Paratelphusa) spinigera*: Alcock, *Cat. Indian Decapod Crust. Indian Mus.*, **1**(2): 72, pl. 11, fig. 53.(West Bengal: Kolkata; Assam: Balaganj, Cachar; Bihar: Darbhanga, Kissenganj: Orissa: Sur Lake, Puri District; Punjab: Safedbein Canal, Jullunder District; Uttar Pradesh: Hardwar, Saharanpur; Kashmir: Khewrah Gorge, Jhelum District).

1970. *Sartoriana spinigera*: Bott, *Abhandl. Sencken.Naturfors. Ges.*, **526**: 39, pl. 4, figs. 35-37; pl. 26, fig. 18.

1999. *Sartoriana spinigera*: Ghosh and Ghatak, *Zool. Surv. India State Fauna Series* **4**: *Fauna of Meghalaya, Part 9*: 575 (Meghalaya – Khasi Hills: Umiam. Garo Hills: Phulbari, Rangai River; Damra bazaar).

1998. *Sartoriana spinigera:* Deb, *Zool. Surv. India State Fauna Series* **3**: *Fauna of West Bengal, Part 10:* 387 (West Bengal – Kolkata, Pulta, Chinsura, Raigunj, Dinajpur, Jalpaiguri, Coochbihar).

1983. *Paratelphusa (Paratelphusa) spinigera:* Dutta, *J. Bombay nat. Hist. Soc.*, **80**(2): 544, fig. 5 (Assam—Goalpara district: Dipo; Kamrup district: Kukurmara beel, Lankeswar dhum near Jalukbari, Boko, Bebejapara near Bozali, Bhulukmara beel near Amingaon, Durmari beel near Chetoli, Gogiakur near River Saulkhua, Mongoldoi, Kali Kuchi, River Kulsi; Nowgong district: Hapakati beel near Morigaon; Karbi-Anglong district: River Jamuna; Cachar district: River Karimganj; Sibsagar district: Nawpukhuri beel, River Namdang near Joysagar, Jorhat, Golaghat, Bokakhat, Gorisagar, Sunari; Lakhimpur district: Pohumara near Singar, North Lakhimpur; Dibrugarh district:).

2000. *Sartoriana spinigera:* Ghosh and Ghatak, *Zool. Surv. India State Fauna Series* **7**: *Fauna of Tripura, Part 4:* 273 (Tripura – North Tripura: Dharmanagar: Birjanagar, Kadamtala, Kanchanpur, Bagon; Bormara; Khusdinpara Chamanu Road; Manubari; Gerania; Champak Nagar. South Tripura: Kamalcherra River, Dullubari; Surmai River, South of Gondacherra; Dhanyasagar, Udaipur. West Tripura: Asrambari).

2008. *Sartoriana spinigera: Zool. Surv. India Fauna of Kopili Hydro Electric Project Site, Wetland Ecosystem Series,* **8**:36, 37 (Assam - River Kopili).

Type locality: Bangladesh: Jessore district.

Habitat: Freshwater Ponds, bill, canals, Paddyfield and wetland area.

Distribution: India: Meghalaya;Uttar Pradesh:(Sharanpur) Uttarakhand (Hardwar) Punjab; Orissa (Puri district); West Bengal:(All districts of the plain area;) Assam: (Cachar; Bihar: Darbhanga,) Bihar (Kissenganj); Tripura, Mizoram, Nagaland, Jharkhand.

Abroad: Bangladesh, Pakistan, Sri Lanka, Myanmar

Conservation Status: It is categorised as Least Concern in IUCN Red list 2013.

Remarks: Its appear as most widely distributed and highly populated crab of Northern part of India. Its edible species and consumed by the local people very frequently.

15. *Sartoriana trilobata* (Alcock, 1909)

1909. *Paratelphusa (Paratelphusa) trilobata:* Alcock, *Rec. Indian Mus.*, **3**: 375 (Assam – Sibsagar).

1910. *Paratelphusa (Paratelphusa) trilobata:* Alcock, *Cat. Indian Decapod Crust. Indian Mus.*, **1**(2): 74, **pl. 11,** fig. 15 (.Assam: Sibsagar).

Type locality: Sibsagar.

Habitat: unknown

Distribution: India – Assam.

Conservation Status: It is categorised as Data Deficient in IUCN Red list 2013.

Remarks: Endemic to Assam. Known from its type locality only. No any further collection of this crab was maid after Alcock, 1909. This species was erected on the base of single female example.

Genus *Sommanniathelphusa* Bott, 1968

1853. *Parathelphusa* H. Milne Edwards, *Ann. Sci. nat. (Zool.)*, 3, **20**: 213 (in partim).

1871. *Paratelphusa*: Wood-Mason, *J. Asiat. Soc. Bengal*, **40**(2): 213 (in partim).

1968. *Sommanniathelphusa* Bott, *Senckenbergiana biol.*, **49**: 407

Type species: *Parathelphusa sinensis* H. Milne Edwards, 1853, by original designation, gender: feminine.

16. *Sommanniathelphusa sinensis* (H. Milne Edwards, 1853)

1853. *Parathelphusa sinensis* H. Milne Edwards, *Ann. Sci. nat. (Zool.)*, 3, **20**: 213.

1983. *Paratelphusa (Paratelphusa) sinensis*: Dutta, *J. Bombay nat. Hist. Soc.*, **80**(2): 540, fig. 4 (Kamrup district: R. Pagladia near Uttarkuchi; Darrang district: Proper Tezpur; Karbi Anglong district: Proper Diphu).

1970. *Sommanniathelphusa sinensis sinensis*: Bott, *Abhandl. Sencken.Naturfors. Ges.*, **526**: 111, pl. 20, figs. 42-44 and pl. 30, fig. 81.

Type locality: China.

Habitat: Freshwater pond and Wet land.

Distribution: India - Assam.

Abroad - China, Hong Kong. So far recorded from Assam of India.

Conservation Status: In IUCN conservation list it it categorized as data deficient.

Remarks: This species is reported first time by Dutta from Different parts of Assam from India. There are no any records of this species after Dutta, 1983.

Family Potamidae Ortmamm, 1896

1838. Thelphusidae MacLeay: [priority suppressed, ICZN plenary powers (Opinion 712].

1896. Potamoninae Ortmann, *Zool. Jb. (Syst.)*, **9**: 445.

1896. Potamidae Ortmann: [spelling corrected from Potamonidae Ortmann, 1896, and name given priority over Thelphusidae under ICZN plenary powers) [Opinion 712].

Subfamily Potaminae Ortmann, 1896

1838. Thelphusidae MacLeay: [priority suppressed ICZN plenary powers (Opinion 712].

1896. Potamoninae Ortmann, *Zool. Jb. (Syst.)*, **9**: 445.

Remarks: The main character of this subfamily is a transverse ridge on the joint of the 7th and 8th thorasic sternite in the longitudinal median line of the abdominal cavity (Yeo and Ng, 2007).

Genus *Acanthopotamon* Kemp, 1918

1910. *Potamon (Acanthotelphusa)*: Alcock, *Cat. Indian Decapod Crust. Indian Mus.*, **1**(2): 61.

1918. *Potamon (Acanthopotamon)* Kemp, *Rec. Indian Mus.*, **14**: 101.

2007. *Acanthopotamon*: Yeo and Ng, *Raffles Bull. Zool., Supplement*, **16**: 274.

Type species: *Paratelphusa martensi* Wood-Mason, 1875, by original designation.

17. *Acanthopotamon fungosum* (Alcock, 1909)

1909a. *Potamon (Paratelphusula) fungosum* Alcock, *Rec. Indian Mus.*, **3**(3): 250 (Cachar).

1910. *Potamon (Acanthopotamon) fungosum*: Alcock, *Cat. Indian Decapod Crust. Indian Mus.*, **1**(2): 65, fig. 12 (Assam: Cachar).

Type locality: Assam.

Habitat: Water falls

Distribution: India - Assam. (Darband Pass, Cachar, Assam);

Abraod: Thailand (Phangnga Province).

Conservation Status: It is categorised as Data Deficient IUCN Red list 2013.

Remarks: This species seems to be rare in India.

18. *Acanthopotamon martensi* (Wood-Mason, 1875)

1875. *Paratelphusa martensii* Wood-Mason, *Proc. Asiat. Soc. Bengal*: 230 (Throughout the Gangetic valley from Hardwar to Jessore(Bangladesh).

1876. *Paratelphusa martensii*: Wood-Mason, *Ann. Mag. nat. Hist.*, 4, **17**: 121, 122 (North India: Hardwar, Purneah, Allahabad).

1893. *Paratelphusa martensii*: Henderson, *Trans. Linn. Soc. Lond. Zool.*, 2, **5**: 386 (Roorke, North-West Provinces).

1910. *Potamon (Acanthotelphusa) martensi*: Alcock, *Cat. Indian Decapod Crust. Indian Mus.*, **1**(2): 68, pl. 11, fig. 52 (West Bengal: Baranagar and Santipur near Kolkata; Bihar: Purnea, Kissengunj, Darbhanga; Uttar Pradesh: Lucknow, Roorke; Abjulgar, Bijnor).

1970. *Acanthopotamon martensi*: Bott, *Abhandl. Sencken.Naturfors. Ges.*, **526**: 145, pl. 38, fig. 22and pl. 45, fig. 20 (Terai, Nishangar; Ganges Banaras).

1998. *Acanthopotamon martensi*: Deb, *Zool. Surv. India State Fauna Series* **3**: *Fauna of West Bengal, Part 10*: 382 (West Bengal - Baranagar near Kolkata; Kasai canal, Midnapore; Teesta River at Jalpaiguri).

1999. *Acanthopotamon martensi*: Ghosh and Ghatak, *Zool. Surv. India State Fauna Series* **4**: *Fauna of Meghalaya, Part 9*: 571(Meghalaya - West Garo Hills: Willamnagore. Muktapur Road, Dawki).

Type locality: Jessore, Bangaladesh.

Habitat: Reiverine crabs.

Distribution: Assam, Meghalaya; Arunachal Pradesh; Bihar; Uttar Pradesh; Uttarakhand; West Bengal; Rajasthan;

Abroad: Bangladesh, Myanmar.

Remarks: Throughout the Gangetic valley down to Calcutta where brackishwater conditions sustained and where it occurred both in fresh and brackishwater like several of its congeners (Wood-Mason, 1875).

Conservation Status: It is categorised as Least Concern in IUCN Red list 2013.

Genus *Alcomon* Yeo and Ng, 2007

2007. *Alcomon* Yeo and Ng, *Raffles Bull. Zool., Supplement*, **16**: 275.

Type species: *Potamon (Geothelphusa) superciliosum* Kemp, 1913 by Yeo and Ng in *Raffles Bull. Zool., Supplement*, **16**: 275 (2007) by original designation.

19. *Alcomon lophocarpus* (Kemp, 1913)

1913. *Potamon (Geotelphusa) adiatretum Var.lophocarpus* Kemp, *Rec. Indian Mus.*, 8: 300, pl. 18, figs. 15-18(Upper Rotung; Egar stream between Renging and Rotung; Lalek stream near Renging; neighbourhood of Rotung; 3 miles south of Yembung; Sipro valley between Janakmukh and Renging; stream near Balek)

2007. *Alcomon lophocarpus*: Yeo and Ng, *Raffles Bull. Zool., Supplement*, **16**:

Type locality: Small stream between 2 and 3 miles of South of Yembung (Arunachal Pradesh)

Habitat: Hill stream

Distribution: Rotung, Renging, Balek area of Arunachal Pradesh.

Conservation Status: It is categorised as Least Concern in IUCN Red list 2013.

Remarks: This species also collected by Kemp, 1913; after that it was not collected till date. It is endemic to Arunachal Pradesh of India.

20. *Alcomon superciliosum* (Kemp, 1913)

1913. *Potamon (Geotelphusa) superciliosum* Kemp, *Rec. Indian Mus.*, 8: 300, pl. 18, figs. 15-18 (Yembung River; Eager stream between Renging and Rotung; stream near Balek).

2007. *Alcomon superciliosum*: Yeo and Ng, *Raffles Bull. Zool., Supplement*, **16**: 303.

Type locality: A stream near Balek (Arunachal Pradesh)

Habitat: Hill stream and River of the altitude of 600 ft to 2000ft.

Distribution: Yembung, Rotung, Balek (Arunachal Pradesh)

Abroad: Myanmar (Mainland)

Remarks: The presnt author collected this species from Namdapha Biosphere Reserve of Arunachal Pradesh and also from Mizoram.

Conservation Status: It is categorised as Data Deficient in IUCN Red list 2013.

Genus *Himalayapotamon* Pretzmann, 1966

1966. *Potamon (Himalayapotamon)* Pretzmann, *Annln. Naturh. Mus. Wien*, **69**:

Type species: *Telphusa atkinsonianum* Wood-Mason, 1871, by original designation, gender: neuter.

21. *Himalayapotamon ambivium* (Alcock, 1909)

1909a. *Potamon (Potamon) atkinsonianum var. ambivium*: Alcock: *Rec. Indian Mus.*, **3**(3): 243 (Dharampur near Simla, 5, 000 ft.).

1910. *Potamon (Potamon) atkinsonianum var. ambivium*: Alcock: 30 (Himachal Pradesh, Dharampur, Simla, 5000 ft.).

Type locality: Dharampur (Himachal Pradesh)

Habitat: Hill stream

Distribution: Himachal Pradesh

Conservation Status: It is categorised as Data Deficient in IUCN Red list 2013.

Remarks: Endemic to Himachal Pradesh; so far no other speciemens were collected beside the type species.

22. *Himalayapotamon atkinsonianum* (Wood-Mason, 1871)

1871. *Telphusa atkinsonianum* Wood-Mason, *J. Asiat. Soc. Bengal*, **40**: 205, pl. 14, figs. 12-16.

1893. *Telphusa atkinsoniana*: Henderson, *Trans. Linn. Soc. Lond. Zool.*, ser. 2, **5**: 385 (Kangra, Simla).

1904. *Potamon (Potamon) atkinsonianum*: Rathbun, *Nouv. Arch. Mus*, sér. 4, **6**: 271[in partim].

1909a. *Potamon (Potamon) atkinsonianum* var. *ventriosum*: *Rec. Indian Mus.*, **3**(3): 244 (Kumaon, about 6000 ft., probably an aberrant individual).

1910. *Potamon (Potamon) atkinsonianum*: Alcock, *Cat. Indian Decapod Crust. Indian Mus.*, **1**(2): 26, pl. 10, fig. 39.

1910. *Potamon (Potamon) atkinsonianum* var. *ventriosum*: Alcock, *Cat. Indian Decapod Crust. Indian Mus.*, **1**(2): 29.

1970. *Potamon atkinsonianum*: Bott, *Abhandl. Sencken.Naturfors. Ges.*, **526**: 140, pl. 37, figs. 14, 15 and pl. 44, fig. 13.

1975. *Potamon atkinsonianum*: Sharma, *J. Bombay nat. Hist. Soc.*, **72**(1): 223 (Kashmir: Poonch Valley).

Type locality: Sikkim.

Habitat: Hill streams connected with Rivers.

Distribution: West Bengal, Sikkim, Himachal Pradesh, Kashmir.

Conservation Status: It is categorised as Least Concern in IUCN Red list 2013.

Remarks: Though this species is reported from Eastern Himalaya and as well as Western Himalaya but this species is actually occurs in Eastern Himalaya only (Brandis,2001), as it is very close to *H. emphyseteum*, there are many chance to miss identify this species with the later ones.

23. *Himalayapotamon babaulti* (Bouvier, 1918)

1918. *Potamon babaulti* Bouvier, *Bull. Mus. Hist. Nat.*, **24**: 392 (Western Himalayas).

Type locality: Western Himalayas.

Habitat: Unknown

Distribution: Only reported from Himachal Pradesh. No collection has been maid after type.

Conservation Status: It is categorised as Data Deficient in IUCN Red list 2013.

Remarks: There is no collection of this species after 1918.

24. *Himalayapotamon emphyseteum* (Alcock, 1909)

1909a. *Potamon (Potamon) atkinsonianum* var. *emphyseteum* Alcock, *Rec. Indian Mus.*, **3**(3): 243 (Punjab Himalayas at Bilaspur and Kangra).

1966. *Potamon (Himalayapotamon) atkinsonianum gordoni* Pretzmann, *Annln. Naturh. Mus. Wien*, **69**: (Himalaya).

1970. *Potamon emphysetum*: Bott, *Abhandl. Sencken.Naturfors. Ges.*, **526**: 141, pl. 37, fig. 17 and pl. 44, fig. 15 (Himachal Pradesh: Dharampur, Simla, Aurkhad, Hari Talyangar).

Type locality: Bilaspur and kangra, Himachal Pradesh

Habitat: Hill stream, Bowri and Rivers.

Distribution: Punjab, Himachal Pradesh, Jammu and Kashmir.

Conservation Status: It is categorised as Least Concern in IUCN Red list 2013.

Remarks: This species is abundant in Western Himalaya.

25. *Himalayapotamon kasaulis* (Pretzmann, 1966)

1966. *Potamon (Himalayapotamon) koolense kasauli* Pretzmann, *Annln. Naturh. Mus. Wien*, **69**: (Himachal Pradesh, former Punjab)).

Type locality: Kasauli (Near Simla), Himachal Pradesh.

Habitat: Not recorded.

Distribution: Only at Kasauli (Himachal Pradesh)

Conservation Status: It is categorised as Data Deficient in IUCN Red list 2013.

Remarks: No further collection after type.

26. *Himalayapotamon koolooense* (Rathbun, 1904)

1904. *Potamon (Potamon) Koolooense* Rathbun, *Nouv. Arch. Mus.*, sér. 4, **6**: 270, pl. 10, fig. 1 (North India: Kooloo Valley).

1910. *Potamon (Potamon) Koolooense*: Alcock, *Cat. Indian Decapod Crust. Indian Mus.*, **1**(2): 24, pl. 10, fig. 38 (Himachal Pradesh: Simla, Dharamsala, 4000-5000 ft.; Uttarakhand: Ramnee (Garwhal), Bhim Tal (Kumaon), Naini Tal; Uttar Pradesh: Hazara; River Ravi, Chamba; Nepal Terai; Afghanistan).

1970. *Potamon koolooense*: Bott, *Abhandl. Sencken.Naturfors. Ges.*, **526**: 143, pl. 38, fig. 19 and pl. 45, fig. 17 (Simla; Tons River, 3600 m; Molta; Kooloo;).

1999. *Potamon koolooense*: Ghosh, H.C and Ghatak, S.S. fauna of Meghalaya. Part 9, 569-570. (Meghalaya: Macesphlang forest hills).

Type locality: Kooloo Valley, Himachal Pradesh.

Habitat: Bowri, Hill Stream.

Distribution: India - Uttar Pradesh, Uttarakhand, Himachal Pradesh; West Bengal (Darjeeling), Meghalya, Sikkim.

Abroad: Afghanistan, Nepal.

Conservation Status: It is categorised as Least Concern in IUCN Red list 2013

Remarks: One of the most common species of Western Himalayas.

27. *Himalayapotamon monticola* (Alcock, 1910)

1910. *Potamon (Potamon) fluviatile* var. *monticola*, Wood-Mason (name only) or *Potamon ibericum* var. *monticola*: 23 (Khasi Hills; Darjeeling Hills).

Type locality: Darjeeling, West Bengal.

Habitat: Rivers and Streams

Distribution: Darjeeling (W.B); Khasi Hills (Meghalya).

Conservation Status: It is categorised as Data Deficient in IUCN Red list 2013.

28. *Himalayapotamon bifarium* (Alcock, 1910)

1910. *Potamon (Potamon) bifarium* Alcock *Cat. Indian Decapod Crust. Indian Mus.*, **1**(2): 30-31, pl.1, fig. 3.

Type locality: Sikkim or Myanmar.

Habitat: Unknown

Distribution: Sikkim/Myanmar

Remarks: The distribution of this species till debatable, as this species was described from a single specimen that was lebbeled as Sikkim or Burma.

Conservation Status: It is categorised as Data Deficient in IUCN Red list 2013.

Genus *Lobothelphusa* Bouvier, 1917

1917. *Hydrothelphusa (Lobothelphusa)* Bouvier, *C. R. Acad. Sci. nat. Paris*, **165**: 620 (in partim).

Type species: *Paratelphusa crenulifera* Wood-Mason, 1875, by subsequent designation by Bott, 1970, Gender: Masculine.

29. *Lobothelphusa wood-masoni* (Rathbun, 1905)

1905. *Potamon (Paratelphusa) woodmasoni* Rathbun, *Nouv. Arch. Mus.*, sér.4,**7**:262, pl.12,fig. 2.

1875. *Paratelpbusa edwardsi* Wood-Mason, Proc. asiat. Soc. Bengal, 231.

1876. *Paratelpbusa edwardsi :* Wood-Mason, Ann. Mag. nat. Hist. (4) 17: 121.

1898. *Parathelphusa edwardsi:* De Man, Ann. Mus. civ. Stör. nat. Genova, 19: 438.

1905. *Potamon (Parathelphusa) woodmasoni* Rathbun, Nouv. Arch. Mus. (4) 7: 262, T. 12 F.12.

1909a. *Paratelphusula milneedwardsi* Alcock, Rec. ind. Mus., 3: 250.

1909b. *Paratelphusula woodmasoni,* Alcock, Rec. ind. Mus., 3: 250.

1910. *Paratelphusa wood-masoni*: Alcock, *Cat. Indian Decapod Crust. Indian Mus.*, **1**(2): 63, pl.11, fig. 50 (Meghalaya - Garo Hills; Assam - Sibsagar).

1966. *Potamon (Spinopotamon) crenuliferum woodmasoni,* Bott, Senckenbergiana biol., 47: 477, Abb. 11.

1970. *Lobothelphusa wood-masoni*: Bott, *Abhandl. Sencken.Naturfors. Ges.*, **526**: 149, pl. 38, fig.26 and pl. 46, fig. 24 (Meghalaya: Garo hills)

Type locality: Garo Hills, Meghalaya.

Habitat: Hill stream, Terrestrial.

Distribution: Assam, Tripura, Meghalya and Mizoram.

Abroad: Bangladesh, Myanmar.

Conservation Status: It is categorised as Least Concern in IUCN Red list 2013.

Remarks: As per IUCN species evaluation committee its categorized as least concern in view of its wide distribution, presumed large population, and because it is unlikely to be declining fast enough to qualify for listing in a more threatened category. But in reality its population is declining and it need to revise its position in conservation categories.

Subfamily Potamiscinae Bott, 1970

In this subfamily there is no any transverse ridge in between the 7th and 8th thorasic sternite in the longitudinal midlines of the abdominal cavity.(Yeo and Ng, 2007)

Genus *Eosamon* Yeo and Ng, 2007

2007. *Eosamon* Yeo and Ng, *Raffles Bull. Zool., Supplement*, **16**: 281.

Type species: *Potamon (Potamon) smithianum* Kemp, 1923, subsequent designation by Yeo and Ng in *Raffles Bull. Zool., Supplement*, **16**: 281.

30. *Eosamon tumidum* (Wood-Mason, 1871)

1871. *Telphusa tumida* Wood-Mason, *J. Asiat. Soc. Bengal*, **40**(2): 453, pl. 27, figs. 6-10.

1910. *Potamon (Potamon) tumidum*: Alcock, *Cat. Indian Decapod Crust. Indian Mus.*, **1**(2): 41, pl. 10, fig. 45 (West Bengal: ? Darjeeling).

1970. *Potamiscus tumidus*: Bott, *Abhandl. Sencken.Naturfors. Ges.*, **526**: 161, pl. 38, fig. 32 and pl. 51, fig. 52.

1998. *Potamiscus tumidus*: Deb, *Zool. Surv. India State Fauna Series* **3**: *Fauna of West Bengal, Part 10*: 387 (West Bengal – Darjeeling district: Darjeeling).

Type locality: Jünnan, Hotha.

Habitat: Hill stream

Distribution: India – Darjeeling District of West Bengal. Himachal Pradesh.

Abroad: China (Yunnan); Myanmar (Myanmar (mainland).

Conservation Status: It is categorised as Data Deficient in IUCN Red list 2013.

Genus Aspermon Yeo and Ng, 2007

31. *Aspermon feae* Rathbun, 1905

1898. *Paratelpbusa feae* DE MAN, *Ann. Mus. civ. Stör. nat. Genova*, (2) 19: 393, T. 4 F. 3.

1905. *Potamon (Parathelpbusa) feae*: Rathbun, Nouv. *Arch. Mus.* (4) 7: 241.

1910. *Potamon (Acanthotelpbusa) feae*: Alcock, *Cat. Indian Decapod Crust. Indian Mus.*, 1 (2): 66, T. 11 F. 51

1913. *Potamon (Acanthotelpbusa) feae*: Kemp, *Rec. ind. Mus.*, 8: 301.

Type locality: Upper Irrawaddy, Myanmar

Habitat: Hill stream and River.

Distribution: India – Assam, Arunachal Pradesh, Mizoram (Present Record).

Abroad: -Myanmar (Upper Irrawaddy River, Teinzo)

Conservation Status: It is categorised as Data Deficient in IUCN Red list 2013.

Genus *Indochinamon* Yeo and Ng, 2007

2007. *Indochinamon* Yeo and Ng, *Raffles Bull. Zool.*, **16**: 282.

Type species: *Potamon villosum* Yeo and Ng, 1998, by subsequent designation by Yeo and Ng in *Raffles Bull. Zool.*, **16**: 282.(2007).

32. *Indochinamon andersonianum* (Wood-Mason, 1871)

1871. *Telphusa andersoniana* Wood-Mason, *J. Asiat. Soc. Bengal*, XL (2): 451, pl. 27, figs. 16-29.

1910. *Potamon (Potamon) andersonianum*: Alcock, *Cat. Indian Decapod Crust. Indian Mus.*, **1**(2): 32, pl. 10, fig. 40.

1970. *Potamon andersonianum*: Bott, *Abhandl. Sencken.Naturfors. Ges.*, 526: 142, pl. 37, figs.16 and pl. 44, fig. 14.

2003. *Potamon andersonianum*: Roy, Ghosh and Ghatak, *Zool. Surv. India State Fauna Series* **10**: *Fauna of Manipur, Part 3*: 122 ((Manipur Hills. No specimen other than the collection of H. H. Godwin-Austein).

Type locality: Upper Hills, Kakhyien Hills, Ponsee.

Habitat: Hill stream

Distribution: India - Manipur. Abroad: Myanmar, China.

Conservation Status: It is categorised as Data deficient in IUCN Red list 2013.

Remarks: Its rare in India.

33. *Indochinamon asperatum* (Alcock, 1909)

1909a. *Potamon (Potamon) andersonianum asperatum* Alcock, *Rec. Indian Mus.*, **3**: 244 (Assam: Ganjam in Cachar Hills, 4000 ft.).

1910. *Potamon (Potamon) andersonianum asperatum*: Alcock, *Cat. Indian Decapod Crust. Indian Mus.*, **1**(2): 35 (Assam: Ganjam, Cachar Hills, 4000 ft.).

1910. *Potamon (Potamon) andersonianum var.asperatum*: Alcock, *Cat. Indian Decapod Crust. Indian Mus.*, **1**(2): 35 (Ganjam Hills, about 1000 ft.).

2007. *Indochinamon asperatum*: Yeo and Ng, Raffles Bull. Zool., Supplement, 16: 304.

Type locality: Assam.

Habitat: Freshwater but microhabitat is unknown.

Distribution: India – Assam

Conservation Status: It is categorised as Data deficient in IUCN Red list 2013.

Remarks: Endemic to Assam.

34. *Indochinamon beieri* (Pretzmann, 1966)

1966. *Potamon beieri* Pretzmann, *Annln. Naturh. Mus. Wien*, **69**: (Himalayas).

1904. *Potamon (Potamon) rangoonense* Rathbun, *Nouv. Arch. Mus. Hist. nat.*, 4, **6**: 279, pl. 11, fig. 2, Abb. 18a-c [Opinion 1640].

1910. *Potamon (Potamon) andersonianum rangoonense*: Alcock, *Cat. Indian Decapod Crust. Indian Mus.*, 1(2): 34, fig. 41.

1970. *Ranguna (Ranguna) rangoonensis*: Bott, *Abhandl. Sencken.Naturfors. Ges.*, **526**: 163, pl. 38, fig. 35 and pl. 47, fig. 31 (Assam; Naga Hills).

Type locality: Sukli, E.side of Dawane Hills, 1200 ft (Myanmar).

Habitat: Hill stream

Distribution: Assam, Nagaland, Mizoram. Abroad: Myanmar

Conservation Status: It is categorised as Data deficient in IUCN Red list 2013.

Remarks: On a recent survey by the present author it seems that this species is quite common in Mizoram and it's a highly economic edible species in that area.

35. *Indochinamon edwardsi* (Wood-Mason, 1871)

1875. *Paratelphusa edwardsi* Wood-Mason, *Proc. Asiat. Soc. Bengal*: 231(Cachar, Sadya and the Garo hills, Naga and Dafla hills).

1876. *Paratelphusa edwardsi*: Wood-Mason, *Ann. Mag. nat. Hist.*, 4, **17**: 121.

Type locality: Cachar, Sadya and the Garo hills, Naga and Dafla hills

Habitat: Hillstream and river

Distribution: India - Assam, Meghalaya and Nagaland.

Abroad: Myanmar

Conservation Status: It is categorised as Data Deficient in IUCN Red list 2013.

36. *Indochinamon manipurense* (Alcock, 1909)

1909a. *Potamon (Potamon) andersonianum manipurense* Alcock, *Rec. Indian Mus.*, **3**: 244 (Manipur Hills).

1910. *Potamon (Potamon) andersonianum manipurense*: Alcock, *Cat. Indian Decapod Crust. Indian Mus.*, **1**(2): 35, pl. **14, fig. 68** (Manipur Hills).

1910. *Potamon (Potamon) andersonianum* var. *manipurense*: Alcock, *Cat. Indian Decapod Crust. Indian Mus.*, **1**(2): 35 (Manipur Hills).

Type locality: Manipur Hills.

Habitat: River and hill stream

Distribution: Manipur.

Conservation Status: It is categorised as Data deficient in IUCN Red list 2013.

Remarks: Endemic to Manipur.

Genus *Larnaudia* Bott, 1966

37. *Larnaudia larnaudii* (A. Milne Edwards, 1869)

1869. *Thelphusa larnaudii* A. Milne Edwards, *Nouv. Arch. Mus.*, **5**: 166, pl. 10, fig. 4.

1893. *Thelphusa larnaudii*: Henderson, *Trans. Linn. Soc. Lond. Zool.*, 2, **5**: 385.

1900. *Potamon larnaudii*: Doflein, *S.-B.math.-phys. Cl. Akad. Wiss. München*, *1900*: 140 [*in partim*] (Simla, Himalaya and Kolkata).

1905. *Potamon (Potamon) larnaudii*: Rathbun, Nouv. Arch. Mus., sér. 4, 6: 275, pl. 10, fig. 7.

1910. *Potamon (Potamon) larnaudii*: Alcock, Cat. Indian. Decapod Crust. Indian Mus., 1(2): 47.

1970. *Larnaudia larnaudii*: Bott, *Abhandl. Sencken.Naturfors. Ges.*, **526**: 175, pl. 39, fig. 50 and pl. 50, fig. 46.

Type locality: Bangkok.

Habitat: Freshwater

Distribution: India – West Bengal (?); Himachal Pradesh.

Abroad: Thailand (Bangkok, Cochinchina, Mois Chero).

Conservation Status: It is categorised as Data Deficient in IUCN Red list 2013.

Remarks: According to Alcock (1910) this species does not occur in British India. However, he included the same in the Catalogue of Indian Decapod Crustacea as other authors reckoned it as Indian.

Genus Potamiscus Alcock, 1909

1909. *Potamon (Potamiscus)* Alcock, *Rec. Indian Mus.*, **3**(3): 250.

1966. *Ranguna* Bott: [Opinion, 1640].

Type species: *Potamon (Potamiscus) annandalei* Alcock, by original designation, gender: masculine.

38. *Potamiscus annandalei* (Alcock, 1910)

1909a. *Potamon annandalii* Alcock, *Rec. Indian Mus.*, **3**(3): 246 (Assam – Nemotha, Cachar).

1910. *Potamon (Potamiscus) annandalei* Alcock, *Cat. Indian Decapod Crust. Indian Mus.*, **1**(2): 57, pl. 3, fig. 10 (Assam: Nemotha., Cachar).

1970. *Potamiscus annandalei*: Bott, *Abhandl. Sencken.Naturfors. Ges.*, **526**: 158, pl. 38, fig. 28 and pl. 46, fig. 26.

Type locality: Nemotha, Cachhar (Assam)

Habitat: Freshwater but Microhabitat unknown

Distribution: India – Assam

Conservation Status: It is categorised as Data deficient in IUCN Red list 2013.

Remarks: Endemic to Assam.

39. *Potamiscus decourcyi* (Kemp, 1913)

1913. *Potamon (Potamiscus) decourcyi* Kemp, *Rec. Indian Mus.*, **8**: 292, pl. 17, figs. 1-3 (Sirpo valley near Renging; Rotung; a few miles south of Kebang; bank of Siyon River below Debuk Damda. All specimens were collected from small hill streams at altitudes of between 1000-1500 ft., under stones).

1970. *Potamiscus decourcyi*: Bott, *Abhandl. Sencken.Naturfors. Ges.*, **526**: 159, pl. 38, fig. 29 and pl. 46, fig. 27.

1999. *Potamiscus decourcyi*: Ghosh and Ghatak, *Zool. Surv. India State Fauna Series* **4**: *Fauna of Meghalaya, Part 9*: 570 (Meghalaya - Khasi Hills: Mawphlang).

Type locality: Sirpo Valley near Renging, Arunachal Pradesh

Habitat: Hill streams (under stones)

Distribution: Arunachal Pradesh, Meghalaya and Mizoram.

Conservation Status: It is categorised as Data deficient in IUCN Red list 2013.

Remarks: So far recorded from Eastern Himalaya only. It's an endemic species to North east Indian hill states.

40. *Potamiscus pealianus* (Wood-Mason, 1871)

1871. *Telphusa pealiana* Wood-Mason, *J. Asiat. Soc. Bengal*, **40**(2): 204, pl. 14, figs. 7-11.

1909a. *Potamon (Potamon) pealianum* var. *antennarium* Alcock, *Rec. Indian Mus.*, **3**: 245 (Two specimens were collected from Sibsagar (Assam) and two from un recorded locality).

1910. *Potamon (Potamon) pealianum antennarium*: Alcock, *Cat. Indian Decapod Crust. Indian Mus.*, **1**(2): 40, pl. 14, fig. 70 (Assam: Sibsagar).

1910. *Potamon (Potamon) pealianum*: Alcock, *Cat. Indian Decapod Crust. Indian Mus.*, **1**(2): 38, pl. 10, fig. 44 (Assam: Sibsagar).

1970. *Potamiscus pealianus*: Bott, *Abhandl. Sencken.Naturfors. Ges.*, **526**: 159, pl. 38, fig. 33 and pl. 47, fig. 29.

Type locality: Assam, Sibsagar.

Habitat: Hill stream

Distribution: Assam, Mizoram. Abroad: Myanmar.

Conservation Status: It is categorised as Data Deficient in IUCN Red list 2013.

Remarks: restricted distribution in two of the Northeasternstates of India and Myanmar.

41. *Potamiscus tumidulum* (Alcock, 1909)

1909 *Potamon (Potamon) tumidulum* ALCOCK, Rec. ind. Mus., 3: 245.

1910 *Potamon (Potamon?) tumidulum,* — ALCOCK, Cat. decap. Crust. ind. Mus., 1 (2): 43, T. 2 F. 6.

Type locality: Sikkim

Habitat: Rivers and Streams

Distribution: Sikkim. Abroad: Myanmar; China.

Conservation Status: It is categorised as Least Concern in IUCN Red list 2013.

Genus *Quadramon* Yeo and Ng, 2007

2007. *Quadramon* Yeo and Ng, *Raffles Bull. Zool., Supplement*, **16**: 292.

Type species: *Potamon (Potamiscus) aborense* Kemp, 1913, by subsequent designation by Yeo and Ng in *Raffles Bull. Zool., Supplement*, **16**: 292 (2007).

42. *Quadramon aborense* (Kemp, 1913)

1913. *Potamon (Potamiscus) aborense* Kemp, *Rec. Indian Mus.*, **8**: 294, pl. 18, figs. 4, 5 (Abor Country: Vicinity of Rotung at altitudes between 1000 and 1300 ft., Egar stream, between Rotung and Sireng stream).

2007. *Quadramon aborense*: Yeo and Ng, *Raffles Bull. Zool., Supplement*, **16**: 292.

Type locality: Rotung, Arunachal Pradesh.

Habitat: Hill stream, behind rocks.

Distribution: India - Arunachal Pradesh, Mizoram

Conservation Status: It is categorised as Data Deficient in IUCN Red list 2013.

Remarks: it's also an endemic form of Eastern Himalaya.

Genus *Teretamon* Yeo and Ng, 2007

2007. *Teretamon* Yeo and Ng, *Raffles Bull. Zool., Supplement*, **16**: 295-296.

Type species: *Potamon (Geotelphusa) adiatretum* Alcock, 1909 by subsequent designation by Yeo and Ng in *Raffles Bull. Zool., Supplement*, **16**: 295-296 (2007).

43. *Teretamon adiatretum* (Alcock, 1909)

1909 *Potamon (Geotelphusa) adiatretum* ALCOCK, Rec. ind. Mus., 3: 250.

1910 *Potamon (Geotelphusa) adiatretum,* — ALCOCK, Cat. ind. decap. Crust. ind. Mus., 1 (2): 59, T. 3 F. 11. (Dafla Hills: Arunachal Pradesh).

Type locality: Mawlamynie (Formerly Known as Moulmein), Myanmar.

Habitat: Unknown

Distribution: India – Arunachal Pradesh

Abroad: Myanmar

Conservation Status: It is categorised as Least Concern in IUCN Red list 2013.

Remarks: It seems to be rare as no any collection were maid after its record from Arunachal Pradesh.

44. *Tiwaripotamon austenianum* (Wood-Mason, 1871)

1871. *Telphusa austeniana* Wood-Mason, *J. Asiat. Soc. Bengal*, **40**(2): 203, pl. 13.

1905. *Potamon (Potamon) austenianus*: Rathbun, *Nouv. Arch. Mus.*, sér. 4, **6**: 287 (Cherra Punji).

1910. *Potamon (Potamon) austenianum*: Alcock, *Cat. Indian Decapod Crust. Indian Mus.*, 1(2): 44 (Meghalaya: Cherrapunji).

1970. *Tiwaripotamon austenianum*: Bott, *Abhandl. Sencken.Naturfors. Ges.*, **526**: 151 (Cherra Punji).

1999. *Tiwaripotamon austenianum*: Ghosh and Ghatak, *Zool. Surv. India State Fauna Series* **4**: *Fauna of Meghalaya, Part 9*: 570 (Meghalaya: Cherrapunji).

Type locality: Cherra Punji (Meghalaya)

Habitat: Hill Stream and River.

Distribution: India - Meghalaya.

Conservation Status: It is acategorised as Data deficient in IUCN Red list 2013.

Remarks: This species is known only from a single locality in India. Further information on its extent of occurrence, ecological role and population trends should be require to conservation of this species.

Genus *Trichopotamon* Dai and Chen, 1984

45. *Trichopotamon sikkimensis* (Rathbun, 1905)

1905. *Potamon (Geotelphusa) sikkimensis*, Rathbun, Nouv. Arch. Du Mus., Paris, sér. 4, 7: 219, vi, pl. 17, fig. 7.

1910. *Potamon (Potamiscus) sikkimense*: Alcock: 56, fig. 48 (West Bengal – Kurseong; Dafla Hills; Rajasthan – Ajmere).

Type locality: Sikkim

Habitat: Hill stream

Distribution: Darjeeling District of West Bengal, Sikkim, Arunachal Pradesh and Rajasthan.

Abroad: Bhutan and Nepal.

Conservation Status: It is categorised as Least Concern in IUCN Red list 2013.

Remarks: Interstingly this species occurs in Eastern Himalaya as well as Rajasthan, dry western part of India.

Discussion

A total of 45 species belonging to 2 families and 20 genera of Freshwater crab have been recorded from different states of Eastern and Western Himalayan part of Indian Territory. The extent of distribution of these species specified areas are summarized in Table 11.1. Among these, 16 species occur in the family Gecarcinucidae, Potamidae represented as 29 species, of this Subfamily Potaminae is consisted as 13 species only and Sub-family Potamiscinae represented as 16 species. From the north-east states, 40 species has been recorded so far, of which, 22 species are recorded from the state of Assam. The remaining species are reported from Meghalaya, Manipur, Nagaland, Tripura, Sikkim and Mizoram. The occurrence of

Table 11.1

Name of the State	Number of Species	Endemic Species
Assam	22	9
Manipur	4	1
Mizoram	11	4
Nagaland	10	3
Sikkim	7	0
Meghalya	10	3
Tripura	3	0
Arunachal Pradesh	6	2
WestBengal(Darjeeling)	9	3
Himachal Pradesh	9	5
Jammu and Kashmir	3	2
Uttarakhand	4	2

higher number of species in these areas may be attributed to the wet climatic conditions of these regions The present study shows that out of 45 species, 20 (44.44 per cent) are distributed exclusively (Endemic) to India Himalayas. Interestingly 15 species are endemic to Eastern Himalayas and only 4 species are endemic to Western Himalayas. One spesies barytelphusa cunicularis (West wood, 1836) are common in both area are also an endemic species of India but its particularly a peninisular species probably recently invaded in Himalayan regions. A list of species distributed in different states, of Indian part of Himalayas are listed here with the number of endemic species harbourded in each state.

Table 11.2: Distribution of Freshwater Crabs in different States of Indian Himalayan Territory

	Distribution in Indian Part of Himalayas
Phylum ARTHROPODA Latreille, 1829	
Subphylum CRUSTACEA Brünnich, 1772	
Class MALACOSTRACA Latreille, 1802	
Order DECAPODA Latreille, 1802	
Suborder PLEOCYEMATA Burkenroad, 1963	
Infraorder BRACHYURA Linnaeus, 1758	
Superfamily GECARCINUCOIDEA Rathbun, 1904	
Family GECARCINUCIDAE Rathbun, 1904	
I. Genus *Barytelphusa*	
1. *Barytelphusa cunicularis* (Westwood, 1836)	Darjeeling of West Bengal, Himachal Pradesh
II. Genus *Globitelphusa*	
2. *Globitelphusa bakeri* (Alcock, 1909)	Assam
3. *Globitelphusa cylindra* (Alcock, 1909)	Assam and Nagaland
4. *Globitelphusa pistorica* (Alcock, 1909)	Assam
III. Genus *Liotelphusa*	
5. *Liotelphusa gagei* (Alcock, 1909)	Darjeeling of West Bengal
6. *Liotelphusa laevis* (Wood-Mason, 1871)	W.B (Darjeeling hills) Meghalaya, Assam, Nagaland, Arunachal Pradesh
7. *Liotelphusa quadrata* (Alcock, 1909)	Assam
Iv. Genus *Maydelliathelphusa*	
8. *Maydelliathelphusa edentula* (Alcock, 1909)	Assam, Nagaland, Mizoram*
9. *Maydelliathelphusa falcidigitis* (Alcock, 1910)	Assam, Meghlaya,Nagaland, Mizoram*
10. *Maydelliathelphusa harpax* (Alcock, 1909)	Assam, Nagaland, Meghalya, Mizoram*
11. *Maydelliathelphusa lugubris* (Wood-Mason, 1871)	Assam, Meghlaya, Nagaland, Tripura, Manipur, Sikkim, Mizoram*, Darjeeling(W.B)
12. *Maydelliathelphusa masoniana* (Henderson, 1893)	Assam, Mehgalya, Himachal Pradesh, J&K
v. Genus *Travancoriana*	
13. *Travancoriana napaea* (Alcock, 1909)	Assam

Contd...

Table 11.2–*Contd...*

	Distribution in Indian Part of Himalayas
vi. Genus *Sartoriana*	
14. *Sartoriana spinigera* (Wood-Mason, 1871)	Assam, Meghlaya, Nagaland, Tripura, Manipur, Sikkim, Mizoram*
15. *Sartoriana trilobata* (Alcock, 1909)	Assam
vii. Genus *Somanniathelphusa*	
16. *Somanniathelphusa sinensis* (H. Milne Edwards, 1853)	Assam

Superfamily POTAMOIDEA Ortmann, 1896
Family POTAMIDAE Ortmann, 1896
Subfamily POTAMINAE Ortmann, 1896

viii. Genus *Acanthopotamon*	
17. *Acanthopotamon fungosum* (Alcock, 1909)	Assam
18. *Acanthopotamon martensi* (Wood-Mason, 1875)	Meghalaya; Assam;
ix. Genus *Alcomon*	
19. *Alcomon lophocarpus* (Kemp, 1913)	Arunachal Pradesh
20. *Alcomon superciliosum* (Kemp, 1913)	Arunachal Pradesh, Mizoram*
x. Genus *Himalayapotamon*	
21. *Himalayapotamon ambivium* (Alcock, 1909)	Himachal Pradesh
22. *Himalayapotamon atkinsonianum* (Wood-Mason, 1871)	Darjeeling (W.B.), Sikkim, Himachal Pradesh, J&K
23. *Himalayapotamon babaulti* (Bouvier, 1918)	Himachal Pradesh
24. *Himalayapotamon bifarium* (Alcock, 1909)	Sikkim(?)
25. *Himalayapotamon emphyseteum* (Alcock, 1909)	Himachal Pradesh, Uttarakhand, J&K
26. *Himalayapotamon kausalis* (Pretzmann, 1964)	Himachal Pradesh
27. *Himalayapotamon koolooense* (Rathbun, 1904)	Darjeeling (W.B.); Meghalaya, Himachal Pradesh, Uttarakhand
28. *Himalayapotamon monticola* (Alcock, 1910)	Darjeeling (W.B.); Khasi Hills (Meghalaya)
xi. Genus *Lobothelphusa*	
29. *Lobothelphusa woodmasoni* (Rathbun, 1905)	Assam, Tripura, Meghlaya, Mizoram*

Subfamily POTAMISCINAE Bott, 1970

xii. Genus *Aspermon*	
30. *Aspermon feae* (de Man, 1898)	Assam, Arunachal Pradesh, Mizoram*
xiii. Genus *Eosamon*	
31. *Eosamon tumidum* (Wood-Mason, 1871)	Darjeeling (W.B), Himachal Pradesh
xiv. Genus *Indochinamon*	
32. *Indochinamon asperatum* (Alcock, 1909)	Assam
33. *Indochinamon beieri* (Pretzmann, 1966)	Assam, Nagaland, Mizoram*

Contd...

Table 11.2–Contd...

	Distribution in Indian Part of Himalayas
34. *Indochinamon edwardsi* (Wood-Mason, 1871)	Assam, Meghalaya, Nagaland (Known Only from Type locality)
35. *Indochinamon manipurense* (Alcock, 1909)	Manipur
36. *Indochinamon andersonianum* (Alcock, 1909)	Manipur
xv. Genus *Lamaudia*	
37. *Lamaudia lamaudi* (A.Milne Edwards, 1869)	Himachal Pradesh
xvi. Genus *Potamiscus*	
38. *Potamiscus annandali* (Alcock, 1909)	Assam
39. *Potamiscus decourcyi* (Kemp, 1913)	Assam, Nagaland, Mizoram*
40. *Potamiscus pealianus* (Wood-Mason, 1871)	Assam, Mizoram*
41. *Potamiscus tumidulus* (Alcock, 1909)	Sikkim
xvii. Genus *Quadromon*	
42. *Quadromon aborense* (Kemp, 1913)	Arunachal Pradesh, Mizoram*
xviii. Genus *Teretamon*	
43. *Teretamon adiatretum* (Alcock, 1909)	Arunachal Pradesh
xix. Genus *Tiwaripotamon*	
44. *Tiwaripotamon austenianum* (Wood Mason 1871)	Meghalaya
xx. Genus *Trichopotamon*	
45. *Trichopotamon sikkimense* (Rathbun, 1905)	Sikkim, Darjeeling(West Bengal)

*Present Record on the basis of a recent survey to Mizoram by the present author.

Threats and Conservation

Freshwater crabs are found in ponds, lakes, streams, rivers and marshes. A few species however, are able to live in brackish water. They subsist on fallen leaves and algae and thereby help in nutrient cycling by consuming detritus of the freshwater ecosystems. They also support small-scale fisheries especially in the rural sectors and thus providing a primary source of protein for the local people. Apart from these, they are an important source of food for a wide range of animals such as fish, amphibians, reptiles and mammals. Freshwater crabs are considered important as environmental markers (especially for rain forests). This necessitates that they are identified correctly. They are also excellent indicators of good water qualitiy as most of them require pristine water for survival. They are important as markers for the study of biogeography, plate tectonics and animal evolution. Some colourful species are also important in aquarium trade. Freshwater crabs are medically important as vectors of the deadly disease Paragonimiasis which affects about 20 million people world-wide. As such, identification of the correct crab vector is important for the control of this disease. They are also considered important in the biological control of pest. Perhaps the most widely used of all the traditional pest management practices is the use of decomposing crabs in the control of rice bugs (as the filling of paddy grain starts, locally available crabs are smashed and put on pointed bamboo sticks in terraced paddy fields). This is practiced throughout the entire state by all communities in Meghalaya. This method is environmentally friendly, as some farmers replace the crab baits as soon they dry up. The crab bait traps can be used in connection with other traditional methods of managing the pest.

Like other parts of the world, freshwater crabs of India are also subjected to tremendous pressure of threats. Major threats to freshwater crabs of India are due to habitat destruction and pollution. Loss of natural forests to land development and agriculture has impacted almost every habitat in which freshwater crabs live. Rapid urbanization, industrialization, poor sloping-land management and unwise land-use change in the high lands continues to be a serious problem resulting to habitat loss and wiping out the freshwater crabs. Only a handful of freshwater crab species have wide distribution and able to tolerate of land-use change. Wide use of pesticides for agriculture is also causing seious concern. At present, their regulation addresses only human safety issues and has no impacts on other non-target organisms or the environments in general. In addition, water quality is also deteriorating very fast even in key natural habitats. Many of the freshwater crabs are extremely sensitive to polluted or silted waters and will not survive when exposed to these features. However, a National Committee for the study of Freshwater Crab may be set up involving Zoological Survey of India towards conservation action.

IUCN has recently included 1280 species of freshwater crabs of the World in the Red List of Threatened Species, of which, 227 has been considered as near threatened, vulnerable, endangered or critically endangered. Further, for another 628 species adequate data is not available to assess their status. Acording to the estimation of IUCN, nearly, two-thirds of freshwater crabs are going to be extinct, with one in every six species particularly vulnerable. So far, from Indian Part of Himalaya, all the 45 species has been enlisted in the IUCN Red data list. Among these, only 13 species enlisted as Least Concern.where as a single species *Liotelphusa quadrata* (Alcock, 1909) are categorized as Vulnarable. Three species namely *Liotelphusa gagei* (Alcock, 1909); *Liotelphusa laevis* (Wood-Mason, 1871) and *Maydelliathelphusa edentula* (Alcock, 1909) are considered as Near Threatened. Surprisingly 26 species are uptil enlisted as Data Deficient categories as there is no collection data or any furthure report of those species since a long period. However, most of the freshwater crabs need to be brought under Rapid Assessmet Survey to ascertain their status in India.

References

Alock, A. 1909a. Diagnoses of new species and varieties of freshwater crabs. Nos. 1-3. *Rec. Indian Mus.*, **3**(3): 243-252.

Alock, A. 1909b. Diagnoses of new species and varieties of freshwater crabs. *Rec. Indian Mus.*, **3(4)**: 375-381.

Alock, A. 1910. Catalogue of the Indian Decapod Crustacea in the collection of the Indian Museum. Part 1. Brachyura. Fasciculus II. The Indian Freshwater Crabs – Potamonidae. Trusees of the Indian Museum, Calcutta, pp. 1-134, pls. 1-11.

Bott, R. 1970. Die Süßwasserkrabben von Europa, Asien, Australien und ihre Stammesgeschichte. Eine Revision der Potamoidea und der Parathelphusoidea. (Crustacea: Decapoda). *Abh. senckenb. Naturforsch. Ges.*, **526**: 3-338**(203)**, figs. 1-8, pls. 1-38. Verlag Waldemar Kramer Frankfurt am Main.

Brandis, D., and Sharma, S. 2005. Taxonomic revision of the freshwater crab fauna of Nepal with description of a new species (Crustacea, Decapoda, Brachyura, Potamoidea and Gecarcinucoidea). *Senkenbergia biologica*, **85 (1)**: 1–30.

Bouvier, E.L. 1918. Sur quelques crustacés décapodes recueillis par M. Guy Babault dans les eaux douces de l'Inde Anglaise. *Bulletin du Muséum national d'Histoire naturelle, Paris [1er série]* **24**: 386–393.

Chopra, B., and Tiwari, K. K. 1947. Decapoda Crustacea of the Patna State, Orissa. *Rec. Indian Mus.*, **45**: 213-224.

Cumberlidge, N. 2008. *Liotelphusa quadrata*. In: IUCN 2013. IUCN Red List of Threatened Species. Version 2013.2. <www.iucnredlist.org>. Downloaded on **19 April 2014**.

Deb, M. 1998. Crustacea: Decapoda: Crabs. *Zool. Surv. India State Fauna Series* **3**: *Fauna of West Bengal, Part* **10**: 345-403.

Dutta, N. K. 1983. Studies on the systematics and distribution of crabs in Assam. *J. Bombay nat. Hist. Soc.*, **80**(2): 539-548, figs. 1-6.

Ghosh, H. C., and Ghatak, S. S. 1999. Crustacea: Decapoda: Potamonidae. *Zool. Surv. India State Fauna Series* **4**: *Fauna of Meghalaya, Part* **9**: 569-576.

Ghosh, H. C., and Ghatak, S. S. 2000. Crustacea: Decapoda: Potamonidae. *Zool. Surv. India State Fauna Series* **7**: *Fauna of Tripura, Part* **4**: 273-275.

Henderson, J. R. 1893. A contribution to Indian carcinology. *Trans. Linn. Soc. Lond. (Zool.)*, ser. 2, **5**: 325-458, pls. 36-40.

Henderson, J. R. 1912. Description of a new species of freshwater crab from southern India. *Rec. Indian Mus.*, **7(11)**: 111-112.

Kemp, S. 1913. Crustacea Decapoda. *Rec. Indian Mus*, **8**: 289-310, pl. 17-31.

Kemp, S. 1924. Crustacea Decapoda of the Siju cave, Garo hills, Assam. *Rec. Indian Mus.*, **26(1)**: 41-48, pl. 3.

Krishnamurthy, P. 1995. Crustacea: Decapoda. *Zool. Surv. India Himalayan Ecosystem Series: Fauna of Western Himalaya, Part 1, Uttar Pradesh*: 23.

Milne Edwards, H. 1837. *Histoire naturelle des Crustacés, comprenant l'anatomie, la physiologie et la classification de ces animaus, Paris.* **2**: 1-532.

Mitra, S., and Devroy, M. K. 2012. On the occurrence and abundance of *Himalayapotapon emphysetum* (Alcock,1909) in bowri system of Himachal Pradesh. *J. Environ., and Sociobiol.*, **9(2)**: 176.

NG, P. K. L., Guinot, D., and Davie, P. J. F. 2008. Systema Brachyura: Part 1. An annotated checklist of extant brachyuran crabs of the world. *Raffles Bull. Zool.*, *Supplement*, **17**: 1-286.

Pretzman, G. 1963. Über einige süd und ost-asiatische Potamoniden. *Ann. Naturh. (Mus.) Hofmus., Wien,* **66**: 361-372.

Pretzman, G. 1966a. Süsswassserkrabben aus dem westlichen Himalaya gebiet. *Annln. Naturh. Mus. Wien,* **69**: 299-303, 4 pls.

Pretzman, G. 1966b. Potamidenaus Asien (*Potamon* Savignyi and *Potamiscus* Alcock) (Crustacea: Decapoda). *Senck. biol.,* **47**: 469-509, 6 pls., 32 figs.

Ramakrishna, G. 1950. Notes on some Indian Potamonid crabs (Crustacea: Decapoda). *Rec. zool. Surv. India,* **48(1)**: 89-92.

Rathbun, M. J. 1904. Les Crabes d'eau douce (Potamonidae). *Nouv. Arch. Mus.,* sér. 4, **6**:225-312, pls. 9-18.

Roy, T. K., Ghosh, S. K., and Ghatak, S. S. 2003. Crustacea: Decapoda: Palaemonidae and Potamonidae. *Zool. Surv. India State Fauna Series* **9**: *Fauna of Sikkim, Part 5*: 117-119.

Roy, T. K., Ghosh, S. K., and Ghatak, S. S. 2004. Crustacea: Decapoda: Palaemonidae and Potamonidae. *Zool. Surv. India State Fauna Series* 10: Fauna of Manipur, Part **3**: 119-123.

Sharma, B. D. 1975. Preferential feeding in captivity by a freshwater crab, *Potamon atkinsonianum* Wood-Mason (Crustacea: Potamonidae) on *Notonecta undulata* (Insecta: Hemiptera). *J. Bombay nat. Hist. Soc.,* **72(1)**: 222-223.

Sharma, K. K., Gupta, R. K., and Langer, S. 2013. Effects of some of the ecological parametrers on freshwater Crab abundance *Paratelphusa masoniana* (Henderson) inhabiting Gho-Manhasan stream, A tributary of River Chenab, Jammu, J&K. *International Journal of Recent Scientific Research,* **4(5)**: 640-644.

Wood-Mason, J. 1871. Contribution to Indian Carcinology. Part 1. Indian and Malayan Telphusidae. *J. Asiat. Soc. Bengal,* **40(2)**: 194-196.

Wood-Mason, J. 1875. On new or little known crustaceans. *Proc. Asiat. Soc. Bengal,* **1875**: 230-232.

Yeo, D. C. J., and Ng, P. K. L. 2003. Recognition of two subfamilies in the Potamidae Ortmann, 1896 (Brachyura: Potamidae) with a note on the genus *Potamon* Savignyi, 1816. *Crustaceana,* **76(10)**: 1219-1235.

Yeo, D. C. J., and Ng, P. K. L. 2007. On the genus "Potamon" and allies in Indo-China (Crustacea: Decapoda: Potamidae). *Raffles Bull. Zool.,* Supplement No., **16**: 273-308.

2015, Animal Diversity, Natural History and Conservation, Vol. 5   Pages 165–176
Editors: V.K. Gupta and Anil K. Verma
Published by: DAYA PUBLISHING HOUSE, NEW DELHI

Chapter 12

Perspective on Fish Diversity

Govind Pandey[1]*, A.B. Shrivastav[2] and S. Madhuri[3]
[1]Professor/Principal Scientist of Pharmacology and Toxicology,
College of Veterinary Science and Animal Husbandry, Rewa
(NDVSU, Jabalpur), M.P., India
[2]Director, Centre for Wildlife Forensic and Health,
Nanaji Deshmukh Veterinary Science University (NDVSU), Jabalpur, M.P., India
[3]Associate Professor/Senior Scientist, College of Fishery Sciences,
Nanaji Deshmukh Veterinary Science University (NDVSU), Jabalpur, M.P., India

ABSTRACT

Unlike mammals or birds, fish are not a clade but they are a paraphylectic collection of taxa, including jawless, skeletal and cartilaginous types (*e.g.*, hagfish, lampreys, sharks and rays, ray-finned fish, coelacanths and lungfish). The lungfish and coelacanths are closer relatives of tetrapods (*e.g.*, mammals, birds, amphibians) than the other fish (*e.g.*, ray-finned fish or sharks). Thus, the last common ancestor of all fish is also an ancestor to tetrapods. However, the paraphyletic groups are no more in modern systematic biology, thus the term 'fish' as a biological group must not be used. In fact, the fish denotes any non-tetrapod craniate (an animal with a skull and mostly with a backbone) that has gills, and whose limbs (if any) are fin-shaped. Although fish usually refer to many aquatic species, but actually all of them are not fish, *viz.*, starfish, shellfish, cuttlefish, jellyfish and crayfish. The jawless fish are the most primitive fish. In fact, the ancestors of cartilaginous fish (having a cartilaginous skeleton) were the bony animals and they developed paired fins firstly. The bony fish are categorized into the lobe finned and ray finned fish. The teleost fish are the most modern (advance). In size, *Paedocypris progenetica* is the smallest fish. The gobies have the shortest life span and they are small coral reef-dwelling fish. Only 58 per cent of extant fish are saltwater, while a disproportionate 41 per cent are freshwater and remaining 1 per cent is

---

* *Corresponding author.* E-mail: drgovindpandey@rediffmail.com

anadromous fish. The groupers are protogynous hermaphrodites, who school in harems of 3 to 15 females. Fish adopt a variety of strategy for nurturing their brood, *e.g.*, shark differently adopts three protocols with brood. Many fishes are food opportunists or generalists as they eat whatever they get easily. The fishes with four eyes have eyes above the top of head which are divided in two different parts, so that they can see below and above the water surface simultaneously. The seahorses are the slowest-moving fishes. The 'toxic fishes' are able to produce strong toxins in their bodies. There are commercial food fish, recreational sport fish, decorative aquarium fish and tourism fish. As the most vertebrate species have brain-to-body weight ratios, fish also have the relative brain weights of vertebrates. Considering all these, this chapter discusses on the diversity of different types of fish.

Keywords: *Fish, Fish Diversity, Fisheries, Types of Fish, Vertebrates.*

Introduction

'Fish' refers any non-tetrapod craniate (*i.e.*, an animal with a skull and in most cases a backbone) that has gills throughout life and whose limbs, if any, are in the shape of fins (Wikipedia- Fish Diversity, 2014). Unlike groupings like birds or mammals, fish are not a single clade but a paraphyletic collection of taxa, including jawless, cartilaginous and skeletal types (*e.g.*, hagfishes, lampreys, sharks and rays, ray-finned fish, coelacanths and lungfish) (Connolly, 2006). In fact, lungfish and coelacanths are closer relatives of tetrapods (*e.g.*, mammals, birds, amphibians) than other fishes, *e.g.*, ray-finned fish or sharks, so the last common ancestor of all fishes is also an ancestor to tetrapods. As paraphyletic groups are no longer recognized in modern systematic biology, the use of term 'fish' as a biological group must be avoided.

Several types of aquatic species generally called as 'fish' are not fish on the basis of definition described above, *e.g.*, shellfish, cuttlefish, starfish, crayfish and jellyfish. In earlier times, the natural historians classified also seals, whales, amphibians, crocodiles, even hippopotamuses, as well as a host of aquatic invertebrates, as fish. However, according the definition of fish, all mammals, including Cetaceans like whales and dolphins, are not fish. In some contexts, especially in aquaculture, the true fish are known as 'finfish' (or fin fish) to differentiate them from other fishes. A typical fish is ecothermic that has a streamlined body for rapid swimming, extracts oxygen from water using gills or uses an accessory breathing organ to breathe atmospheric oxygen, has two sets of paired fins, usually one or two (rarely three) dorsal fins, an anal fin and a tail fin, has jaws, has skin that is usually covered with scales, and lays eggs. Each of these criteria has exception, *e.g.*, tuna, swordfish and some species of sharks show some warm-blooded adaptations; they can heat their bodies significantly above ambient water temperature. The streamlining and swimming performance varies from fish (*e.g.*, tuna, salmon and jacks that can cover 10 to 20 body-lengths/second) to species (*e.g.*, eels and rays that swim no more than 0.5 body-lengths/second) (Wikipedia- Fish Diversity, 2014). Several groups of freshwater fish extract oxygen from air and water using a variety of different structures. The lungfish have paired lungs similar to those of tetrapods; gouramis have a structure called 'labyrinth organ' that performs a similar function; while many catfishes, *e.g.*, corydoras extract oxygen via the intestine or stomach (Nelson, 2006).

The fish have many shapes and sizes, ranging from the huge 16 m (52 ft) whale shark to the tiny 8 mm (0.3 inch) stout infant fish. The body shape and the arrangement of the fins are highly variable, covering such seemingly un-fishlike forms as seahorses, puffer fish, angler fish and gulpers. Similarly, surface of the skin may be naked (as in moray eels), or covered with scales of a variety of different types generally defined as placoid (typical of sharks and rays), cosmoid (fossil lungfish and coelacanths), ganoid (various fossil fish but also living gars and bichirs), cycloid, and ctenoid (these last two are found on most bony fish). There are even fish that live mostly on land. Mudskippers feed and interact with one another on mudflats and go underwater to hide in their burrows (Wikipedia- Fish Diversity, 2014). The catfish, *Phreatobius cisternarum* lives in underground, phreatic habitats, and a relative lives in waterlogged leaf litter (Aird, 2007; Campbell and Reece, 2005).

With the above background, therefore, this chapter deals about the diversity of fish with more common types. Although most fish species have probably been discovered and described, about 250 new ones are still discovered every year. According to FishBase, 32,100 species of fish had been described by September, 2011, which is more than the combined total of all other vertebrates: mammals, amphibians, reptiles and birds (Wikipedia- Fish Diversity, 2014).

Fish Diversity

Fish are very diverse and are categorized in many ways, as discussed below (Wikipedia- Fish Diversity, 2014):

1. Diversity Depending on Species

(a) Cartilaginous Fish

These fishes have a cartilaginous skeleton; however, their ancestors were bony animals and they were the first fish to develop paired fins. The cartilaginous fish do not have swim bladders. Their skin is covered in placoid scales (dermal denticles) which are as rough as sandpaper. Because these fishes do not have bone marrow, the spleen and special tissues around the gonads produce red blood cells (RBCs). Their tails can be asymmetric, with the upper lobe longer than the lower lobe. Some cartilaginous fishes possess an organ called "Leydig's organ", which also produces RBCs. There are over 980 species of cartilaginous fish, including shark, ray and chimaera.

(b) Bony Fish

They include the lobe finned fish and the ray finned fish. The lobe finned fish is the class of fleshy finned fishes, consisting of lungfish and coelacanths. They are bony fish with fleshy, lobed paired fins, which are joined to the body by a single bone. These fins evolved into the legs of the first tetrapod land vertebrates, amphibians. The ray finned fishes are so-called because they possess lepidotrichia or 'fin rays'; their fins being webs of skin supported by bony or horny spines (rays). Their three types are chondrosteans, holosteans and teleosts. The chondrosteans and holosteans are primitive fishes sharing a mixture of characteristics of teleosts and sharks. In comparison with the other chondrosteans, the holosteans are closer to the teleosts and further from sharks.

(c) Jawless Fish

They are most primitive fish. There is current debate over whether these are really fish at all. They have no jaw, no scales, no paired fins and no bony skeleton. Their skin is smooth and soft to touch, and they are very flexible. Instead of a jaw, they possess an oral sucker. They use this to fasten on to other fish, and then use their rasp-like teeth to grind through their host's skin into viscera. They inhabit both fresh and salt water environments. Some are anadromous, moving between both fresh and salt water habitats.

The extant jawless fish are either lamprey or hagfish. The juvenile lamprey feed by sucking up mud containing microorganisms and organic debris. The lamprey has well developed eyes, while the hagfish has only primitive eyespots. The hagfish coats itself and carcasses it finds with noxious slime to deter predators, and periodically ties itself into a knot to scrape the slime off. It is the only invertebrate fish and the only animal which has a skull but no vertebral column (Campbell and Reece, 2005). It has four hearts, two brains and a paddle-like tail (Aird, 2007).

(d) Teleosts

They are the most advanced or 'modern' fishes. They are overwhelmingly the dominant class of fishes with nearly 30,000 species, covering about 96 per cent of all the extant fish species. They are ubiquitous throughout freshwater and marine environments from the deep sea to the highest mountain streams. They include nearly all the important commercial and recreational fishes. The teleosts have a movable maxilla and premaxilla and corresponding modifications in the jaw musculature. These modifications make it possible for teleosts to protrude their jaws outwards from the mouth. The caudal fin is homocercal, meaning the upper and lower lobes are about equal in size. The spine ends at the caudal peduncle, distinguishing this group from those in which the spine extends into the upper lobe of the caudal fin. Some examples of teleost fish are rose fish, swordfish, eel and seahorse.

2. Diversity Depending on Size

The smallest fish species is *Paedocypris progenetica*. It is a type of minnow which lives in the dark-coloured peat swamps of the Indonesian island of Sumatra. The females of this species have a standard length of 7.9 mm (0.31 inch) at maturity (Kottelat *et al.,* 2005). Until recently, this was the smallest of all known vertebrates. However, recently a minute Papua New Guinea frog, *Paedophryne amauensis*, with a standard length of 7.7 mm (0.30 inch) was discovered. The slender Indonesian fish may still be the smallest vertebrate by weight. Male individuals of the anglerfish species *Photocorynus spiniceps* are 6.2 to 7.3 mm long at maturity, and thus could be claimed as an even smaller species. However, these males do not survive on their own merits but only by sexual parasitism on the larger female. Another very small fish is the stout infant fish, a type of goby. According to the Guinness Book of World

Records, the sinarapan (a goby fish, found in Philippines) is the world's smallest commercially harvested fish (Foot, 2000).

The sinarapan has an average length of 12.5 mm (0.49 inch), and is threatened by overfishing. The largest fish is the whale shark. It is a slow moving filter feeding shark with a maximum published length of 20 m (66 ft) and a maximum weight of 34 tonnes. The whale sharks can live up to 70 years. The heaviest bony fish is the ocean sunfish. It can weigh up to 2,300 kg (5,100 lb). It is found in all warm and temperate oceans. The longest bony fish is the king of herrings. Its total length can reach 11 m (36 ft) and it can weigh up to 272 kg (600 lb). It is a rarely seen oarfish found in all the world's oceans, at depths of between 20 m (66 ft) and 1,000 m (3,300 ft).

3. Diversity Depending on Life Span

Some of the shortest-lived species are gobies, which are small coral reef-dwelling fish. Some of the longest-lived are rockfish.

The shortest lived is the seven-figure pygmy goby, which lives for at most 59 days. This is the shortest lifespan for any vertebrate (Depczynski and Bellwood, 2005). The short lived fish have particular value in genetic studies on aging. The ram cichlid is used in laboratory studies because of its ease of breeding and predictable aging pattern (Herrera and Jagadeeswaran, 2004). The longest-lived fish is the 205 years reported for the rough eye rockfish, *Sebastes aleutianus*, which is found offshore in the North Pacific at 25 to 900 m (14-490 fathoms). This fish exhibits negligible senescence (Cailliet *et al.*, 2001).

The maximum reliably reported age for a goldfish is 41 years. The longest living commercial fish may be orange roughy, with a maximum reported age of 149 years. One of the longest living sport fish is the Atlantic tarpon, with a maximum reported age of 55 years. Some of the longest living fish are living fossils, *e.g.*, green sturgeon. This species is among the longest living species found in freshwater, with a maximum reported age of 60 years. This is also among the largest fish species found in freshwater, with a maximum reported length of 2.5 m (8.2 ft) and a maximum reported weight of 159 kg. Another living fossil is the Australian lungfish. The Australian lungfish has hardly changed for 380 million years. The oldest fish in captivity that has lived in an aquarium for 75 years is an Australian lungfish.

4. Diversity Depending on Habitat

There is 10,000 times more saltwater in the oceans than there is freshwater in the lakes and rivers. However, only 58 per cent of extant fish species are saltwater. A disproportionate 41 per cent are freshwater fish (remaining 1 per cent is anadromous). This diversity in freshwater species is, perhaps, not surprising, since the thousands of separate lake habitats promote speciation. Fish can also be demersal or pelagic. The demersal fish live on or near the bottom of oceans and lakes, while the pelagic fish inhabit the water column away from the bottom. Habitats can also be vertically stratified. The epipelagic fish occupy sunlit waters down to 200 m (110 fathoms), the mesopelagic fish occupying deeper twilight waters down to 1,000 m (3,300 ft), and the bathypelagic fish inhabiting the cold and pitch black depths below. Most oceanic species (78 per cent or 44 per cent of all fish species) live near the shoreline. These coastal fish live on or above the relatively shallow continental shelf. Only 13 per cent of all fish species live in the open ocean, off the shelf. Of these, 1 per cent is epipelagic, 5 per cent are pelagic and 7 per cent are deep water.

Fish are found in nearly all natural aquatic environments. Most fish, whether by species count or abundance, live in warmer environments with relatively stable temperatures. However, some species survive temperatures up to 44.6°C (112.3°F), while others cope with colder waters; there are over 200 finfish species south of the Antarctic Convergence. Some fish species tolerate salinities over 10 per cent. The world's deepest living fish, *Abyssobrotula galatheae*, a species of cusk eel, lives in the Puerto Rico Trench at a depth of 8,372 m (27,467 ft). At the other extreme, the Tibetan stone loach lives at altitudes over 5,200 m (17,100 ft) in the Himalayas. Some marine pelagic fish range over vast areas, *e.g.*, blue shark fish that lives in all oceans. At the other extreme are fish confined to single, small living spaces, such as isolated cave fish like Lucifuga in the Bahamas and Cuba, or equally isolated desert pupfish living in small desert spring systems in Mexico and the southwest USA, or bythitid vent fish like *Thermichthys hollisi*, living around thermal vents 2,400 m (1,300 fathoms) down.

5. Diversity Depending on Breeding Behaviour

The groupers are protogynous hermaphrodites, who school in harems of 3 to 15 females. When no male is available, the most aggressive and largest females shift sex to male, probably as a result of behavioural triggers. In very deep waters, it is not easy for a fish to find a mate. There is no light, so some species depend on bioluminescence. Others are hermaphrodites, which doubles their chances of producing both egg and sperm when an encounter does occur. The female anglerfish releases pheromones to attract tiny males. When a male finds her, he bites on to her and never lets go. When a male of the anglerfish species, *Haplophryne mollis,* bites into the skin of a female, he releases an enzyme that digests the skin of his mouth and her body, fusing the pair to the point where two circulatory systems join up. The male then atrophies into nothing more than a pair of gonads. This extreme sexual dimorphism ensures that, when the female is ready to spawn, she has a mate immediately available. Some sharks, *e.g.*, hammerheads are able to breed parthogenetically. The female groupers change their sex to male if no male is available. Male toadfish 'sing' at up to 100 decibels with their swim bladders to attract mates.

6. Diversity Depending on Brooding Behaviour

The fish adopt a variety of strategies for nurturing their brood, *e.g.*, the shark variously follows three protocols with its brood. Many sharks, including lamniformes are ovoviviparous, bearing their young after they nourish themselves after hatching and before birth, by consuming the remnants of the yolk and other available nutrients. Some, like hammerheads are viviparous, bearing their young after nourishing hatchlings internally, analogously to mammalian gestation. Finally catshark and others are, oviparous, laying their eggs to hatch in the water. Some animals, predominantly fish (*e.g.*, cardinalfish) practice mouth brooding, caring for their offspring by holding them in the mouth of a parent for extended periods of time. The mouth brooding has evolved independently in different families of fish. Others, like seahorse males, practice pouch-brooding (analogous to Australia's kangaroos), nourishing their offspring in a pouch in which the female lays them. A female *Cyphotilapia frontosa* mouth brooding fry can be seen looking out her mouth. The chain catshark (an oviparous) is laying its eggs to hatch in the water. The great white shark (an oviparous) is gestating eggs in the uterus for 11 months before giving birth. The scalloped hammerhead (a viviparous) bears its young after nourishing hatchlings internally.

7. Diversity Depending on Feeding Behaviour

There are three basic methods by which food is gathered into the mouths of fish: by suction feeding, by ram feeding and by manipulation or biting. Mostly, all fish species use one of these styles and most use the two. The early fish lineages had inflexible jaws limited to little more than opening and closing. Modern teleosts have evolved protusible jaws (*e.g.*, protusible jaw of sling jaw wrasse) that can reach out to engulf prey. Its mouth extends into a tube half as long as its body, and with a strong suction it catches prey. The equipment tucks away under its body when it is not in use. In practice, the feeding modes lie on a spectrum, with suction and ram feeding at the extremes. Many fish capture their prey using both suction pressures combined with a forward motion of the body or jaw.

The cookie cutter shark is a small dogfish which derives its name from the way it removes small circular plugs, looking as though cut with a cookie cutter, from the flesh and skin of cetaceans and larger fish, including other sharks. The cookie cutter attaches to its larger prey with its suctorial lips, and then protrudes its teeth to remove a symmetrical scoop of flesh (Martins and Knickle, 2009).

Most fish are food opportunists, or generalists. They eat whatever is most easily available. For example, the blue shark feeds on dead whales and nearly everything else that wriggles: other fish, cephalopods, gastropods, ascidians, crustaceans. The ocean sunfish prefer jellyfish. Other fish have developed extreme specializations. The silver arowana (monkey fish) can leap 2 m out of the water to capture prey. They usually swim near the surface of water waiting for potential prey. Their main diet consists of crustaceans, insects, smaller fishes and other animals that float on the water surface, for which its draw-bridge-like mouth is exclusively adapted for feeding. The remains of small birds, bats and snakes have also been found in their stomachs. The archerfish prey on land-based insects and other small animals by literally shooting them down with water droplets from their specialized mouths. They are remarkably accurate; the adults almost always hit the target on the first shot. They can bring down insects, *e.g.*, grasshoppers, spiders and butterflies on a branch of an overhanging tree, 3 m above the water's surface. This is partially due to good eye sight, but also the ability to compensate for light refraction when aiming. The triggerfish also use jets of water, to uncover sand dollars buried in sand or overturn sea urchins. The doctor fish (nibble fish) live and breed in the outdoor pools of some. Turkish spas, where they feed on the skin of patients with psoriasis. The fish are like cleaner fish, in that they only consume the affected and dead areas of the skin (eat parasites off other fish), leaving the healthy skin to recover.

8. Diversity Depending on Toxicity

The toxic fish produce strong poisons in their bodies. Both poisonous fish and venomous fish contain toxins, but deliver them differently. The venomous fish bite, sting, or stab, causing an envenomation. They do not necessarily cause poisoning if they are eaten, since the digestive system often destroys the venom. By contrast, the digestive system does not destroy poisonous fish toxins, making them poisonous to eat. The most poisonous fish is the puffer fish. It is the second most poisonous vertebrate after the golden dart frog. It paralyses the diaphragm muscles of human victims, who can die from suffocation. In Japan, the skilled chefs use parts of a closely related species, the blowfish to create a delicacy called 'fugu', including just enough toxin for that 'special flavour'. The spotted trunkfish is a reef fish which secretes a colourless ciguatera toxin from glands on its skin when touched. The toxin is only dangerous when ingested, so there is no immediate harm to divers. However, the predators as large as nurse sharks can die as a result of eating a trunkfish. The giant moray is a reef fish at the top of the food chain. Like many other apex reef fish, it is likely to cause ciguatera poisoning if eaten.

The outbreaks of ciguatera poisoning in the 11th to 15th centuries from large, carnivorous reef fish, caused by harmful algal blooms, could be a reason why Polynesians migrated to Easter Island, New Zealand and possibly Hawaii (Rongo *et al.*, 2009). In a study done in 2006, it was found that there are at least 1200 species of venomous fish. There are more venomous fish than venomous snakes. In fact, there are more venomous fish than the combined total of all other venomous vertebrates (Smith and Wheeler, 2006).

The venomous fish are found in almost all habitats around the world, but mostly in tropical waters. They wound over 50,000 people every year. They carry their venom in venom glands and use various delivery systems, *e.g.*, spines or sharp fins, barbs, spikes and fangs. The venomous fish tend to be either very visible, using flamboyant colours to warn enemies, or skillfully camouflaged and may be buried in the sand. Apart from the defense or hunting value, the venom helps bottom dwelling fish by killing the bacteria that try to invade their skin. The most venomous known fish is the reef stonefish. It has a remarkable ability to camouflage itself amongst rocks. It is an ambush predator that sits on the bottom waiting for prey to approach. Instead of swimming away if disturbed, it erects 13 venomous spines along its back. For defense, it can shoot venom from each or all of these spines. Each spine is like a hypodermic needle, delivering the venom from two sacs attached to the spine. The stonefish has control over whether to shoot its venom, and does so when provoked or frightened. The venom results in severe pain, paralysis and tissue death, and can be fatal if not treated. Despite its formidable defenses, the stonefish have predators. Some bottom feeding rays and sharks with crushing teeth feed on them, as does the Stokes' sea snake.

The beautiful lionfish (a coral reef fish) is also venomous. Unlike stonefish, lionfish can only release venom when something strikes its spines. Although not native to the USA coast, the lionfish have appeared around Florida, and have spread up the coast to New York. They are attractive aquarium fish, sometimes used to stock ponds, and may have been washed into the sea during a hurricane. The lionfish can aggressively dart at scuba divers and attempt to puncture their facemask with their venomous spines. The stargazer, *Uranoscopus sulphureus*, is also a venomous fish. The stargazer buries itself and can deliver electric shocks as well as venom. It is a delicacy in some cultures (cooking destroys venom), and can be found for sale in some fish markets with the electric organ removed. They have been called 'the meanest things in creation'. The stingray envenomation can occur to people who wade in shallow water and tread on them. It can be avoided by shuffling through the sand or stamping on bottom, as the rays detect this and swim away. The stinger usually breaks off in the wound. It is barbed, so it easily penetrates but not so easily be removed. The stinger causes local trauma from the cut itself, pain and swelling from the venom and possible later infection from bacteria. Rarely, severed arteries or death can result. Treatment for venom stings normally includes the application of heat, using water at temperatures of about 45°C (113°F), since the heat breaks down most complex venom proteins.

9. Diversity Depending on Locomotion

The slowest-moving fishes are the seahorses. The slowest of these, the dwarf seahorse attains about 5 ft/hour. Among the fasted sprinters are the Indo-Pacific sailfish and the black marlin. Both have been recorded in a burst at over 110 km/hour (68 mph). For the sailfish, that is equivalent to 12 to 15 times their own length per second. The wahoo is perhaps the fastest fish for its size, attaining a speed of 19 lengths per second, reaching 78 km/hour (48 mph). The short fin mako sharks are fast enough and agile enough to chase down and kill an adult swordfish, but they do not always win. Sometimes in the struggle with a shark, a swordfish can kill it by ramming it in the gills or belly. The short fin mako's speed has been recorded at 50 km/hour (31 mph), and there are reports that it can achieve bursts of up to 74 km/hour (46 mph). It can jump up to 9 m (30 ft) in the air. Due to its speed and agility, this high-leaping fish is sought as game worldwide. This shark is highly migratory. Its exothermic constitution partly accounts for its relatively great speed. The Atlantic bluefin tuna is capable of sustained high speed cruising, and maintains high muscle temperatures, so it can cruise in relatively cold waters.

A number of species jump while swimming near the surface, skimming the water. The flying fish have unusually large pectoral fins, which enable the fish to take short gliding flights above the surface of the water, in order to escape from predators. Their glides are typically around 50 m (160 ft), but they can use updrafts at the leading edge of waves to cover distances of at least 400 m (1,300 ft). The flying fish was able to stay aloft for 42 seconds by beating the surface of the water with its caudal (tail) fin. The mudskipper is a type of walking fish. The walking fish are often amphibious and can travel over land for extended periods of time. They are able to spend longer times out of water. These fish may use a number of means of locomotion, including springing, snake-like lateral undulation and tripod-like walking. The mudskipper is probably the best land-adapted of contemporary fish, and is able to spend days moving about out of water and can even climb mangroves, although to only modest heights. There are some species of fish that can walk along the sea floor but not on land, *e.g.*, flying gurnard fish.

10. Diversity Depending on Vision

The four-eyed fish have eyes raised above the top of the head and divided in two different parts, so that they can see below and above the water surface at the same time. These fishes actually have only two eyes, but their eyes are specially adapted for their surface-dwelling lifestyle. The eyes are positioned on the top of the head, and the fish floats at the water surface with only the lower half of each eye under water (Nelson, 2006).

The two halves of four-eyed fish are divided by a band of tissue, and the eye has two pupils connected by part of the iris. The upper half of the eye is adapted for vision in air, the lower half for vision in water. The lens of the eye also changes in thickness top to bottom to account for the difference in the refractive indices of air versus water. These fish spend most of their time at the surface of the water. Their diet mostly consists of the terrestrial insects which are available at the surface. Many species of fish can see the ultraviolet end of the spectrum, beyond the violet.

The two stripe dam shelfish, *Dascyllus reticulatus*, has ultraviolet-reflecting colouration which they appear to use as an alarm signal to other fish of their species (Losey, 2003). The predatory species can not see this if their vision is not sensitive to ultraviolet. There is further evidence for this view that some fish use ultraviolet as a 'high-fidelity secret communication channel hidden from predators', while yet other species use ultraviolet to make social or sexual signals (Siebeck *et al.*, 2010). The mesopelagic fishes live in deeper waters, in the twilight zone down to depths of 1000 m, where the amount of sunlight available is not sufficient to support photosynthesis. These fishes are adapted for an active life under low light conditions. The barrel eyes are a family of small, unusual-looking mesopelagic fishes, named for their barrel-shaped, tubular eyes which are generally directed upwards to detect the silhouettes of available prey. They have large, telescoping eyes which dominate and protrude from the skull. These eyes generally gaze upwards, but can also be swiveled forwards in some species. Their eyes have a large lens and a retina with an exceptional number of rod cells and a high density of rhodopsin (visual purple pigment); there are no cone cells. The barrel eye species, *Macropinna microstoma*, has a transparent protective dome over the top of its head, somewhat like the dome over an airplane cockpit, through which the lenses of its eyes can be seen. The dome is tough and flexible, and presumably protects the eyes from the nematocysts (stinging cells) of the siphonophores from which it is believed that the barrel eye steals the food (Robison and Reisenbichler, 2008).

The four-eyed fish feeds at the surface of the water with eyes that allow it to see both above and below the surface at the same time. The two stripe damselfish can signal secret alarms by reflecting ultraviolet to other fish of its species. The barrel eye has barrel-shaped, telescopic eyes which are directed upwards, but can also be swiveled forward. The flashlight fish use a retro- reflector behind the retina with photophores to detect the eye shine in other fishes.

11. Diversity Depending on Use by Humans

The fish are sought by humans for their value as commercial food fish, recreational sport fish, decorative aquarium fish, and in tourism, attracting snorkelers and SCUBA divers. Throughout human history, important fisheries have been based on forage fish. The forage fish are small fish that are eaten by larger predators. They usually school together for protection. The typical ocean forage fish feed near the bottom of the food chain on plankton, often by filter feeding. They include the family Clupeidae (*e.g.*, herrings, sardines, menhaden, hilsa, shad and sprats fish), as well as anchovies, capelin and halfbeaks. Important herring fisheries have existed for centuries in the North Atlantic and North Sea. Likewise, important traditional for anchovy and sardine fisheries have operated in the Pacific, Mediterranean and southeast Atlantic. The world annual catch of forage fish in recent years has been around 25 million tonnes, or one quarter of the world's total catch.

Higher in the food chain, Gadidae (*e.g.*, cod, pollock, haddock, saithe, hake and white fish) also supports important fisheries. Concentrated initially in the North Sea, Atlantic cod was one of Europe's oldest fisheries, later extending to Grand Banks (Armstrong *et al.,* 2004), declining numbers led to international 'cod wars' and eventually virtual abandonment of these fisheries.

The Alaska pollock supports an important fishery in Bering sea and north Pacific, yielding about 6 million tonnes, while the cod amounts to about 9 million tonnes. The yellowfin tuna are now fished as a replacement for the depleted Southern bluefin tuna. The Atlantic cod fisheries have collapsed. The koi and goldfish have been kept in decorative ponds for centuries in China and Japan. The food fish, oily fish, white fish, farmed fish and the fish used for medicinal purposes are also consumed by the humans. The recreational and sport fishing is big business. Some of the more popular recreational and sport fish include bass, marlin, porgy, shad, mahi-mahi, smelt whiting, swordfish and walleye. Fish keeping is another popular pastime, and there is a large international trade for aquarium fish. Snorkeling and SCUBA diving attract millions of people to beaches, coral reefs, lakes and other water bodies to view fish and other marine life.

12. Miscellaneous Diversity

The fish hold the records for relative brain weights of vertebrates. Most vertebrate species have similar brain-to-body weight ratios. The deep sea bathypelagic cusk-eel, *Acanthonus armatus,* has the smallest ratio of all known vertebrates. At other extreme, the elephant nose fish (African freshwater fish) has the largest ratio of all known vertebrates. The *Sarpa salpa* (a species of bream) is recognizable by the golden stripes running the length of its body and can induce LSD-like hallucinations, if it is eaten. These widely distributed coastal fish became a recreational drug during the Roman Empire and are called 'the fish that make dreams' in Arabic. Other hallucinogenic fish are *Siganus spinus*, called 'the fish that inebriates' in Reunion Island, and *Mulloidichthys samoensis*, called 'the chief of ghosts' in Hawaii.

References

Aird, W.C. 2007. *Endothelial Biomedicine.* Cambridge University Press.

Armstrong, M.J., Gerritsenb, H.D., Allenc, M., McCurdya, W.J., and Peel, J.A.D. 2004. Variability in maturity and growth in a heavily exploited stock: cod (*Gadus morhua* L.) in the Irish sea. *J. Marine Sci.,* **61(1)**: 98-112.

Cailliet, G.M., Andrews, A.H., Burton, E.J., Watters, D.L., Kline, D.E., and Ferry-Graham, L.A. 2001. Age determination and validation studies of marine fishes: do deep-dwellers live longer ? *Exp. Gerontol.,* **36**: 739-764.

Campbell, N.A., and Reece, J.B. 2005. *Biology,* 7th Ed. Benjamin Cummings, San Francisco CA.

Connolly, R. 2006. *Phycodurus eques.* www.iucnredlist.org.

Depczynski, M., and Bellwood, D.R. 2005. Shortest recorded vertebrate lifespan found in a coral reef fish. *Current Biol.,* **15(8)**: R288-R289.

Foot, T. 2000. *Guinness Book of World Records 2001.* Guinness World Records Ltd.

Herrera, M., and Jagadeeswaran, P. 2004. Annual fish as a genetic model for aging. *J. Gerontol. Series A: Biol. Sci. Med. Sci.,* **59**: B101-B107.

Kottelat, M., Britz, R., Tan, H.H., and Witte, K.E. 2005. *Paedocypris,* a new genus of Southeast Asian cyprinid fish with a remarkable sexual dimorphism, comprises the world's smallest vertebrate. *Proceedings of the Royal Society,* **B 273**: 895-899.

Losey, G.S. Jr. 2003. Crypsis and communication functions of UV-visible coloration in two coral reef damselfish, *Dascyllus aruanus* and *D. reticulatus. Ani. Behav.,* **66(2)**: 299-307.

Martins, C., and Knickle, C. 2009. *Megamouth Shark-Parasites. Florida Museum of Natural History.* http://www.flmnh.ufl.edu/fish/Gallery/descript/Megamouth/megamouth.htm.

Nelson, J.S. 2006. *Fishes of the World.* John Wiley and Sons, Inc.

Robison, B.H., and Reisenbichler, K.R. 2008. *Macropinna microstoma* and the paradox of its tubular eyes. *Copeia,* **4**: 780-784.

Rongo, T., Bush, M., and van Woesik, R. 2009. Did ciguatera prompt the late Holocene Polynesian voyages of discovery? *J. Biogeogr.,* **36(8)**: 1423-1432.

Siebeck, U.E., Parker. A.N., Sprenger, D., Mathger, L.M., and Wallis, G. 2010. A species of reef fish that uses ultraviolet patterns for covert face recognition. *Current Biol.,* **20(5)**: 407-410.

Smith, W.L., and Wheeler, W.C. 2006. Venom evolution widespread in fishes: a phylogenetic road map for the bioprospecting of Piscine venoms. *J. Hered.,* **97(3)**: 206-217.

Wikipedia- Fish Diversity. 2014. www.google.com.

2015, Animal Diversity, Natural History and Conservation, Vol. 5   *Pages* **177–211**
Editors: **V.K. Gupta and Anil K. Verma**
Published by: **DAYA PUBLISHING HOUSE, NEW DELHI**

Chapter 13

DNA Fingerprinting in Conservational Research on Ichthyofauna

Tanmay Mukhopadhyay and Soumen Bhattacharjee*
Cell and Molecular Biology Laboratory, Department of Zoology,
University of North Bengal, P.O. North Bengal University,
Raja Rammohunpur, Siliguri, District Darjeeling – 734 013,
West Bengal, India

ABSTRACT

Genetic variability contributes to the diverse forms of life on earth. Moreover, biological heterogeneity is essential to maintain the integrity of the biological systems that leads to the survival and adaptation of the whole population within that system. Therefore, the loss of genetic diversity as well as genetic integrity reduces the potentiality/capability of a population for adaptation and thus increases the risk of extinction. The recent concern and awareness about conserving nature and biodiversity as a whole have catalyzed into a massive global research efforts in the field of biodiversity and conservation genetics. With the refinement of molecular tools and DNA fingerprinting technologies, researchers found a wider way to resolve the empirical questions of conservation concern. Over exploitation, habitat alteration, inbreeding and loss of genetic diversity are the primary threats behind the considerable depletion of almost all economically important global fish populations. Therefore, fishery scientists have recognized conservational fish genetics as a guide to manage the natural fish stocks to derive sustainable benefit from the standpoint of economy, social and cultural aspects through multidisciplinary scientific concepts of population genetics, ecology and evolution.

* *Corresponding author.* E-mail: soumenb123@rediffmail.com, sbhnbu@gmail.com

The first step of fingerprinting technologies in the field of conservation was introduced with the advent of protein polymorphism analysis which was followed by mitochondrial DNA (mtDNA) sequence analyses, nuclear and mtDNA-based Restriction Fragment Length Polymorphism (RFLP), Random Amplification of Polymorphic DNA (RAPD), Amplified Fragment Length Polymorphism (AFLP) and microsatellite markers. Genomic approaches like whole genome sequencing and Single Nucleotide Polymorphism (SNP) analysis opened a promising door to look at the genome of a species with economical, ecological, and evolutionary and conservation relevance. Moreover, these genome-wide fingerprinting approaches give a detailed and relevant picture in the level of functionally important genetic variation as well as target sequence variations in conservation context. In summary this review provides an overview of the biodiversity and how this widespread biodiversity can be conserved with the help of molecular techniques with a special focus on fishery research to understand and evaluate genetic data.

Keywords: *Genetic diversity, Molecular markers, DNA fingerprinting, Fish conservation.*

Introduction

Biodiversity is the umbrella term and it covers the variety of life forms and it includes the entire biological hierarchy from single molecules to entire ecosystems, or the entire taxonomic hierarchy from alleles to kingdoms. It includes all the genotypes, individuals, species, populations and higher logical classes and all of the different members of all those classes. It also describes all the variability found within the living interactions and processes at all these levels of organization within the hierarchy (Sarkar and Margules, 2002). Therefore such a wide ranging description of the term "Biodiversity" makes it a paradigm to define. In 1996, Gaston defined Biodiversity as "the variability among living organisms from all sources, including terrestrial, marine and other aquatic ecosystems and the ecological complexes of which they are a part; this includes diversity within species, between species and of ecosystems". Williams *et al.* (1994) used the term Biodiversity to signify the "variety of life" or "God's creation" which encompasses the irreducible and incredible complexity of the totality of life forms.

Genetic differences among the individuals of a species are the basis of the phenotypic variation within that species. Genetic diversity may be defined as the heritable variations within and between populations of organisms. Knowledge of the genetic structure of a species is essential for assessing the genetic status and subsequently implementing conservation programme of the genetic resources as well as selective improvement. Genetic variation is a key tool for assessing the biological potential of an organism. A large population with high level of genetic variation may be better capable of dealing with changes with its surroundings such as fluctuations in physical and environmental factors, epidemics etc. Whereas the small populations have experienced a reduction in the level of genetic variation because of genetic bottlenecks (inbreeding depression and genetic drift), founder effects and many other stochastic processes. So it is due to loss of genetic variation or the heterozygosity, the smaller population become continuously homogenous and has a great chance to be extinct ultimately. However, to ensure the adaptation, expansion, and reestablishment in natural populations, maintenance of genetic variation in endangered species is very essential (Alam *et al.,* 2010). Natural population may fluctuate greatly in their sizes. The loss of heterozygosity is proportional to the reciprocal of the population size. Generations with small population sizes will dominate the effect on loss of heterozygosity. This is analogous to the unbalanced sex-ratio in small population leading to loss of heterozygosity (Allendorf and Luikart, 2007). So, the information on the genetic structure of cultivable fish is very essential for optimizing identification of potential brood stock, stock enhancement, selective breeding programmes, management for sustainable yield, and conservation of the genetic diversity and biodiversity as well (Haniffa *et al.,* 2007).

Therefore, fishery scientists and ichthyologists have recognized conservational fish genetics as a guide to manage the natural fish stocks to derive sustainable benefit from the standpoint of economy, social and cultural aspects through multidisciplinary scientific concepts of population genetics, conservation, ecology and evolution. The first step of fingerprinting technologies in the field of conservation genetics was introduced with the introduction of allozyme analysis which was followed by mitochondrial DNA (mtDNA) sequence analyses, nuclear and mtDNA-based Restriction Fragment Length Polymorphism (RFLP), Random Amplification of Polymorphic DNA (RAPD), Inter Simple Sequence Repeat (ISSR), Amplified Fragment Length Polymorphism (AFLP) and microsatellite markers. Genomic approaches like whole genome sequencing and Single Nucleotide Polymorphism (SNP) analysis opened a promising door to look at the genome of a species with economical, ecological, and evolutionary and conservation relevance. Moreover, these genome-wide fingerprinting approaches give a detailed and relevant picture in the level of functionally important genetic variation as well as target sequence variations in conservation context. In summary this review provides an overview of the biodiversity and how this widespread biodiversity can be conserved with the help of molecular techniques with a special focus on fishery research to understand and evaluate genetic data.

History of the Term "Biodiversity"

The term biodiversity was coined by Walter G Rosen at some point during the organisation of the 21-24 September, 1986 "National Forum on Biodiversity" held in Washington DC, under the auspices of US National Academy of Sciences and Smithsonian Institution (Takacs, 1996). In his book Takacs (1996) has also pointed out that in 1998, the term *"biodiversity"* did not appear as a key word in Biological Abstracts and *"Biological Diversity"* appeared only once till date. In 1993, *" biodiversity"* appeared seventy-two times and *"biological diversity"* nineteen times. The first journal with *"biodiversity"* in its title, *Canadian Biodiversity*, appeared in 1991; a second, *Tropical Biodiversity* surfaced in 1992; *Biodiversity Letters* and *Global Biodiversity* followed in 1993.

The UN Conference on Environment and Development or UNCED held in Rio de Janeiro in June 1992 (United Nation Environment Programme, 1992), declared preservation of biodiversity as a major element of 'sustainable development' and acknowledged it as a new type of natural resource. The entire gamut of "Biodiversity" provides raw material for productive activities ranging from genetic engineering to medical industries. Therefore, biodiversity has appeared to be a systematically surveyed and prospected resource (Reid *et al.,* 1993). The definition and significance of 'Biodiversity' has given rise to a discourse and this discourse is originated from ecology, evolutionary biology and conservation but this discourse has widened rapidly within a short time. Eventually the concern of Biodiversity relates to different realms of human practices and the actual facts can be defined in several contexts (Haila and Kauki, 1994) (Figure 13.1).

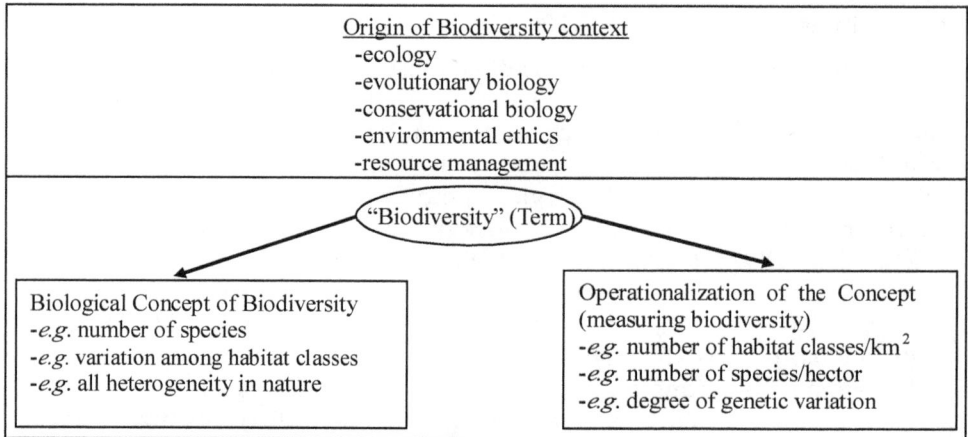

<u>Origin of Biodiversity context</u>
-ecology
-evolutionary biology
-conservational biology
-environmental ethics
-resource management

"Biodiversity" (Term)

Biological Concept of Biodiversity
-*e.g.* number of species
-*e.g.* variation among habitat classes
-*e.g.* all heterogeneity in nature

Operationalization of the Concept (measuring biodiversity)
-*e.g.* number of habitat classes/km^2
-*e.g.* number of species/hector
-*e.g.* degree of genetic variation

Figure 13.1: The Formation of Biodiversity under different Contexts (Adapted from Haila and Kauki, 1994).

The Elements of Biodiversity

Composition, structure and function are the three primary attributes of biodiversity which are further classified as a nested hierarchal pattern that incorporates elements of each attribute and summarizes a plethora of universal and phenomenological observation of nature and natural entities (Figure 13.2). Therefore, it is essential to ascertain a clear distinction between the patterns of biodiversity and explanations of their significances. Biodiversity can be categorized at four levels in a biological hierarchal pattern: (1) genetic diversity refers to the sum total of information present in the genes of individual organisms of a species; (2) species diversity is the number and frequency of organisms in a given area, such as the area occupied by a biological community; (3) ecosystem diversity is related to the variety of ecological processes, communities, and habitats within a region; and (4) landscape diversity is the spatial heterogeneity of the various land uses and ecosystems within a larger region (Norse, 1986) (Figure 13.2).

Genetic variation is the raw materials of evolution. A change in genetic composition of population and species is the primary mechanism of evolutionary change within species. The pool of genetic diversity of species can exist at three levels - genetic diversity within individuals (or heterozygosity), genetic diversity among

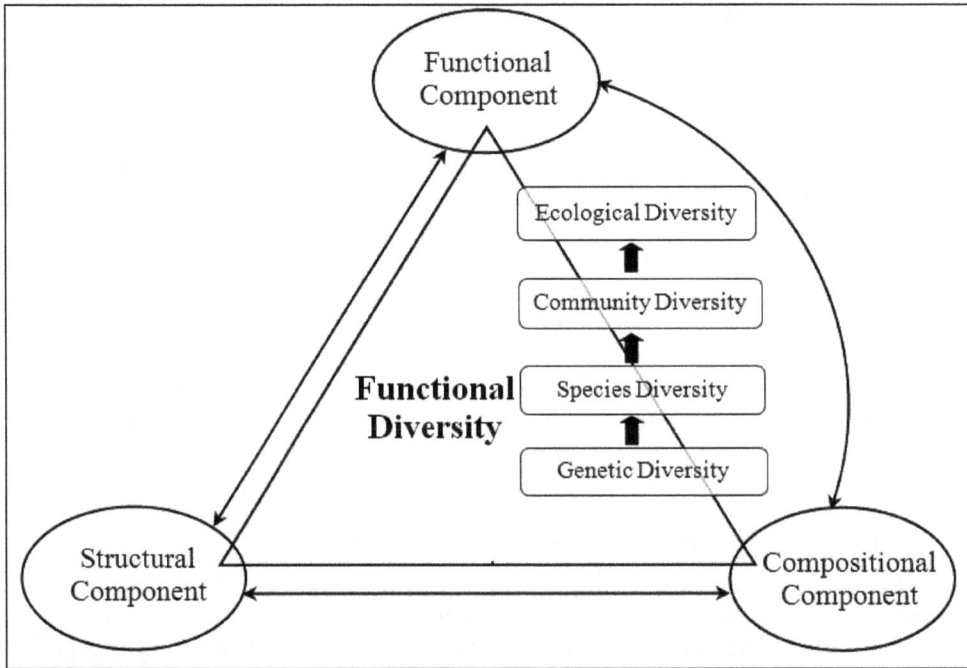

Figure 13.2: The Components and Hierarchy of a Functional Biodiversity System.

different individuals within a population and genetic variations among different populations. Thus, genetic diversity contributes to all the diverse forms of life on earth. The loss of genetic diversity reduces the capability for adaptation and increases the risk of extinction (Frankham, 1995; Caughley and Gunn, 1996; Avise and Hamrick, 1996; Landweber and Dobson, 1999). Wimsatt (1974) characterized "biodiversity" as "descriptive complexity"; and patterns and processes of biological realms are marked by variability and complexity at every level of organization. The conservation of biodiversity is proportionally related in some way to the preservation and maintenance of biological integrity. The biodiversity-biointegrity relationship implies that the increased diversity at every level of biological organisation leads to increased complexity and stability, and which leads to integrity and persistence of the biological system (McCann, 2000).

The main criterion that lie behind the diversity is functional (molecules within genome looks different, individuals within population looks different, populations within community looks different etc.). But the structural and evolutionary criteria are readily visible behind the biodiversity hierarchies. This can be tied into two metaphoric viewpoints: (a) in *Resilience Metaphor*, 'diversity' can be viewed as a buffer against unexpected disturbances and perturbations and (b) in *Selection Metaphor*, 'diversity' can be viewed as a raw material for adaptive change and evolution (Grene, 1987).

Relevance of Biodiversity Conservation

Biodiversity is a combination of heterogeneities and these heterogeneities are very essential for maintaining the integrity of biological systems. Therefore it is important to ascertain or specify the focus of the conservation concern of the biodiversity in the mixture of biological heterogeneities. The focus of the conservation concern must have a goal to maintain our planet suitable for human livelihood and preserve the aesthetic beauty of the nature, so as the human being with variable nature could perceive it. All the phenomena of biological diversity are not equally important for conservation and thus human concern trimmed the range of heterogeneity in two extremes that is important in the conservation discourse on biodiversity. One extreme may involve the smallest scale, *i.e.,* life processes within individuals, but this discourse only continues to function as long as there are organisms. On the other extreme, phenomena like extinctions and the depletion of ecological systems on the bio-geographic level are of great concern, because those were nearly stagnant for a very large extent of time (Haila and Kauki, 1994).

Origin of the Ichthyofaunal Conservation

The 'diversity crisis' was the main force that pushed ichthyologists to think about the conservation of ichthyofaunal diversity. The threat of the diversity crisis focused interest in the mechanism of extinction and in population viability and this approach drew upon population genetics, population ecology and conservation genetics together. With the increasing global population, the demand for high quality protein is also increasing, especially from aquatic resources. Due to over-exploitation, habitat alteration, introduction of exotic fish species and overharvesting towards certain sizes or age classes, there is considerable depletion in fish populations almost all economically important species. So an urgent need for conservation of fish genetics has been recognised by fishery scientists and aquaculturist worldwide (Lakra *et al.,* 2007). Fish habitats are destroyed as a consequence of many stochastic and environmental factors, such as deforestation, watershed erosion, and silting processes. Agricultural runoffs, pesticides, fertilizers, sewage, and chemical pollutants have added additional stresses towards declining fish populations. Construction of dams also creates barriers to the natural dispersal pathways of migratory fishes, and eliminates opportunities for gene flow among populations. Introduction of exotic fish species may result in hybridization between exotic and naturally occurring species which, in turn, leads to alterations of the genetic composition, which may lead to breakdown of locally adapted genetic architecture of native populations. Moreover, exotic species compete with native species for nest and food; they often reproduce and quickly replace native fishes (Banerjee *et al.,* 2008). The basic aim of natural fisheries management should be aimed at deriving a long term sustainable benefit through multidisciplinary scientific concepts involving population biology, ecology, evolution, genetics, social, political and economic aspects to manage and domestication of the natural fish stocks. The main objective of conservation is to maintain the genetic integrity and diversity of the species in their natural habitats. Therefore a proper documentation of genetic variability and diversity is of critical significance for sustainable fishery management with long term potential impact.

Moreover, some theoretical problems are there and some critical questions have to be answered before prioritizing any conservation practices. The problematic area and questions are as follows:

1. Is it possible to identify any critical species or species group of as a bio-indicator in ascertaining biological diversity? (Pearson and Cassola, 1992).

2. What are the minimum thresholds of diversity before ecological systems lose their functional integrity or evolutionary adaptability? This is the question of redundancy of 'diversity' or 'heterogeneity'. Is there any lower limit to diversity to maintain the integrity? (Walker, 1992).

Conservation Genetics and Conservation Genomics

The most powerful catalyst for increased research efforts in the field of conservation has been advances in genetic and molecular technologies, leading to an increasingly wider variety of molecular methodologies for application in conservation genetic and population genetic studies. To date, genetic and molecular methods have been applied vastly in conservation biology primarily as selectively neutral molecular tools for resolving the empirical questions of conservation and evolutionary relevance (Primmer, 2009).

Role of Molecular Genetic Method in conservation

Frankel and Soule (1980) reviewed the application of population genetic and evolutionary genetic theory and the methodology to address the issues of conservation concern has a long history. The first step of molecular biological technologies in the field of conservation genetics was taken in the 1960s with the introduction of protein polymorphism analysis (Ridgeway, 1962), followed by the mitochondrial DNA (mtDNA), Restriction Fragment Length Polymorphism (RFLP) methodologies (Bowen and Avise, 1990), Randomly Amplified Polymorphic DNA (Williams *et al.*, 1990), and most recently microsatellite (Taylor *et al.*, 1994) and Inter Simple Sequence Repeat markers (Zietkiewicz *et al.*, 1994). Each of these molecular-marker was seen as a revolution at the time when it was developed, due to the new possibilities that each new development created compared to previously available markers (*e.g.*, high-quality DNA vs. low-quality DNA; high polymorphism vs. Low polymorphism, recombining vs. non-recombining, neutral vs. possibly selected) (De Young and Honeycutt, 2005).

Therefore, the applications of particular types of genetic markers are becoming more and more specialised to achieve a particular goal to answer/solve the specific questions of conservation concern (Schlotterer, 2004). To date, the main use of genetic markers in conservation biology has been as molecular tools for resolving numerous questions, including: estimating parentage or relatedness in *ex situ* breeding programs (McLean *et al.*, 2008), estimating the effective population size (Vera *et al.*, 2008), detecting population bottlenecks as well as genetic drift (Luikart, 1998), determining population structure for definition of management or identify the evolutionary significant units (ESUs) (Fraser and Bernatchez, 2001), and detecting hybridization (Randi and Lucchini 2002; Barilani, 2005). Furthermore the analysis of neutral markers can be used to assess the evolution of a particular loci in response to the changes in environmental processes and this in turn could be used to assess the potential for populations to adapt the future changes of the environment (Bonin *et al.,* 2007).

Genomic Approaches in Conservation Biology

The genomic era was started when the genome of the evolutionary, ecologically and commercially important model organisms were successfully sequenced (Kohn *et al.,* 2006). This genomic information has been used for the study of organism having great economic value and as well as of their relationship with other important taxa. Vasemagi *et al.* (2005a) studied the evolutionary genetics of wild Atlantic salmon population with the genomic information available from the Atlantic salmon (*Salmo salar*) and Rainbow trout (*Oncorhynchus mykiss*).

There are several benefits of focusing the conservation towards genomic approach. Firstly, it provides a raw material to develop a larger number of molecular markers of a wide variety of organisms and also expected to solve the uncertainties reflecting in population structure towards a specific conclusion (Koskinen *et al.,* 2004). Secondly, the single-nucleotide polymorphism (SNPs) marker have the advantages to develop highly sophisticated marker system enabling analyses of even more highly degraded DNA that is hardly possible with PCR-based markers such as microsatellites (Sanchez and Endicott, 2006). Thirdly, this genomic information has enabled a new way to analyse the conservation genetic problems with the help of neutral markers, thus enhancing conservation of functionally relevant genetic variation (Kohn *et al.,* 2006).

The plethora of information provided by the various genome projects, advances in functional genomics and transgenic analysis have added great resources in the field of biotechnology and genetic engineering. These resources open a wider way in the field of fisheries and aquaculture to achieve several goals like increased food production, development of new natural resources, awareness about diminishing genetic diversity and to halt the deleterious effect of modern civilization on our environment. Melamed *et al.* (2002) described some biotechnological approaches to fisheries and aquaculture science (Figure 13.3).

Expressed sequence tag (EST) libraries are the valuable resource for the study of conservation and evolutionary genetics. The EST has numerous homologous sequences, which could be aligned with overlapping sequences to identify polymorphic sites and simultaneously could develop the SNP marker using bioinformatic tools by CASCADE databases, called *in silico* SNP mining pipeline (Guryev *et al.,* 2005). Those ESTs having microsatellite sequences at the untranslated region has been used to develop PCR primers using simple bioinformatic tools to detect gene polymorphisms (Vasemagi *et al.,* 2005a). Vasemagi *et al.* (2005b) used this approach to develop seventy-five polymorphic EST-associated microsatellite markers in five salmonid species. More recently genomic technological advances like next generation sequencing technologies and "deep sequencing" or "ultra-high throughput sequencing" technologies (Mardis, 2008; Holt and Jones, 2008) provide a wider path to understand the empirical questions regarding the conservation genetics of model as well as non-model organisms.

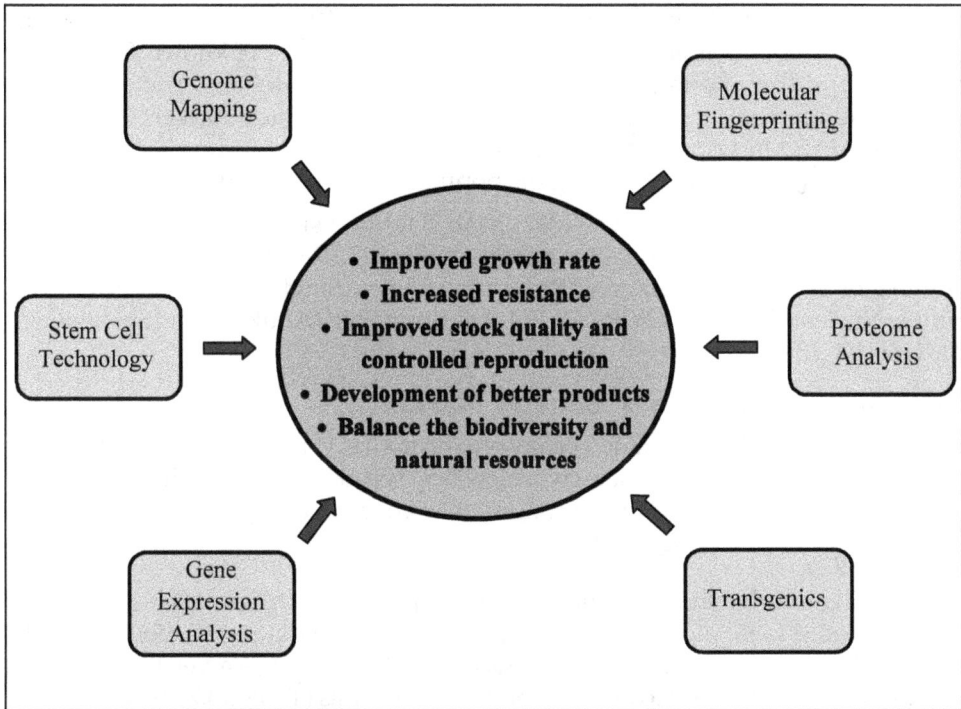

Figure 13.3: An Overview of some of the Platform Technologies (Square boxes) in Functional Genomics, and their Potential Applications to Aquaculture (Round box) (Adapted from Melamed *et al.*, 2002).

Molecular Markers in Population Biology Studies

Over the last few decades the key innovations in the field of molecular biology has made a great impact on the population biological analyses. The gene and genotypic variations within an individual are influenced by migration, genetic drift, population size natural selection, founder effect and historical events. Population genetics studies investigate the link between the demographic features and molecular genetic variations (Sunnucks, 2000). Therefore by measuring the genetic variation with different population genetic models, we can make an inference about the population biology of a species. The genetic markers are all heritable characters and also reflect changes in DNA sequences. Therefore, in organisms the genetic markers can be used to analyse the variation for the alleles per locus.

Types of Molecular Markers

Molecular markers can be classified into Type I and Type II markers on the basis of gene function. Type I markers are associated with genes of known function, while type II markers are associated with unknown genomic function as well as regions. Based on this classification allozyme markers are Type I markers because the protein they encode has known functions. RAPD and ISSR markers are Type II markers because RAPD and ISSR bands are amplified from anonymous genomic regions with arbitrary primers via the polymerase chain reaction (PCR). Microsatellite markers are also Type II markers unless they are associated with genes of known function. In general, type II markers such as RAPDs, ISSR, microsatellites, and AFLPs are considered non-coding and therefore selectively neutral. Such markers have found widespread use in population genetic studies to characterize genetic divergence, genetic diversity within and among the populations or species (O'Brien, 1991).

Altukhov and Salmenkova (2002) have classified six different types of DNA polymorphisms used for population genetic studies, *viz.,* RFLP, Minisatellites (Varying Number of Tandem Repeats, VNTR), Microsatellites (Single Tandem Repeats, STR; Simple Nucleotide Repeats, SSR), RAPD, AFLP and SNP (Altukhov and Salmenkova, 2002) (Table 13.1).

Criteria for Choosing a Genetic Marker

To study population genetics of any organisms it is very much essential to choose a genetic marker with appropriate characteristics. The main issues for choosing genetic marker to solve a given question of conservation concern are briefly described below:

Sensitivity

In order to solve the particular question regarding the population genetics the marker should have a correct sensitivity. Based on the data available during the studies (either too many or too little), the accumulated data have given the idea about what sorts of markers has to be used for proper analysis (Table 13.2).

Multi-locus or Single-Locus

Usually there is desirable but incompatible relationship between the practicality and accuracy of genetic markers. This dichotomous relationship has been found in the Multi-locus (RAPD, ISSR, and AFLP) techniques and Single locus (microsatellite) techniques (Table 13.2). Multi-locus markers are technically convenient but imprecise and it has some drawback including the variations they detect can be non-heritable. A fundamental limitation is dominant inheritance, because DNA fragments can be scored only as present or absent (1 or 0), in contrast to co-dominant inheritance where each of the two alleles at a locus in an individual can be identified and thus analysed more precisely (Burt *et al.,* 1996). By contrast, single-locus markers are far more flexible and informative because they can be analysed as genotypic arrays, as alleles with frequencies and as gene genealogies (Sunnucks *et al.,* 1997). Sometimes multi-locus RAPD bands can be cloned and converted to single locus co-dominant marker (Fa *et al.,* 2010).

Gene genealogies and Frequencies

Mitochondrial or nuclear DNA sequences affected and evolved by means of mutation rates, population parameters and natural selections. Genealogies describe the population processes, phylogeographic phenomenon, speciation and evolutionary relationships. Molecular phylogenies can help to untangle the current structure from the effects of historical events (Templeton, 1998). Therefore, markers that capable of analysing gene genealogies have great advantages over other markers.

Table 13.1: Types of DNA Polymorphism (Adapted from Altukhov and Salmenkova, 2002)

Variation Type	Cause of Polymorphism	Fields of Application
Restriction fragment length polymorphism (RFLP)	Nucleotide differences in restriction sites	Population genetics, systematics, phylogeny, genetic mapping, QTL mapping
Minisatellites (varyingnumber of tandem repeats, VNTR)	Varying number of tandemly repeated nucleotide DNA sequences with repeat size of 10–100 nucleotides	Population genetics (single-locus polymorphisms, multilocus polymorphisms), for estimation of kinship and pedigree analysis, for studying induced mutation process.
Microsatellites (single tandem repeats, STR; simple nucleotide repeats, SSR)	Varying number of short repeated nucleotide DNA sequences with repeat size of 1–6 nucleotides	Population genetics, evolutionary, demographic, and ecological genetics; identification of kinship and population assignment, genetic mapping
Randomly amplified polymorphic DNA fragments (RAPD)	Nucleotide differences in the primer binding sites	Population genetics, systematics, Phylogeny studies. Identification of plant cultivars and animalbreeds, genetic and QTL mapping
Amplified fragment length polymorphism (AFLP)	Nucleotide differences in restriction sites and flanking sites	Population genetics, systematics, phylogeny, identification of individual genotypes, analysis of kinship and pedigrees, genetic and QTL mapping
Single-nucleotide polymorphism (SNP)	Substitutions of single nucleotides in a DNA sequence	Evolutionary and population genetics; genetic mapping; particularly often used in studying SNP associations with diseases

Organelle and Nuclear DNA

The eukaryotic cells obtain nuclear DNA biparentally but mitochondrial or chloroplast DNA are uniprental (Table 13.2). This mode of transmission and the differences in the pattern of evolution helps to understand the population biology and evolutionary history of organelle DNA and nuclear DNA gene genealogies. Mitochondrial DNA has a lower N_e (effective number of alleles) than nuclear markers so mitochondrial DNA markers are used to identify the taxa (Sunnucks, 2000).

The technical advancement of molecular markers has led to the blossoming of genetic analysis of populations. However, genetic markers have been indiscriminately used for population genetics to conservation genetics studies and thus sometimes potentially lead to inappropriate interpretations. This is due to the two contradictory differences in the study question (historic/contemporary genetic variation), population size (big/small populations), time window (over past/recent events), and purpose (understanding evolutionary forces/exploring effects of diversity on survival (Wan *et al.*, 2004). Population genetics emphasizes the role of different evolutionary forces that play over time and conservation genetics highlights the effects of genetic structure to preserve endangered species. Both of which can be deduced with the help of appropriate molecular markers.

Application of Molecular Markers in Fishery Science

Molecular genetic markers have been used as a powerful and high resolution molecular genetic tools to detect the genetic variability as well as uniqueness of individuals, populations and species and have a revolutionised analytical power to explore the genetic diversity (Lakra *et al.*, 2007). The data gathered from genetic diversity study will help to implement the conservation and management programmes towards natural resources. This genetic diversity and variation within/between population can provide clues to the population's life histories, divergence and degree of evolutionary isolation. Measuring genetic diversity in wild fish populations or aquaculture stocks is essential for interpretation, understanding and effective management of these populations or stocks in the wild (Okumus and Ciftci, 2003). Parker *et al.* (1998) suggested that the appropriate techniques for a particular type of question is dependent on (1) the extent of genetic polymorphism required to best answer the question, (2) the analytical or statistical approaches available for the technique's application, and (3) the pragmatics of time and costs of materials (Parker *et al.*, 1998). Different types of molecular markers available for studying the fish population may basically be divided into 2 types of markers *viz.* Protein and DNA and there are three general classes of molecular markers used in population genetic and phylogenetic studies: (1) allozymes, (2) mitochondrial DNA and (3) nuclear DNA (Parker *et al.*, 1998).

The different types of molecular markers used in fishery science for the study of population genetics and phylogenetic studies are given below:

Allozyme

The term "allozyme" refers to different allelic forms of nuclear-encoded enzymes, whereas "isozyme" is a more general term referring to different biochemical forms of

Table 13.2: Major Attributes of Molecular Markers Commonly Used in Fisheries and Aquaculture Genetics (Modified from O'Connell and Wright, 1997; Parker et al., 1998; Sunnucks, 2000)

Marker	Genome	PCR Assay	Expression	Loci	No. of Loci Readily Available	Ovarall Variation	Population Structure	Individual Identification	Gene Mapping
Allozyme	Nuclear	No	Co-dominant	Single	Moderate	Low	Moderate/High	Low	Low
mtDNA RFLP	Organeller	No	Varies/Co-dominant	Single	Single	Low (Variable-Moderate)	Moderate/High	Inappropriate	Low
mtDNA Sequence	Organeller	Yes	Varies/Co-dominant	Single	Single	Variable (Low-High)	Moderate/High	Cumbersome	Low
RAPD/AFLPs	Nuclear	Yes	Dominant	Multi-locus	Many	High	Low	Moderate	Moderate RAPD/ High AFLP
Multilocus mini-/microsatellite 'fingerprints'	Nuclear	No	Dominant	Multi-locus	Many	High	Low	Moderate	–
Nuclear RFLPs	Nuclear	No	Co-dominant	Single	Many	Variable	High	Moderate/High	High
Minisatellite	Nuclear	Few	Co-dominant	Single	Moderate	High	High	High	High
Microsatellite	Nuclear	Yes	Co-dominant	Single	Many	High	Moderate/High	Low	Low

an enzyme identified by electrophoresis. Allozyme electrophoresis denotes the technique for identifying the genetic variation at the level of enzymes, which are directly encoded by DNA. The allelic variants give rise to protein variants known as *allozymes* that differ slightly in electrical charges and can be separated by enzyme electrophoresis. Allozymes are co-dominant markers because it inherited as Mendelian fashion (both alleles are expressed in homozygous as well as heterozygous condition) and characters passed from parent to offspring in a predictable manner. The data obtained from allozyme variation provides single locus genetic variation and this locus specific genetic data allow us to answer many basic questions regarding the diversity available in fish populations (May, 2003).

Each of these allozymes is a product of a unique allele whose amino acid sequence is slightly different from each other. In diploid organisms, there will be a combination of two alleles at each locus, designated as F (fast) and S (slow) or as "a" and "b" to distinguish between them. Therefore, the genotype at a gene locus coding for an enzyme can be deduced for each individual in the sample from the number and position of the band observed on the electrophoretic gels. If the enzyme is a monomer (the complete enzyme consists of only one polypeptide) the electrophoretic band pattern shown in Figure 13.4a would result for homozygotes and heterozygotes. An enzyme whose quaternary structure is a dimer (the complete enzyme consists of two polypeptides) would show the electrophoretic band pattern demonstrated in Figure 13.4b. In an individual, heterozygous at a dimeric enzyme locus for slow (a) and fast (b) alleles, three enzyme associations will be evident (aa, ab, bb). The heterozygote will show up twice as dark because both forms of each polypeptide will randomly associate, meaning that there is only one combination that will yield aa or bb, but there are two combinations which will yield ab. The electrophoretic pattern for a tetrameric enzyme is given in Figure 13.4c (Bader, 1998).

The allozyme electrophoresis technique is relatively cost effective, comparatively specialized equipment are required and the general applicability of this technique has made this the most studied form of molecular variation (Park and Moran, 1995). This technique is very much advantageous for screening of a large number of loci often more than 30-40 and no marker development phase is necessary and the analysis can start for any species immediately after the samples have been collected from the field (Okumus and Ciftci, 2003). Furthermore, comparable data from previous studies

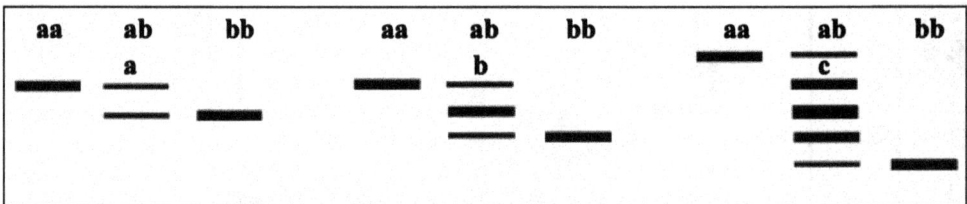

Figure 13.4: Electrophoretic Pattern for Allozymes.

a) Monomeric enzyme; b) Dimeric enzyme and c) Tetrameric enzyme. Homozygotes are represented by aa, bb and heterozygotes are represented by ab (adapted from Bader, 1998).

and a wealth of standard statistical procedures make allozymes appealing for studies of both fine and broad scale genetic variation (Weir, 1990). Besides these advantages there are some limitations. One of the major limitations of this has been the inability to read genotypes from small quantities of tissue, which makes allozymes inapplicable for small organisms (*e.g.* larvae) and also in endangered species; further more the major drawback is that is difficult to overcome is that only a small fraction of enzyme loci appear to be allozymically polymorphic in many species (Okumus and Ciftci, 2003). In addition, a given change in nucleotide sequence may not result in a change in amino acid at all, and thus would not be detected by protein electrophoresis. Furthermore, a change in the DNA that results in a change in an amino acid may not result in a change in the overall charge of the protein and, therefore, would also not be detected (Park, 1994).

In spite of these limitations, allozyme analyses have had a more profound effect on fisheries research and management than most all other genetic tools. There are widespread applications of allozymes in fisheries, these include allozyme variation in Brown trout, *Salmo trutta* from Central Spain (Machordom *et al.,* 1999), in *Chiloglanis paratus* and *C. pretoriae* from the Limpopo River system, Southern Africa (Engelbrecht and Mulder, 2000), Johnston's topminnow, *Aplocheilichthys johnstoni*, population from the Zambezi River system (Steenkamp *et al.,* 2001), Australian rainbow trout, *Oncorhynchus mykiss* (Farrington *et al.,* 2004), Spotted Murrel, *Channa punctatus* from South Indian river system (Haniffa *et al.,* 2007), three populations of the Armored Catfish *Hypostomus regani* from Brazil (Zawadzki *et al.,* 2008), *Nandus nandus* populations from Bangladesh (Zohora *et al.,* 2010). The allozyme electrophoresis study can also be used for the study of population structure and genetic diversity analysis in different contexts, the previous studies included from genetic analysis among different populations of Atlantic Salmon (Cordes *et al.,* 2005), left eyed founder *Paralichthys olivaceus* (Feng *et al.,* 2007), Endemic and Endangered Yellow Catfish, *Horabagrus brachysoma* (Muneer *et al.,*2007). This technique can be used for the molecular phylogenetic analysis of fish species (Matsuoka, 2003).

Mitochondrial DNA (mtDNA) and Restriction Fragment Length Polymorphism (RFLP)

A small portion of (< per cent 1) of the DNA of eukaryotic cells is non-nuclear; it is located within organelles in the cytoplasm called mitochondria. The major features of mtDNA: (a) in general maternally inherited haploid single molecule; (b) the entire genome is transcribed as a unit; (c) not subject to recombination and provides homologous markers; (d) mainly selectively neutral and occurs in multiple copies in each cell; (e) replication is continuous, unidirectional and symmetrical without any apparent editing or repair mechanism; and (f) optimal size, with no introns present (Billington, 2003). Like nuclear DNA the genome includes coding and non-coding regions and later evolves much faster than coding regions of DNA. One consequence of maternal transmission is that the effective population size for mtDNA is smaller than that of nuclear DNA, so that mtDNA variation is a more sensitive indicator of population phenomena such as bottlenecks and hybridizations (Avise, 1994).

The rapid rate of evolution, the maternal mode of inheritance and the relatively small size of mtDNAs make the RFLP analysis of this molecule one of the methods of choice for many population studies (Ferguson *et al.,* 1995). RFLP reveals polymorphism in an individual defined by restriction fragment sizes of distinctive lengths produced by a specific restriction endonuclease. Variation at mtDNA may be analysed mainly with two different approaches. (a) RFLP analysis of whole purified mtDNA obtained from fresh tissue (usually liver or gonad) by digesting it with restriction endonucleases; (b) RFLP analysis or DNA sequencing of small segments of the mtDNA molecule obtained by means of PCR amplication (Billington, 2003). These techniques yields maximum resolution and information that has made examination of mtDNA variations considerably easier and faster (Ferguson *et al.,* 1995). RFLP analysis in mtDNA studies have several advantages, including quite high levels of detectable variation, evolving of a high mtDNA, high genetic drift and low gene flow, possibility of reconstructing the phylogenetic history of mtDNA and thus populations, analysis of very small tissues (Hansen and Mensberg, 1998; Nielsen *et al.,* 1998; Hansen *et al.,* 2000).

Several sets of "universal primers" have been developed to analyse the same mtDNA segments in a variety of fish species. In initial studies with fish carried out with salmonids, universal primers for analysing the mtDNA ND-1 and ND-5/6 segments were used (Cronin *et al.,* 1993). The D-loop region of the mtDNA is practically the only non-coding region in the entire mtDNA of vertebrates. Some markers have targeted the D-loop region is highly variable in mammals but for some fish species little polymorphism is shown in the D-loop region (Okumus and Ciftci, 2003). Studies with a number of fish species *Salmo trutta* (Hall and Nawrocki, 1995); *S. salar* (McConnell *et al.,* 1995); *Anguilla anguilla* (Daemen *et al.,* 1996); *Hirundichthys affinis* (Gomes *et al.,* 1999); *Liporinus elongates* (Martins *et al.,* 2003) have actually shown less variability in the non-coding D-loop region than elsewhere in the mitochondrial genome. Thus the cytochrome b, cytochrome oxidase gene, 12S and 16S rRNA genes and dehydrogenase genes may be examined more profitably Atlantic cod (Carr and Marshall, 1991); *Tor tambroides* Valenciennes, (Esa *et al.,* 2008); Mekong giant cat fish (Nakorn *et al.,* 2006); six flatfish species (Kartavtsev *et al.,* 2007); anemone fishes (Ghorashi *et al.,* 2008) and *Barilius bendelisis* (Sah *et al.,* 2011 see Figure 13.5). MtDNA is also useful for barcoding of different fish species like *Channa marulius* (Lakra *et al.,* 2011). Genetic divergence study of six *Labeo* species from nine Indian River system has been also carried out with mtDNA Cytb gene sequence analysis (Luhariya *et al.,* 2012). Mitochondrial DNA (mtDNA) sequence variations have also been studied in two highly endangered fish *Chitala chitala* and *Anaecypris hispanica* (Mandal *et al.,* 2011 and Alves *et al.,* 2001). The mtDNA have a number of applications in fisheries biology, management and aquaculture and has attracted a lot of attention in many species, especially for population and evolutionary studies to provide critical information for use in the conservation and rehabilitation programmes.

Random Amplified Polymorphic DNA(RAPD)

RAPD is a random amplification of anonymous loci by PCR. It has several advantages and has been quite widely employed in fisheries studies. The method is

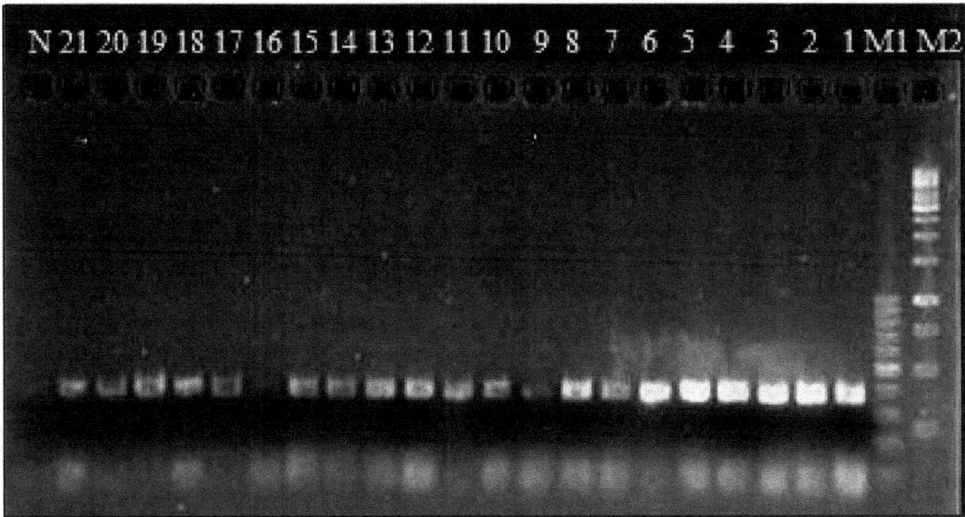

**Figure 13.5: Amplified Cytochrome b Gene in
Three Populations of *Barilius bendelisis.*
(M₂) 1 Kb ladder, (M₁) 100bp ladder and (N) Negative control (Sah *et al.,* 2011).**

simple, rapid and cheap, it reveals high polymorphism, only a small amount of DNA is required and most importantly, no prior knowledge of the genetic make-up of the organism in question is required (Hadrys *et al.,* 1992). RAPD markers allow creation of genomic markers from species of which little is known about target sequences to be amplified. The technique is a based on the PCR amplification of discrete regions of genome with short oligonucleotide primers of arbitrary sequence. A small amount of genomic DNA, one or more oligonucleotide primer (usually about 10 base pair in length), free nucleotides and polymerase with a suitable reaction buffer are major requirements. The main drawback with RAPDs is that the resulting pattern of bands is very sensitive to variations in reaction conditions, DNA quality, and the PCR temperature profile (Williams *et al.,* 1990; Welsh and McClelland, 1990). Even if the researcher is able to control the major parameters, other drawbacks of RAPD will remain: homozygous and heterozygous states cannot be differentiated and the patterns are very sensitive to slight changes in amplification conditions, giving problems of reproducibility (Ferguson *et al.,* 1995).

Random Amplified Polymorphic (RAPD) markers (type II markers) have been used for phylogenetic studies for species and subspecies identification of fish, for gynogenetic fish identification, analysis of inter-specific geneflow, analysis of paternity and kinship relationships and for gene mapping studies in fish. DNA polymorphisms have been extensively employed as a means of assessing genetic diversity in aquatic organisms. RAPD fingerprinting offers a rapid and efficient method for generating a new series of inheritable DNA markers in fishes (Foo *et al.,* 1995; Ali *et al.,* 2004). Two modifications of detecting RAPD markers have been described as DNA Amplification Fingerprinting (DAF) and Arbitrarily Primed Polymerase Chain Reaction (AP-PCR). DAF uses short random primers of 5-8 bp and visualizes the relatively greater number of amplification products by Polyacrylamide Gel Electrophoresis (PAGE) and silver staining. AP-PCR uses slightly longer primers and amplification products are radioactively labelled and also resolved by polyacrylamide gel electrophoresis (Hadrys *et al.,* 1992).

Assessment of genetic diversity of Tilapia stock under aquacultural practices in Fiji has been done RAPD markers to ascertain polymorphism and impacts of current management practices on genetic diversity and gene introgression among the tilapia stocks (Appleyard and Mather, 2002). Similar studies in flounder *Paralichthys olivaceus,* in China, has been done to investigate gradually diminishing physiological adaptability, slower growth, higher mortality and lower reproduction in the cultured stock in comparison to those of natural ones in spite of extensive fishery management efforts (Feng *et al.,* 2007). RAPD technique was used to examine the genetic variability and status of Brazilian endemic fish, *Brycon lundii, Astyanax altiparanae* and a migratory fish *Prochilodus marggravii*. The studies suggested the occurrence of a distinct structured populations and important implications for the conservation of the genetic variability of distinct natural stocks (Wasko and Galetti, 2002; Leuzzi *et al.,* 2004; Hatanaka and Galetti, 2003). Genetic stock structure of IMC *Labeo rohita* was investigated with the help of RAPD in Bangladesh (Islam and Alam, 2004). More recently, in a similar study in Bangladesh, the genetic structure has been ascertained in freshwater mud eel, *Monopterus cuchia,* and a baseline for any conservation approach towards threatened or near-threatened species (Alam *et al.,* 2010). RAPD has also been used for species identification. Species identifications of the Pacific lamprey *Entosphenus tridentatus* from four other Japanese lampreys, *Lethenteron japonicum,* *L. kessleri,* and two undescribed *Lethenteron* species, were carried out on the basis of PCR-based RAPD (Yamazaki *et al.,* 2005). The interspecific genetic variability and genetic relatedness among five Indian sciaenids namely *Otolithes cuvieri, Johnieops sina, Johnieops macrorhynus, Johnieops vogleri* and *Protonibea diacanthus* was studied for the first time using RAPD marker (Lakra *et al.,* 2007). A RAPD-based molecular identification and assessment of intraspecific genetic similarity in two ornamental fishes, *Badis badis* and *Dario dario* provided an excellent guideline for taxonomic identification of the fishes (Brahmane *et al.,* 2008). We have also carried out RAPD-PCR analysis to delineate the genetic structure of *Badis badis* from three major river streams of the Terai region of West Bengal, India (Mukhopadhyay and Bhattacharjee, 2013) (Figure 13.6).

Moreover, this technique can be used for phylogenetics, linkage group identification, chromosome and genome mapping, analysis of interspecific gene flow and hybrid speciation, and analysis of mixed genome samples (Hadrys *et al.,* 1992).

Inter Simple Sequence Repeat (ISSR)

The Inter Simple Sequence Repeat (ISSR) technique is another PCR-based method that uses microsatellites (usually 15-25 bp long) as primers in a single-primer PCR reactions targeting amplification of multiple genomic loci of DNA segment present at an amplifiable distance in between two identical microsatellite repeat regions (inter-SSR sequences) oriented in opposite direction (Reddy *et al.,* 2002). The microsatellite repeats used as primers can be di-nucleotide, tri-nucleotide, tetranucleotide or penta-

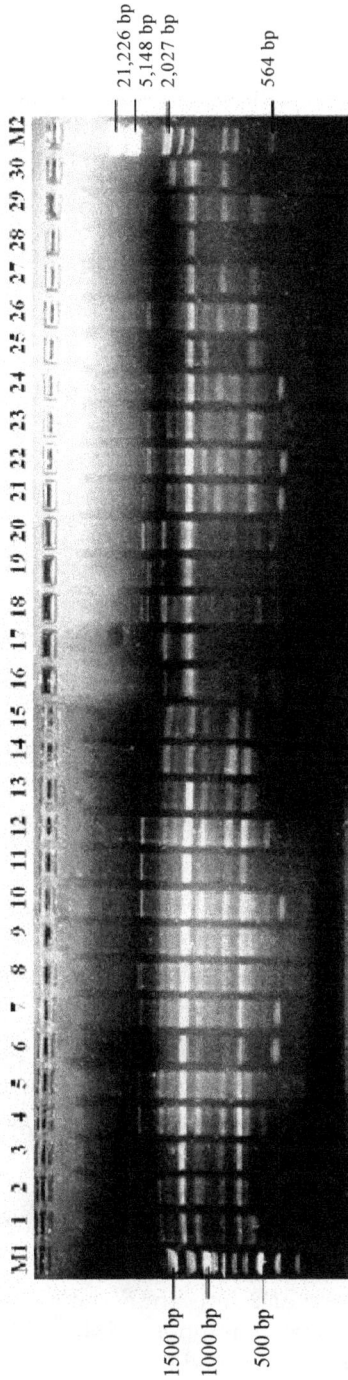

Figure 13.6: Random Amplified Polymorphic DNA Fragment Patterns Generated from Five Individuals of Six different Riverine Spots using OPA16 Primer.

M1=100 base pair DNA ladder, Lane 1-5= Balason River, Palpara, Matigara; 6-10= Mahananda-Panchanol River Junction; 11-15= Mahananda River, Champasari; 16-20= Mahananda Barrage, Fulbari; 21-25= Panchanol River; 26-30= Balason River, Tarabari; M2=Lambda DNA EcoRI/HindIII double digest.

nucleotide repeat sequence. ISSRs have high reproducibility possibly due to the use of longer primers (16–25 mers) as compared to RAPD primers (10- mers) which permits the subsequent use of high annealing temperature (45– 60°C) leading to higher stringency (Reddy *et al.,* 2002).

The primers used can be either unanchored (Meyer *et al.,* 1993; Gupta *et al.,* 1994; Wu *et al.,* 1994) or more usually anchored at 3' or 5' ends with 1 to 4 degenerate bases extended into the flanking sequences of either direction (Zietkiewicz *et al.,* 1994) (Figure 13.7). ISSRs segregate mostly as dominant markers (Gupta *et al.,* 1994; Wang *et al.,* 1998). However, they have also been shown to segregate as co-dominant markers in some cases thus enabling distinction between homozygotes and heterozygotes (Wu *et al.,* 1994; Akagi *et al.,* 1996; Wang *et al.,* 1998; Sankar and Moore, 2001).

ISSRs have been successfully used in many organisms to estimate the extent of genetic diversity at inter- and intra-specific level. Over the last few years ISSR markers have been widely used in the population genetic studies and stock identification in fishes including Japanese flounder (Liu *et al.,* 2006), *Brycon* (Antunes *et al.,* 2010), *Botia superciliaris* (Liu *at al.,* 2012), four Tilapia species (*Oreochromis niloticus, Oreochromis aureus, Tilapia zillii* and *Sarotherodon galilaeus*) (Saad *et al.,* 2012), grouper, cobia and red coral trout (Chiu *et al.,* 2012).

Amplified Length Polymorphism (AFLP)

AFLPs are dominant markers and the alleles are visualized in agarose or polyacrylamide gels. It combines the strengths of RFLP and RAPD markers and overcome their problems. The approach is PCR amplification based and requires no probe or previous sequence information as needed by RFLP. It is reliable because of high stringent and reproducibility of the amplification in contrast to RAPD's problem of low reproducibility and stringency (Okumus and Ciftci, 2003). AFLP analysis is based on selective amplification of fragments obtained by restriction of the genomic DNA. This method involves three stages: (a) DNA digestion (typically by two restriction enzymes) and binding sticky fragment ends with oligonucleotides adapters by a ligase; (b) selective PCR amplification of the set of restriction fragments; and (c) electrophoretic analysis of the amplified fragments. The nucleotide sequence of the adapter and an adjoining restriction site serves as a target for the annealing of the primer; the primer sequence complementary to the target is elongated at the 3' end by several arbitrary nucleotides. This permits to selectively amplify only the fragments whose restriction sites are flanked by the corresponding complementary nucleotides. Using this method and RAPD, particular sets of fragments (fingerprints) can be produced by PCR without previously knowing their nucleotide sequence. The fingerprint polymorphism is determined by the polymorphism of restriction sites and flanking nucleotides and is manifested as the presence or absence of particular bands in the gel. The method has high resolution and, in contrast to RAPD, good repeatability (Mueller and Wolfenbarger, 1999).

AFLP seems to be much more efficient than the microsatellite loci in discriminating the population differentiation in whitefish (*Coregonus clupeaformis*) (Campell *et al.,* 2003). The researchers also noted that the AFLP resulted in higher level of success in all levels of stringency and the likelihood difference between

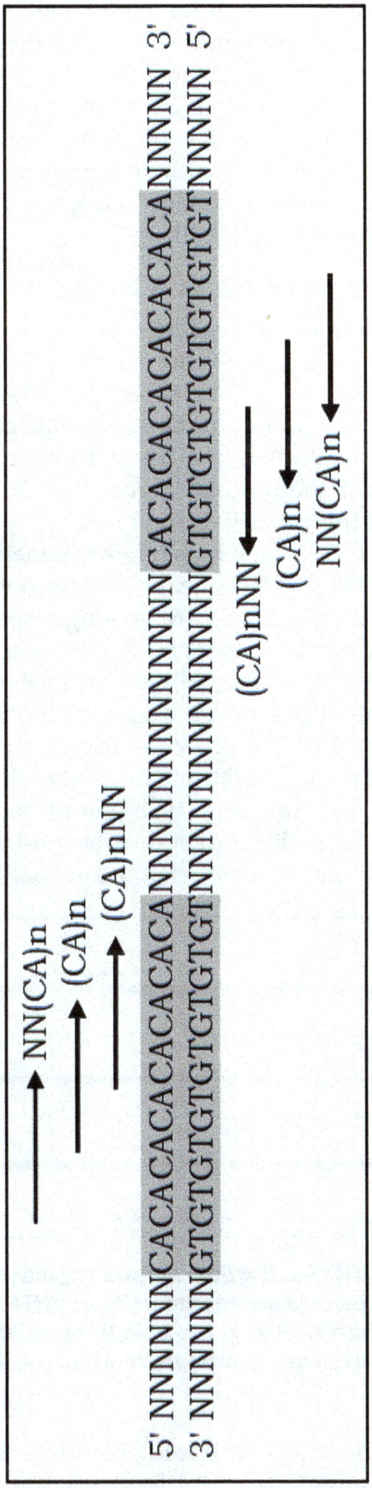

5' NNNNNNCACACACACACACACACACACACACANNNNNN 3'
3' NNNNNNGTGTGTGTGTGTGTGTGTGTGTGTNNNNNN 5'

NN(CA)n

(CA)n

(CA)nNN

(CA)nNN

(CA)n

NN(CA)n

Figure 13.7: A Schematic Representation of ISSR-PCR with Unanchored, 5' and 3' Anchored Primers. The shaded base pairs indicate the microsatellite region and arrows indicate the individual primers.

populations can be studied appropriately through AFLP marker. Similar to RAPD, AFLP analysis allows the screening of many more loci within the genome in a relatively within a short time and in an inexpensive manner. AFLP markers shows dominant nature of inheritance therefore it limits their utility for application in population genetics. Sometimes, AFLP markers have been used for intra and inter population mtDNA analysis. The AFLP methodology is also difficult to analyze due to the large number of unrelated fragments are visible on the electrophorectic gel along with the polymorphic fragments (Okumus and Ciftci, 2003).

Variable Number Tandem Repeats (VNTRs)

The nuclear genome of eukaryotes including fishes contains segments of DNA that are repeated tens or even hundreds to thousands of time, this are known as Variable Number of Tandem Repeat or VNTR. These repeated sequences are the most important class of repetitive DNAs found in the eukaryotic genome (O'Reilly and Wright, 1995). The tandem repeat vary in number at different loci and in different individuals dispersed throughout the genome (Figure 13.8). Based on the size of the repeat unit two main classes of this repetitive and highly polymorphic DNAs have been distinguished: a) *Minisatellite DNA*, which refers to genetic loci with repeats of smaller length (9-65 bp) concentrated near the centromeric region, and b) *Microsatellite DNA*, in which the repeat unit is only 2 to 8 (1-6) bp long dispersed throughout the genome. The term VNTR is frequently used for both mini- and microsatellite DNAs (Magoulas, 1998). Most importantly, microsatellites are much more numerous in the genome of vertebrates. Minisatellites are also classified as multilocus and single-locus. These two markers are alternatively known by a number of synonyms. Minisatellites: DNA fingerprinting and VNTR; microsatellites: simple sequence (Tautz, 1989), short tandem repeats (STRs) (Craig *et al.,* 1988); simple sequence repeats (SSRs) (Orti *et al.,* 1997). Mini- and microsatellite markers are the most important population genetic tools and increasingly applied in fisheries and aquaculture studies.

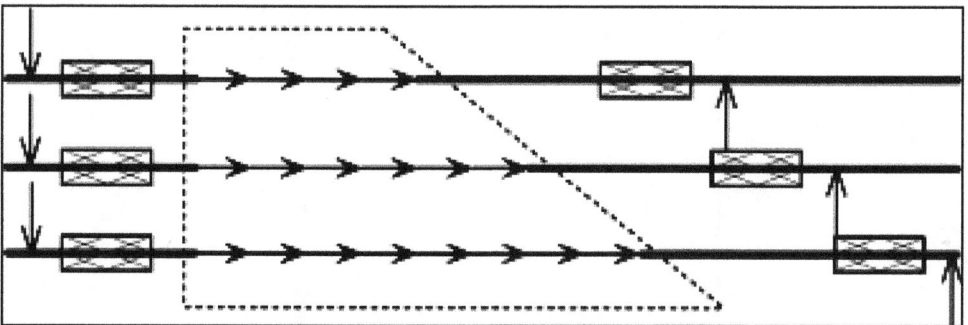

Figure 13.8: Polymorphisms in VNTRs. Vertical arrows represent restriction enzyme recognition sites; horizontal arrows are repeat units in VNTR; the dotted shape marks out core sequences of the VNTR loci; the bold lines indicate regions flanking core sequences; the shaded boxes show primers binding the flanking sequences.

Minisatellites

They refer to sequences which are composed of tandem repeats of 9 to 65 base pairs and have a total length ranging from 0.1 to 7kb (Jeffreys *et al.,* 1985). Many minisatellite loci are highly variable and useful in parentage analysis or for marking individual families. In addition to high level of polymorphism, there are major advantages such as generation of many informative bands per reaction and high reproducibility. Taggart and Ferguson (1990) first developed the single locus minisatellite marker for Atlantic salmon and tilapia species (Taggart and Ferguson, 1990). Unfortunately the highly variable loci are less useful for discriminating populations unless large sample sizes are used. Large numbers of alleles can also lead to difficulties in scoring and interpretation of data. Another limitation is the complex mutation processes (O'Reilly and Wright, 1995).

Despite the initial technical problems associated with using minisatellite probes, they have proved very successful in detecting genetics variations within and between fish populations (*e.g.* Taggart *et al.*, 1995 and Taylor, 1995), micro-geographical (between tributaries of a river) population differences (Galvin *et al.*, 1995) and reproductive success of farm escapees (Ferguson *et al.*, 1995). Recently minisatellites have been applied in fisheries for genetic identity, parentage, forensics, identification of varieties, estimate mating success and conforming gynogenesis (Jeffreys *et al.*, 1985).

Microsatellites

Microsatellites are thought to occur approximately once every 10 kbp, while minisatellite loci occur once every 1500 kbp in fish species, which suggests that microsatellites may be more useful for genome mapping (O'Connell and Wright, 1997). Microsatellites are co-dominant, inherited in a Mendelian fashion and tandem arrays of very short repeating motifs of 2-8 DNA bases that can be repeated up to ~100 times at a locus. They are among the fastest evolving genetic markers, with 10^{-3}-10^{-4} mutations/generation and their high polymorphism, and PCR based analysis has made them one of the most popular genetic markers (Goldstein *et al.*, 1995). Some microsatellite loci have very high numbers of alleles per locus (>20), making them very useful for applications such as parent-offspring identification in mixed populations, while others have lower numbers of alleles and may be more suited for population genetics and phylogeny (O'Connell and Wright, 1997). Primers developed for one species will often cross-amplify microsatellite loci in closely related species (Okumus and Ciftci, 2003).

The much higher variability at microsatellites results in increased power for a number of applications (Luikart and England, 1999). Only small amounts of tissue are required for typing microsatellites and these markers can be assayed using non-lethal fin clips and archived scale samples, facilitating retrospective analyses and the study of depleted populations (McConnell *et al.*, 1995). One of the main problems is the presence of so-called "null alleles" and this null alleles occur when mutations take place in the primer binding regions of the microsatellite locus, *i.e.* not in the microsatellite DNA itself (O'Reilly and Wright, 1995; Jarne and Lagoda, 1996). Even though microsatellites have already proven to be powerful single locus markers for a variety of genetic studies, the need to develop species-specific primers for PCR amplification of alleles can be expensive. However, primers developed to amplify markers in one species may amplify the homologous markers in related species as well (Morris *et al.*, 1996). Another important disadvantage of microsatellite alleles is that amplification of an allele via PCR often generates a ladder of bands (1 or 2 bp apart) when resolved on the standard denaturing polyacrylamide gels. These accessory bands (also known as stutter or shadow bands) are thought to be due to slipped-strands impairing during PCR (Tautz, 1989) or incomplete denaturation of amplification products (O'Reilly and Wright, 1995). The practical outcome of PCR stutter is that it may cause problems scoring alleles.

The main application of microsatellite in fisheries and aquaculture are phylogenetics and phylogeography (Hansen, 2002), population genetic structure (Nielsen *et al.,* 1999), conservation of biodiversity and effective population size (Reilly *et al.,* 1999), hybridization and stocking impacts (Hansen, 2002; Ruzzante *et al.,* 2001), inbreeding (Tessier *et al.,* 1997), domestication, quantitative traits (Jackson *et al.,* 1998), studies of kinship and behavioural patterns (Bekkevold *et al.,* 2002). Microsatellites are also becoming increasingly popular in forensic identification of individuals, and determination of parentage and relatedness, genome mapping, gene flow and effective population size analysis (Withler *et al.,* 2004).

Single Nucleotide Polymorphisms (SNPs)

Single nucleotide polymorphisms or SNPs describes the polymorphisms caused by point mutation in a single nucleotide base within a genome. These substitutions ultimately give rise to different alleles of alternative bases nucleotide position within a locus. SNPs has growing recent attention in the field of conservation genetics because they represent the most abundant polymorphisms in any organisms genome either coding or non-coding (Morin *et al.,* 2004; Liu and Cordes, 2004). A SNP present within a locus can produce as many as two alleles, each containing one of two possible base pairs at the SNP site. More than two SNP alleles would mean a second mutation at the same site which is usually rare. Therefore, SNPs have been regarded as bi-allelic markers. SNP markers are inherited as a co-dominant markers. Several approaches have been used for SNP discovery including Single-Strand Conformation Polymorphism (SSCP) analysis, heteroduplex analysis, and direct DNA sequencing (Hecker *et al.,* 1999). DNA sequencing has been the most accurate and most used approach for SNP analysis.

Concluding Remarks

The main purpose of this review is to provide an overview of the term Biodiversity and how this widespread Biodiversity can be conserved or at least put to a sustainable use. The discourse intends to relate biodiversity conservation with recent molecular biological developments with a special emphasis on the molecular and genetical markers in Fishery science and conservation. The questions being answered were what biodiversity is? How it can be conserved? What are the molecular markers? What are their main advantages and disadvantages? Where they can be used in fishery sciences? These rapidly developing markers can identify closely related species, populations/stocks, genetic strains, families and individuals. It is becoming increasingly important for fisheries and aquaculture stock managers to understand and evaluate genetic data. The case of biodiversity is only special because its conservation will protect the variety and variability of the patterns and processes of life forms. The issue of biodiversity is much broader than merely the question of extinction and endangered species. Ultimately the key to protecting 'biodiversity' is maintaining heterogeneity which is a necessary characteristic of ecological systems. In small structured populations (*e.g.* most endangered taxa), these markers may become linked to effectors of fitness, and the persistence of molecular polymorphisms may be a consequence of genetic hitchhiking. Molecular markers have proven their usefulness in solving many difficult taxonomic problems with endangered species, in designing and monitoring captive breeding programmes and understanding breeding systems, in detecting the geographical structure of genetic diversity, in managing gene flow, and in understanding factors contributing to fitness and to design an appropriate conservation programme.

References

Akagi, H., Yokozeki, Y., Inagaki, A., Nakamura, A., and Fujimura, T. (1996): A co-dominant DNA marker closely linked to the rice nuclear restorer gene, Rf-1, identified with inter-SSR fingerprinting. *Genome*, **39**: 1205–1209.

Alam, Md. S., Islam, Md. S., and Alam, Md. S. 2010. DNA Fingerprinting of the Freshwater Mud Eel, *Monopterus cuchia* (Hamilton) by Randomly Amplified Polymorphic DNA (RAPD) Marker. *International Journal of Biotechnology and Biochemistry*, **6**:271-278.

Ali, B. A., Huang, T. H., Qin, D. N., and Wang, X. M. 2004. A review of random amplified polymorphic DNA (RAPD) markers in fish research. *Reviews in Fish Biology and Fisheries*, **14**:443-453.

Allendorf, F. W., and Luikart, G. 2007. *Conservation and the Genetics of Population*. Blackwell Publishing, UK.

Altukhov, Yu. P and Salmenkova, E. A. 2002. DNA Polymorphism in Population Genetics. *Russian Journal of Genetics*, **38 (9)**: 989–1008

Alves, M. J., Coelho, H., Collares-Pereira, M. J., and Coelho, M. M. 2001. Mitochondrial DNA variation in the highly endangered cyprinid fish *Anaecypris hispanica*: importance for conservation. *Heredity*, **87**:463-473.

Antunes, R. S. P., Gomes1, V. N., Prioli, S. M. A. P., Prioli, R. A., Julio Jr, H. F., Prioli, L. M., Agostinho, C. S., and Prioli, A. J. 2010. Molecular characterization and phylogenetic relationships among species of the genus *Brycon* (Characiformes: Characidae) from four hydrographic basins in Brazil. *Genetics and Molecular Research*, **9 (2)**: 674-684.

Appleyard, S. A., and Mather, P. B. 2002. Genetic characterization of cultured Tilapia in Fiji using allozymes and random amplified polymorphic DNA. *Asian Fisheries Science*, **15**:249-264.

Avise, J. C. 1994. *Molecular Markers, Natural History and Evolution*. Chapman and Hall, New York.

Avise, J.C., and Hamrick, J. L. (eds.) 1996. Conservation and Genetics: Case Histories from Nature. Chapman and Hall, New York,.

Bader, J. M. 1998. Measuring Genetic Variability in Natural Populations by Allozyme Electrophoresis. *Proceedings of the 19th Workshop/Conference of the Association for Biology Laboratory Education (ABLE)* pp 25-42.

Banerjee, T, Raj, K.D., and Misra, V. 2008. Conservation of natural fish population. *Proccedings of Taal2007:The 12th World Lake Conference.* pp: 562-567.

Barilani, M. 2005. Detecting hybridization in wild (*Coturnix c. coturnix*) and domesticated (*Coturnix c. japonica*) quail populations. *Biological Conservation,* **126**: 445–455.

Bekkevold, D., Hansen, M. M., and Loeschcke, V. 2002. Male reproductive competition in spawning aggregations of cod (*Gadus morhua,* L.). *Molecular Ecology,* **11**: 91-102.

Billington, N. 2003. Mitochondrial DNA. E. M. Hallerman (Ed.), *Population genetics: principles and applications for fisheries scientists.* American Fisheries Society, Bethesda, Maryland: 59-100.

Bonin, A., Nicole, F., Pompanon, F., Miaud, C., and Taberlet, P. 2007. Population adaptive index: a new method to help measure intraspecific genetic diversity and prioritize populations for conservation. *Conservation Biology,* **21(3)**: 697-708.

Bowen, B. W., and J. C. Avise. 1990. Genetic structure of Atlantic salmon and Gulf of Mexico populations of sea bass, menhaden, and sturgeon—Influences of zoogeographic factors and life-history patterns. *Marine Biology,* **107**: 371–381.

Brahmane, M. P., Mitra, K., and Misra, S. S. 2008. RAPD fingerprinting of the ornamental fish *Badis badis* (Hamilton 1822) and *Dario dario* (Kullander and Britz 2002) (Perciformes, Badidae) from West Bengal, India. *Genetics and Molecular Biology,* **31**:789-792.

Burt, A., Carter, D. A., Koenig, G. L., White, T. J., and Taylor, J. W. 1996. Molecular markers reveal cryptic sex in the human pathogen *Coccidioides immitis. Proc. Natl. Academy of Science U. S. A.* **93**: 770–773.

Campbell, D., Duchesne, P., and Bernatchez, L. 2003. AFLP utility for population assignment studies: analytical investigation and empirical comparison with microsatellites. *Molecular Ecology,* **12**: 1979–1991.

Carr, S. M., and Marshall, H. D. 1991. Detection of intraspecific DNA sequence variation in the mitochondrial cytochrome *b* gene Atlantic cod (*Gadus morhua*) by the polymerase chain reaction. *Canadian Journal of Fish Aquatic Science,* **48**: 48-52.

Caughley, G., and Gunn, A. 1996. Conservation Biology in Theory and Practice. Blackwell Science. Cambridge.

Chiu,T. H., Su, Y. C., Lin, H. C., and Hsu, C. K. 2012. Molecular electrophoretic technique for authentication of the fish genetic diversity. Gel Electrophoresis-advanced techniques. In Tech Publisher, pp. 83-96.

Cordes, J.F., Perkins, D.L., Kincaid, H.L., and May, B. 2005. Genetic analysis of fish genomes and populations: allozyme variation within and among Atlantic salmon from Downeast rivers of Maine. *Journal of Fish Biology,* **67**: 104–117.

Craig, J., Fowler, S., Burgoyne, L. A., Scott, A.C., and Harding, H. W. J. 1988. Repetitive deoxyribonucleic acid (DNA) and human genome variation: a concise review relevant to forensic biology. *Journal of Forensic Science*, **33**: 1111-1126.

Cronin, M. A., Spearman, W. J., Wilmot, R. L., Patton, J. C., and Bickham, J. W. 1993. Mitochondrial variation in chinook (*Oncorhynchus tshawytcha*) and chum salmon (*O. keta*), detected by restriction enzyme analysis of polymerase chain reaction (PCR). *Canadian Journal of Fish Aquatic Science*, **50**: 708–715.

Daemen, E., Volckart, F., Hellemans, B., and Ollevier, F. 1996. The genetic differentiation of the European eel(*Anguilla anguilla* L.) on the European continental shelf. *Royal Belgium Academy of Sciences*, **65(1)**: 39-42.

DeYoung, R. W., and Honeycutt, R. L. 2005. The molecular toolbox: genetic techniques in wildlife ecology and management. *Journal of Wildlife Management*, **69**: 1362–1384.

Engelbrecht, G. D and Mulder, P. F. S. 2000. Allozyme variation in *Chiloglanis paratus* and *C. pretoriae* (Pisces, Mochokidae) from the Limpopo River system, Southern Africa. *Water*, **26**:111-114.

Esa, Y. B., Siraj, S. S., Daud, S. K., Rahim, K. A. A., Japning, J. R. R., and Tan, S. G. 2008. Mitochondrial DNA Diversity of *Tor tambroides* Valenciennes (Cyprinidae) from Five Natural Populations in Malaysia. *Zoological Studies*, **47(3)**: 360-367.

Fa, L.S., Jie, T.S., and Qi, C.W. 2010. RAPD-SCAR markers for genetically improved new gift Nile Tilapia (*Oreochromis niloticus niloticus* L.) and their application in strain identification. *Zoological Research*, **31(2)**:147-153.

Farrington, L.W., Austin, C. M., Burridge, C. P., Gooley, G. J., Ingram, A., and Talbot, B. 2004. Allozyme diversity in Australian rainbow trout, *Oncorhynchus mykiss*. *Fisheries management and ecology*, **11**:97-106.

Feng, Y., Peijun, Z., Keeling, Y., and Jianhai, X. 2007. Genetic variation of natural and cultured stocks of *Paralichthys olivaceus* by allozyme and RAPD. *Chinese Journal of Oceanology and Limnology*, **25**:78-84.

Ferguson, A., Taggart, J. B., Prodohl, P. A., McMeel, O., Thompson, C., Stone, C., McGinnity, P., and Hynes, R. A. 1995. The application of molecular markers to the study and conservation of fish populations, with special reference to *Salmo*. *Journal of Fish Biology*, **47**: 103-126.

Foo, C. L., Dinesh, K. R., Lim, T. M., Chan, W. K., and Phang, V. P. E. 1995. Inheritance of RAPD markers in Guppy fish, *Poecilia reticulate*. *Zoological Science*, **12**: 535-541.

Frankel, O. H., and Soule, M. E. 1980. *Conservation and Evolution*. Cambridge University Press. Cambridge.

Frankham, R. 1995. Conservation genetics. *Annual Review of Genetics*, **29**: 305–327.

Fraser, D. J., and Berntachez, L. 2001. Adaptive evolutionary conservation: towards a unified concept for defining conservation units. *Molecular Ecology*, **10**: 2741–2752.

Galvin, P., McKinnell, S., Taggart, J. B., Ferguson, A., O'Farrell, M., and Cross, T. F. 1995. Genetic stock identification of Atlantic salmon using single locus minisatellite DNA probes. *Journal of Fish Biology*, **47**: 186-199.

Gaston, K. J. 1996. What is Biodiversity. In *Biodiversity: a biology of numbers and diference*, K.J Gaston (Ed.). Blackwell Science Ltd. Oxfrod,U.K.

Ghorashi, S. A., Fatemi, S. M., Amini, F., Houshmand, M., Salehi Tabar, R., and Hazaie, K. 2008. Phylogenetic analysis of anemone fishes of the Persian Gulf using mtDNA sequences. *African Journal of Biotechnology*, **7 (12)**:2074-2080.

Goldstein, D. B., Linares, A. R., Cavalli-Sforza, L. L., and Feldman, M. W. 1995. An evaluation of genetic distances for use with microsatellite loci. *Genetics*, **139**: 463-471.

Gomes, C., Oxenford, H. A., and Dales, R. B. G. 1999. Mitochondrial DNA D-loop variation and implications for stock structure of the four-wing flyingfish *Hirundichthys affinis,* in the central western atlantic. *Bulletin of Marine Science*, **64(3)**: 485-500.

Grene, M. 1987. Hierarchies in biology. *American Science*, **75**:504-510.

Gupta, M., Chyi, Y. S., Romero-Severson, J., and Owen, J. L. (1994): Amplification of DNA markers from evolutionarily diverse genomes using single primers of simple-sequence repeats. *Theoretical Applied Genetics*, **89**: 998–1006.

Guryev, V., Berezikov, E., and Cuppen, E. 2005. CASCAD: a database of annotated candidate single nucleotide polymorphisms associated with expressed sequences. *BMC Genomics*, **6**:10-13.

Hadrys, H., Balick, M., and Schierwater, B. 1992. Applications of random amplified polymorphic DNA (RAPD) in molecular ecology. *Molecular Ecology*, **1**:55–63.

Haila, Y and Kauki, J. 1994.The phenomenon of biodiversity in conservation biology. *Annals Zoolici Fennici*, **31**:5-18.

Hall, H. J., and Nawrocki, L. N. 1995. A rapid method for detecting mitochondrial DNA variation in the brown trout, *Salmo trutta. Journal of Fish Biology*, **46**: 360–364.

Haniffa, M. A., Nagarajan, M., Gopalakrishnan, A., and Musammilu, K. K. 2007. Allozyme variation in a threatened freshwater fish Spotted Murrel (*Channa punctatus*) in the South Indian River system. *Biochemical Genetics*, **45**: 363-374.

Hansen, M. M. 2002. Estimating the long-term effects of stocking domesticated trout into wild brown trout(*Salmo trutta*) populations: An approach using microsatellite DNA analysis of historical and contemporary samples. *Molecular Ecology*, **11**:1003-1015.

Hansen, M. M., and Mensberg, K. L. D. 1998. Genetic differentiation and relationship between genetic and geographical distance in Danish sea trout (*Salmo trutta* L.) populations. *Heredity*, **81**: 493–504.

Hansen, M.M., Ruzzante, D.E., Nielsen, E.E., and Mensberg, K.L.D. 2000. Microsatellite and mitochondrial DNA polymorphism reveals lifehistory dependent interbreeding between hatchery and wild brown trout (*Salmo trutta* L.). *Molecular Ecology*, **9**: 583–594.

Hatanaka, T., and Galetti, P. M. 2003. RAPD markers indicate the occurrence of structured populations in a migratory freshwater fish species. *Genetics and Molecular Biology*, **26**: 19-25.

Hecker, K.H., Taylor, P.D., and Gjerde, D.T. (1999) Mutation detection by denaturing DNA chromatography using fluorescently labeled polymerase chain reaction products. *Analytical Biochemistry*, **272(2)**, 156-164.

Holt, R.A., and S.J.M. Jones. 2008. The new paradigm of flow cell sequencing. *Genome Research*, **18**: 839–846.

Islam, M. S., and Alam, M. S. 2004. Randomly amplified polymorphic DNA analysis of four different populations of the Indian major carp, *Labeo rohita* (Hamilton). *Journal of Applied Ichthyology*, **20**: 407–412.

Jackson, T. R., Ferguson, M. M., Danzmann, R. G., Fishback, A. G., Ihssen, P. E., and Crease, T. J. 1998. Identification of two QTL influencing upper temperature tolerance in three rainbow trout (*Oncorhynchus mykiss*) half-sib families. *Heredity*, **80**: 143-151.

Jarne, P., and Lagoda, P. J. L. 1996. Microsatellites, form molecules to populations and back. *Trends in Ecology and Evolution*, **11**: 424–429.

Jeffreys, A. J., Wilson, V., and Thein, S. L. 1985. Hyper variable minisatellite regions in human DNA. *Nature*, **314**: 67-73.

Kohn, M. H., Murphy, J. W., Ostrander, E. A., and Wayne, R. K. 2006. Genomics and conservation genetics. *Trends in Ecology and Evolution*, **21(11)**: 629-637.

Koskinen, M. T., Hirvonen, H., Landry, P. A., and Primmer, C. R. 2004. The benefits of increasing the number of microsatellites utilized in genetic population studies: an empirical perspectives. *Hereditas*, **141(1)**:61-67.

Lakra, W. S., Mohindra, V., and Lal, K. K. 2007. Fish Genetics and conservation research in India: Status and perpectives. *Fish Physiology and Biochemistry*, **33**: 475-487.

Lakra, W. S., Verma, M. S., Goswami, M., Lal, K. K., Mohindra, V., Punia, P., Gopalakrishnan, A., Singh, K. V., Ward, R. D., and Hebert, P. 2011. DNA barcoding of Indian marine fishes. *Molecular Ecology Resources*, **11**:60-71.   .

Landweber, L.F., and Dobson, A.P. (eds.) 1999. Genetics and the Extinction of Species: DNA and the Conservation of Biodiversity. Princeton University Press.New Jersey.

Leuzzi, M. S. P., de Almeida, F. S., Orsi, M. L., and Sodre, L. M. K. 2004. Analysis by RAPD of the genetic structure of *Astyanax altiparanae* (Pisces, Characiformes) in reservoirs on the Paranapanema River, Brazil. *Genetics and Molecular Biology*, **27**: 355-362.

Liu, H., Xiong, F., Duan, X., and Chen, D. 2012. Genetic Diversity of *Botia Superciliaris* populations in the upper reaches of the Yangtze River revealed by ISSR markers. International Conference on Environment, *Chemistry and Biology IPCBEE, IACSIT*

Press, Singapore. **49**.6: 22-30.

Liu, Y. G., Chen, S. L., Li, J., and Li, B. F. 2006. Genetic diversity in three Japanese flounder (*Paralichthys olivaceus*) populations revealed by ISSR markers. *Aquaculture,* **255**: 565–572.

Liu, Z.J., and Cordes, J.F. 2004. DNA marker technologies and their applications in aquaculture genetics. *Aquaculture,* **238**, 1-37.

Luhariya, R.K., Lal, K.K., Singh, R.K., Mohindra, V., Punia, P., Chauhan, U.K., Gupta, A., and Lakra, W.S. 2012. Genetic divergence in wild population of *Labeo rohita* (Hamilton, 1822) from nine Indian rivers, analyzed through MtDNA cytochrome b region. *Molecular Biology Reports,* **39**:3659–3665.

Luikart, G, England, P. R. 1999. Statistical analysis of microsatellite DNA data. *Trends in Ecology and Evolution,* **14**: 253-256.

Luikart, G. 1998. Usefulness of molecular markers for detecting population bottlenecks via monitoring genetic change. *Molecular Ecology,* **7**: 963–974.

Machordom, A., Marini, J. L. G., Sanz, N., Almodovar, A., and Pla, C. 1999. Allozyme diversity in brown trout (*Salmo trutta*) from Central Spain: Genetic consequences of restocking. *Freshwater Biology,* **41**:707-717.

Magoulas, A. 1998. Application of molecular markers to aquaculture and broodstock management with special emphasis on microsatellite DNA. *Cahiers Options Mediterrannes,* **34**: 153- 168.

Mandal, A., Mohindra, V., Singh, R.K., Punia, P., Singh, A.K., and Lal, K.K. 2011. Mitochondrial DNA variation in natural populations of endangered Indian Feather-Back Fish, *Chitala chitala. Molecular Biology Reports.* DOI 10.1007/s11033-011-0917-9.

Mardis, E.R. 2008. The impact of next-generation sequencing technology on genetics. *Trends in Genetics,* **24**: 133–141.

Martins, C., Wasko, A.P., Oliveira, C., and Foresti, F. 2003. Mitochondrial DNA variation in wild populations of *Leporinus elongatus* from the Paraná River basin. *Genetics and Molecular Biology,* **26 (1)**:33-38.

Matsuaoka, N. 2003. Molecular phylogeny and allozyme variation of the five common fish species of the suborder Percoidei. *Bull fac agriculture and life science hirosaki university,* **17(5)**: 17-22.

May, B. 2003. *Allozyme variation.* E. M. Hallerman (Ed.), Population genetics: principles and applications for fisheries scientists. American Fisheries Society, Bethesda, Maryland: 23-36.

McCann, K. S. 2000. The diversity-Stability debate. *Nature,* **405**:228-233.

McConnell, S. K., O'Reilly, P., Hamilton, L., Wright, J. N., and Bentzen, P. 1995. Polymorphic microsatellite loci from Atlantic salmon (*Salmo salar*) – genetic differentiation of North American and European populations. *Canadian Journal of Fisheries and Aquatic Science,* **52**: 1863– 1872.

McLean, J. E., Seamons, T. R.; Dauer, M. B., Bentzen, P., and Quinn, T. P. 2008. Variation in reproductive success and effective number of breeders in a hatchery population of steelhead trout (*Oncorhynchus mykiss*): examination by microsatellite-based parentage analysis. *Conservation Genetics*, **9**: 295–304.

Melamed, P., Gong, Z., Fletcher, G., and Hew, C.L. 2002. The potential impact of modern biotechnology on fish aquaculture. *Aquaculture*, **204**: 255–269.

Meyer, W., Mitchell, T. G., Freedman, E. Z., and Vilgays, R. (1993): Hybridization probes for conventional DNA fingerprinting used as single primers in the polymerase chain reaction to distinguish strains of *Cryptococcus neoformans*. *Journal of Clinical Microbiology*, **31**: 2274–2280.

Morin, P.A., Luikart, G., Wayne, R.K., and the SNP working group. 2004 SNPs in ecology, evolution and conservation. *Trends in Ecology and Evolution*, **19(4)**, 208-216.

Morris, D. B., Richard, K. R., and Wright, J. M. 1996. Microsatellites from rainbow trout (*Oncorhynchus mykiss*) and their use for genetic study of salmonids. *Canadian Journal of Fisheries and Aquatic Science*, **53**: 120–126.

Mueller, U.G., and Wolfenbarger, L.L. 1999. AFLP Genotyping and Fingerprinting, *Trends Ecology and Evolution*, **14**: 389–394.

Mukhopadhyay, T., and Bhattacherjee, S. 2013. Assessment of intra-population diversity of the threatened ornamental fish *Badis badis* (hamilton-buchanan, 1822) from the terai region of West Bengal. *NBU Journal of Animal Science*, *7: 13–23*.

Muneer, P. M. A., Gopalakrishnan, A., Lal, K. K., and Mohindra, V. 2007. Population Genetic Structure of Endemic and Endangered Yellow Catfish, *Horabagrus brachysoma*, Using Allozyme Markers. *Biochemical Genetics*, **45**: 437-445.

Nakorn, U., Suknanomon, S., Nakijima, M., Taniguchi, N., Kamonrat, W., Poompuang, S., and Nguyen, T. T. T. 2006. MtDNA diversity of the critically endangered Mekong giant catfish (*Pangasianodon gigas* Chevey, 1913) and closely related species: implications for conservation. *Animal Conservation*, **9**:483–494.

Nielsen, E. E., Hansen, M. M., and Loeschcke, V. 1999. Genetic variation in time and space: Microsatellite analysis of extinct and extant populations of Atlantic salmon. *Evolution*, **53**: 261-268.

Nielsen, E. E., Hansen, M. M., and Mensberg, K. L.D. 1998. Improved primer sequences for the mitochondrial ND1, ND3/4 and ND5/6 segments in salmonid fishes: Application to RFLP analysis of Atlantic salmon. *Journal of Fish Biology*, **53**: 216-220.

Norse, E. A. 1986. Conserving biological diversity in our national forests. The Wilderness Society, Washington, DC.

O'Brien, S. J. 1991. Molecular genome mapping: lessons and prospects. *Current Opinion in Genetics and Development*, **1**: 105–111.

O'Connell, M., and Wright, J. M. 1997. Microsatellite DNA in fishes. *Reviews in Fish Biology and Fisheries*, **7**: 331-363.

O'Reilly, P., and Wright, J. M. 1995. The evolving technology of DNA fingerprinting and its application to fisheries and aquaculture. *Journal of Fish Biology*, **47**: 29-55.

Okumus, I and Ciftci, Y. 2003. Fish population genetics and molecular markers: II-molecular markers and their applications in fisheries and aquaculture. *Turkish Journal of Fisheries and Aquatic Sciences*, **3**: 51-79.

Orti, G., Pearse, D. E., and Avise, J. C. 1997. Phylogenetic assessment of length variation at a microsatellite locus. *Proceedings of the National Academy of Sciences*, **94**: 10745-10749.

Park, L. K. 1994. *Introduction to DNA basics*. L. K. Park, P.l Moran, and R. S. Waples (Eds.), Application of DNA Technology to the Management of Pacific Salmon. Proceedings of the Workshop,Seattle, Washington.

Park, L. K., and Moran, P. 1995. *Developments in molecular genetic techniques in fisheries*. G.R. Carvalho, and T.J. Pitcher (Eds.), Molecular Genetics in Fisheries. Chapman and Hall, London.

Parker, P. G., Snow, A. A., Schug, M. D., Booton, G. C., and Fuerst, P. A. 1998. What molecules can tell us about populations: choosing and using a molecular marker. *Ecology*, **79(2)**: 361-382.

Pearson, D. L., and Cassola, F. 1992. World-wide species richness patterns of tiger beetles (Coleoptera: Cicindelidae): Indicator txon for biodiversity and conservation studies. *Conservation Biology*, **6**: 376-391.

Primmer, R. C. 2009. From Conservation Genetics to Conservation Genomics. *Annals of New York Academy of Science*, **1162**: 357–368.

Randi, E., and Lucchini, V. 2002. Detecting rare introgression of domestic dog genes into wild wolf (*Canis lupus*) populations by Bayesian admixture analyses of microsatellite variation. *Conservation Genetics*, **3**: 31–45.

Reddy, M. P., Sarla, N., and Siddiq, E. A., 2002. Inter simple sequence repeat (ISSR) polymorphism and its application in plant breeding. *Euphytica*, **128**: 9–17.

Reid, W. V., Laird, S. A., Meyer, C. A., Gamez, R., Sittenfeld, A., Janzen, D. H., Gollin, M. A., and Juma, C. 1993. Biodiversity Prospecting: Using genetic resources for sustainable development.World Resource Institute, Washington DC.

Reilly, A., Elliott, N. G., Grewe, P. M., Clabby, C., Powell, R., and Ward, R. D. 1999. Genetic differentiation between Tasmanian cultured Atlantic salmon (*Salmo salar* L.) and their ancestral Canadian population: comparison of microsatellite DNA and allozyme and mitochondrial DNA variation. *Aquaculture*, **173**: 459–469.

Ridgeway, G. J. 1962. The application of some special immunological methods to marine population problems. *American Naturalist*, **96**: 219–224.

Ruzzante, D. E., Hansen, M. M., and Meldrup, D. 2001. Distribution of individual inbreeding coefficients,relatedness and influence of stocking on native anadromous brown trout (*Salmo trutta*) population structure. *Molecular Ecology*, **10**: 2107-2128.

Saad, Y. M., Rashed, M. A., Atta, A. H., and Ahmed, N. E., 2012. Genetic diversity among some *Tilapia* species based on ISSR Markers. *Life Science Journal*, **9(4)**: 4841-4846.

Sah, S., Barat, A., Pande, V., Sati, J., and Goel, C. 2011. Population Structure of Indian Hill Trout (*Barilius bendelisis*) Inferred from Variation in Mitochondrial Dna Sequences. *Advances in Biological Research*, **5 (2)**: 93-98.

Sanchez, J. J., and P. Endicott. 2006. Developing multiplexed SNP assays with special reference to degraded DNA templates. *Nature Protocol*, **1**: 1370–1378.

Sankar, A. A., and Moore, G. A. (2001): Evaluation of inter-simple sequence repeat analysis for mapping in *Citrus* and extension of genetic linkage map. *Theoretical Applied Genetics*, **102**: 206–214.

Sarkar, S., and Margules, C. 2002. Operationalizing biodiversity for conservation planning. *Journal of Bioscience(Suppl. 2)*, **27**:299-308.

Schlotterer, C. 2004. The evolution of molecular markers—Just a matter of fashion? *Nature Review of Genetics*, **5**: 63–69.

Steenkamp, M. K. J., Engelbrecht, G. D and Mulder, P. F. S. 2001. Allozyme variation in a Johnston's topminnow, Aplocheilichthys johnstoni, population from the Zambezi River system. *Water*, **27**:53-55.

Sunnucks, P. 2000. Efficient genetic markers for population biology. *Tree*, **15**: 199-203.

Sunnucks, P., De Barro, P.J., Lushai, G., Maclean, N., and Hales, D. 1997. Genetic structure of an aphid studied using microsatellites: cyclic parthenogenesis, differentiated lineages and host specialization. *Molecular Ecology*, **6(11)**:1059-73.

Taggart, J. B., and Ferguson, A. 1990. Hypervariable minisatellite DNA single locus probes for Atlantic salmon, *Salmo salar* L. *Journal of Fish Biology*, **37**: 991-993.

Taggart, J. B., Verspoor, E., Galvin, P. T., Moran, P., and Ferguson, A. 1995. A minisatellite DNA marker for discriminating between European and North American Atlantic salmon (*Salmo salar* L.). *Canadian Journal of Fisheries and Aquatic Science*, **52**: 2305-2311.

Takacs, D. 1996. *The idea of biodiversity: philosophies of paradise.* Johns Hopkins University Press.

Tautz, D. 1989. Hypervariability of simple sequences as a general source for polymorphic DNA markers. *Nucleic Acids Research*, **17(16)**: 6463-6471.

Taylor, A. C., Sherwin, W. B., and Wayne, R. K. 1994. Genetic variation of microsatellite loci in a bottlenecked species: the northern hairy-nosed wombat Lasiorhinus krefftii. Molecular Ecology, *3(4):277-290.*

Taylor, E. B. 1995. Genetic variation at minisatellite DNA loci among North Pacific populations of steelhead and rainbow trout (*Oncorhynchus mykiss*). *Journal of Heredity*, **86**: 354–363.

Templeton, A.R. 1998. Nested clade analysis of phylogenetic data: testing hypotheses about gene flow and population history. *Molecular Ecology*, **7**:381-397

Tessier, N., Bernatchez, L., and Wright, J. M. 1997. Population structure and impact of supportive breeding inferred from mitochondrial and microsatellite DNA analyses in land-locked Atlantic salmon *Salmo salar* L. *Molecular Ecology*, **6**: 735–750.

Vasemagi, A., Nilsson, J., and Primmer C.R. 2005a. Expressed sequence tag (EST) linked microsatellites as a source of gene associated polymorphisms for detecting signatures of divergent selection in Atlantic salmon (*Salmo salar* L.). *Molecular Biology and Evolution*, **22**: 1067–1076.

Vasemagi, A., Nilsson, J., and Primmer C.R. 2005b. Seventy-six EST-linked Atlantic salmon (*Salmo salar* L.) microsatellite markers and their cross-species amplification in five other salmonids. *Molecular Ecology Notes*, **5**: 282–288.

Vera, J. C., Wheat, C. W., Fescemyer, H. W., Frilander, M. J., Crawford, D. L., Hanski, I., and Marden, J. H. 2008. Rapid transcriptome characterization for a nonmodel organism using 454 pyrosequencing. *Molecular Ecology*, **17(7)**:1636-1647.

Walker, B. H. 1992. Biodiversity and ecological redundancy. *Conservation Biology*, **6**: 18-23.

Wan, Q.H., Wu, H., Fujiharaz, T., and Fang, S.G. 2004. Which genetic marker for which connservation genetics issue? *Electrophoresis*, **25***: 2165–2176

Wang, G., Mahalingan, R., and Knap, H. T. (1998): (C-A) and (GA) anchored simple sequence repeats (ASSRs) generated polymorphism in soybean, *Glycine max* (L.) Merr. *Theoretical Applied Genetics*, **96**: 1086–1096.

Wasko, A. P., and Galetti, P. M. 2002. RAPD analysis in the Neotropical fish *Brycon lundii*: genetic diversity and its implications for the conservation of the species. *Hydrobiologia*, **474**: 131–137.

Weir, B. S. 1990. Genetic data analysis. Sinauer, Sunderland, Massachusetts, USA.

Welsh, J., and McClelland, M. 1990. Fingerprinting genomes using PCR with arbitrary primers. *Nucleic Acids Research*, **18**: 7213-7218.

Williams, J. G. K., Kubelik, A. R., Livak, K. J., Rafalsk, J. A., and Tingey, S. V. 1990. DNA polymorphisms amplified by arbitrary primers are useful as genetic markers. *Nucleic Acid Research*, **18**: 6531-6535.

Williams, P. H., Gaston, K. J., and Humphries, C. J. 1994. Do conservationists and molecular biologists value differences between organisms in the same way? *Biodiversity Letters*, **2**: 67–78.

Wimsatt, W. C. 1974. Complexity and organization. *Proc. Meetings Philos. Sci. Ass.* **1972**: 67-86.

Withler, R. E., Candy, J. R., Beacham, T. D., and Kristina Miller, K. M. 2004. Forensic DNA analysis of Pacific salmonid samples for species and stock identification. *Environmental Biology of Fishes*, **69**:275–285.

Wu, K., Jones, R., Dannaeberger, L., and Scolnik, P. A. 1994. Detection of microsatellite polymorphisms without cloning. *Nucleic Acids Research*, **22**: 3257–3258.

Y. P. Kartavtsev, Y.P., Park, E.T.J.,Vinnikov, E.K.A., Ivankov, E.V.N., Sharina, E.S.N., and Lee, E.J.S. 2007. Cytochrome b (Cyt-b) gene sequence analysis in six flatfish species (Teleostei, Pleuronectidae), with phylogeneticand taxonomic insights. *Marine Biology*, **152**:757–773.

Yamazaki, Y., Fukutomi, N., Oda, N., Shibukawa, A., Niimura, Y., and Iwata, A. 2005. Occurrence of larval Pacific lamprey *Entosphenus tridentatus* from Japan, detected by random amplified polymorphic DNA (RAPD) analysis. *Ichthyology Research*, **52**: 297–301.

Zawadzki, C. H., Renesto,E., Peres, M. D., and Paiva, S. 2008. Allozyme variation among three populations of the armored catfish Hypostomus regani (Ihering, 1905) (Siluriformes, Loricariidae) from the Paraná and Paraguay river basins, Brazil. *Genetics and Molecular Biology*, **31** (3): 767-771.

Zietkiewicz, E., Rafalski, A., and Labuda, D. (1994): Genome fingerprinting by simple sequence repeat (SSR) – anchored polymerase chain reaction amplification. *Genomics*, **20**: 176–183.

Zohora, N., Khan, M.M.R., Ahammad, A.K.S., and Hasan, M. 2010. Morphological and allozyme variation of three wild meni (*Nandus nandus*, hamilton) populations in Bangladesh. *International Journal of Biological Research*, **1(5)**: 33-40.

2015, Animal Diversity, Natural History and Conservation, Vol. 5   *Pages 213–225*
Editors: V.K. Gupta and Anil K. Verma
Published by: DAYA PUBLISHING HOUSE, NEW DELHI

Chapter 14

Aspects of Dietary Tactics and Choices in Little Brown Doves (*Streptopelia senegalensis senegalensis*) from Jammu (J&K), India

Deep Novel Kour[1]* and D.N. Sahi[1]
[1]Postgraduate Department of Zoology,
University of Jammu, Jammu – 180 006, J&K, India

ABSTRACT

The present study reflects an attempt to collect data on the diverse aspects of feeding activity of Little Brown Doves including observations on the type of foraging habitat, feeding postures employed, feeding tactics, feeding sessions, feeding preferences, time spent on feeding activity, size of feeding flocks, feeding associations with other birds and economic importance of feeding behaviour to agriculture. In the study area, Little Brown Doves were recorded to be strictly grainivorous with the preference of weed seeds in their diet as weed seeds constituted 36.1 per cent of their diet. However, cereals, oil seeds and pulses were reported to be consumed only after weed seeds, thereby accounting for 23.17 per cent, and 11.96 per cent and 7.3 per cent of the total food of Little Brown Doves respectively.

Keywords: Flocks, Foraging, Agriculture, Feeding Tactics.

* *Corresponding author.* E-mail: deepnovel10@yahoo.co.in

Introduction

Knowledge about diets is fundamental for understanding species' niches, roles in communities and potential impacts on other species (Moegenburg and Levey, 2003; French and Smith, 2005) and food is often considered to be the most important factor affecting the production of off-spring (Lack, 1968; Martin, 1987). In India, not many studies have been conducted for columbids in general and Little Brown Doves (*Streptopelia senegalensis senegalensis*) in particular. Columbids are the most unique among birds in their production of crop milk and in their drinking habits. They are world-wide in distribution, with different species found in the various zoogeographical regions of the world. Goodwin (1983) described the Columbidae as a large and successful family of birds. The present communication encompasses the main objective to gather firsthand information pertaining to the various dietary items and dietary preferences of a common and resident columbid *i.e.* Little Brown Dove from the diverse feeding habitats of the study area as virtually nothing has been reported pertaining to the aforementioned aspects of the bird from the study area.

Study Area

The study was carried out from June, 2009 to June 2010 in Jammu, winter capital of the state of Jammu and Kashmir (India). The city of Jammu sprawls on a hillock, on both the banks of picturesque River Tawi in the foothills of Himalayas and has a geographical area of 3097 sq.km. Besides, it lies between 32° 27′ and 33° 50″ North latitudes and 74° 19″ and 75° 20″ east longitudes. Altitudinally, it extends from 250 to 410 m above the mean sea level. The climatic conditions in and around the study area are dry sub-humid to arid. There are four well marked seasons in a year namely winter, summer, monsoon and autumn.

The study area possessed dominant plant species like *Acacia nilotica* (Babul), *Acacia modesta* (Kramishatrav), *Dalbergia sissoo* (Sheesham), *Morus alba* (Shahtoot), *Eucalyptus tereticornis* (Safeda), *Mangifera indica* (Aam), *Zizyphus* spp.(Ber), *Butea monosperma* (Pallash), *Adhatoda vasica*, *Grewia optiva* (Dhamin), *Emblica officinalis* (Amla), *Cannabis sativa* (Bhang), *Ficus bengalensis* (Barghad), *Ficus religiosa* (Peepal), *Calotropis procera* (Akvan) etc. Agriculture included predominantly of *Oryza sativa* (Rice), *Triticum aestivum* (Wheat), *Zea mays* (Maize), vegetables and fruit trees.

Methodology

The data was collected during a continuous period of one year from June, 2009 to May, 2010. Periodic surveys were performed in the area under inquisition from 0630 to 1200 h in the morning and 1300 to 1900 h in evening during summer and 0730 to 1200 h in morning and 1400 to 1830 h in evening during winter. In addition to it, several erratic excursions were also conducted during different hours of the day, before sunrise to sunset. The birds were observed with naked eye and through binoculars (Bushnell 7 X 50 U. S. A. made) whenever found necessary to record the data from quite a long distance in order to avoid any interference to birds due to the presence of observers. Photographic evidence was collected with the aid of Canon EOS camera fitted with 300 mm zoom lens, Digital Camera (Sony) fitted with 14.1

megapixel lens with an optical zoom of 4X. Besides, videos recordings were done with the aid of 800 X Digital 200M/Optical 20 X video camera.

The observations were taken on their foraging behaviour using "**Focal Animal Sampling**" technique (Altmann, 1974). The data was collected when Little Brown Doves were actively feeding in the feeding habitats. To identify the food items, samples were collected from the study site besides analysing the gut contents of accidently killed specimens in the field. Furthermore, the seasonal changes in the feeding habits of doves were also studied. To study seasonal variation in feeding methods, the study period was divided into four stages:

(a) December to March

(b) March to June

(c) July to September

(d) October to November

The collected dead adult birds were dissected and their crops and gizzards were taken out and opened to analyse the diverse food contents. The contents were put in screen sieves, washed and placed on a blotting paper and dried for 10-15 minutes depending upon room humidity at room temperature. Different food items were sorted out and identified. Each food item was weighed on a digital balance. After recording the weights, the unidentified seeds, etc. were preserved dry for subsequent studies.

When the gut contents of birds could not be studied immediately, guts (crops and gizzards) were taken out and each was tied at both ends with a cotton thread. The guts were labelled and put in 10 per cent formaldehyde solution. On the subsequent day, the preserved guts were de-formalized by keeping them in water for about 15 minutes and then examined.

In addition to it, feeding ecology of Little Brown Doves with reference to community ecology was also studied by analysing their intra and interspecific foraging associations. They were considered to be associated with other birds if the former were within 2-3 m of foraging birds while studying the interspecific interactions. In addition to it, intra specific interactions were also studied by recording the interactions among the members of the flocks.

Results and Discussion

Feeding Behaviour of Little Brown Doves

Little Brown Dove was noticed to be a permanent resident in the study area throughout the year feeding either singly or in pairs and frequently in uncultivated grassy fields to consume large quantities of weed seeds. Although, in breeding season, only one member of the pair was discerned to feed as the other one was found to devote its time to incubation. Furthermore, for most of the times, Little Brown Doves were noted foraging along the road sides pecking the wild seeds with Indian Ring Necked Doves. They were also found to search the defecated seeds from cattle dung but they were never seen to damage the standing crops during the study period. Mathew *et al.* (1980) reported the feeding behaviour of Little Brown Doves in Andhra

Pradesh and Kerala (India) and recorded them to feed in open fields, fallows and cart tracks.

Feeding Associations

During the present course of study, Little Brown Doves were spotted to feed intraspecifically with the birds like Indian Blue Rock Pigeons (*Columba livia*), Indian Ring Necked Doves (*Streptopeia decaocto decaocto*) and Indian Spotted Doves ((*Streptopelia chinensis suratensis*) whereas interspecifically, it was seen to feed with House Crow (*Corvus splendens*), House Sparrow (*Passer domesticus*), Red Wattled Lapwing (*Vanellus indicus indicus*), Bank Myna (*Acridotheres ginginianus*), Indian Myna (*Acridotheres tristis tristis*) and Rose Ringed Parakeet (*Psitaculla krameri*) as displayed in Table 14.1. Saini (1988) noted Little Brown Doves to glean the grains in close association with Indian Ring Necked Doves and Indian Blue Rock Pigeons in Punjab.

Table 14.1: List of Birds Sharing the Feeding Sites with Little Brown Doves in the Study Area

Sl.No.	Name of the Bird	Order	Family
1.	Indian Blue Rock Pigeons (*Columba livia*)	Columbiformes	Columbidae
2.	Indian Ring Necked Doves (*Streptopeia decaocto decaocto*)	Columbiformes	Columbidae
3.	Indian Spotted Doves (*Streptopelia chinensis suratensis*)	Columbiformes	Columbidae
4.	Indian House Crow (*Corvus splendens splendens*)	Passeriformes	Corvidae
5.	Red Wattled Lapwing (*Vanellus indicus indicus*)	Ciconiiformes	Charadriidae
6.	Red Vented Bulbul (*Pycnotus cafer cafer*)	Passeriformes	Pycnonotidae
7.	House Sparrow (*Passer domesticus*)	Passeriformes	Ploceidae
8.	Bank Myna (*Acridotheres ginginianus*)	Passeriformes	Sturnidae
9.	Common Myna (*Acridotheres tristis tristis*)	Passeriformes	Sturnidae
11.	Rose Ringed Parakeet (*Psitaculla krameri manillensis*)	Psicattiformes	Psittacidae

Feeding Sessions

In the summer months, Little Brown Doves initiated singing early in the morning before the sunrise and the birds gradually started congregating in the open soon after sunrise. The birds were recorded to devote their time in feeding usually in the groups ranging from 2-36 individuals till 10.00 to 11.30 hours and thereafter as the temperature was reported to increase, they were observed to take shelter in the shady trees keeping on singing during noon. In the afternoon, again active feeding was re-initiated. In winters as the days become much shorter, the birds were observed to carry on feeding throughout the day though intermittently.

Food of Adult

The dietary items preferred by Little Brown Doves are depicted in Table 14.2 in which cereals were found to compose 23.17 per cent of the total food whereas pulses, oilseeds and weed seeds represented 6.78 per cent, 11.96 per cent and 35.13 per cent respectively of the total diet intake of Little Brown Doves. The other dietary items like unidentified plant matter and animal matter (in the form of molluscan shells) were reckoned to constitute 6.95 per cent and 2 per cent of food only. Besides, grit made a considerable contribution to the food in the form of 13.02 per cent (Figure 14.1).

Table 14.2: List of Dietary Items Consumed by Little Brown Doves in the Study Area during 2009 and 2010

Category of Food Item with Per cent Contribution	Common Name of Food Item	Scientific Name	Family
Cereals (23.17 per cent)	Paddy	*Oryza sativa*	Poaceae
	Wheat	*Triticum aestivum*	Poaceae
	Lesser Millet	*Pennisetum typhoideum*	Poaceae
Pulses (6.78 per cent)	Green Gram	*Phaseolus aureus*	Fabaceae
	Black Gram	*Phaseolus mungo*	Fabaceae
Oil Seeds (11.96 per cent)	Mustard	*Brassica* sp.	Brassicaceae
	Til	*Sesamum indicum*	Pedaliaceae
Weed Seeds (35.13 per cent)	Gullidanda	*Phalaris minor*	Gramineae
	Bathu	*Chenopodium album*	Chenopodiaceae
	Jangli Palak	*Rumex* sp.	Polygonaceae
	Hulhul	*Cleome viscosa*	Cleomaceae
	Itsit	*Trianthema monogyna*	Aizoaceae
	Swank	*Echinochloa crusgalli*	Poaceae
	Salara	*Celosia argentea*	Amaranthaceae
	Metha	*Trigonella foenumgraecum*	Fabaceae
	Gha	*Poa annua*	Poaceae
	Takri Ghas	*Digitaria* sp.	Poaceae
	Unidentified weed seeds	–	–
Unidentified Vegetative Matter (6.95 per cent)	–	–	–
Animal Matter (2 per cent)	–	–	–
Grit (13.02 per cent)	–	–	–

Gut analysis (n=30) reflected that the weed seeds were preferred by Little Brown Doves over other categories of food (Table 14.2). Saini (1988) also putforth the same observations. However, Frith *et al.* (1976) reported weed seeds to form a considerable portion of the diet of *Streptopelia senegalensis* but Rana (1976) noted cereals especially pearl millet as the major food of Little Brown Doves. Among the category of weed seeds, seeds of Gullidanda, Bathu, Jangli Palak, Hulhul, Itsit, Swank, Salara, Metha, Gha, Takri Ghas and some unidentified weed seeds were eaten by Little Brown Doves. Mathew *et al.* (1980) recognised wild seeds of *Croton sparsiflorus* and *Paspalum*

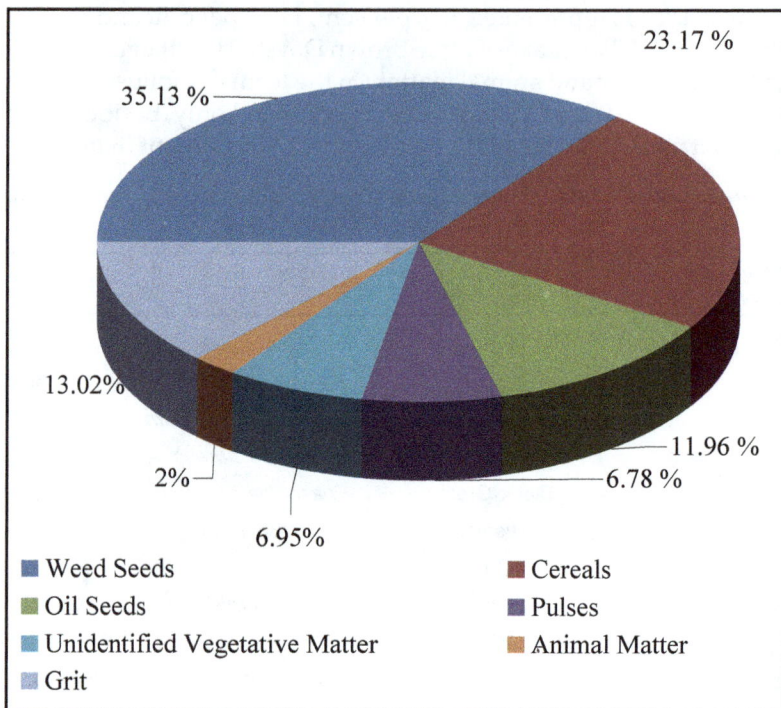

Figure 14.1: Percentage of different Dietary Components in the Food of Little Brown Doves.

scrobiculatum to be the main preferences of Little Brown Doves accompanying with cultivated grains like paddy and ragi. Rowan (1983) and Adang *et al.* (2008) reported *Brachiaria lata* as the most important food item with respect to frequency of occurrence in *S. senegalensis.*

Among the cereals which were identified to be the second most preferred food item by the Little Brown Doves, Paddy, Wheat and Bajra were consumed. Mustard seeds and Til constituted the category of oil seeds being consumed by Little Brown Doves and the least consumption was noticed in case of pulses represented by Green Grams and Black Grams. However, contrary to present results, Rana (1976) reported that Little Brown Doves fed upon considerable proportion of insect food (45.20 per cent) during breeding season. Moreover, Dean (1979) revealed predominance of seeds of commercial crops *i.e.* sunflower and maize in the diet of *Streptopelia senegalensis.* However, no such dominance of the aforementioned seeds was revealed during the study.

Seasonal Changes in Food in Case of Little Brown Doves

Besides, identifying the various food items of Little Brown Doves in the study area, the present study also encompassed the detailed examination of the seasonal variations in the relative percentage of food as presented in Table 14.3. However, no specimens could be collected in the months of December and January.

Table 14.3: Relative Percentage of different Food Items in the Guts of Little Brown Doves, *Streptopelia senegalensis senegalensis* during the study period *I.e.* June, 2009 to May, 2010

Food Items	2009							2010				
	June (n=6)	July (n=3)	Aug (n=2)	Sept (n=5)	Oct (n=3)	Nov (n=0)	Dec (n=0)	Jan (n=2)	Feb (n=2)	Mar (n=3)	Apr (n=2)	May (n=3)
CEREALS												
Paddy (*Oryza sativa*)	0.68	0.35	0.28	1.03	0.74	0	0	0.14	0.07	0	0	0.02
Wheat (*Triticum aestivum*)	0.20	0.20	0.98	0.48	0.06	0	0	0.08	0.57	0.33	1.34	0.06
Lesser Millet (*Pennisetum typhoideum*)	3.04	2.47	1.04	2.89	1.23	0	0	2.12	0.68	1.09	0.67	0.37
Total Cereals	**3.92**	**3.02**	**2.3**	**4.4**	**2.03**	**0**	**0**	**2.34**	**1.32**	**1.42**	**2.01**	**0.45**
PULSES												
Green gram (*Phaseolus aureus*)	0.12	0.01	0.09	0.24	0.37	0	0	0	0	0	0	0.25
Black Gram (*Phaseolus mungo*)	0	0	0.11	0.19	0.49	0	0	0.04	0.01	0	0	0
Crotolaria sp.	1.13	0	0	0	0	0	0	0.18	1.23	1.35	1.00	0
Total Pulses	**1.25**	**0.01**	**0.2**	**0.43**	**0.86**	**0**	**0**	**0.22**	**1.24**	**1.35**	**1**	**0.25**
OIL SEEDS												
Mustard (*Brassica* sp.)	0	0	0	0	0	0	0	2.35	1.98	1.49	0.28	0.08
Til (Sesame)	0.35	0.23	0.14	1.57	1.23	0	0	1.07	0.47	0.58	0.13	0
Total Oil seeds	**0.35**	**0.23**	**0.14**	**1.57**	**1.23**	**0**	**0**	**3.42**	**2.45**	**2.07**	**0.41**	**0.08**
UNCULTIVATED (WILD) PLANTS SEEDS OF WHEAT												
Gullidanda (*Phalaris minor*)	0	0.26	0	0	0	0	0	0	0	0.47	0.89	0
Bathu (*Chenopodium album*)	0	0	0	0	0	0	0	0	0.24	0.67	0.39	0
Jangli Palak (*Rumex* sp.)	0.16	0	0	0	0	0	0	0	0.78	1.34	0.52	0.10

Contd...

Table 14.3– Contd...

Food Items	2009							2010				
	June (n=6)	July (n=3)	Aug (n=2)	Sept (n=5)	Oct (n=3)	Nov (n=0)	Dec (n=0)	Jan (n=2)	Feb (n=2)	Mar (n=3)	Apr (n=2)	May (n=3)
UNCULTIVATED (WILD) PLANTS SEEDS OF PADDY												
Hulhul (*Cleome viscosa*)	0.46	0.67	0.86	1.01	0	0	0	0	0	0.45	0.39	0.24
Itsit (*Trianthema monogyna*)	1.38	1.77	2.46	0.94	0.43	0	0	0.34	0.39	0.22	0	0
Swank (*Echinochloa crusgalli*)	0.08	0.13	0.10	0	0	0	0	0	0	0	0	0
Salara (*Celosia argentea*)	0.41	0.36	0.23	0.07	0	0	0	0.07	0.01	0.26	0.39	0.03
Metha (*Trigonella foenumgraecum*)	0.08	0.01	0	0	0	0	0	0	0	0	0	0
Gha (*Poa annua*)	0	0	0	0	0	0	0	0	0	0.18	0.10	0.07
Takri ghas (*Digitaria* sp.)	0	0	0	0	0	0	0	0.11	0.15	0.20	0.04	0
Unidentified uncultivated plant seeds	0.35	0.57	0.79	0.19	3.06	0	0	1.35	1.59	0.34	1.00	4.97
Total uncultivated (Wild) Plants seeds	**2.92**	**3.77**	**4.44**	**2.21**	**3.49**	**0**	**0**	**1.87**	**3.16**	**4.13**	**3.72**	**5.41**
Unidentified Vegetative Matter	0.52	0.38	1.56	0.26	0.85	0	0	0.58	0.73	0.22	0.50	1.33
Anima Matter	0.10	0.46	0.32	0.10	0.08	0	0	0.03	0.09	0.17	0.25	0.39
Grit	0.81	2.08	1.06	0.88	1.44	0	0	1.30	0.86	0.55	1.99	2.05

It is evident from Table 14.3 that weed seeds form the predominant food type throughout the year. Unidentified weed seeds were the most abundant in the months of May (4.97 per cent), followed by October (3.06 per cent), February (1.59 per cent), January (1.35 per cent) and April (1 per cent). However, the other months of June to September and March reflected less than 1 per cent of contribution to food *i.e.* 0.35 per cent, 0.57 per cent, 0.79 per cent, 0.19 per cent and 0.34 per cent respectively. In the month of August, seeds of Itsit were consumed in maximum quantity *i.e.* 2.46 per cent of the food which was followed by the month of July and June each contributing to 1.77 per cent and 1.38 per cent of food respectively. However, the months of September, October, January, February and March represented 0.94 per cent, 0.43 per cent, 0.34 per cent, 0.39 per cent and 0.22 per cent of contribution respectively.

In addition to it, Hulhul registered its considerable presence in the guts in the months of September, August and July, thereby composing 1.01 per cent, 0.86 per cent and 0.67 per cent of total food respectively. Moreover, least consumption was reported in the months of June (0.46 per cent), March (0.45 per cent), April (0.39 per cent) and May (0.24 per cent). Salara was also noted in the guts throughout the year except the months of October, November and December. The maximum consumption was found in the months of June (0.41 per cent) followed by April (0.39 per cent), July (0.36 per cent), March (0.265) and August (0.23 per cent). Although, in the months of September, January and February the contribution was less than 0.10 per cent.

In the course of study, seeds of Jangli Palak were reported to be present in the guts in the months of June (0.16 per cent), February (0.78 per cent), March (1.34 per cent), April (0.52 per cent) and May (0.10 per cent). On the other hand, Gullidanda marked its presence only in the months of July (0.26 per cent), March (0.47 per cent) and April (0.89 per cent). Similarly, seeds of Bathu were also discerned only in the three months in a year *i.e.* February to March, rendering a representation of 0.24 per cent, 0.67 per cent and 0.39 per cent of the total food respectively. Also, the months of January to April witnessed the presence of seeds of Takri Ghas which accounted for 0.11 per cent, 0.15 per cent, 0.20 per cent and 0.04 per cent of the total food respectively. Seeds of Gha, however, were eaten maximally in the months of March (0.18 per cent), followed by April (0.10 per cent) and May (0.07 per cent). Swank was taken as food only in the months of June to August, thereby reflecting a percentage of 0.08 per cent, 0.13 per cent and 0.10 per cent respectively whereas Metha was observed to contribute 0.08 per cent and 0.01 per cent in the months of June and July respectively (Figure 14.3).

As far as the seasonal changes in the consumption of cereals were concerned, it is clear from Table 14.3 that Lesser Millet (Bajra) was consumed almost throughout the year. About 3.04 per cent of Lesser Millet was eaten by Little Brown Doves in June which is the maximum quantity followed by the months of September (2.89 per cent), July (2.47 per cent), January (2.12 per cent), October (1.23 per cent), March (1.09 per cent) and August (1.04 per cent). The months of February, April and May recorded a proportion of >1 per cent. Besides, wheat represented the food maximally in the months of April (1.34 per cent), August (0.98 per cent) and February (0.97 per cent). The other months reflected the contribution of wheat in >0.50 per cent proportion. Paddy also constituted 0.68 per cent of food in the month of June, 0.35 per cent in July,

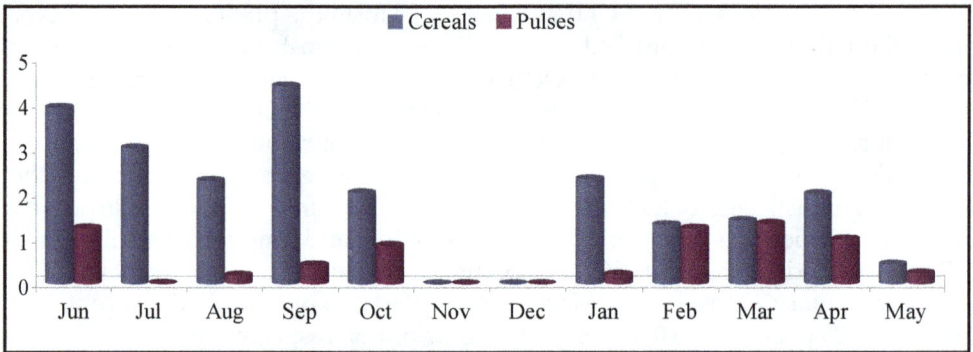

Figure 14.2: Graphical Representation of Monthly Contribution of Cereals and Pulses in the Total Diet of Little Brown Doves.

0.28 per cent in August, 1.03 per cent in September, 0.74 per cent in October, 0.14 per cent in January, 0.07 per cent in February and a meagre proportion of 0.02 per cent in May (Figure 14.2).

Oil seeds were discerned to be preferred as food by Little Brown Doves after cereals and in oil seeds, Til seeds were reported to be consumed by Little Brown Doves in the months of June to October and January to March in an amount of 0.35 per cent, 0.23 per cent, 0.14 per cent, 1.57 per cent, 1.23 per cent, 1.07 per cent, 0.47 per cent, 0.58 per cent and 0.13 per cent respectively. Moreover, the months of January to May displayed the presence of Mustard seeds in the guts of Little Brown Doves which further represented 2.35 per cent, 1.98 per cent, 1.49 per cent, 0.28 per cent and 0.08 per cent of total food respectively (Figure 14.3).

In the category of pulses, October (0.37 per cent) was detected as the month depicting the highest contribution of Green grams and July (0.01 per cent) was the month in which Green Grams contributed minimally to the total food. In the other months, the range of the percentage was noticed to be 0.09 per cent -0.25 per cent. Besides, Black Grams contributed 0.11 per cent in the month of August, 0.19 per cent in September, 0.49 per cent in October, 0.04 per cent in January and 0.01 per cent in February (Figure 14.2).

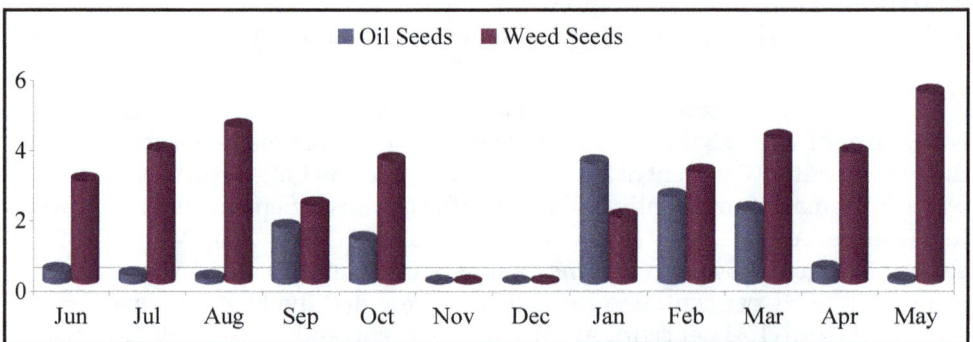

Figure 14.3: Graphical Representation of Monthly Contribution of Oil Seeds and Weed Seeds in the Total Diet of Little Brown Doves.

Unidentified vegetative matter, animal matter and grit reflected the same trend of the distribution in the different months of the year, although the percentage differed among these food items. August and May were discerned to be the months when the unidentified vegetative matter made its appearance in the maximum amounts in the guts of Little Brown Doves *i.e.* 1.56 per cent and 1.33 per cent respectively. 0.85 per cent, 0.73 per cent, 0.58 per cent, 0.52 per cent and 0.50 per cent of the unidentified vegetative matter was contributed in the months of October, February, January, June and April respectively. In the other months (whenever it was recorded), it reflected >0.50 per cent of contribution (Figure 14.4). Animal matter was recorded during June to October and then again from January to May. In the month of July, highest proportion of animal matter was consumed *i.e.* 0.46 per cent and in the month of January, lowest contribution of 0.03 per cent was deduced. The range varied from 0.08 per cent -0.39 per cent in the remaining months (Figure 14.4). On the other hand grit was noted in maximum amount in the months of July (2.08 per cent) and May (2.05 per cent). The month of March emerged out as the period of low consumption of grit in the form of only 0.55 per cent. The months of June, August, September, October, January, February and April represented 0.81 per cent, 1.06 per cent, 0.88 per cent, 1.44 per cent, 1.30 per cent, 0.86 per cent and 1.99 per cent respectively (Figure 14.5).

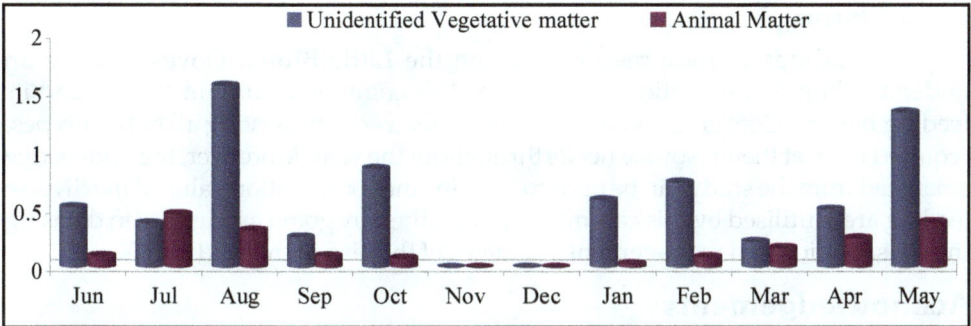

Figure 14.4: Graphical Representation of Monthly Contribution of Unidentified Vegetative Matter and Animal Matter in the Total Diet of Little Brown Doves.

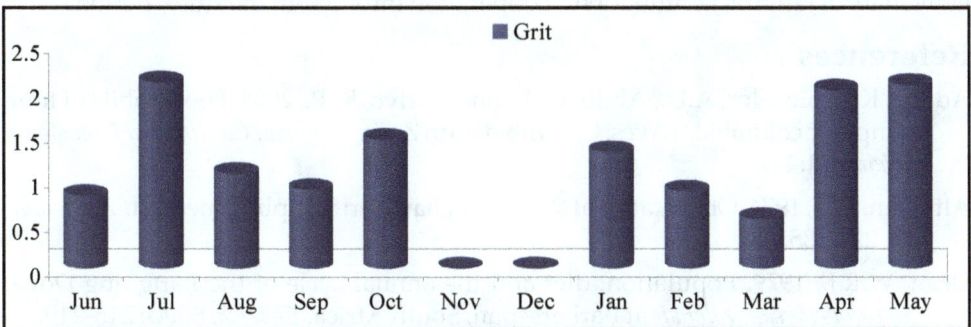

Figure 14.5: Graphical Representation of Monthly Contribution of Grit in the Total Diet of Little Brown Doves.

Thus, from the foregoing account, it is clear that weed seeds were the most preferred food item followed by cereals and oil seeds whereas Pulses were observed to be the least preferred food item by Little Brown Doves. Thus, the order of preference of the major food categories in case of Little Brown Doves can be enumerated as under:

Weed seeds > Cereals > Oil Seeds > Pulses.

Economic Status

Little Brown Doves can be regarded as exclusively grainivores birds. The molluscan shells represented the only animal component forming less than 0.50 per cent of the food. Weed seeds, cereals, pulses and oil seeds formed the exclusive diet of Little Brown Doves. During the study period, Little Brown Doves were not found to damage either the standing crops or sown crops in the agricultural fields but were detected to mainly glean the fallen grains. Additionally, these doves were viewed to consume significant amounts of weed seeds. Though major proportion of these weed seeds was found crushed in the gizzards, though chances of escaping digestion also prevail. Therefore, the role of Little Brown Doves as distributors of weeds requires experimental testing. The present study documents that in relation to agriculture, Little Brown Doves have neutral status.

Conclusion

The findings of the present study on the Little Brown Doves provide an understanding of the relationship between this common columbid species and its feeding habitat. Conservation of a species depends on preserving all habitat types, required to meet their resource needs throughout the year. Moreover, the knowledge emanated from the study can be used to redefine the conservation value of the diverse feeding areas utilised by this columbid species, thereby going a long way in devising the conservation and management strategies of the bird in the study area.

Acknowledgements

The authors would like to thank Department of Science and Technology, Govt. of India for rendering the financial assistance in the form of INSPIRE Fellowship to carry out the study. Besides, thanks are also due to Head, Department of Zoology, University of Jammu, Jammu for his cooperation throughout the study period.

References

Adang, K. L., Ezealor, A.U., Abdu, P. A., and Yoriyo, K. P., 2008. Food habits of four sympatric columbids (Aves: Columbidae) in Zaria, Nigeria. *Continental J. Biological Sciences,* **1**: 1-9.

Altmann, J. C., 1974. Observational study of behaviour: Sampling method. *Behaviour,* **49**: 227-285.

Dean, W.R.J., 1979. Population, diet and the annual cycle of the Laughing Dove, *Strpetopelia senegalensis* at Barbersspan, South Africa. *Ostrich,* **50(4)**: 215-219.

French, A. R., and Smith, T. B., 2005. Importance of body size in determining dominance hierarchies among diverse tropical frugivores. *Biotropica,* **37**: 96-101.

Goodwin, D., 1983. Pigeons and Doves of the World.3 1 edition, Ithaca, Cornell University Press and British Museum (Natural History), pp.363.

Lack, D., 1968. Ecological Adaptations for Breeding in Birds. Methuen, London.

Martin, T.E., 1987. Food as a limit on breeding birds: a life history perspective. *Ann Rev Ecol. Syst.*, **18**:453–487

Mathew, D. N., Narendran, T. C., and Zacharias, V. J., 1980. A comparative study of the feeding habits of the certain species of Indian birds affecting agriculture. *Journal of Bombay Natural History Society*, **75**: 1178-1197.

Moegenburg, S. M., and Levey, D. J., 2003. Do frugivores respond to fruit harvest? An experimental study of short - term responses. *Ecology*, **84**: 2600-2612.

Rana, B.D., 1976. Observations on the food of the Indian Ring Dove, *Streptopelia decaocto* and Little Brown Dove, *Streptopelia senegalensis*. *Z. Angew Zool.*, **63(1)**: 25-30.

Rowan, M. K., 1983. The Doves, Parrots, Louries and Cuckoos of Southern Africa, *Academic Press, London*, pp. 429.

Saini, H.K., 1988. *Feeding ecology and population dynamics of doves and pigeons of Punjab*. Ph. D. Thesis. College of Basic Sciences and Humanities, Punjab Agricultural University, Ludhiana.

2015, Animal Diversity, Natural History and Conservation, Vol. 5   Pages 227–250
Editors: V.K. Gupta and Anil K. Verma
Published by: DAYA PUBLISHING HOUSE, NEW DELHI

Chapter 15

Conservation Land: An Integrated Conceptual Framework Towards *in situ* Biodiversity Conservation Programmes in India

Soumyajit Chowdhury[1,2]
[1]Department of Zoology,
V.J.R. College, Kolkata – 700 032, West Bengal, India
[2]Centre for Biodiversity and Ecological Studies,
Kolkata – 700084, West Bengal, India

ABSTRACT

Apart from the traditionally maintained forests and other ecosystems by local communities, biodiversity conservation at ecosystem level in India has taken many different shapes for the past few decades. Recognized as a megadiversity country with four biodiversity hot spots, India with its immense bioresources still lacks a proper network and classification of such *in situ* conservation programmes. In the present text, a concept termed "Conservation Land' is proposed by the author, which incorporates all of the major (with some minor) *in situ* conservation approaches taken in the country on a national or international basis. The umbrella term 'conservation land-use' deals with a flexible and simple classification based on 'ecosystem-approach' that may identify potential 'gap areas' in the management programmes of the country; helping further the policy and decision makers as well as conservation biologists of the country towards better management and conservation of the threatened habitats and bioresources of India.

Keywords: Biodiversity, Conservation land, Ecosystem approach, in situ Conservation, India.

---

* *Corresponding author.* E-mail: wildlifesc@gmail.com

Introduction

In the present era, wild habitats and ecosystems are being disrupted to varying extremes by multifarious anthropogenic activities at the global scale. Conservationists throughout the globe are much concerned about the threatening situations prevailing in different terrestrial and aquatic biomes, that are increasingly leading towards the change in climatic conditions as well as depletion in natural resources. Alarmed by these threats, several conservation strategies and programmes have been adopted in different countries of the world to sustain the natural resources and green environment.

India, being recognized as a country of megadiversity for biological resources, has been in the process of conserving the same for decades through a variety of *in situ* management programmes. With a view to keep the natural ecosystems and other landscapes of natural importance unaltered, such *in situ* management programmes are either run by the Government or non-Government initiatives in different parts of the country. However, much of the general people and even amateurs in different sectors but related to wildlife and biodiversity conservation, are aware of such implemented programmes as well as the way they interact and act in our country. Understanding the gap in knowledge regarding the network of *in situ* conservation programmes and the importance of management at the landscape level in India, the author of the present text has proposed a framework under the concept of "Conservation Land". The latter may help in better understanding of the different *in situ* conservation programmes running in India and also a way of integrating them towards better protection and maintenance of the different ecosystems of the country.

1. Understanding the Levels of Biodiversity

The term 'biodiversity' is formed by a contraction of the term *biological diversity*. The term 'biodiversity' was coined by Walter G. Rosen in 1985 (UNEP, 1995), which is commonly used to describe the number, variety and variability of living organisms. 'Biological diversity' can be defined as the variability among living organisms from all sources including *inter alia*, terrestrial, marine and other aquatic ecosystems and the ecological complexes of which they are part; this includes diversity within species, between species and of ecosystems (The Convention on Biological Diversity, 2013a).

At broader scale, biodiversity includes three fundamentally different but closely and hierarchically related levels of biological organization. These are genetic-, species- and ecosystem-level diversities.

a) Genetic Diversity

It refers to the variation of genes within and between populations of organisms. The levels of genetic diversity range from the variations in the sequence of four base-pairs in the nucleic acids to the amount of DNA per cell and chromosome structure and number. Mutations at the genetic and chromosomal level gives rise to new genetic variations in individuals, and can be spread through the population by recombination in organisms practicing sexual reproduction. The significance of genetic variation is thus to enable both natural evolutionary change and artificial selective breeding to occur (Krishnamurthy, 1998).

b) Species Diversity

As the living world is mostly considered in terms of species, biodiversity is very commonly used as a synonym of *species diversity*. It refers to the variety of species within a region. Such diversity could be measured on the basis of two components: (i) Species richness, which describes the number of species inhabiting any point in space and time, and (ii) Species evenness, which takes into account the relative abundance of the species within the community and degrees by which a few of the species may be numerically dominant. Species diversity is significant as the species are the primary focus of evolutionary mechanisms, and the origination and extinction of species are the principal agents in governing biological diversity in most senses.

c) Ecosystem Diversity

In an ecosystem, there may exist different landforms, each of which supports different and specific vegetation. Ecosystem diversity in contrast to genetic and species diversity is difficult to measure since the boundaries of the communities which constitute the various sub-ecosystems are elusive. Ecosystem diversity could be best understood if one studies the communities in various ecological niches within the given ecosystem; each community is associated with definite species complexes. These complexes are related to composition and structure of the biodiversity.

2. Why the 'Ecosystem Approach' has Become a Central Strategic Plan Towards Biodiversity Conservation?

Ecosystem can be defined as an assemblage of biotic communities and their environment. According to the Convention on Biological Diversity (CBD), ecosystem can be defined as a dynamic complex of plant, animal and microorganism communities and their non-living environment interacting as a functional unit. An ecosystem shows its dynamic dimensions ranging from the biosphere to the biomes, thereby forming a distinct element in biodiversity and ecological studies throughout the globe (Convention on Biological Diversity, 2013a). Ecosystems are self-organized with their complex food-webs and biogeochemical cycles, thereby operating as unique functional units in nature. Maintaining prominent spatial boundaries, ecosystems in most cases however, exchange materials, energy and even biotic forms with adjacent ecosystems (Jorgensen, 2008) – accordingly resulting in a mosaic of interconnected systems within a broader landscape (Chowdhury, 2013a). *Landscape diversity* has been considered as the fourth form of biodiversity by several workers (Wilson, 1988; Szaro and Shapiro, 1990). It is also designated by Scheiner (1992) as the *Pattern Diversity* because of the pattern contributed to the landscape by repeated habitat components.

Emerging earlier as a central principle in the implementation of the Convention on Biological Diversity, the concept of '*ecosystem approach*' was adopted as the primary framework for action under the Convention on November, 1999 at Jakarta (2nd meeting of CBD). The concept was subsequently referred to in the elaboration and implementation of the various thematic and cross-cutting programmes of work, and in the guidelines that were developed as part of these programmes of work. Furthermore, the ecosystem approach is incorporated in the goals and activities of each of the Convention's current work programmes, and is also playing the central role in the Strategic Plan of the Convention (Secretariat of the Convention on Biological Diversity, 2004).

Although being considered as a central concept, 'ecosystem approach' has always been found difficult to define in a simple manner. The term, however, refers to the diversity at the ecosystem (or habitat) level. When compared to *genetic diversity* (within-species diversity) and *species diversity* (between-species diversity), *ecosystem diversity* places itself on a broader perspective in the hierarchy by explaining the habitat diversities, ecosystem types and ecological processes occurring within each type (Krishnamurthy, 1998). According to CBD, the ecosystem approach is a strategy for the integrated management of land, water and living resources that promotes conservation and sustainable use in an equitable way.

At a global or regional scale, broadly similar communities like forests, grasslands, wetlands, oceans and others are assembled together to generate definite management strategies. By utilizing its dynamic character, the concept of ecosystem diversity can be variably applied at different scales for adopting such strategies. Biodiversity studies at ecosystem level moreover aims to emphasize easier recording and monitoring of the change, trend and effect of natural and human activities on ecosystems than on individual populations of species (Krishnamurthy, 1998). In other words, different species in an ecosystem survive by interacting with other organisms and their environment and in turn influenced by essential ecosystem functions. Ecosystem approach therefore recognizes the conservation of diversity at species and genetic levels too. Furthermore, the approach also recognizes humans, along with their cultural diversity, as an integral part of ecosystem (Secretariat of the Convention on Biological Diversity, 2004).

3. Applying Ecosystem Approach in Biodiversity Conservation

With increased anthropocentric demands, the present era illustrates rapid habitat degradation and environmental pollution with consequent threats to biological diversities at global and regional scales. Loss of genetic, species or ecosystem diversity may disrupt the natural processes that are vital for maintaining life-supporting environment on Earth (Krishnamurthy, 1998). The Convention on Biological Diversity ponders to achieve three primary objectives so as to maintain the biodiversity resources on Earth, *viz.* (i) Conservation of biological diversity, (ii) Sustainable use of biological diversity and (iii) Equitable sharing of benefits generated from the utilization of genetic resources. The application of ecosystem approach will help to maintain a balance of the three objectives of the Convention.

'Ecosystem-based' approaches emphasize on strategies involving conservation of habitat or ecosystem diversities and in turn the species associated with them and their genetic diversities, thereby providing significant information on ecosystem services. This mode of conservation aims to maintain representative areas of ecosystems or important habitat sites through (a) a network of protected areas or (b) control on land use (Jorgensen, 2008). The present-day concept of conservation at the landscape level furthermore strengthens the ecosystem-approach (Chowdhury, 2013a).

According to CBD, an ecosystem approach is based on the application of appropriate scientific methodologies that are focused on different levels of biological organization; the latter encompass the essential structure, processes, functions and interactions among organisms and their environment. The term 'ecosystem', as defined by CBD refers to any functional unit at any scale and stressing on the determination of the scale of analysis and action by the problem being addressed. Thus the focus of ecosystem approach on the structure, processes, functions and interactions is consistent with the definition and explanation of ecosystem by the CBD. However, in order to respond to uncertainties in ecosystem processes, ecosystem approach must be adaptive. Besides, this mode of approach does not preclude other management and conservation practices (*viz.* biosphere reserves, protected areas, single-species conservation programmes, etc) as well as other approaches carried out under existing national policies and legislative frameworks. Rather ecosystem approach integrates all these practices and strategies to deal with complex situations. The need of the moment is to facilitate the implementation of the ecosystem approach as the primary framework for balancing the three objectives of the Convention. (Secretariat of the Convention on Biological Diversity, 2004)

4. Biodiversity Conservation in India

4.1. Biogeography and its Zonation in India

Because of its dual origin, partly from Eurasia and partly from the Gondwana land, India shows diverse climatic and topographic conditions resulting in varied ecosystems and habitats like forests, grasslands, deserts, wetlands and marine ecosystems. The country has a *moist tropical – temperate – tundra series* at its northern end, while at the other end there is a *dry tropical – grassland – shrubland – desert series*. The tundra biome type found here, however, is not the true tundra, but an alpine tundra originating due to the great heights reached by the Himalayas during the Pleistocene epoch. The Indian desert is of recent origin and is a secondary formation due to the shift in the direction of monsoon. The grasslands, similarly, are also secondary and either biotic or edaphic in origin (Mukherjee, 1996).

Within India, 10 *biogeographic zones* (with 26 *biotic provinces*) have been recognized according to the classification proposed by Rodgers and Panwar (1988) (Table 15.1). This scheme gives a more detailed view of the different ecosystem conditions and their spatial distributions in the country (Figure 15.1), compared to the earlier schemes proposed by Udvardy (1975) and MacKinnon (1997) that deal at broader scales with the World and south-east Asia respectively.

4.2. Biodiversity Conservation Approaches in India

A *conservation programme* can be defined as the planned management of natural resources so as to retain the natural balance, diversity and evolutionary change in the environment (Lincoln *et al.,* 1982). Conservation programmes are generally carried out as *in situ* and *ex situ* policies. The *in-situ* conservation stresses on conserving the ecosystem and the different habitat types within, without modifying the environment for the flora and fauna. According to the Convention on Biological Diversity, such an approach remains engaged in establishing 'a system of protected areas or areas

Table 15.1: Biogeographic Zones of India with Biotic Provinces (after Rodgers and Panwar, 1988)

Sl.No.	Biogeographic Zones	Biotic Provinces
1.	Trans-Himalayan	Ladakh
2.	Himalayan	North West Himalaya, West Himalaya, Central Himalaya, East Himalaya
3.	Desert	Kutch, Thar
4.	Semi-Arid	Punjab, Gujarat-Rajwara
5.	Western Ghats	Malabar Coast, Western Ghat Mountain
6.	Deccan Peninsula	Deccan Plateau South, Cenral Plateau North, Eastern Plateau, Chota-Nagpur, Central Highlands
7.	Gangetic Plain	Upper gangetic Plain, Lower Gangetic Plain
8.	North-East India	Brahmaputra Valley, Assam Hills
9.	Islands	Andaman Islands, Nicobar Islands, Lakshadweep Islands
10.	Coasts	West Coast, East Coast

where special measures are need to be taken to conserve biological diversity and develop, where necessary, guidelines for the selection, establishment and management of such areas' (Convention on Biological Diversity, 2013b). The *ex-situ* conservation is employed primarily for complementing *in-situ* measures, and focuses on conserving target species of flora and fauna (and their body representatives like gametes) outside their natural environments and preferably in the country of origin of such components in several ways (*viz.* Botanic and Zoological Gardens, Sperm and Ovum Banks, Seed Banks, etc.); and including measures to recover and rehabilitate threatened species and to reintroduce them as well into their natural environment under appropriate conditions (Convention on Biological Diversity, 2013c, Krishnamurthy, 1998). Both these strategies are well applied to Indian scenario to various extents and forms aiming towards (i) prevention of the loss of genetic diversity, (ii) recovering from species loss and prevention of species extinction and (iii) promoting sustainable utilization of natural resources by protecting an ecosystem from existing or prospective threats. The modern concept of conservation thus is briefed as the wise maintenance and utilization of the earth's resources.

4.3. Ecosystem-Based Conservation Approaches in India: The *in situ* Programmes

The ecosystems in India are degraded to different degrees primarily because of (i) deforestation, (ii) overgrazing, (iii) agriculture, (iv) mining, and (v) urbanisation. Each of these sources is an aggregate of further finer causes that in turn lead to a huge array of factors that have to be mitigated, as far as practicable, so as to conserve the ecosystems and their related services and hence reducing the deterioration in environmental quality (Chowdhury, 2013a)

India is one of the 17 megadiversity countries with four biodiversity hot-spots (Table 15.2). The hotspot areas, along with other areas of biodiversity importance are

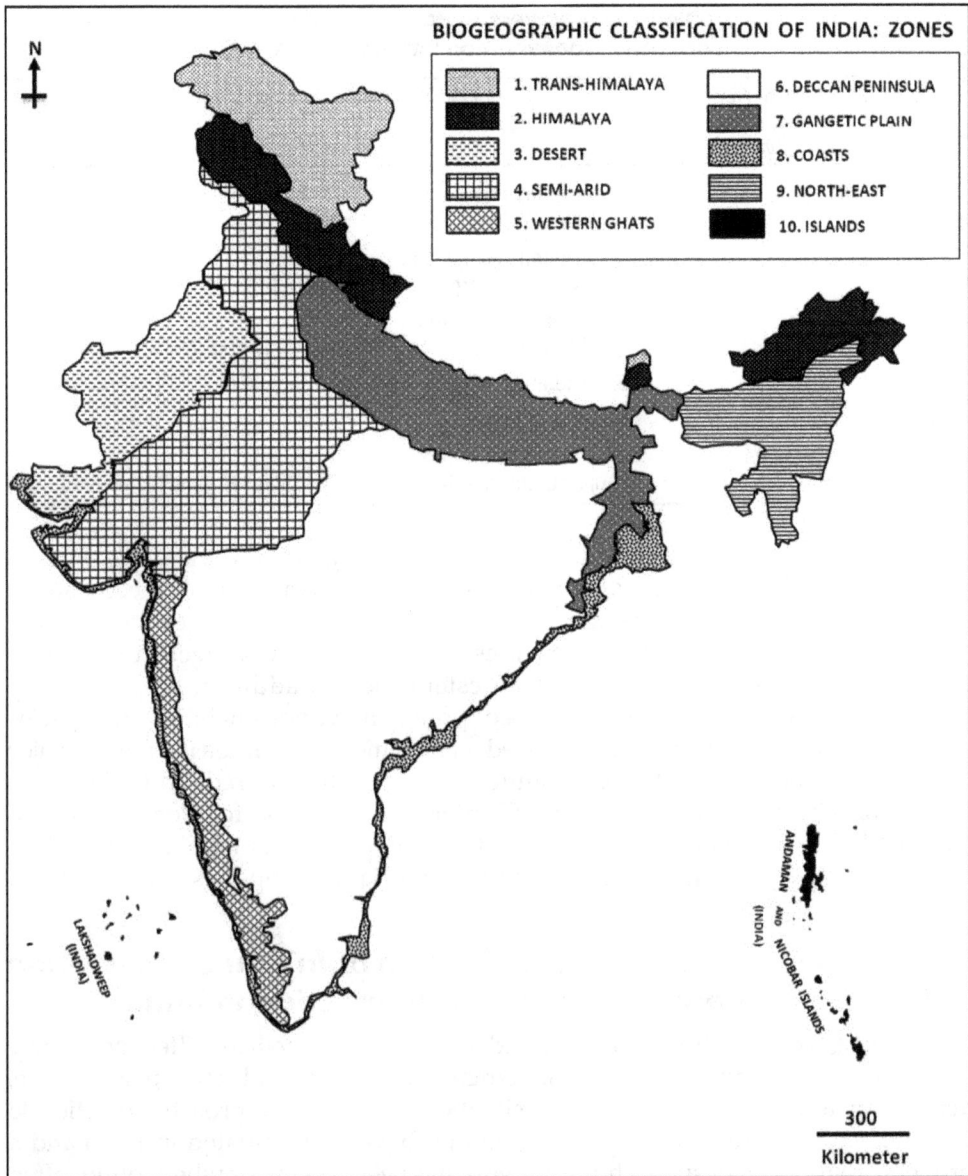

Figure 15.1: Biogeographic Zones of India (after Rodgers, Panwar and Mathur, 2000).

facing tremendous biotic impact with fast dwindling of their resources. Several wild species of plants and animals are threatened. In conjunction with the traditional approaches (like maintenance of community-conserved areas, sacred groves, etc.), several modern approaches have been initiated in India, especially in the form of conservation of ecoregions through enactment of comprehensive legislation and initiation of action programmes (Chowdhury, 2013a).

**Table 15.2: Biodiversity Hot Spots in India
(according to Conservation International, 2013)**

Sl.No.	Biodiversity Hot Spot Category (Asia-Pacific Region)	Hot Spots in India and their Distribution
1.	Himalaya	a. Eastern Himalayas [West Bengal, Sikkim, Assam, Arunachal Pradesh]
		b. Western Himalayas [Kumapn-Garhwal, Northwest Kashmir]
2.	Indo-Burma	a. North-Eastern India [Meghalaya, Assam, Tripura, Nagaland, Mizoram,Manipur]
		b. Andaman Islands
3.	Sundaland	Nicobar Islands
4.	Western Ghats and Sri Lanka	Western Ghats [Gujarat, Maharashtra, Goa, Karnataka, Kerala, Tamil Nadu]

Through various ecosystem-based conservation programmes, *in situ* conservation strategies in India have been planned for sustainable management of ecosystems and natural habitats and populations of species existing therein. For the past few decades several National Parks, Sanctuaries, Biosphere Reserves, Tiger and Elephant Reserves, Ramsar sites, etc. have been established in addition to community conservation using traditional practices of protecting various habitats and species therein. Moreover, for achieving improved management skills India is a part to major international conventions like Convention on International Trade in Endangered Species of wild fauna and flora (CITES) International Union for Conservation of Nature (IUCN), International Convention for the Regulation of Whaling, UNESCO-World Heritage Committee and Convention on Migratory Species (CMS) (MOEF, 2013a).

5. The Need of a Proper Classification of *In Situ* Conservation Programmes and their Mode of Interaction in India

In India, biodiversity conservation and ecosystem management differ very widely often within the country particularly in terms of their criteria and action plans. Hence, depending upon the needs of a specific situation, no uniform approach is practicable (Basu *et al.*, 2006). Several *in-situ* programmes have been initiated in India and a proper classification of such a huge network has been lacking or when found, often confusing particularly in relation to the ways they are implemented in the country.

As for example, apart from the Protected Areas (like Sanctuaries, National Parks, Conservation and Community Reserves), the ENVIS Centre on Wildlife and Protected Areas has brought different conservation programmes in India (like Biosphere Reserves, Tiger Reserves, Elephant Reserves, Ramsar Wetland Sites and World Heritage Sites) under the term 'Conservation Area' (ENVIS Centre on Wildlife and Protected Areas, 2011a). But the term has been defined by Sarkar *et al.* (2007) as "any geographical unit at which a conservation plan for biodiversity is being implemented".

Their study has aimed towards spatial configuration of conservation areas in a conservation area network in India, as has been previously used by conservation planners in many countries like Australia, Canada, Papua New Guinea and South Africa (Sarkar *et al.,* 2007). The same term has also been found to be used by the UNESCO World Heritage Centre to indicate a broad natural landscape with a definite geographic attribute like the proposed Bhitarkanika Conservation Area in Orissa (UNESCO World Heritage Centre, 1992-2014). Hence the use of such a term for integrating different *in situ* conservation programmes in the country seems confusing and unsuitable as well.

A proper classification through an integrated network of such programmes in India is the need of the hour. This may further illustrate the ways of action implemented in the country and also the prospective approaches to be initiated in the future.

6. Conservation Land: A Concept Proposed Towards Integration and Classification of *In Situ* Biodiversity Conservation Programmes in India

Under the present scenario of rapid habitat disruptions in India, a growing realization is that managing the different *in-situ* conservation systems separately may not prove effective in conserving biodiversity in the long run. However, as mentioned earlier, a term which can contain all such *in situ* conservation programmes in the country as an umbrella concept has been lacking. The term 'Conservation Land' is thus proposed keeping in mind of the central theme of ecosystem approach. The term in turn allows integrated ecosystems approach for networking the different *in-situ* programmes in India under the umbrella term '*Conservation Land-use*'. Such an approach may further encourage towards strengthening long-term conservation of bioresources and governance issues in India by way of applying the ecosystem approach (Chowdhury, 2013b).

A classification of 'conservation land' has also been proposed to incorporate the different land-use programmes used by major ecosystem-based *in situ* conservation approaches operating in the country (Figure 15.2). The umbrella class 'Conservation Lands-Use' is broadly categorized into *Ecosystem based management* and *Species specific management* at *Level 1 of classification,* based on the primary target of conservation and management. The *Level 2* categories are further grouped more or less in terms of the type of conservation programmes that run in the country as well as the system of governance in managing such ecosystems, illustrated in turn at *Level 3 of classification.* Unlike Level 2 categories like 'Protected Area Network' and 'Biosphere Reserve' that are declared by the Government of India, other categories at this level like 'Wetland', 'Community-based Conservation Area' and 'Special Area' are introduced in the present scheme as hypothetical or textual concepts so as to provide a simple understanding of the multifaceted *in-situ* conservation programmes launched in India under National and International initiatives. 'Special Area' includes some miscellaneous conservation lands of significance, which are not found to fit to the programmes under the other categories. The types of conservation lands legally protected under different Indian laws are indicated in the classification scheme. Other categories may also have legal status but at individual levels, which are outside

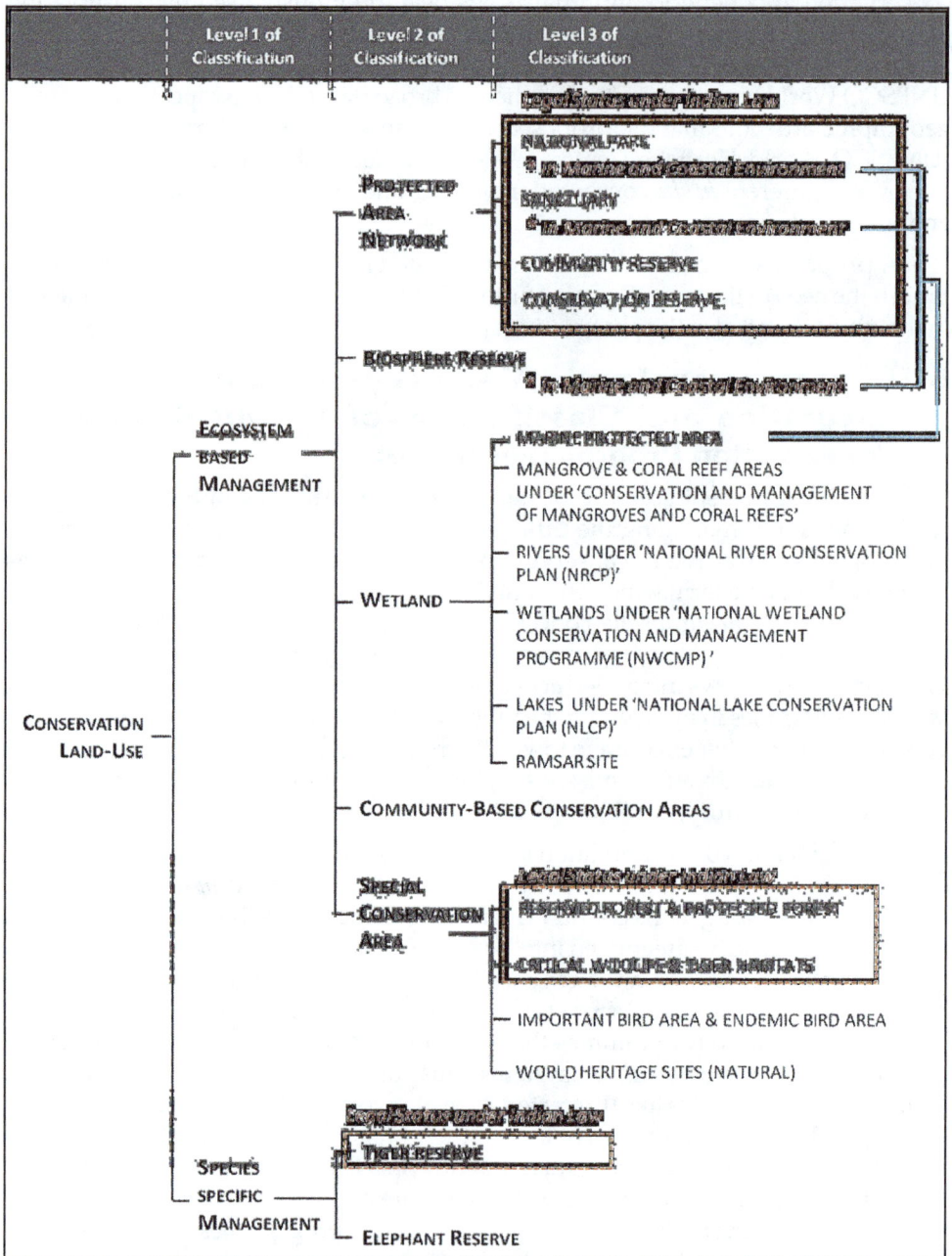

Figure 15.2: The Scheme of Classification of 'Conservation Land-Use' Concept for the *in situ* Conservation Programmes in India.

the scope of the present context and hence not indicated in the present classification (Figure 15.2).

7. Explanation of the Categories Under Conservation Land-Use Classification Scheme

7.1. Ecosystem-Based Management

This category includes all such *in situ* programmes that focus on conserving the ecosystem (or the landscape) as a whole, and not highlighting a definite species.

7.1.1. Protected Area Network

Protected Areas (PAs) are areas of natural habitats or ecosystems that are protected under *in situ* conservation systems (Krishnamurthy, 1998). IUCN defines a protected area as 'a clearly defined geographical space, recognized, dedicated and managed, through legal or other effective means, to achieve the long-term conservation of nature with associated ecosystem services and cultural values' (Dudley, 2008). In brief, CBD defines PA as 'a geographically defined area which is designated or regulated and managed to achieve specific conservation objective' (Convention on Biological Diversity, 2013a). Generally aimed to conserve and nurture biological diversity, PAs also shows importance (a) in producing ecosystem services that are beneficial to local, national and global economies and (b) as a natural habitat for a significant part of world's population that depend on such areas for their livelihood (Jensen, 2013).

Not following the IUCN classification proper, India has built its own nomenclature and classification of PAs. The Protected Area Network comprises of National Parks, Sanctuaries, Community Reserves and Conservation Reserves (MOEF, 2013a). Serving distinct functions, each of these categories has a specific origin and governance pattern. PAs like National Parks and Sanctuaries were described and legalized much earlier in the Wild Life (Protection) Act, 1972. But categories like the Community and Conservation Reserves have recently been introduced and legalised under the Wild Life (Protection) Amendment Act, 2002 for broadening the concept of PAs and encouraging greater coverage and involvement of local people (Krishnan *et al.,* 2012).

The integrity of 668 PAs extending over 1, 61,221.57 sq. km (4.90 per cent of the total geographic area) of the country is secured mainly through the preparation of management plans and also through provision of support for surveillance and monitoring activities by the Ministry of Environment and Forest (MOEF), Government of India. PAs in India comprise of 102 National Parks, 515 Sanctuaries, 47 Conservation Reserves and 4 Community Reserves encompassing terrestrial, coastal and marine biomes (Basu *et al.,* 2006; MOEF, 2013a).

A. Sanctuary

A *Sanctuary* is an area of adequate ecological, faunal, floral, geomorphological, natural or zoological significance that is declared for the purpose of protecting, propagating or developing wildlife or its environment. A sanctuary is notified by the State Government and protected by the Forest Department. No alterations of the boundaries of a sanctuary shall be made except on a resolution passed by the

Legislature of the State. Certain rights of people living inside a sanctuary could be permitted, or the same could be acquired after providing adequate compensation to the person concerned. Grazing (or movement) of livestock and fishing are not prohibited within a sanctuary, but may be regulated by the Chief Wildlife Warden. Livestock when allowed to graze within the sanctuary as well as kept within five kilometers of a sanctuary may be immunized against communicable diseases, if prescribed by the Chief Wildlife Warden. However, hunting, exploitation or removal of any wildlife and forest produce or its habitat is prohibited within the sanctuary (WPA, 1972; MOEF, 2013a).

B. National Park

A National Park is defined similarly to a sanctuary. Similar provisions apply but the primary difference between a sanctuary and a national park lies in the vesting of rights of people living inside. Hunting and cattle grazing along with the exploitation or removal of any wildlife or its habitat within a national park are prohibited. Moreover, only a completely unencumbered area, in which all rights have become vested in the state government, can be declared as a national park. Once established, its boundaries may not be altered except through a resolution passed by the state legislature (WPA, 1972; MOEF, 2013a).

C. Conservation Reserve

A Conservation Reserve is any area owned by the State Government, particularly adjacent the National Parks and Sanctuaries that is declared by the same for the purpose of linking one Protected Area with another. Such reserves are declared by the State Government for the purpose of protecting landscapes, seascapes, flora and fauna including their habitat, but after having consultations with the local communities. The rights of the people living inside a Conservatio Reserve are not affected (WPAA, 2002; MOEF, 2013a).

D. Community Reserve

A Community Reserve is any private or community land declared by the State Government that is not comprised within a National Park, Sanctuary or a Conservation Reserve, where an individual or a community has volunteered to conserve wildlife and its habitat. These Reserves are declared for the purpose of protecting fauna, flora and traditional or cultural conservation values and practices. The rights of the people living inside a Community Reserve are not affected (WPAA, 2002; MOEF, 2013a).

7.1.2. Biosphere Reserve

For developing relationship between man and nature towards wise use of natural resources, the Man and the Biosphere (MAB) Programme was initiated by UNESCO in the early 1970s as an Intergovernmental Scientific Programme (UNESCO, 2013). This framework recognizes the major ecoregions as well as natural and cultural landscapes of the world within 'Biosphere Reserve Network' (GOI, 2007). Ideally developed as large planning areas biosphere reserves act as integrated ecosystems. Such a reserve area contains legally protected *core zones* and well-defined *buffer and transition zones* maintaining human settlement and resource exploitation area. Such an approach aims to merge conservation of natural resources and socio-economic

Figure 15.3: Chintamoni Kar Bird Sanctuary in Kolkata, West Bengal is an Urban Green Island which Harbours a Rich Variety of Bird and Insect Fauna, with Numerous Native Floral Species of Medicinal Importance. The area provides a stressed repository of biodiversity among the rapidly changing landscape induced by urbanization. (*Picture: Soumyajit Chowdhury*)

development with maintenance of cultural values associated with the area and its people (GOI, 2007; Rodgers and Panwar, 1988).

In India, The National Biosphere Reserve Programme was initiated in 1986. The National Man and Biosphere (MAB) Committee of India identifies and recommends potential sites for designation as BRs; while their management is the responsibility of the concerned State or Union Territory, with necessary technical expertise provided by the Central Government. Eighteen (18) BRs have been declared by the Government of India and all incorporate existing major PAs (MOEF, 2013a). Nine of the eighteen BRs are designated as part of the World Network of Biosphere Reserves under UNESCO's Man and the Biosphere Programme (UNESCO, 2013b).

7.1.3. Wetland

Wetlands are defined as lands transitional between terrestrial and aquatic ecosystems where the water table is usually at or near the surface or the land is covered by shallow water (Mitch and Gosselink, 1986). As wetlands are greatly important in maintaining a healthy environment and their ecosystem services are increasingly understood, they are recognized for protection in various ways throughout the globe. India, with its annual rainfall of over 130 cm, varied topography

Figure 15.4: Gorumara National Park in Jalpaiguri, West Bengal is One of Best Protected and Managed PAs of India. The park harbours Indian One-horned Rhinoceros in good numbers, along with Elephants, Gaur, Tiger, Leopard and numerous bird species. (*Picture: Soumyajit Chowdhury*)

and climatic regimes support diverse and unique wetland habitats (Prasad *et al.,* 2002). Separate provisions are waiting for specific legal instrument for wetland conservation in the country. The legal framework for their conservation and management is however, provided by several legislations relevant to wetland conservation like the Water (Prevention and Control of Pollution) Act, 1974 (as amended upto 1988), the Environment (Protection) Act, 1986, the Wildlife (Protection) Act, 1972 (as amended upto 2002), the Indian Fisheries Act, 1857, Municipal Solid Wastes (Management and Handling) Rule, 2000, etc. (MOEF, 2007). They are conserved and managed by different national and international conservation programmes, as classified (Figure 15.2) and explained in the following.

A. Marine Protected Area

Marine and costal biodiversity can be protected effectively throughout the globe by the declaration and establishment of Marine Protected Areas (MPAs). An MPA can be defined as any area of the intertidal or subtidal terrain, together with its overlying water and associated flora, fauna, historical and cultural features, which has been reserved by law or other effective means to protect part or all of the enclosed environment (Kelleher, 1999). Although there is no specific provision for the term "Marine Protected Area" itself under any legislation in India, such a programme includes several Protected Areas (National Parks and Sanctuaries) that fall in whole

Figure 15.5: Sundarbans, a Unique Landscape in Eastern India, is Managed and Conserved at Various Levels by different National and International Conservation Programmes. The Sundarbans Biosphere Reserve is home to the largest chunk of mangrove forest in India as well as the only mangrove-dwelling tiger population (*Panthera tigris tigris*) of the world. The latter is protected and delimited under Sundarban Tiger Reserve (STR). The BR also holds one National Park and two Wildlife Sanctuaries. Being considered ecologically important and sensitive, the mangrove ecosystem thriving here is under intensive conservation programmes of MOEF. The wetland area as a whole, being considered threatened and sensitive to anthropogenic stresses, is also managed under National Wetland Conservation Programme. The landscape has also been recognized as a Natural World Heritage Site. (*Picture: Soumyajit Chowdhury*)

or in part within swath of 500 m from the high tide line and to marine environment of the country. There are 31 MPAs in India, covering a total area of 627121 hectares (Singh, 2003; Rajagopalan, 2008).

B. Mangroves and Coral Reef Areas: Conservation and Management of Mangroves and Coral reefs in India

Marine ecosystems, especially the mangroves and coral reefs are globally considered as highly productive as well as biodiverse ecosystems. They are threatened as well due to increasing incidences of marine and coastal pollution and overexploitation of their resources. In India, mangroves and coral reefs were declared as ecologically sensitive areas under the Environment Protection Act, 1986 which prohibits their exploitation, followed by a Coastal Regulation Zone (CRZ) Notification, 1991 prohibiting developmental activities and disposal of wastes in such ecosystems. During 1986-87, MOEF initiated a scheme on conservation and management of mangroves and coral reefs. A National Committee was thus constituted, which recommended 38 mangrove areas (in 10 provinces covering 4639 sq. km) and four coral reef areas (in four provinces covering 2375 sq. km) in the country for intensive conservation (Singh, 2003; MOEF Press Release, 2008).

C. Rivers under National River Conservation Plan (NRCP)

Rivers in India are the major water sources and like the other water bodies they are threatened to various degrees from pollution caused by increasing anthropogenic activities adjoining the former. The National River Conservation Directorate, functioning under the Ministry of Environment and Forests is engaged in implementing the River [and Lake] Action Plans under the National River Conservation Plan (NRCP) [and National Lake Conservation Plan (NLCP)] by providing financial assistance to the State Governments. NRCP aims to improve the water quality of the rivers through the implementation of pollution abatement works. So far a total of 41 rivers have been covered under the programme (MOEF, 2013b).

D. Wetlands under 'National Wetland Conservation and Management Programme (NWCMP)

Wetlands offer multifarious benefits towards maintenance of bioresources and environmental health as well as to human benefit. Often being ignorant about its services, wetlands in India are mostly threatened by various biotic and abiotic pressures. MOEF in 1985-86 initiated the National Wetland Conservation Programme (NWCP) in collaboration with state governments for the conservation and wise-use of wetlands in the country. The objectives of the programme is to prevent degradation of wetlands due to encroachment, siltation, weed infestation, catchment erosion, surface run-off carrying pesticides and fertilizers from the agricultural field, and discharge of domestic sewage and effluents, which resulted in deterioration of water quality, prolific weed growth decline of biodiversity and other associated problems. **115 wetlands** have been identified by the Ministry, which require urgent conservation and management initiatives under NWCMP (MOEF, 2010).

E. Lakes under National Lake Conservation Plan (NLCP)

In India, lakes are either natural or man-made based on their origin. However, compared to the innumerable man-made lakes in different parts of the country, natural lakes are few in number; most of the latter are located in the Himalayan region and the floodplains of the Indus, Ganges and Brahmaputra rivers. Although aware of their immense ecological and anthropocentric importance, most lakes (like other wetlands) in India are in different states of degradation (MOEF, 2010). To facilitate conservation of lakes, the National Lake Conservation Programme (NLCP) was initiated by the MOEF in 2001. Considering different activities to be required for the conservation of lakes, mostly for the urban ones with greater exposure to pollution, the lakes were separately identified and brought under NLCP, and alienated from their former NWCP. The objective of the programme is to restore and conserve the urban and semi-urban lakes of the country degraded due to wastewater discharge into the lake and other unique freshwater ecosustems, through an integrated ecosystem

Figure 15.6: Gulf of Kachchh Marine National Park in Jamnagar, Gujarat is the First Marine National Park in India Declared for Conserving the Coral Reefs and Mangrove Ecosystems in the Southern Part of the Gulf of Kachchh.
(*Picture: Soumyajit Chowdhury*)

approach. A total of 61 lakes (and lake systems) in India have been brought under NLCP till 2013 (MOEF, 2013b).

F. Ramsar Site

The Ramsar Convention on Wetlands is an intergovernmental treaty signed in Ramsar, Iran in 1971. The treaty deals with conservation aspects and wise-use of inland waters and the near-shore coastal areas through a framework for national action and international cooperation. The identified wetlands under this programme are designated as Ramsar sites and included in the Ramsar List of Wetlands of International Wetlands (MOEF, 2007; MOEF, 2013b). In India, 26 sites have been designated as Ramsar sites till date (MOEF, 2013b).

7.1.4. Community-based Conservation Area

From time immemorial, local communities in different parts of the world have been engaged in managing various natural ecosystems (like forests, wetlands, etc.) and wildlife populations for significant ecological, social and economic reasons. These traditionally conserved areas by the local communities are commonly known as 'community-based conservation areas' (Future Generations, 2013; Pathak, 2009). India, likewise, have been in the process of conserving forests, wetlands, trees, birds and other animals from historic times through various traditional and ingenious

Figure 15.7: East Kolkata Wetlands in Kolkata, West Bengal is the only Ramsar Site in the World Practicing Fisheries and Agriculture Traditionally using City Sewage. This wetland is also supervised under the NWCMP. The site harbours a rich array of resident and migratory birds, some unique mammals like Marsh Mongoose as well as diverse plant and other animal groups. (*Picture: Soumyajit Chowdhury*)

ways. Such *community-based conservation areas* are green islands for several bioresources and traditional knowledge banks, situated amid rapidly changing landscape pattern of the localities. Such threatened ecosystems are presently considered as 'conservation-important areas' and managed under different categories by various levels of governance in the country (Krishnan, 2012; WNEP-WCMC. 2013).

7.1.5. Special Conservation Area

Apart from the huge network of Protected Areas, Biosphere Reserves, Wetlands, Community-based Conservation Areas and Species Reserves maintained across different Indian biomes to conserve the natural ecosystems and species therein, several small-scale national and international programmes manage many natural habitats to encourage conservation. Such diverse areas are hereby called Special Conservation Areas (Chowdhury, 2013c). They have gained legal status as National programmes under various Indian laws and in turn include the following categories.

A. Reserved Forest (RF) and Protected Forest (PF)

Apart from the variously aforementioned protected forest habitats in India, some other forest areas are accorded certain degree of protection by declaring them as Reserved Forest (RF) or Protected Forest (PF). The Indian Forest Act, 1927 introduced the terms and also explains the way of notifying these forest areas by the State Government as well as exercise of the rights related to the same.

Any forest-land or waste-land which is the property of the State Government, or over which the Government has proprietary rights, or any part or whole of the forest-produce to which the Government is entitled, may be declared a *Reserved Forest* by the State Government. Any claim to rights to pasture or to forest-produce is passed by the Forest Settlement Officer, whereby an order is issued to admit or reject the same in whole or in part.

Any other forest-land or waste-land not included in a reserved forest, but a property of the State Government, or over which the Government has proprietary rights, or any part or whole of the forest-produce to which the Government is entitled, may be declared a *Protected Forest* by the same. However, unlike the RFs, PFs may not abridge or affect any existing rights of forest communities to activities required to sustain their livelihood partially or entirely from the forest resources or products.

B. Critical Wildlife and Tiger Habitats

Critical Wildlife Habitats (CWHs) are areas of National Parks and Sanctuaries that are to be kept inviolate and shall not be subsequently diverted by the State or Central Government or any other entity for other uses. Such CWHs are defined under Section 2(b) and sub-sections (1) and (2) of Section 4 of the Forest Rights Act, 2006. Such an area will be notified by the MOEF after consultation with an Expert Committee based on scientific and objective criteria.

Critical Tiger Habitats (CTHs) are the core areas of National Parks and Sanctuaries under Tiger Reserve Programmes that are required to be kept as inviolate for the purpose of tiger conservation. Such CTHs are defined under Section 38V.4 (i) of the Wild Life (Protection) Amendment Act, 2006. Such areas will be notified by the State Government in consultation with an Expert Committee, without affecting the rights of the Scheduled Tribes or such other forest dwellers (WPAA, 2006).

C. Important Bird Area (IBA) and Endemic Bird Area (EBA)

International programmes like Important Bird Areas (IBA) and Endemic Bird Areas (EBA) by BirdLife International (BirdLife International, 2013) identify key sites of conservation and conservation priorities in the country. Many-a-times these areas find themselves as a part of larger conservation network like a National Park, Sanctuary, Tiger Reserve or a Biosphere Reserve and in the way may gain a degree of protection.

D. World Heritage Sites (Natural)

With an aim to identify, protect and preserve the cultural and natural heritage around the world that are considered to be of outstanding value to humanity, UNESCO (United Nations Educational, Scientific and Cultural Organization) encourage countries to sign the World Heritage Convention. The international treaty, called the *Convention concerning the Protection of the World Cultural and Natural Heritage* was adopted by UNESCO in 1972 (UNESCO World Heritage Centre, 2008). India is a part of this international conservation programme as a State Party for the purpose of improved management of the identified heritage sites. A total of six Natural Heritage

Sites have been inscribed on the World Heritage List from the country, namely Kaziranga National Park and Manas Wildlife Sanctuary (Assam), Keoladeo National Park (Rajasthan), Nanda Devi and Valley of Flowers National Parks (Uttaranchal), Sundarbans National Park (West Bengal) and Western Ghats (UNESCO World Heritage Centre, 1992-2014; ENVIS Centre on Wildlife and Protected Areas, 2011a).

7.2. Species-Based Management

In support of conservation of huge network of ecosystems and biomes in India at landscape/ecosystem levels through *'ecosystem-approach'*, there is biodiversity management at species level. Such *'species-based management'* strategies are centrally sponsored special flagship programmes targeted towards conservation of large mammalian species (like tigers and elephants), that are highly threatened especially due to their illicit hunting and habitat fragmentation. Such programmes have been implemented at landscape level to form special conservation areas - designated here as *'species reserves'* - that are planned for recovery of the species populations and conservation of their habitats. Such reserves in India have taken the schemes of *Tiger Reserves* and *Elephant Reserves* (Krishnan *et al.,* 2012; MOEF, 2013a).

7.2.1. Tiger Reserve

The Project Tiger was launched by the Government of India in 1973 in order to save the declining population of the threatened Bengal Tiger (*Panthera tigris tigris*) in the country. With an initial number of nine reserves declared in 1973-74, the number increased to 43 as on February, 2014 (ENVIS Centre on Wildlife and Protected Areas, 2011b). Created on a 'core-buffer' strategy, Tiger Reserves (TRs) have recently gained legal protection under Wild Life (Protection) Amendment Act, 2006 along with other previously declared Protected Areas (*viz.* National Park, Sanctuary, Conservation Reserve and Community Reserve).

7.2.2. Elephant Reserve

To protect the elephants (*Elephas maximus*) and their habitats and corridors from increasing man-elephant conflict in India, Project Elephant was launched by the Gvernment of India. The Project was launched as a Centrally Sponsored Scheme in 1992, under which financial and technical support are being provided to major elephant bearing states in the country. Till now, 26 Elephant Reserves (ERs) have been formally notified by various State Goovernments. However, consent for establishment of six more ERs has been approved by MOEF; although the concerned State Governments are yet to notify the same (MOEF, 2014).

8. Discussion

Under the present circumstances of rapid habitat disruption in India, the concept of 'conservation land' may prove effective in real-life conservation management programmes by the Government and different Non-Government agencies. Moreover, a proper land-use scheme under the proposed concept may also prove beneficial to the citizens and amateurs for understanding the present scenario of the management programmes running in the country.

The significance of the concept can thus be summarized as:

1. Simple understanding of the different *in situ* conservation strategies in India,

2. The interlinking and networking among the different strategies, and

3. Flexibility of the general classification scheme with possibilities of incorporations of newer approaches implemented or modifications and development of existing ones in the country, but without disturbing the basic structure.

The potential 'gap areas' thus can be highlighted from the proposed classification under the 'Conservation Land-Use' umbrella scheme, that in turn may help the policy and decision makers as well as conservation biologists of the country towards better management and conservation of the threatened habitats and bioresources of India.

References

Basu, R.C., Khan, R.A., and Alfred, J.R.B. (Ed). 2006. *Environmental Awareness and Wildlife Conservation*. Director, Zoological Survey of India, Kolkata, 1-302 pp. +20 plates.

BirdLife International. 2013. Country Profile: India. Available from: http://www.birdlife.org/datazone/country/India. Accessed on 5 January, 2013.

Chowdhury, S (Ed.). 2013a.　Ecosystem Diversity: Understanding biodiversity and conservation approaches at the ecosystem level, with an emphasis on Indian ecoregions. Available from: http://www.biomeindia.org/ecosystem-diversity.html. Accessed on: 20 December, 2013.

Chowdhury, S (Ed.). 2013b. Conservation Lands: A concept proposed for in situ conservation network programmes in India and its classification under the "Conservation Land-Use" scheme. Available from: http://www.biomeindia.org/conservation-lands.html. Accessed on: 20 December, 2013.

Chowdhury, S (Ed.). 2013c. Special Conservation Area: An introduction to the proposed concept and its categories in India. Available from: http://www.biomeindia.org/sc-area.html. Accessed on: 20 December, 2013.

Convention on Biological Diversity. 2013a. Article 2: Terms of Use. Available from: http://www.cbd.int/convention/articles/default.shtml?a=cbd-02. Accessed on: 20 December, 2013.

Convention on Biological Diversity. 2013b. Article 8. In-situ Conservation. Available from: http://www.cbd.int/convention/articles/default.shtml?a=cbd-08. Accessed on: 20 December, 2013.

Convention on Biological Diversity. 2013c. Article 9. Ex-situ Conservation. Available from: http://www.cbd.int/convention/articles/default.shtml?a=cbd-09.Accessed on: 20 December, 2013.

Dudley, N. (Ed.) 2008. *Guidelines for Applying Protected Area Management categories*. Gland, Switzerland: IUCN, x+86 pp.

ENVIS Centre on Wildlife and Protected Areas, 2011a. Conservation Areas of India. Available from: http://wiienvis.nic.in/Database/ConservationAreas–844.aspx. Accessed on: 20 December, 2013.

ENVIS Centre on Wildlife and Protected Areas, 2011b. Tiger Reserve Database. Available from: http://wiienvis.nic.in/Database/Tiger–Reserve_Database_7850.aspx. Accessed on: 28 February, 2014.

Future Generations. 2013. Community-based Conservation. Available from: http://www.future.org/applied-research/community-based-conservation. Accessed on: 10 Dec, 2013.

GOI (Government of India). 2007. *Protection, Development, Maintenance and Research in Biosphere Reserves in India: Guidelines and Proformae.* Ministry of Environment and Forests, New Delhi, 35 pp.

Jensen, L. (Ed) 2013. *The Millennium Development Goals Report 2013.* United Nations, New York, 68 pp.

Jorgensen, S.E. 2008. *Encyclopedia of Ecology Vol. 1 and 2.* Elsevier B.V. The Netherlands, 3839 pp.

Krishnamurthy, K.V. 1998. *An Advanced Textbook on Biodiversity: Principles and Practice.* Oxford and IBH Publishing Co. Pvt. Ltd. New Delhi, 260 pp.

Krishnan, P., Ramakrishnan, R. Saigal, S., Nagar, S. Faizi, S., Panwar, H.S., Singh, S., and Ved, N. 2012. *Conservation Across Landscapes: India's Approaches to Biodiversity Governance. United Nations Development Programme.* New Delhi, India.

Lincoln, R.J., Boxshall, G.A., and Clark, P.F. 1982. *A Dictionary of Ecology, Evolution and Systematics.* Cambridge University Press, Cambridge, 371 pp.

MacKinnon, J. 1997. *Protected Areas System Review of the Indo-Malayan Realm.* Asian Bureau for Conservation and World Conservation Monitoring Centre/World Bank Publication. Canterbury (UK).

Mitsch, W. I., and Gosselink, I.G. 1986. *Wetlands.* Van Nostrand Reinhold, New York, 539 pp.

Press Information Bureau (GOI, MOEF). 2008. Conservation Programme for Mangroves and Coral Reefs. Available from: http://pib.nic.in/newsite/PrintRelease.aspx. Accessed on: 20 December, 2013.

MOEF. 2007. *Conservation of Wetlands in India: A Profile (Approach and Guidelines).* Conservation Division- I. Ministry of Environment and Forests, Government of India, New Dellhi, 42 pp.

MOEF. 2010. *Conservation and Management of Lakes – An Indian Perspective.* National River Conservation Directorate, Ministry of Environment and Forests, Govt. of India, New Delhi, 102 pp.

MOEF. 2013a. Protected Area Network in India. Available from: http://moef.nic.in/public-information/protected-area-network. Accessed on 24 December, 2013.

MOEF. 2013b. *Annual Report 2012-13.* Environmental Infirmation System (ENVIS), Ministry of Environment and Forests, Government of India, 464 pp.

MOEF. 2014. Project Elephant. Available from: http://envfor.nic.in/division/ introduction-4. Accessed on: 10 January, 2014.

Mukherjee, B. 1996. *Environmental Biology.* Tata McGraw-Hill Publishing Company Limited. New Delhi, 691 pp.

Pathak, N. (Ed). 2009. *Community Conserved Areas in India – A Directory.* Kalpavriksh, Pune/New Delhi, 812 pp.

Prasad, S.N., T.V. Ramachandra, N. Ahalya, T. Sengupta, A. Kumar, A. K. Tiwari, V.S. Vijayan and L. Vijayan. 2002. Conservation of Wetlands of India – a review. *Tropical Ecology,* **43**(1): 173-186.

Rajagopalan, R. 2008. *Samudra Monograph: Marine Protected Areas in India.* International Collective In Support of Fishworkers. Chennai, India, 69 pp.

Rodgers, W. A., and Panwar, H.S. 1988. *Planning a Wildlife Protected Areas Network in India. Vol 1 and 2.* Dept of Environment, Forests, and Wildlife/Wildlife Institute of India report. Wildlife Institute of India, 341+261 pp.

Rodgers, W. A., Panwar, H.S., and Mathur, V.B. 2000. Wildlife Protected Area Network in India: A review, Executive Summary. Wildlife Institute of India, Dehradun.

Sarkar, S., Mayfield, M., Cameron, S., Fuller, T., and Garson, J. 2007. Conservation area networks for the Indian region: Systematic methods and future prospects. *Himalayan Journal of Sciences,* **4**(6):27-40.

Scheiner, S.M. 1992. Measuring pattern diversity. *Ecology,* **73**:1860-1867.

Secretariat of the Convention on Biological Diversity. 2004. *The Ecosystem Approach, CBD Guidelines.* Secretariat of the Convention on Biological Diversity. Montreal, 49 pp.

Singh, H.S. 2003. Marine Protected Areas in India. *Indian Journal of Marine Sciences,* **32** (3): 226-233.

Szaro, R.C., and Shapiro, B. 1990. *Conserving our Heritage: America's Biodiversity.* The Nature Conservancy, Arlington, VA, 16 pp.

Udvardy, M. D. F. 1975. *A classification of the biogeographical provinces of the world.* IUCN Occasional Paper no. 18. Morges, Switzerland: IUCN, 49 pp.

UNESCO. 2013a. Man and the Biosphere Programme. Available at: http:// www.unesco.org/new/en/natural-sciences/environment/ecological-sciences/man-and-biosphere-programme/. Accessed on: 18 January, 2013.

UNESCO. 2013b. Asia and the Pacific (124 biosphere reserves in 23 countries) Last Update: June 2013. Available at: http://www.unesco.org/new/en/natural-sciences/environment/ecological-sciences/biosphere-reserves/asia-and-the-pacific/. Accessed on: 18 January, 2013.

UNESCO World Heritage Centre, 1992-2014. India. Available from: http:// whc.unesco.org/en/statesparties/in. Accessed on: 18 January, 2014.

UNESCO World Heritage Centre, 2008. *World Heritage Information Kit.* UNESCO World Heritage Centre, Paris, 30 pp.

UNEP, 1995. *Global Biodiversity Assessment.* Cambridge University Press, Cambridge.

Wilson, E.O. 1988. The Current State of Biological Diversity. In: *Biodiversity*, (Eds.) E.O. Wilson and F.M. Peters. National Academy Press, Washington DC, p. 3-18.

WNEP-WCMC. 2013. Indigenous and Community Conserved Areas (ICCAs). Available from: http://www.unep-wcmc.org/indigenous-and-community-conserved-areas_263.html. Accessed on: 20 December, 2013.

WPA, 1972. The Indian Wild Life (Protection) Act, 1972 (as amended upto 1993).

WPAA, 2002. The Wild Life (Protection) Amendment Act, 2002: No. 16 of 2003. Ministry of Law and Justice (Legislative Department), New Delhi.

2015, Animal Diversity, Natural History and Conservation, Vol. 5   Pages 251–266

Editors: V.K. Gupta and Anil K. Verma

Published by: DAYA PUBLISHING HOUSE, NEW DELHI

Chapter 16

Assessing Management Effectiveness and its Conservation in Hadagarh Wildlife Sanctuary, Odisha, India

S.D. Rout

Department of Wildlife and Biodiversity Conservation,
North Orissa University, Takatpur,
Baripada – 757 003, Odisha, India

ABSTRACT

The Hadagarh Wildlife Sanctuary is located in Keonjhar, Mayurbhanj and Balasore districts of Odisha. The landscape with miscellaneous vegetation and smaller hills contains variety of flora and fauna The sanctuary has well demarcated core as well as buffer zone. The overall objectives of wildlife sanctuaries was to deal with the conservation of biodiversity, dependencies of local people upon the sanctuary and maintenance of associated cultural values. Considering the above facts the sanctuaries has been receiving financial assistance to execute different projects, programmes and activities inside and outside the sanctuary from the Ministry of Environment and Forests (MoEF), Government of India and State Government since its inception. Using a host participatory evaluation methods, conservation and management success of Hadagarh Sanctuary was evaluated. As per the objectives the total activities of the sanctuary have categorized under six major head of programmes such as Sanctuary Infrastructure Development and Maintenance, Forest Conservation Activities, Wildlife Habitat Management, Wildlife Conservation, Livelihood option and Community Development Activities

* *Corresponding author.* E-mail: srusti_d_rout@rediffmail.com

and Tourism development and are ranked on the basis of scoring points. The major recommendations for achieving long term sustainability of this programme are described for implementation of sustainable development of Sanctuary.

Keywords: *Hadagarh, Management effectiveness, Conservation, Wildlife Sanctuary, Odisha.*

Introduction

Biodiversity is the variety and variability of life on Earth. The earth supports an incredible array of biodiversity. Unfortunately, the earth's biodiversity is disappearing, with an estimated 1,000 species per year becoming extinct. Conserving biodiversity is especially crucial in developing countries where people's livelihoods are directly dependent on natural resources such as forests, fisheries and wildlife. With the increasing pressure and demand for land and unplanned land use of various landscapes, for majority of the biodiversity, protected areas have become the main abode for survival. Protected areas are tracts of forest, mountains, wetlands, grasslands, deserts, lakes, rivers, coral reefs, and even oceans that are managed to maintain the Earth's biodiversity. Because habitat loss is a serious threat to biodiversity, establishing protected areas as safe havens for biodiversity is critical to ensuring the survival of the Earth's natural heritage. Establishing a protected area requires more than simply setting aside a track of land for "protection." A protected area requires a management plan to ensure that biodiversity is in fact protected. To be effective, the plan must address the various threats to the area and the biodiversity it supports. Threats to any given protected area might include the conversion of habitat to agriculture, unsustainable extraction of resources, illegal logging and poaching, the introduction of invasive or alien species, pollution, and climate change. As emphasized by the Convention on Biological Diversity, any strategy to slow the loss of biodiversity and to enhance its contribution to development must integrate three essential elements: conservation of biodiversity, sustainable use of its components and the equitable sharing of benefits.

Considering the above facts the sanctuaries has been receiving financial assistance to execute different projects, programmes and activities inside and outside the sanctuary from the Ministry of Environment and Forests (MoEF), Government of India and State Government since its inception. However, yet we do not whether the project inputs has actually contributed towards achieving the stated objectives. Hence, the present study on effectiveness of management is an attempt to fulfill this gap by understanding the cause effect relationships. The overall goal of the evaluation is to assess the extent of physical output and outcomes achieved *vis a vis* the envisaged ones. Thus the main objectives of this evaluation are:

1. To assess the overall outcome and impact,
2. To suggest the improvements in the quality of implementation, and
3. To determine continued relevance of the scheme in the context of Environmental Policy.

There are several studies have been carried out both at national (Singh, 1999; Ganguly *et al.,* 2003; Mishra, 2010; Rout, 2011) and international levels (Bruner *et al.,* 2001; Farrell and Marion, 2001; Goodman, 2003; Hockings, 2003) on the evaluation and assessment of PAs, but there is no comprehensive study that has analysed the impacts of project or sanctuaries inputs and the desired conservation outputs with respect to sanctuaries of Odisha. Assessing the management effectiveness of a PA system can enable policymakers to develop strategic, system wide responses to pervasive management problems. Therefore, the present evaluation is significant from the point of view of i) tracking the progress towards achieving project objectives, ii) providing timely information for decision making to improve project efficiency and effectiveness, and iii) identifying problems as they come up and before they run into crisis.

Study Area Description

The Hadagarh sanctuary lies between 21°12' to 21°23' North latitude and 86°12' 30" to 86°21' 30" East longitude in Kenosha, Mayurbhanj and Balasore districts of Odisha state. The sanctuary has well demarcated core as well as buffer zone. Core area constitute Boula R.F. and Satkosia R.F. (part) having 179.310 Km^2 and Buffer area constitute Satkosia R.F. (part), Nada R.F., Atei R.F., Santoshpur R.F., Kuldiha R.F and Gadasahi P.F. having 220.442 Km^2 (Figure 16.1). It comprises part of Satkosia RF of Mayurbhanj and Boula RF of Keonjhar district. The clusters of hills extend through the western and southern tract of sanctuary (Rout, 2013). The valley is occupied by Hadagarh reservoir and its catchments. The highest peak being 1861 ft. is the Boula Pahad. The major portion of the Sanctuary under plain tract is around reservoir and its catchments. The weather is more pleasant with an average annual rainfall of 120 cm and the maximum temperature during the hot weather rises up to 42°C while the minimum during winter falls as low as 6°C.

Water is the most important requirement for any Wildlife habitat and its importance is significant when the tract falls in a seasonal rainfall area like Hadagarh. The spatial distribution of water sources is more important than the total quantity of water. Hadgarh has a network of perennial and seasonal water sources. There are numerous seasonal and perennial nalas inside the sanctuary. Besides this, Salandi River and the Hadgarh reservoir

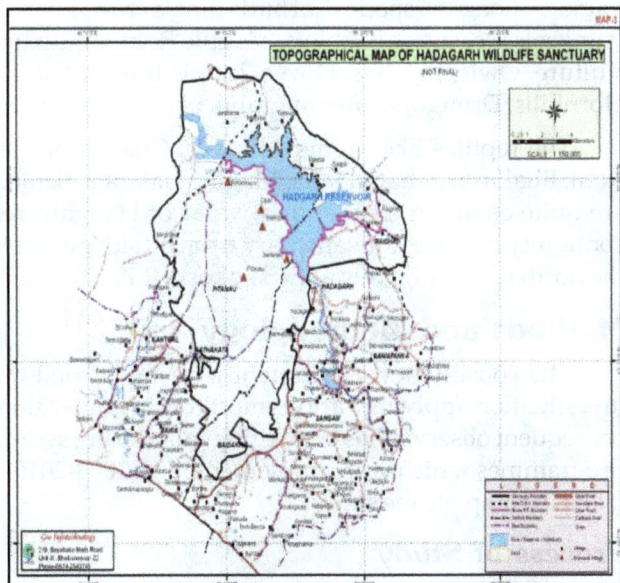

Figure 16.1: Map Shows the Study Area.

is the largest perennial source of water for wildlife. However during pinch period and summer months, some perennial water holes are required in interior areas such as Sangam, Bahia, Baniapanka and Kathakata etc. to avoid migration of wild animals in search of water.

According to the Biogeographic classification by Rodgers and Panwar (1988) the Hadagarh sanctuary falls in to three hierarchical levels of planning units as follows (i) Biogeographic Zone- Decan Peninsula, (ii) Biogeographic Province-Chhotanagpur and (iii) Biogeographic Region – Mahanadian. Similarly, Hadagarh is composed of 2 main type of vegetation such Dry Peninsular Sal Forests and Northern dry mixed deciduous forests(Champion and Seth, 1968).

The vegetation is tropophilous having some xerophytic character in many of its species. The main portion of sanctuary comprises of mixed deciduous forests with miscellaneous species. A few pockets in Boula R.F. have miscellaneous forests with almost no sal. It is observed that forests in Hadagarh Sanctuary are generally poor in comparison to adjoining Sal forest in Similipal Biosphere Reserve of Mayurbhanj district. The associates of sal such as Harida, Bahada, Anla, Arjuna, Asan, Jamu, Mahua, Simili, Karada, Senha, Char, Bahlia, etc. are found. Climbers found are *Bauhinia vahalii, Millettia auriculata* and *Butea superba* in Hadagarh Sanctuary. Some common grasses found are spear grass, Sabai and broom grass in southern slopes of the hill along the boundary of Boula R.F. between Kantipal and Bohidakhia along Mayurbhanj boundary. Bamboo is almost absent in the Sanctuary.

Hadagarh Wildlife Sanctuary is famous for migrating elephants and resident population of Wild boar, Bears, Hare and Sambars. In addition to it Langur, Macaques, Tiger, Leopard, Jackal, Hyaena, Jungle cat, Mongoose, Otter, Fishing cat, Chital, Barking deer, Flying squirrel, Porcupine, Rodents, Pangolin, Snakes are commonly found. Among 156 species \of birds Jungle crow, Jungle myna, Crow phesants, Bulbull, Parakeets, Peacocks, Babblers, Magpie Robin, Woodpeckers, Koel, Bee eater, Swifts, Vulture, Owls, Pigeons, Doves, Jungle bush quail, Patridge, Egrets, Ducks, Teals, Hornbills, Drongo, winter migrants like Ducks are also found.

The reptiles like Monitor lizards, Chamelions are also found. Among fishes, Rohi, Bhakura, Pohara, Balia, Magura, Jalanga, Seula, Kau, Singi, Kantia, Mahurali are quite common in river and water of Hadgarh reservoir. Hadagarh forest has contiguity between the sanctuary from all sides except southern and Eastern side. In the north it is contiguous with Satakosia R.F, Noto R.F. upto Similipal Tiger Reserve.

Methods and Methodology

The consultancy assignment is being carried out through social survey and investigation applying various monitoring and evaluation tools and techniques and consequent observations on existing monitoring system for around 33 (Annexure – I) programmes which were implemented in 2009-2010. The following steps are to be taken to complete the study.

Process of Study

A host of participatory evaluation methods (Burgess, 1982; Mukherjee, 1995) were administered throughout the evaluation process. The work required collection of primary as well as secondary information. Before undertaking the fieldwork for data collection the entire study framework and the methodology has been thoroughly discussed with the Divisional Forest Officer, Keonjhar Wildlife Division, Anandapur and the Range officer, Hadagarh.

Data Collection

The study would require primary as well as secondary data. The primary data pertaining to the project were collected from officials and villagers. The secondary data were collected from the documents like basic project proposal, progress report documents, action plan, annual report and audit report. The collection of data involve the following steps like interview with key persons, questionnaire survey, village meeting, participant observation, case study documentation, site level observation and photography.

Data process and Analysis

The data that pertaining to the individuals programmes were analysed and discussed in the following subheads such as objectives, implementation process, technological support, benefits derived, issues identified, rating of component, and overall rating.

In addition to these points some special points were considered. The content of questionnaires, interviews, discussion with people were analysed. The overall performance of the activities undertaken in Hadagarh Range was evaluated based on scoring a number of parameters such as attainment of objectives, achievements of outputs and activities, implementation approach and quality, livelihood and community welfare, education and awareness generation, research management interface and programme sustainability on a scale of 10 points ranging from Outstanding: (8-10); Very Good : (5-<8); Good: (3-<5); Poor (<3) and finally calculating the overall rating. The rating points were given based on the responses and personal observations.

Results

Threats to Hadagarh Wildlife Sanctuary

Hadagarh has tolerated many ravages of destruction on account of population pressure, illicit felling, domestic livestock grazing, encroachment and shifting cultivation, activities of mining, hunting, forest fire, NTFP collection, insect attacks and other pathological problems. These threats creating the measure problem for the conservation of Sanctuary.

Critical Issues of Hadagarh Wildlife Sanctuary

There are 15 villages comprising of 1302 families with a population of about 6538 people fully dependent upon Hadagarh Sanctuary for their livelihood. These villages have been established by the displaced people of Hadagarh dam project by way of encroachment of Boula R.F. in early seventies. The villagers maintain huge number of domestic animals to support them. In Satkosia RF of Mayurbhanj district another 3 villages exists. These encroachments need to be evicted since they occupy the fertile valleys and plains inside the sanctuary. Once evicted these areas can be converted into excellent grasslands and meadow with fodder trees on periphery. Till the relocation of the encroachers outside the sanctuary is finalized they can be ringed out with trenches and stone wall in order to prevent further encroachments. At present these encroachers maintain their livelihood by cultivation, fishing, smuggling of timber, hunting, collection of MFPs and selling of firewood etc. As per field inspection they are not putting much effort in agriculture and keeping these lands fallow except in few low lying pockets.

Illegal firewood and minor forest produce collection are major problems that need to be curbed. Villagers living along the boundary of the Sanctuary depend fully on sanctuary for their fuel wood and small timber requirements. Moreover, a significant section of population has taken up cutting and selling of firewood, small sizes, bellies, and planks as their occupation. Almost all encroacher and settlers inside sanctuary are engaged in timber business and illegal fishing. Villagers like Kantipal, Kathakata, Chhenapadi, Bidyadharpur etc. are the marketable points for timber and MFPs. The labourers engaged in mining operation by FACOR, IMFA and OMC in Boula Mines area go for illicit felling for meeting their small wood, firewood requirement from Sanctuary area. Organized illicit fellings are habitually done by the villagers on the periphery of Boula RF. Organized gang smuggling of timber is rampant in Raighati, Sankatapalia and Boula mines area. Since there are very few protection staff deployed inside the sanctuary smuggling of valuable timber has become uncontrolled. Lack of adequate field staff, lack of arms and ammunitions, lack of communication facilities, viz VHF, vehicle etc. have worsened the protection situation. Only two forest section offices are functioning on the fringe of sanctuary with nine Forest Guards. Illegal removal of timber not only results in colossal loss of revenue and destruction of the vegetative cover but also adversely affects the future stock due to irregular felling, high stumps, felling damage, and removal of immature and promising stems.

Inputs in Hadgarh Wildlife Sanctuary

Financial Investment, Planning and Management

An amount of Rs. 36.48lakh was spent during the year 2009-2010 to carry out the activities envisaged in the management action plan. Funds flow was driven by annual plan through Non-plan, State-plan and central plan. Though the State Government has been able to prepare and submit the annual plan to Govt. of India regularly, there has been problem of late receipt of funds for the field implementation. This process has been compromised on the likely impact and outcome of the programme and activities.

Implementation Approach

Implementation of various activities of the Sanctuary was done through already existing staff as well as other institution staff for training and capacity building programme, and establishing a coordination mechanism for the overall supervision. While doing so, the guidelines issued by Government of India and State Government have been scrupulously followed by the executing agencies. While carrying out the activities, the provisions of Wildlife Protection Act (1972) have also been followed strictly.

Institutional Building

Coordination Committees

Three Coordination committees namely the State level Steering Committee, Local level Coordination Committee, and the Research Committee has been set up for the overall supervision and monitoring of the implementation of the Hadagarh Wildlife Sanctuary activities, and also to monitor the progress towards achieving the set objectives. The state level steering committee is headed by the Chief Secretary of the state to oversee the programme. Local level Co-ordinator Committee is headed by the Regional Chief Conservator of Forests, who co-ordinates activities of this division, and recommends suitable management interventions for incorporation in the Management Action Plan. Research Committee is headed by the Principal Chief Conservator of Forests (Wildlife) and Chief Wildlife Warden of the concerned state for smooth conduct of research. The other members are Director, Botanical Survey of India, Zoological Survey of India, representatives of local University are carried out with identified research proposals to help management.

Village Level Institutions

Institution building at the grass root level was the most important aspect for sustainability of Sanctuary. This was long stretched concurrent activity throughout the project period involving a series of meetings and awareness workshops. Participatory tools were used to identify the existing institutions in different villages and Eco-development Committee (EDC), Self Help Group (SHG), Vana Surakhya Samiti (VSS) and Forest Protection Committees.

Capacity Building of Staff and Community

Field staff and local people were the key actors in the sanctuary. The PA staffs responsible for translating the management policies into action have traditionally been competent in activities related to PA protection, habitat management, forest conservation, livelihood option and eco-restoration work. They may have some idea about the village communities and their resource dependencies. But their overall activities to sanctuary ecosystem has not favourable and this issues still remain a weak area. Similarly, the local people may be very rich with respect to their traditional knowledge. But there is still a lot of scope to upgrade the understanding of the local communities about latest issues and happenings in the area of conservation and development. They may also not be aware about the available opportunities for their local development as well as access to these opportunities of late, due to socio-political factors and the impact of market forces, village communities have become highly disorganized and divided and their traditional institutional structures have become almost crumpled. Most of the NGOs are also quite active and competent in human development issues. But only a few NGOs are well organized to take up the new responsibility of undertaking sanctuary conservation development initiatives. Therefore, capacity building and empowerment of all these three critical actors was very important for such a programme that focused primarily on the twin objectives of conservation of biodiversity and compatible conservation friendly local development.

Quantitative Changes

The broader objective of biodiversity conservation can not be achieved unless the damaged habitat is repaired and biotic pressure is reduced. For reducing the biotic pressure protection has to be strengthened, and at the same time dependency of the local people upon the sanctuary has to be reduced. An effort has been taken to improve the condition in all respects by the local communities and has been success. Illegal cutting of trees and cases of forest fires has also been reduced considerably as the VSS, EDCs, Elephant driving squads and Green Brigades volunteers are actively involved in forest protection and strengthening the intelligence network. Forest cover has improved as well as the poaching of animal has decreased indicating an improvement of the general health of the forest ecosystem.

Qualitative Changes

A number of impacts that can not be documented in quantitative term s can be seen because of the proper initiatives. To identify these impacts we may need further scientific investigation. However, during the meetings and interactions with the staff, community, and local NGOs, the following qualitative impacts were quite evident and thus documented. The following activities have been evaluated with objectives, benefit, overall rating and issues during the evaluation period.

1. Sanctuary Infrastructure Development and Maintenance

Objective

The main objective of this programme is to develop the infrastructure inside the sanctuary area for better management and protection and to provide good communication inside sanctuary.

The following activities have been carried out during 2009-10 under the head of Sanctuary Infrastructure Development and Maintenance:

- ☆ The Beat house of Baniapanka, Kathakata, Hatadihi and Hadagarhhas maintained in this period by replacing the asbestos, white washing and repairing of the floor.
- ☆ The camp shed at Hadagarh and Baidakhia beat house has maintained by replacing the broken asbestos and white washing.
- ☆ To facilitate boating for tourist one J.T at Hadagarh Salandiriver has been constructed and boat has repaired.
- ☆ 2 nos. staff quarter, 2 nos. of garage and boundary wall of the camp shed has been constructed at Salapada to provide amenities to the office staffs.
- ☆ A watch tower at Bahia has constricted to watch the anti forest activities.
- ☆ To demarcate the sanctuary area 50 nos. pillar has posted around Boula RF.
- ☆ The forest road from Pitanau – Dalki and Sajanapal to Chakratritha has been maintained for better communication.

Benefit

Local villagers were benefited by getting employment during the construction of the above activities; the staffs also get amenities by means of accommodation and communication. The construction of Forest road has also attracted tourism activities.

Overall Rating

Very good – 7

Issues

No signboards are fixed aside the forest road and sanctuary area disseminating awareness information for tourist.

2. Forest Conservation Activities

Objective

The main objective of this programme is to Conserve and improve the biodiversity, aesthetic and geo-morphological values of sanctuary by management of sites, habitats, control of forest fire and illicit felling.

The following activities have been carried out during 2009-10 under the head of Forest Conservation activities:

- ☆ Awareness meetings on fire and wildlife protection has been organised in the villages of the sanctuary. Banner and Leaflets for fire protection awareness and wildlife protection were printed and distributed among the local people for generate mass awareness.
- ☆ Fire watchers were engaged to control fire in forest.
- ☆ Fire line has been created at Baniapanka and Boula RF to resist the fire havoc.
- ☆ To develop a habit of fire protection among the villages incentive has been given to 3 nos. VSS and 32 SSG. Separate incentive has been provided to award the best fire protection group.

Benefit

Engagement of fire watcher is creating employment opportunity for the unemployed youth of the villages. Local people are participating in fire control activities.

Overall Rating

Very good – 7

Issues

No signboards are fixed aside the forest road and sanctuary area disseminating awareness information for tourist.

3. Wildlife Habitat Management Activities

Objective

The main objective of these activities is to Conserving viable population of wildlife in their natural habitat with flagship species conservation approach and conserve, protect and improve wildlife habitat with special emphasis on the core area/reservoir/corridor and buffer areas.

The following activities have been carried out during 2009-10 under the head of Wildlife Habitat Management activities:

☆ Development of Medow done at Pitanau, Dalki, Bahia, Kantipal and Phuljhar

☆ 5 nos. of salt lick has been maintained at Pitanau, Bahia, Sajanapal

☆ Weed eradication work done at Pitanau

Benefit

Through habitat management the movement of wildlife towards nearby villages has been reduced.

Overall Rating

Very good – 7

Issues

Periodical maintenance of salt lick is needed to keep it usable for wild animal. More cultivation of grasses species may be done in the meadow.

4. Wildlife Conservation Activities

Objective

The main objective of these activities is to conserve viable population of wildlife in general and elephant/tiger in particular in their natural habitat with flagship species conservation approach.

The following activities have been carried out during 2009-10 under the head of Wildlife Conservation activities:

☆ Engagement of anti poaching watcher

☆ Management of elephant depredation by engaging Elephant watcher

☆ Engagement of Elephant tracker to track elephant movement.

☆ Construction of elephant proof stone wall at Siadimalia to stop elephant to enter into the villages and crop fields.

Benefit

☆ Local people are getting employment opportunities as anti poaching squad, elephant squad and elephant tracker.

☆ Reduce human- wildlife conflict

Overall Rating

Out standing – 9

Issues

More anti poaching squad needed to stop the poaching activities inside the forest area.

5. Livelihood Option and Community Development Activities

Objective

The main objective of these activities is to facilitate the community with services to win their confidence and support for forest and wildlife management activities as well as Create awareness among local people regarding the necessity to conserve and protect wildlife habitat.

The following activities have been carried out during 2009-10 under the head of Livelihood Option and Community Development Activities:

- ☆ Construction of one drying yard at Dangachua
- ☆ Organization of training programme at Tanla
- ☆ Water facility done at Dhenka beat house
- ☆ Organisation of cattle immunization camp at Kathakata, Bahia, Pitanau and Dalki
- ☆ Organization of health camp at Pitanau, Dalki, Salapada, Kathakata and Bahia

Benefit

- ☆ Local communities are getting services in the inaccessible area.
- ☆ Through these activity Forest Department win the confidence of the villages and are getting their cooperation in forest management.

Overall Rating

Good – 5

Issues

More awareness to stop the illegal fire activities inside the forest is required.

6. Tourism Development Activities

Objective

The main objective of these activities is to Conserve and develop important wildlife habitat along with places of religious/ecological importance for the purpose of eco-tourism and disseminate the idea of conservation among visitors.

The following activities have been carried out during 2009-10 under the head of Tourism Development Activities:

- ☆ Boat repairing done at Baitarani river
- ☆ Construction of J.T at Hadagarh Salandi river
- ☆ Sinage for eco-tourist Installation of 2nos. sign board at Hatadihi and Chhenapadi

Benefit

- ☆ Local communities are getting services in the inaccessible area.

Overall Rating

Good – 5

Issues

Though there are ample scopes for eco-tourism development but less attempt has been taken to attract more and more tourist to the sanctuary. No information as well as awareness sign boards has installed inside the sanctuary.

Physical Activities, Overall Outcome and Impacts

Activities implemented in Sanctuary towards achieving the broader objectives of the programme can broadly be classified as Sanctuary Infrastructure Development and Maintenance, Forest Conservation Activities, Wildlife Habitat Management, Wildlife Conservation, Livelihood option and Community development Activities and Tourism Development. The rating score of each category of activities was rated on a 10 point scale ranging from ' highly unsatisfactory=1' to 'highly satisfactory = 10'. Each parameter was scored separately and finally, an overall rating of 6.6 was arrived at on the scale of 10 point (Table 16.1).

Table 16.1: Evaluation of Conservation and Management Effectiveness of Hadagarh Wildlife

Sl.No.	Parameters	Score Obtained
	Sanctuary	
1.	Sanctuary Infrastructure Development and Maintenance	**07**
2.	Forest Conservation Activities	**07**
3.	Wildlife Habitat Management	**07**
4.	Wildlife Conservation	**09**
5.	Livelihood option and Community development Activities	**05**
6.	Tourism Development	**05**
	Overall Score	**6.6**

Recommendations

The major recommendations have been recommended for achieving long term sustainability basing on the conservation actions for villages inside the Hadagarh Sanctuary and Outside the Sanctuary.

A. Conservation Action for Villages inside Sanctuary

All villages inside must be brought under eco-development program and for this the priority villages are the ones located within the identified Critical Areas for conservation. In each village the following actions needs to be undertaken:

1. **Formation of Village Forest Committee/Eco Development Committees:** In each village VFC/EDC must be formed which would include members form all castes living in the village. They should be made responsible for all the protection related, resource sharing and management aspects with the overall monitoring of all matters to be taken up by the Forest Department.

2. **Improving livelihoods**: Improve agriculture through improvement in water resource, soil moisture and other allied resources. Introduce and encourage use of alternate energy like bio gas, solar lights, solar pressure cooker and so on. Try and experiment should be taken for other livelihood options like poultry and livestock farming.

3. **Resources to be made available within their village environs** : Plantation of required/preferred species like medicinal plants, fruit bearing plants must be done along the village fringes (degraded presently) and also along the farm areas as part of farm forestry.

4. **Use of Grass species**: Grass from road sides and pilgrim paths should be allowed to be taken in a regulated and controlled manner that would in turn ensure reduction in fire occurrences, improve natural regeneration by way of the seeds being able to fall onto the soil instead of on the grass bed formed due to non removal, and above all gain peoples' support towards conservation.

5. **Prevention of Fire**: The VFCs/EDCs with the help of the other villagers are to be involved in prevention of fire during dry season in addition to taking all precaution, which includes make fire lines and fire breakers at appropriate locations and sites and also involve in creating awareness on the negative impact of fire to the locals people and tourists.

6. **Awareness creation on sustainable use and environment and its restoration:** A systematic and regular awareness programs should be taken up mainly on the aspects of sustainable use of the resources, sharing it equally, managing it and above all about the significance of restoration and its related issues.

7. **Conservation Action formulation:** The actions needed to conserve the sites of the sanctuary with formation of Village Forest Committees (VFC), restoration, removal of invasive and weeds, soil and moisture conservation, fodder improvement and management, arranging the health camp, veterinary camps, encourage farm forestry activities, provide/encourage alternate energy, fire control, eco-development program and awareness program.

B. Conservation Actions for Villages outside the Sanctuary

The conservation strategies to be drawn for the villages located along the fringes, outside the boundary, *i.e.*, 1km area from the PA boundary, which is the buffer are different in many aspects as it would be a partnership of three groups, the villagers or Panchayat of the concern village, Forest Department and the Non Government Organization. The Forest Department being the main stakeholder of the forested areas, villagers or Panchayat being owners of the revenue lands, village commons (grazing lands, forest lands) and private lands including the agriculture lands, within the village limits and NGOs would be the main catalyst and helping in planning various interventions and restoration along with setting proper and appropriate institution for managing each and every aspects that would lead to conservation of the PA and ensure sustainability of all resources that are used by the locals and their livestock. We would achieve this with the following major actions:

1. **Restoration of forests in the fringe areas under JFM arrangements:** It is very crucial that the reserved forestland in the surrounding of the villages that are located along the boundary outside be restored through the combined effort of the villagers, Forest Department and the NGOs, which is possible under Joint Forest Management (JFM) arrangements. Once this is done then based on the existing status of the forest land plan appropriate restoration according to the need of the people and set up various necessary institutional mechanisms that would aid in the proper and sustainable management of the area of concerned. This would not only help the people but also render linkages (corridors) with the forest of the PA for all animals that would help in increasing the productivity of their agriculture in addition to improving the environmental conditions. All this would ultimately reduce the dependency of the people on the Sanctuary.

2. **Support regeneration of village grasslands and other commons to meet fodder needs:** It is very important to improve the productivity and potential of the village grass or grazing lands and other common lands in order to increase the fodder availability. Site specific interventions like improving moisture regime, soil nutrients, and grazing or fodder use and management system, in addition to selecting species for restoration according to the need of the people that should take into consideration aspects like, seeds of native species of high nutritive value, seeds of fast growing species, mixture of seeds of perennial and annual grass species, seeds of native legumes for fixing the75nitrogen and so on. All this would ensure a sustainable use of the resource in addition to reducing the grazing pressure in the forests along the boundary of the Area.

3. **Interventions in animal husbandry:** It is very critical to improve the quality of the livestock so that the livestock numbers can be reduced, introduce systems like rotational grazing and cut and carry, chaff cutters and so on.

4. **Improve livelihoods to increase cash incomes:** It is very important to improve the livelihoods of the local people in the form of increase in cash incomes. This should be done by undertaking on farm interventions like improving the moisture and water availability, soil conditions, growing good and high yield local crop varieties. Further, introduction of poultry farming, livestock farming and so on can also be tried that would ensure cash incomes.

5. **Building and linking community institutions aimed at protection of the entire landscape:** A well planned and proper institutional mechanism with regard to management, use and protection of the forest and the other natural resource must be built up along with appropriate step being made to link these community institutions with the other stake holders like the revenue, agriculture, animal husbandry, forest and other related departments, and all the connected NGOs. This would ensure the protection of the entire landscape within which the PA is located.

6. **Awareness creation on sustainable use and environment and its restoration:** A systematic and regular awareness programs and education for different age groups in these villages should be taken up mainly on the aspects of sustainable use of the resources, sharing it equally, managing it and above all about the significance of restoration and its related issues like what, when, where and how to restore. This should be substantiated with similar programs on the significance of the forest and the biodiversity present in it by highlighting various benefits the local communities get and other related aspects.

Acknowledgements

I express my sincere thanks to Regional Chief Conservator of Forests, Rourkela for providing this opportunity. I am very much thankful to Divisional Forest Officer, Keonjhar Wildlife Division, Anandapur and Range Officer, Hadagarh for their co-operation and logistic arrangement in conducting this exercise. My thanks to other Forest Department personnel, local people for their co-operation during the evaluation. I am also thankful to Sri T.K. Behera, for his support and necessary help during field work and preparation of this manuscript.

References

Bruner, A.G., Gullison, R.E., Rice, R.E., and da Fonseca,G.A. B., 2001. Effectiveness of parks in protecting tropical biodiversity. *Science,* **291**: 125-128.

Burgess, R.G. (ed.). 1982. *Field Research: A Source Book and Field Manual.* George Allen and Unwin, London, U.K.

Champion, H.G., and Seth, S.K., 1968. *A Revised Survey of the Forest Types of India.* Manager of Publication, Government of India. Delhi.

Farrell, T.A., and Marion, J.L., 2001. Identifying and assessing ecotourism visitors impacts at eight protected areas in Costa Rica and Belize. *Environmental Conservation,* **28**: 215-225.

Ganguly, A., Gokhale, Y., and Gadgil, M., 2003. Developing responsive indicators for the Indian Biosphere Reserve Programme. *J. Bombay Nat. Hist. Soc.,* **100(2 and 3)**: 214-225.

Goodman, P.S., 2003. Assessing management effectiveness and setting priorities in protected areas in KwaZulu-Natal. *Bioscience,* **53**:843-850.

Hockings, M., 2003. Systems for assessing the effectiveness of management in protected areas. *Bioscience.* **53**:823-832.

Mishra, B.K., 2010. Conservation and management effectiveness of Similipal Biosphere Reserve, Orissa, India. *Indian Forester,* **136(10)**: 1310-1326.

Mukherjee, N., 1995. *Participatory Rural Appraisal and Questionnaire Survey: Comparative Field Experience and Methodological Innovations.* Concepts Publishing Co., New Delhi, India.

Rodgers, W.A., and Panwar, H.S., 1988. *Planning a Wildlife Protected Area Network in India. Vol*s. 1 and 2, Wildlife Institute of India, Dehradun.467 pp.

Rout, S.D., 2011. Final Report on Monitoring and Evaluation of SimilipalBiosphere Reserve Programme. Submitted to Field Director, Similipal Tiger Reserve, Baripada.

Rout, S.D., 2013. Report on Monitoring and evaluation of Hadagarh Wildlife Sanctuary for the year 2009-2010. Submitted to Divisional Forest Officer, Keonjhar Wildlife Division, Anandapur.

Singh, S., 1999.Assessing management effectiveness of wildlife protected areas in India. *Parks,* **9**: 234-249.

2015, Animal Diversity, Natural History and Conservation, Vol. 5   Pages 267–279
Editors: V.K. Gupta and Anil K. Verma
Published by: DAYA PUBLISHING HOUSE, NEW DELHI

Chapter 17

Biodiversity of Chakrashila Wildlife Sanctuary, Western Assam, India: Major Issues in Conservation

Bijita Barman and Susmita Gupta*
Department of Ecology and Environmental Science,
Assam University, Silchar – 788 011, Assam, India

ABSTRACT

Biodiversity is the degree of variations of life forms within biome or ecosystems. Northeastern India is very rich in biodiversity, harboring many endemic plant and animal species. It falls in Eastern Himalayan region. Knowledge of the fauna of this region is poor. Most of the information available is on the larger vertebrates that are easily observed and inventoried. The smaller mammals, reptiles, amphibians, and fishes have been neglected and the most abundant taxonomic group, the insects, have been virtually ignored. Chakrashila Wildlife Sanctuary is the home of the endangered primate Golden langur (*Trachypithecus geei*) in India. The sanctuary is also rich in flora and fauna. The water bodies of the Sanctuary attract many migratory birds in winter season each year. The Sanctuary is identified as Key Biodiversity Area for addressing the conservation of biodiversity of the Sanctuary. Very few studies on biodiversity of this Sanctuary have been carried out so far. The growing population in the surrounding villages is having immense pressure on the resources of the sanctuary finally leading to habitat degradation and biodiversity loss. Thus the habitat of the Sanctuary has to be conserved for the conservation of genetic diversity.

Keywords: *Biodiversity, Eastern Himalayan region, Chakrashila Wildlife Sanctuary.*

* *Corresponding author.* E-mail: susmita.au@gmail.com

Introduction

The Biodiversity can be defined as the variability among living organisms from all sources including terrestrial, marine and other aquatic ecosystems, and the ecological complexes of which they are part; this includes diversity within species, between species, and of ecosystems. The term biodiversity was first coined by the entomologist E.O. Wilson in 1986. Biodiversity is the foundation of life on Earth. It is crucial for the functioning of ecosystems which provide us with products and services without which we couldn't live. Oxygen, food, freshwater, fertile soil, medicines, shelter, protection from storms and floods, stable climate and recreation - all have their source in nature and healthy ecosystems. But biodiversity gives us much more than this. We depend on it for our security and health; it strongly affects our social relations and gives us freedom and choice. Biodiversity is extremely complex, dynamic and varied like no other feature of the Earth. Innumerable plants, animals and microbes physically and chemically unite the atmosphere (the mixture of gases around the Earth), geosphere (the solid part of the Earth), and hydrosphere (the Earth's water, ice and water vapour) into one environmental system which makes it possible for millions of species, including human being.

India lies within the Indomalaya ecozone and contains three biodiversity hotspots (Conservation International, 2007). India, with 2.4 per cent of the World's area, has over 8 per cent of the World's total biodiversity, making it one of the 17 megadiversity countries in the world (Mittermeier *et al.*,1997). This status is based on the species richness and levels of endemism recorded in a wide range of taxa of both plants and animals. This diversity can be attributed to the vast variety of landforms and climates, resulting in habitats ranging from tropical to temperate and from alpine to desert. India is also considered one of the world's eight centre's of origin of cultivated plants. Being a predominantly agricultural country India also has a mix of wild and cultivated habitats, giving rise to much specialised biodiversity, which is specific to the confluence of two or more habitats (MoEF and Kalpavriksh, 2004). According to FSI (2002), forest cover has been assessed to be 20.55 per cent of the country's geographical area. Forest cover information of 589 of the 593 districts showed that Madhya Pradesh has the maximum forest area of 77,265 sq km, followed by Arunachal Pradesh (68,045 sq km).

The north-east Indian biogeographic zone is most significant as it represents the transition zone between the Indian, Indo-Malay and Indo-Chinese biogeographic regions, as well as a meeting-place of Himalayan Mountains with those of Peninsular India. The region acts as a biogeographic gateway for plant migration. In India, apart from the Western Ghats, Northeast India is one of the 34 biodiversity 'hotspots' in the world. The biodiversity hotspots contain around 50 per cent of the world's endemic plant species and 42 per cent of all terrestrial vertebrates, but have lost around 86 per cent of their original habitat (Mittermeier *et al.*, 2004). The Northeastern Region of India contains more than one-third of the country's total biodiversity (www.biodiversityhotspots.org). The northeast India is a part of the Eastern Himalayan region which is part of two larger hotspots: the Indo-Burma and the Himalaya Hotspots (CEPF, 2005). This means that the region contains exceptional levels of plant endemism (at least 1,500 endemic species) and has lost 70 per cent or more of its original habitat (Myers *et al.,* 2000). Thus the region has been in focus and a priority for leading conservation agencies of the world. The eastern Himalayan region has been in the spotlight as a part of 'crisis Ecoregions'(Hoekstra *et al.,* 2005); 'Biodiversity Hotspots' (Myers *et al.,* 2000; Mittermeier *et al.,* 2004); 'Endemic Bird Areas' (Stattersfield *et al.,* 1998); 'Mega Diversity Countries' (Mittermeier *et al.,* 1997) and 'Global 200 Ecoregions' (Olson and Dinerstein, 2002). There are 99 protected areas covering 15 per cent area of the Eastern Himalaya. More than 7000 species of plants, 175 species of mammals, and over 500 species of birds have been recorded in the Eastern Himalayas alone (WWF and ICIMOD, 2001). The richness of the region's avifauna largely reflects the diversity of habitats associated with a wide altitudinal range. The Eastern Himalaya and the Assam plains have been identified as an Endemic Bird Area by the Royal Society for Protection of Birds, (ICBP, 1992). The region's lowland and montane moist to wet tropical evergreen forests are considered to be the northernmost limit of true tropical rainforests in the world (Proctor *et al.,* 1998).

According to the State of Forest Report 2011, released by the Forest Survey of India (FSI), the Assam state of India has a total of 27,673 sq.km. forest cover. This area is 24.58 per cent of the total geographical area of the state. Assam shares 3.2 per cent of India's total forest cover (State of Forest Report, 2011). The state of Assam is blessed with great rivers, hills and forests. In these one can find exotic flora and fauna and hence Assam is a delight for the various naturalists. The state boasts of five national parks, 17 wildlife sanctuaries (www.assamportal.com). Some of the endangered animals like the Indian rhinoceros and golden langur are found in these forests and is a safe haven for these wild beings. Chakrashila Wildlife Sanctuary is one of the 17th Wildlife Sanctuaries in Assam.

Methodology

For this study both primary and secondary data have been collected. The secondary data includes various research papers of different authors and news paper clippings. Various WebPages also have been collected for the collection of information. Several field visits during 2010 to 2013 have been made for generating primary data.

Results and Discussion

Geography and Climate of Chakrashila Wildife Sanctuary

The Wildlife Sanctuary is spread over the Dhubri and Kokrajhar districts of Assam (26°15'–26°26'N, and 90°15'– 90°20'E) (Figure 17.1). The sanctuary covers an area of 45.568 km² (4556.8 hectares) and is located at a distance of 6 km from the Kokrajhar town. The Chakrashila was initially declared as a reserve forest by the Government of Assam in 1966 and finally on 14th July 1994 it was given the status of a wildlife sanctuary. The Boundary of the Sanctuary includes Salakati crossing and Tarang river in the north, in the south-NH 37 and Bahalpur crossing, in the east-Chapar Tea Estate and parts of Bhairab Pahar Reserve forest and in the west- Gaurang river. The natural attributes of the zone of influence (ZI) of the Sanctuary includes Gaurang, Tarang and Kaljani river flowing through the ZI; five large and medium water bodies in the south western portion. These are Dhir, Diplai, Dakra, Sakwa-

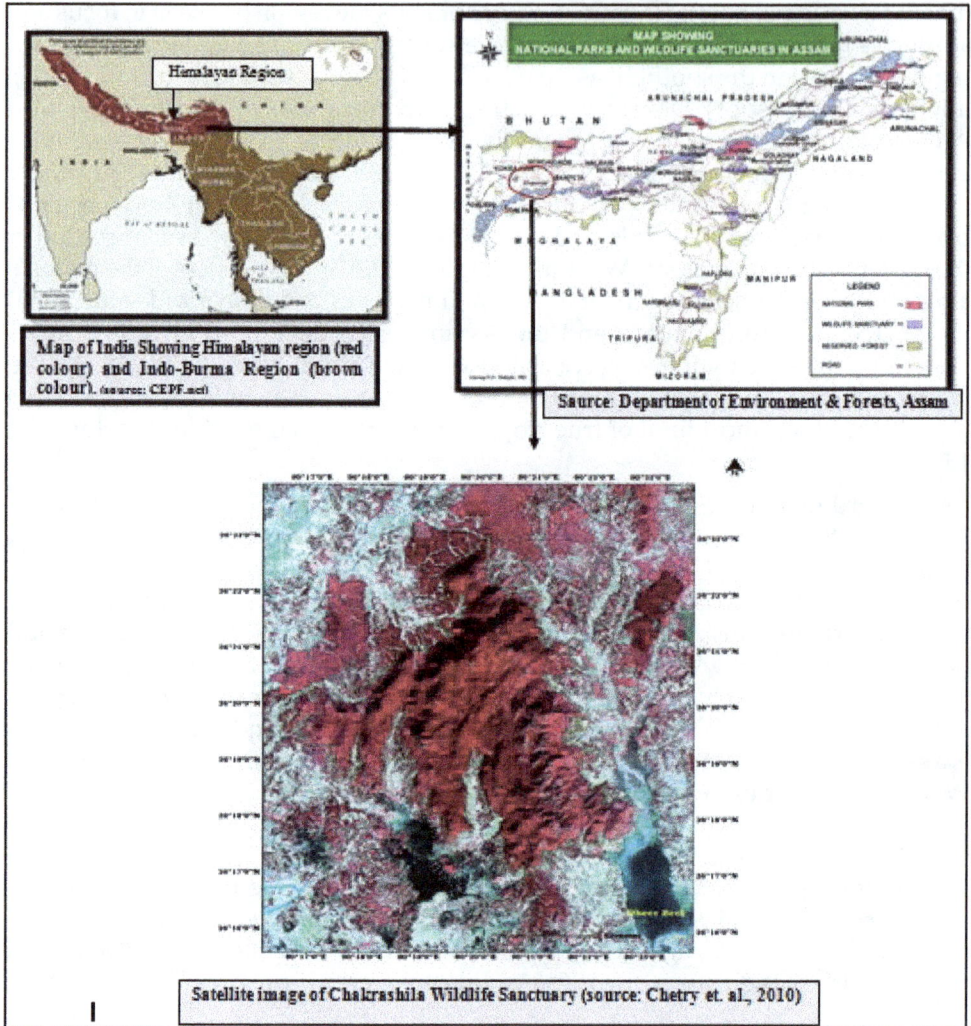

Map of India Showing Himalayan region (red colour) and Indo-Burma Region (brown colour). (source: CEPF.net)

Source: Department of Environment & Forests, Assam

Satellite image of Chakrashila Wildlife Sanctuary (source: Chetry et. al., 2010)

Figure 17.1: The Satellite Imaginary of Chakrashila Wildlife Sanctuary and it's Position in Assam and India.

Sakwi and Hashdoba beel. Chakrashila Wildlife Sanctuary has dry winter and hot summer followed by heavy rainfall. Annual rainfall is between 200 to 400 cm. Air temperature throughout the year generally varies between 8°C to 30°C. Relative humidity is very high with an average of 86 per cent in summer.

Water Bodies of the Sanctuary

There are three large lakes on the southern periphery of the Sanctuary, *i.e.* Dhir, Diplai and Dakra beel, which contribute a lot and become an integral part of the rich bird diversity of the sanctuary. The Asian Wetland Bureau, and the International Waterfowl and Wetlands Research Bureau have identified the first two lakes as internationally important sites for protection of a number of species. Dhir beel located at the base of Chakrashila Wildlife Sanctuary is a proposed Ramsar Site get connected with the Brahmaputra river during monsoon. All these wetlands together constitute more than 800 ha. Bird Life International declared the Diplai Beel an Important Bird Area (IBA) site in 2004. The Asian Water Fowl Census, conducted by the Wetlands International in 1996, recorded 37 species of birds, including nine endangered species, in this wetland. In 2006 Environment activist group Nature's Beckon spotted an extremely vulnerable species of turtle – Black Spotted Pond Turtle (*Geoclemys hamiltoni*) at the Diplai Beel. This species of turtle is included in the Appendix 1 of CITES, in Schedule 1 of the Indian Wildlife Protection Act and it is one of the vulnerable species as per the International Union for Conservation of Nature (IUCN) (The Assam Tribune, 2006).Both wetlands, Dhir and Diplai, attract a lot migratory birds in winter including near Threatened Ferruginous Duck and the Vulnerable Baer's Pochard (*Aythya baeri*). On January 11, 1991 during the mid-winter Annual Waterfowl Census, 565 Ferruginous Ducks were counted in the Diplai beel and 500 in the Dhir beel on January 9, 1991 and 328 on January 8, 1992 (Choudhury, 2000).According to Lopez and Mundkur (1997), 19,828 birds of 43 species were counted in 1994 in Dhir beel during Annual Waterfowl Count. In 1995, the population increased to 26,433 although number of species remained same. The population declined in 1996 to only 7,102 and the number of species seen was 41. In the Diplai beel, in 1994, the total number of waterfowl was 3,224 of 37 species. For Dakra beel, in 1996 a total of 2,409 birds of 20 species were counted.

There are several small springs for quenching the thirst of the wild animals of this Hilly forest. But the two major perennial springs in the sanctuary are Howhowi Jhora and Bamuni Jhora, which flow over the rocks, sparkling and spattering throughout the year, adding to the scenic beauty of the Sanctuary. Besides these two there are many small and medium streams in the Snactuary like, Jornagra stream, Mauriagaon stream, Bhalukjhora stream, Bakuamari stream, Phundibari stream etc.

Flora and Fauna

33.04 per cent of the geographical area of Kokrajhar District is under forest land (State Biodiversity Strategy and Action Plan, Assam, 2009). The forests in the Chakrashila Wildlife Sanctuary are of dense semi-evergreen and moist deciduous type, with patches of grasslands and scattered bushes. Sal tree (*Shorea robusta*) is a dominant tree in this forest. The riparian vegetation of the Sanctuary includes big trees of Sal (*Sorrea robusta*), himolu (*Bombax cieba*),Amari (*Amoora wallichi*), Bhumura (*Terminalia belerica*), Bandardima (*Dysoxylum procerum*), hilikha (*Terminalia chebuja*), poma (*Cedrela toona*), bogiopoma (*Chikrasia tablaris*), Jamuk(*Syzygium cuminni*), Kadam (*Anthocephalus cadamba*), Moj (*Albizzia lucida*), Udal (*Sterculia villosa*), outenga (*Dillenia indica*), Paroli (*Sterosperwum personatum*), Gomari (*Gmelina arborea*), Bhatghila (*Oroxylum indicum*), Bonsum (*Phoebe goalparens*), Dimoru (*Ficus hispida),* and shrubs includes *Phlogacanthus thyrsiflorus, Pandanus odorifer,* etc. (*www.assaminfo.com*)**.** 32 species of ferns also found here, of which more than 30 per cent have medicinal and commercial value.

The Chakrashila wildlife sanctuary is the home for different kinds of animals and birds (Figure 17.2–17.7). There are about 30 species of mammals, 11 species of reptiles, 14 species of amphibians and 60 species of fishes in the Sanctuary. Crocodiles, Alligators, Indian Short-tailed Mole (*Tulpa micrura micrura*), Indian Flying Fox (*Pteropus giganteus*), Short Nosed Fruit Bat (*Cynopterus brachyotis*), Indian False Vampire bat (*Megaderma lyra*), Indian Pipistrelle (*Pipistrellus coromandra*), Asiatic Jackal (*Canis aureus*), Bengal Fox (*Vulpes bengalensis*), Assamese Macaque (*Macaca assamensis*), Porcupine (*Hystricomorph hystricidae*), Sambar (*Cervus unicolor* niger), Golden Langur (*Trachypithecus geei*), Rhesus Macaque (*Macaca mulatta*), Chinese Pangolin (*Manis pentadactyla*), Crab-eating Mongoose (*Herpestes urva*), Tiger (*Panthera tigris*), Leopard (*P. pardus*), Leopard Cat (*Prionailurus bengalensis*), Barking Deer (*Muntiacus muntjak*), Gaur (*Bos frontalis*), Hoary-bellied Himalayan Squirrel (*Callosciurus pygerythrus*), Particoloured Flying Squirrel (*Hylopetes alboniger*) and Golden Jackal (*Canis aureus*). A variety of reptiles and amphibians are also found in this site such as Kraits (*Bungarus* sp.), Cobras (*Naja* sp.), Python (*Python molurus*) and Monitor Lizard (*Varanus* sp.), etc. (*www.assaminfo.com*).

A total of 213 bird species are known to occur in the Sanctuary and the surrounding waterbodies (Barua, 1995), some of which are listed as endangered in IUCN Red Data List. Some of the important and commonly seen bird species are Black Francolin (*Francolinus francolinus*), Jungle Bush Quail (*Perdicula asiatica*), Red Jungle fowl (*Gallus gallus*), Lesser Whistling Duck (*Dendrocygna javanica*), Cotton Teal/Indian Pygmy-goose (*Nettapus coromandelianus*), Little Grebe (*Tachybaptus ruficollis*), Asian Openbill (*Anastomus oscitans*), Woolly-necked Stork (*Ciconia episcopus*), Cinnamon Bittern (*Ixobrychus cinnamomeus*), Indian Pond Heron (*Ardeola grayii*), Cattle

© Bijita Barman's Photography

Figure 17.2: Vegetation and Stream of Chakrashila Wildlife Sanctuary.

Figure 17.3: Forest Cover of the Chakrashila Wildlife Sanctuary.

Figure 17.4: Perennial Stream Flowing through Chakrashila Wildlife Sanctuary.

Figure 17.5: Golden Langur with Child.

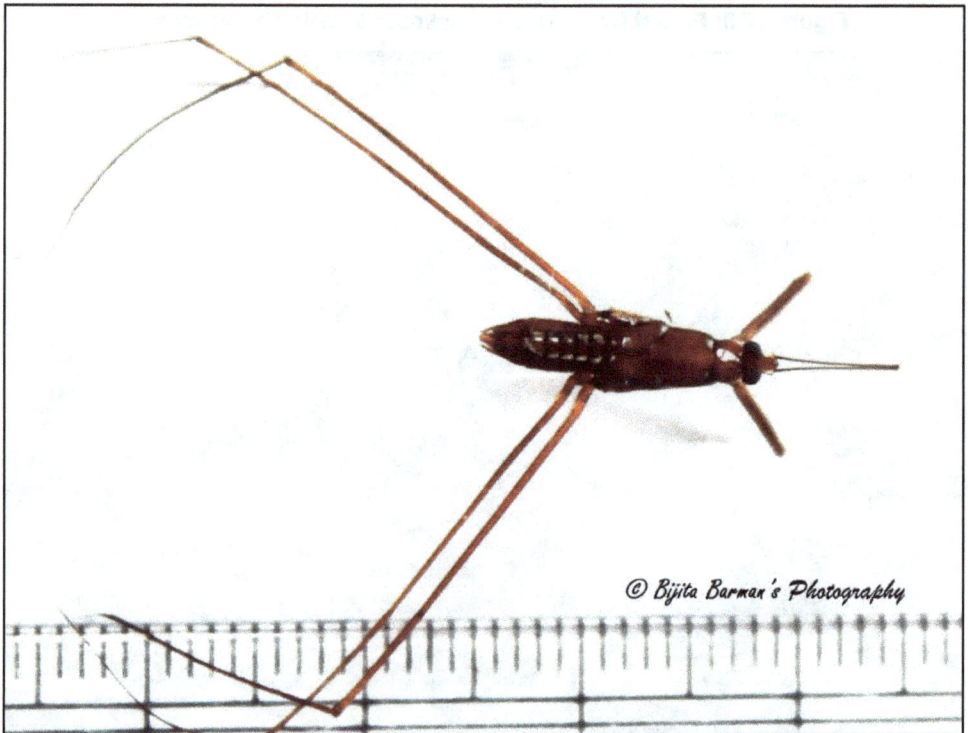

Figure 17.6: Water Strider Available in the Streams of CWS.

Figure 17.7: Dragonfly Nymph in the Lower Stream of CWS.

Egret (*Bubulcus ibis*), Purple Heron (*Ardea purpurea*), Great Egret (*Ardea alba*), Intermediate Egret (*Egretta intermedia*), Indian Cormorant (*Phalacrocorax fuscicollis*), Red-necked Falcon (*Falco chicquera*), Black Kite (*Milvus migrans*), Brahminy Kite (*Haliastur indus*), Indian White-backed Vulture (*Gyps bengalensis,* Critically endangered), Crested Serpent Eagle (*Spilornis cheela*), Crested Goshawk (*Accipiter trivirgatus*), Eurasian Sparrowhawk (*Accipiter nisus*), Greater Spotted Eagle (*Aquila clanga*, Vulnerable), Brown Crake (*Amaurornis akool*), Ruddy-breasted Crake (*Porzana fusca*), Common Moorhen (*Gallinula chloropus*), etc.

However, some globally threatened species are still seen, namely Greater Adjutant (*Leptoptilos dubius*), Lesser Adjutant (*L. javanicus*) and two *Gyps* species of vultures. Among the Near Threatened species, there are Darter (*Anhinga melanogaster*), Ferruginous Duck (*Aythya nyroca*), Greater Greyheaded Fish-Eagle (*Ichthyophaga ichthyaetus*), Red-headed Vulture (*Sarcogyps calvus*), Great Pied Hornbill (*Buceros bicornis*) and Pallid Harrier (*Circus macrourus*).

The Sanctuary also supports a pristine habitat for butterflies. A total of 154 species representing 6 families were recorded within a small area of 45 m^2 of the Sanctuary. Among the butterflies Danaid Eggfly (*Hypolimnas misippus*) is enlisted in Schedule-II of the Indian Wildlife (Protection) Act, 1972 (Chaudhury and Ghosh, 2009). 60 different species of insects are also found in Chakrashila wildlife sanctuary.

The streams also support many numbers of aquatic insects like water bugs (Gerridae- *Ptilomera assamensis, Limnogonous nitidus*; Veliidae- *Rhagovelia obesa*) water beetles (Dytiscidae, Gyrinidae), mayflies (Baetidae- *Offadense* sp.), Caddisflies (Hydropsychidae- *Hydropsyche bidens*), springtails (Embryonidae), etc. These also support various species of fishes, shrimps, etc.

Golden Langur: The Flagship Species

The Sanctuary is especially famous for the endangered Golden Langur. The species is a Schedule I species in the Indian Wildlife Protection Act (1972). It is also listed on CITES, Appendix I. The Golden Langur is found only in the hills of Assam and Bhutan. Chakrashila is the second protected habitat for the Goden Langur. The species has a black face, and a very long tail measuring up to 70-100 cm in length that helps in balancing the body for their arboreal habit. The coat color changes from golden to creamy white in the summer and to a red color in the winter season. During the rainy season it obtains water from dew and rain drenched leaves. It is herbivorous in diet and feeds on fruits, leaves, seeds, buds and flowers. Generally they live in troops of about 8 individuals (at times, more) with several females to each adult male (ZSI, 2010).

There are 501 individuals of the langur in the Sanctuary (Ghosh, 2008) and 5,599 individuals in India (Horwich *et al.*, 2013). The population trend of Golden Langur is declining, for its habitat destruction and the consequent restriction of its populations in fragmented forest pockets.

In India, only a small portion of the Golden langurs' range is protected (40 sq. km in Manas NP and 45 sq. km in Chakrashila WLS) while a substantial population is distributed in different Reserve Forests, Proposed Reserve Forests and in non-forested areas of Dhubri, Bongaigaon and Kokrajhar districts of Assam. Recent estimates compiled with satellite images revealed that 30 per cent of these forest habitats of the Golden langur were lost during the last 10-12 years in India (Forest Survey of India, 1997) resulting in severe fragmentation and degradation of the habitats. The populations that live in these fragmented Reserve Forests and Proposed Reserve Forests are virtually trapped, isolated from the main breeding population and vulnerable to demographic and genetic factors.

In recent year a record of organochlorine insecticide poisoning in three Golden Langurs in the sanctuary was found to be a major threat. In Kokrajhar District there is a rubber plantation covering about 750 bighas (100ha approx.) of land bordering CWS on the north eastern side. A strip of agricultural land of about 100m wide separates the CWS from the rubber plantation. Golden Langurs frequently visit the plantation area in search of food. They are adapted to eat leave and rubber seeds. According to Pathak (2011) the lesions in the liver, kidneys and intestine of the animals suggested prolonged exposure to organochlorine insecticides by ingestion of contaminated leaves of the rubber plants over a long period of time.

Biodiversity Loss and Conservation

The tribes that inhabit the fringe villages of Chakrashila are the Rabha and Bodo. Besides them there are some Garo and the Rajbanshi tribals, along with some

Muslim families as new entrants to the villages. Agriculture is the main occupation of the villagers, with paddy as the main crop. The income levels of the villagers are low, and they depend upon the surrounding forest resources in order to meet most of their daily requirements, such as raw material for houses, agricultural and musical implements; and for food, fuel and fodder. Most of the protein in their diet comes from the forest areas in the form of fish, snails and insects. There is a heavy dependence on the perennial springs of the forest for irrigation and potable water (Datta, 2002).

The denudation of forests began here due to the extreme poverty of the villagers. In order to earn a daily living, the villagers worked for the affluent merchants who hailed from different districts of Assam. They used the local villagers' services as labourers for extraction of firewood and valuable timber from these forests. As a result of indiscriminate smuggling of Sal and other valuable trees, more than 5 sq km area of the thick forest has been completely denuded. The rise in forest denudation led the villagers to encroach deeper and deeper into the forest. This in turn caused further drastic shrinkage of the forests. Again the degradation led to a scarcity of biomass for the local villagers. In March, 2011, a large area near the sanctuary has been cleared by felling valuable trees, including *sal*. The area is around 2km from the Chakrashila forest bungalow and 5km from Kokrajhar town (The Telegraph, 2011). It is an open secret that this sort of land clearing is patronized by political leaders who set up tea and rubber gardens in the area illegally and the fresh clearing of forest land is part of their action plans to grab more land. Chakrashila was a happy hunting ground for local landlords (Zamidars) and others even after it was declared as a reserve forest in 1966. Poaching of birds, mainly waterfowl was also prevalent in the area before it was declared as wildlife sanctuary. This indiscriminate hunting led to the local extinction of a few species such as the Swamp Francolin *Francolinus gularis*.

Though the site falls in the Assam Plains Endemic Bird Area (Stattersfield *et al.*, 1998) the terrain of the Sanctuary is hilly, while the surrounding areas are plain, having a mixture of beels and wet grasslands. Three species have been listed in this EBA: Manipur Bush-Quail (*Perdicula manipurensis*), Marsh Babbler (*Pellorneum palustre*) and Black-breasted Parrotbill (*Paradoxornis flavirostris*). Only Marsh Babbler is likely to occur here as some good patches of grassland are still found. Hence the area demands protection and conservation for endemic species.

Conclusion

The protection of the Sanctuary is of great importance as the Wildlife Sanctuary falls on Eastern Himalayan region and in the biodiversity hotspot Himalaya (CEPF, 2005) which is a data deficient region due to lack of information on current status of diversity and distribution of species (Allen *et al.*, 2010). Again the Wildlife Sanctuary is introduced as a Key Biodiversity Area (KBA) by CEPF (2005) to introduce a protective measurement to the area for the endemic and the most vulnerable species that inhabit the sanctuary. Inspite of different conservational efforts the sanctuary is still under threat. It can be only addressed through the concerted efforts of the villagers, forest departments, politicians and people of the state.

References

Allen, D.J., Molur, S., and Daniel, B.A. (Compilers) 2010. The Status and Distribution of Freshwater Biodiversity in the Eastern Himalaya. Cambridge, UK and Gland, Switzerland: IUCN, and Coimbatore, India: Zoo Outreach Organisation.

Barua, M., 1995. Bird observations from Chakrashila Wildlife Sanctuary and adjacent areas, *Newsletter for Birdwatchers*, **35(5)**: 93-94.

CEPF, 2005. Ecosystem Profile: Indo-Burman Hotspot, Eastern Himalayan Region. WWF, US-Asian Program/CEPF.

Choudhury, A. U., 2000. The Birds of Assam. Gibbon Books and World Wide Fund for Nature Northeast Regional Office, Guwahati. Pp. 240.

Choudhury, K., and Ghosh, S., 2009. Butterflies of Chakrashila Wildlife Sanctuary, Assam. *Indian Forester,* **135(5)**:714-720.

Conservation International, 2007. The Biodiversity Hotspots. (http://www.conservation.org/where/priority_areas/hotspots/Pages/hotspots_main.aspx).

Datta, S., 2002. 'An NGO Initiated Sanctuary: Chakrashila, India'. In A. Kothari, N. Pathak, R.V. Anuradha, and B. Taneja, Communities and Conservation: Natural Resource management in South and Central Asia (New Delhi, Sage Publications and UNESCO, 1998).

Forest Survey of India, 1997. The State Forest Report. Ministry of Environment and Forest, Government of India, Varun Offset Printers, Dehradun.

Forest Survey of India, 2002. State of Forest Report 2001, Forest Survey of India, Ministry of environment and Forest, Dehra Dun, India, 130pp.

Ghosh, S., 2008. Report on Population Estimation of Golden Langur (Southern Population) in Chakrashila Wildlife Sanctuary and Reserve Forests Under Kokrajhar, Dhubri and Bongaigaon Districts of Assam. Report to the Assam Forest Department, Guwahati, Assam.

Hoekstra, J.M., Boucher, T.M., Ricketts, T.H., and Roberts, C., 2005. Confronting a biome crisis: Global disparities of habitat loss and protection. *Ecological Letters*, **8**: 23-29.

Horwich, R. H., Das, R., and Bose, A., 2013. Conservation and the Current Status of the Golden Langur in Assam, India, with Reference to Bhutan. *Primate Conservation,* **(27)**:1-7.

ICBP 1992. Putting biodiversity on the flap- Cambridge, U.K.: International Council for Bird Preservation.

Lopez, A., and Mundkur, T. (eds.) 1997. The Asian Waterfowl Census: 1994-1996. Results of the Coordinated Waterbird Census and Overview of the Status of Wetlands in Asia. Wetlands International, Kaula Lumpur. Pp. 118.

Mittermeier, R.A., Gil P.R., and Mittermeier, C.G.., 1997. *Megadiversity: Earth's Biologically Wealthiest Nations.* Conservation International, Cemex.

Mittermeier, R.A., Gil, P.R., Hoffmann, M., Pilgrim, J., Brooks, T., Mittermeier, C.G., Lamoreux, J., and da Fonseca, G.A.B., 2004. *Hotspots revisited*. Cemex: Mexico City, Mexico.

MoEF and Kalpavriksh, 2004. Nat. Biodiversity Strategy and Action Plan,India:Final Tech.Report of the UNDP/GEF Sponsored Project. MoEF,Govt.of India, and Kalpavriksh, New Delhi/Pune.

Myers, N., Mittermeier, R.A., Mittermeier, C.G., da Fonseca, G.A.B., and Kent. J., 2000. Biodiversity hotspots for conservation priorities. *Nature*, **403**:853-858.

Olson, D.M., and Dinerstein, E., 2002. The Global 200;Priority ecoregions for global conservation. *Annals of Missouri Botanical Garden*, **89**: 199-224.

Pathak, D.C., 2011. Organochlorine insecticide poisoning in Golden Langurs (*Trachypithecus geei*). *JoTT*, **3(7)**: 1959–1960.

Proctor, K. H., Haridasan, K., and Smith, G. W., 1998. How far does lowland tropical rainforest go? *Global Ecology and Biogeography Letters,* **7**:141-146.

State Biodiversity Strategy and Action Plan, Assam, 2009. Forest Eco-System. (http://asmenvis.nic.in).

State of Forest Report, 2011. India State of Forest Report 2011. Forest Survey of India, Ministry of Environment and Forests, Govt. of India, Dehra Dun.

Stattersfield, A. J., Crosby, M. J., Long, A. J., and Wege, D. C., 1998. Endemic Bird Areas of the World: Priorities for Biodiversity Conservation.BirdLife Conservation Series No. 7. Bird Life International, Cambridge, U.K.

The Assam Tribune, 2006.Vulnerable turtle spotted at Diplai Beel. By A Staff Reporter, March 4. (http://environmentportal.in/content/228363/vulnerable-turtle-spotted-at-diplai-beel).

The Telegraph, 2011. Encroachment threat to Chakrashila. March 1 (http://www.telegraphindia.com/1110302/jsp/northeast/story_13653466.jsp).

WWF and ICIMOD. 2001. Ecoregion-Based Conservation in the Eastern Himalaya: Identifying Important Areas for Biodiversity Conservation. Kathmandu: WWF Nepal.

ZSI, 2010. ZSI e-NEWS The Monthly Electronic Newsletter of Zoological Survey of India.pp.17.

2015, Animal Diversity, Natural History and Conservation, Vol. 5   *Pages 281–303*

Editors: V.K. Gupta and Anil K. Verma

Published by: DAYA PUBLISHING HOUSE, NEW DELHI

Chapter 18

In situ and *Ex situ* Conservation of Red Panda in Darjeeling District, West Bengal, India

Jayanta Kumar Mallick (Retd.)

Wildlife Wing (Headquarters),
Forest Directorate, Government of West Bengal

ABSTRACT

The nominate race of Red Panda *Ailurus fulgens fulgens* F.G. Cuvier, 1825, endemic to Eastern Himalayas, is a flagship species for *in situ* and *ex situ* conservation. In West Bengal, two disjunct populations (Linear distance ca. 110 km) are confined mostly to the core area of two protected areas (National Parks) of Darjeeling District [Singalila (78.60 km²) on the north-west and Neora Valley (88 km²) on the north-east] and rarely in the surrounding Reserve Forests (buffer). They prefer the undisturbed temperate upper hill forests near a waterbody at an altitudinal range of 2,100-3,600 m, particularly the dense evergreen canopy and thick undergrowth (visibility not more than 5–10 m, often less than 2 m) of *Abies, Acer, Juniperus, Lithocarpus, Magnolia, Quercus, Rhododendron, Yushania and Arundinaria.* Four captive-bred red pandas of Padmaja Naidu Himalayan Zoological Park, Darjeeling, have been successfully reintroduced in Singalila in two phases. Due to anthropogenic threats like poaching, conversion of natural high forests into plantations, wanton exploitation of the natural resources, encroachment, expansion of tourism facilities, lack of coordination between the forest department and other external agencies and have caused decline in its population and preferred habitats. Strengthening of the forest protection force, provision of equipments, extension of contiguous suitable habitats, joint protected area management by the park managers and the stakeholders, species-specific conservation initiatives will assist in long term survival of this vulnerable species.

Keywords: Conservation, Habitat, Survey, Sighting, Signs, Threats.

* *Corresponding author.* E-mail: jayantamallick2007@rediffmail.com

Introduction

Red Panda, *Ailurus fulgens* F.G. Cuvier, 1825 (type locality "East Indies") [synonyms *ochraceus* Hodgson 1847 (type locality "sub-Himalayas") and perhaps *refulgens* Milne-Edwards 1874 (type locality unknown)] is the only species of the family Ailuridae Gray, 1843 and a terminal relic of once flourishing group, a living fossil, which is not closely related to any other extant species. It is both extremely significant biologically and of high conservation value. Visually, it is a very attractive (most strikingly patterned and vividly coloured of all mammals- chestnut, chocolate and white) animal, but, behaviourally, rather unusual member of Carnivora, which has adapted to a very specialized niche in the broadleaf and conifer ecoregion with narrow ecological requirements and become a specialised herbivore with a low nutrient diet, *i.e.* the bamboo (mostly *Arundinaria maling* and *A. aristata*) leaves and during the monsoon it is supplemented by fruits of various trees and shrubs as well as bamboo shoots. Because of its specialized diet and narrow range of habitat, this charismatic species is considered an indicator of the ecosystem health in Eastern Himalayas. Moreover, red panda has a high mortalily rate due to insufficient maternal care of vulnerable young, high susceptibility to canine distemper virus and other bacteria and ectoparasites, deleterious effects of inbreeding, very low reproductive or protracted growth rate and very small litter size compared to other carnivores. It has also become vulnerable (IUCN Red List) to extinction in the wild because its population had already declined by 40-50 per cent over the last 50 years due to tremendous anthropogenic threats (deforestation, developmental activities, encroachment, habitat loss and fragmentation, poaching, capture and illegal trade).

Therefore, red panda is considered a flagship species in the Himalayan region and designated for *in situ* and *ex situ* conservation. Habitat protection and supporting wild populations or *in situ* conservation, is the most effective management technique, use of reintroduction of captive-bred animals as a tool of conserving species is becoming increasingly popular since many wild populations and their habitat are continuing to decline. Here, *ex situ* conservation is also important for maintenance of a species for eventual release back into the wild to restore or supplement wild populations, as has been experimented in Padmaja Naidu Himalayan Zoological Park (PNHZP), Darjeeling.

Distribution and Range

Red Panda is mostly found in the high altitude temperate forests. The possible exception to this is a small population which has been reported to occur in the semi-tropical forests of Meghalaya. Out of two subspecies (the western, nominate form, *A. f. fulgens,* and the eastern Chinese form, *A. f. styani*), the former race is endemic to Eastern Himalayas, the range forming a crescent from Nepal, Tibet, Bhutan, north-eastern India (northern West Bengal, Sikkim, Arunachal Pradesh and Meghalaya) up to western Yunnan province in China and northern Myanmar. They split around 3 million years ago when torrential river flows cut the Eastern Himalaya forming the Brahmaputra Gorge.

In northern West Bengal, Red Panda is recorded from the high forests of Darjeeling (Agrawal *et al.,* 1992; Anon, 1967, 2008, 2009, 2010; ATREE, 2008; Bahuguna, 1993,

1995; Bahuguna *et al.*, 1998; Bahuguna and Mallick, 2010; Biswas and Ghose, 1982; Biswas *et al.*, 1999; Chaudhuri and Sarkar, 2003; Choudhury, 2001; Chowdhury, 1983; Ghose *et al.*, 2007, Jha, 2011, 2012; Mallick, 2010a, b, 2012; Mukhopadhyay, 1996; O'Malley, 1907; Pradhan, 1998, 1999, 2006; Pradhan *et al.*, 2000, 2001; Prince, 2003; Saha and Singhal, 1996, 1999; Saha *et al.*, 1992; Sclater, 1891; Sharma, 1990; Sharma *et al.*, 2005; Singhal and Mukhopadhyay, 1998; Tikadar, 1983; UNESCO, 2009, Williams, 2004; Wright, 2004). It is now confined only to the core areas of two National Parks (NPs), Singalila and Neora Valley, in Darjeeling District. Red Panda also inhabited Senchal Wildlife Sanctuary (Darjeeling District) in the recent past. In 2008, a Red Panda was found at Kainjaley, 47 km from Darjeeling, which is located outside its known range in Darjeeling. Prior to notification of Neora Valley NP, Red Panda was also recorded in Mouchowki forests (9.57 km²) under East Nar block, at an altitude of 1,311 m, in the lower Neora Valley NP. Red panda has never been known in the entire West Nar block of this Lower Range.

Study Significance

Despite its ecological significance and taxonomic uniqueness, red pandas have been studied much more in captivity than in the wild. Hence, very little is known about this rare, shy, reclusive and secretive species in its natural habitat. Not much information is available on the site-specific conservation status of this species. Being a traditional trek route, Singalila is popular, but until a biological expedition in December 1982, Neora Valley was poorly known.

It is neither familiar to the general public nor many professional biologists. Although it is listed as a Schedule- I (Endangered) species in the Indian Wildlife (Protection) Act, 1972 and an Appendix I species under the Convention on International Trade in Endangered Species of Wild Fauna and Flora (CITES), reversing its risk of extinction in the wild is the most challenging task of the protected area managers, which can only be achieved through intensive joint forest management.

Study Areas

Singalila, located on the northwestern border with Sikkim and Nepal, at an altitudinal range of 1,500-3,660 m, was declared as NP over 78.60 km² in 1992 (Figure 18.1). It lies between 27°13' and 27°14'N and between 88°01' and 88°07'E. The buffer area extends over 96.28 km² of contiguous forests of Darjeeling forest division. The park has temperate and sub-alpine climates. In temperate areas, the mean summer temperatures range as high as 17C, in winter as low as -1C. In the sub-alpine zone, the temperatures are lower. The annual rainfall is up to 330 cm with maximum rainfall occurring during June-August. Humidity is very high throughout the year (average 83-96 per cent). There is regular snowfall above 3,300 m at Sandakphu.

Figure 18.1: Singalila National Park.

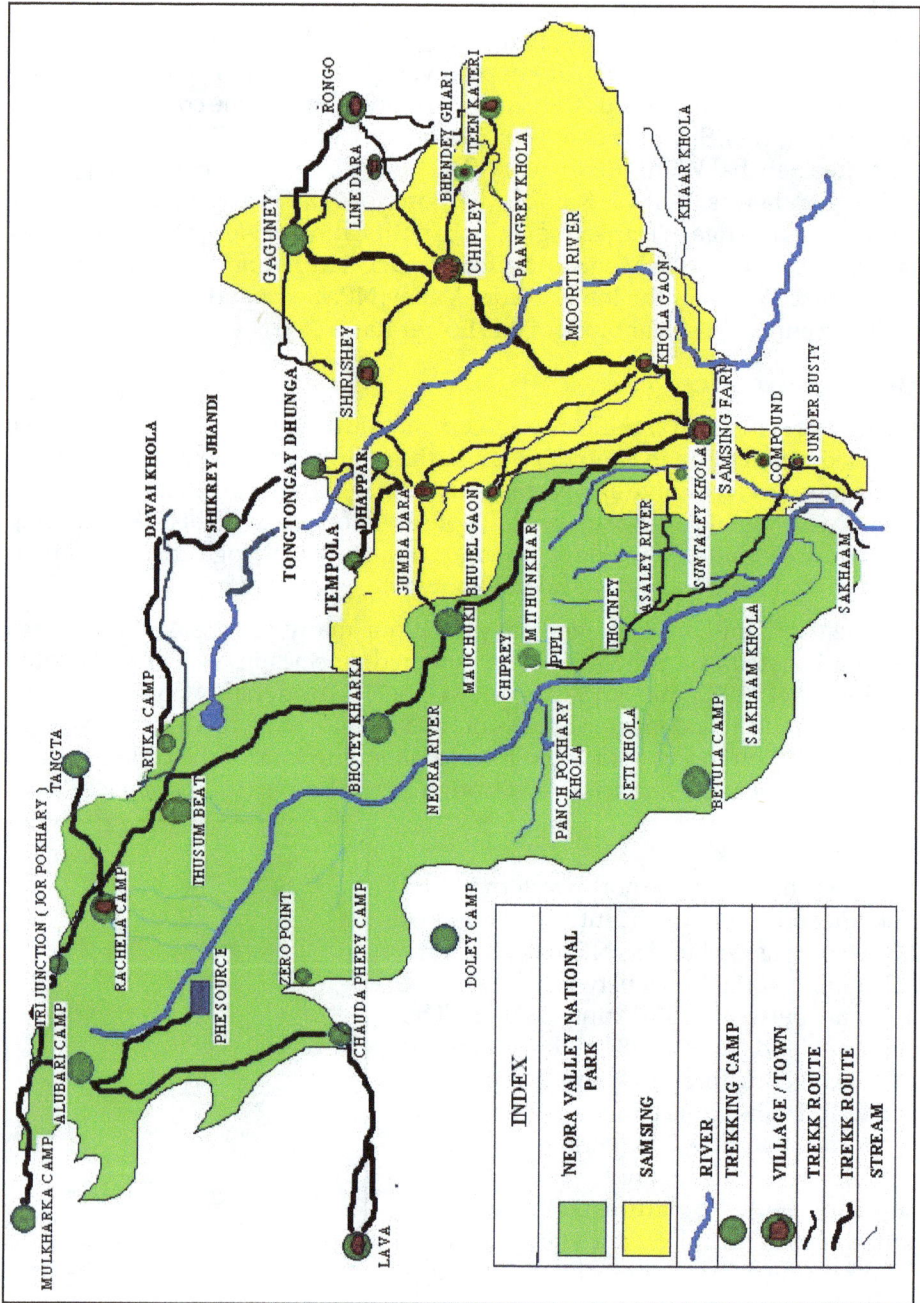

Figure 18.2: Neora Valley National Park and Surrounding Forests.

Neora Valley NP (88 km²) was notified as a protected area in April 1986 and finally gazetted in December 1992. It was short-listed as a World Heritage Site on 26 May 2009 (UNESCO World Heritage Centre 2009). Being located in the Kalimpong subdivision, it lies within 26°52'–27°7'N, 88°45'–88°50'E (Figure 18.2). It is contiguous with Sikkim and Bhutan at its northern and north-eastern boundaries respectively. The southern boundaries are adjoining to the forests of Jalpaiguri district which have connectivity with the Chapramari WLS and the Gorumara NP. The eastern and western sides are reserve forests under Kalimpong Forest Division. Neora Valley is divided into two ranges, Upper (Headquarters: Lava, the western entry point) and Lower (Headquarters: Samsing, the eastern entry point). The park has a wide altitudinal range (183–3,170 m). The highest point is Rachila *danda* (peak). The climatic condition also varies from tropical or sub-tropical in its lower range to temperate in its higher range.

Methods

A study was conducted in Darjeeling Himalayas, based on a literature review, questionnaire survey and forest trail survey for direct sighting and signs like faecal pellets during 2009-2010. The questionnaire survey was conducted on a set *pro forma* among forest officials and camp staff, researchers, tourists/guides, the forest and fringe villagers including Eco-development and Forest Protection Committee members. Colour photographs of eight species of small carnivores recorded in the region, namely Clouded Leopard *Neofelis nebulosa*, Marbled Cat *Pardofelis marmorata*, Hog Badger *Arctonyx collaris*, Masked Palm Civet *Paguma larvata*, Yellow-throated Marten *Martes flavigula*, Beech Marten *Martes foina*, Stripe-backed Weasel *Mustela strigidorsa* and red panda, were shown to the villagers for correct identi-fication of red panda. The field stations were visited for interviewing the staff posted there.

During the ground survey conducted during the pre-monsoon (February–May) and post-monsoon periods (October–January) and in the rainy season during clement weather, the dense forest on rugged terrain, particularly in Neora Valley NP, prevented straight-line transects. The trail-cum-trekking routes in the parks were walked in search of live red Pandas and their evidences.

During the study, approximately 576 hours, equivalent to 72 person-days (excluding inclement weather conditions [stormy, rainy, foggy, frosty and snowy], unproductive traversing and holidays) were used for habitat evaluation, direct sighting, collection of its faeces and questionnaire survey. In all, 188 people of various categories like researchers, forest staff and officers having working experience in the study area, villagers living in and around the parks and tourists/guides) responded to the questionnaire survey.

Results and Discussion

Habitat Use (Figures 18.3–18.4)

In Singalila NP, red panda is relatively abundant within 2,800 and 3,600 m, where its narrow preferred range is 2,800-3,100 m (Pradhan *et al.*, 2001). Here the 2,600–2,800 m zone is dominated by the Oak Forest and that of 2,800–3,000 m by

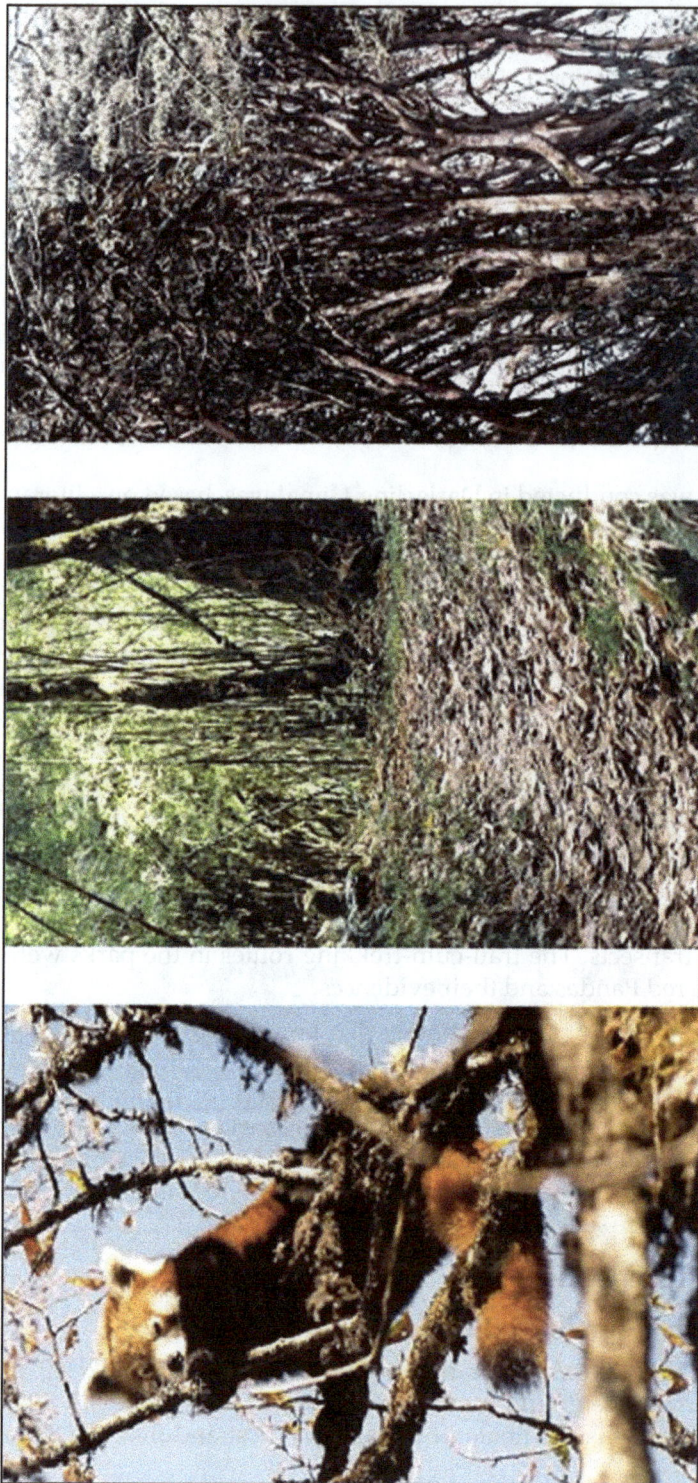

Figure 18.3: Neora Valley NP: (a) Red panda resting on tree; (b) Bamboo habitat; (c) Rhododendron habitat (Sourabh Basu).

Broad Leaf Deciduous Forest. On the eastern side of Singalila ridge, the Oak Forest is composed of an upper canopy of *Lithocarpus pachyphylla* (81.2), *Litsea sericera* (20.9) and *Magnolia campbelli* (17.1) with the sub-canopy dominated by *Litsea elongata* (55.0), *Schefflera impressa* (36.4), *Rhododendron arboreum* (32.0) and *R. griffithiana* (17.2) while the most common shrubs found were *Arundinaria maling, Vitex erubescena, Daphne cannabina,* and *Piptanthus nepalensis.* The Broadleaf Deciduous Forest is dominated by *Sorbus cuspidata* (49.6), *Acer* sp. (25.1), *Litsea sericera* (17.0), and *Quercus pachyphylla* (10.8) with the sub canopy being composed of *Symplocos* sp. (68.8), *Rhododendron arboreum* (33.3), *Osmanthus suavis* (23.47), *Rhododendron falconeri* (21.4), *Viburnum erubescens* (10.3), and *Meliosma delleniaefolia* (8.9) and the common shrubs were *Arundinaria maling, A. aristata, Vitex erubescense, Daphne cannibina, Rosa sericera, Piptanthus nepalensis,* and *Eltzostia* sp.

Encounter rate (n=32) of both red panda and pellet groups in Singalila NP was highest in the broad leaf deciduous forest followed by subalpine and oak forest close to a waterbody (Pradhan *et al.,* 2001). The tree species used frequently and which formed important components of red panda habitat were *Abies densa, Betula utilis, Sorbus cuspidata, Quercus pachyphylla, Magnolia campbellii* and *Rhododendron* spp.

Pradhan *et al.* (1998) registered sightings from the core zone at Gairibans, Kalpokhri, Sandakphu, Phalut, Moley (Sabarkum) and the upper reaches of Gorkhey bordering Sikkim. Sighting was not recorded from lower Gorkhey, Rammam, Siri, Samanden, Rimbik and Gurdung areas in the buffer zone. The lowest altitude from where red panda was reported was Upper Phedi (2,400 m) in the buffer areas. However, Bahuguna *et al.* (1998) confirmed its presence at Kankibong at an altitude of 2,100 m.

Williams (2004) in his study along the Indo-Nepal border recorded three sightings- one at 2,857 m, when a red panda first ran into the Nepalese side of the ridge and then scampered across the border road into Singhalila NP. The second sighting occurred at approximately 2,442 m, 0.5 km below Gairibans on the Jamuna-Gairibas road, when a red panda shifted from a *Lindera pulcherrima* into a *Rhododendron arboreum* and scampered away over a large boulder. The third sighting took place at 2,685 m, directly below Kaiyakatta, 200 m southwest of Kaiyakatta creek. The red panda jumped down from its resting spot on the stump of a twisted *Rhododendron grande* and sped into the mist. He had also sighted two red pandas surrounding Kaiyakatta border area. Therefore, maximum number of sightings took place in and around Kaiyakatta in between 2,000 m and 2,850 m. Though the red pandas are usually solitary, three more adult red pandas were seen together on a tree (*Rhododendron* sp.) during the mating season in late December (winter), whereas two cubs with a female red panda were seen in October and one cub with mother in November.

In Neora Valley NP, the tropical lower hill forest, mostly deciduous and semi-evergreen, extends up to about 750 m. Sub-tropical middle hill forest is found over 750–1,700 m with abundant evergreen trees, dense canopy and undergrowth. Temperate upper hill forests (1,700–3,200 m) are also distinguished by the predominance of Lauraceae, *Quercus lamellosa, Q. lineata, Q. spicata, Elaeocarpus lanceaefolius, Echinocarpus* and *Acer campbelli* (Buk-Oak mixed vegetation) and

Lithocarpus (High-level Oak) forests. The lower limit of upper hill forest, *i.e.* Lauraceous (*Machilus–Michelia*) forest occurs between 1,650 m and 2,100 m. Deep valleys have diverse shrubby (understorey) species forming mixed and mesophyll communities. Pure stands of the dwarf bamboo *Yushania* (= *Arundinaria*) *maling* occur in small stretches of almost flat lands, particularly in the middle hills (Singhal and Mukhopadhay, 1998).

Here, red Panda was once recorded as low as 1,170 m, but now it is found only above 2,100 m in the upper Neora Valley NP (Ghose *et al.,* 2007). The minimum density of vegetation is greater than 40 per cent canopy coverage except at Rachila Chawk, which was deforested in 1879 and brought under a regeneration programme in 1996 and 1997 (Singhal and Mukhopadhyay, 1998). In the Buk-Oak mixed forest at 2,100–2,400 m, *Quercus lamellosa* dominates, with associates like *Q. lineata, Q. spicata, Castanopsis tribuloides, Acer campbelli, Machilus odoratissima, M. gammieana* and *Elaeocarpus, Michelia excelsa* and *Bucklandia.*

In the High-level Oak forest over 2,400–2,750 m, mainly in Rashet and Rachila blocks, *Lithocarpus pachyphylla* predominates, with common associates like *Q. lamellosa, Acer campbelli* and *Magnolia campbelli. Yushania maling* is found scattered all over *Quercus* and *Lithocarpus* forests.

Further higher zone (coniferous forest) harbours pure patches of *Tsuga* and undergrowth of bamboos and *Rhododendron.*

Mallick (2010) recorded 34 sightings of red panda (31 in upper Valley and three in contiguous Kalimpong Forest Division) during a period of 11 years, *i.e.* 1999–2009 (Table 18.1). In 2009, red pandas were sighted on eleven occasions. There were three dur-ing pre-monsoon (May), four during rainy season (June–July) and four during post-monsoon (October–January). Red pandas were sighted mostly in the early morning and late afternoon. On all occasions, only a single red panda was sighted.

Red panda faeces (no other form of sign was recorded) were found at six locations. In all six, red panda was also sighted: Upper and Lower Choudapheri, Zero Point, Alubari, Rhenock and Rachila. Faeces were mostly on *Abies densa* and *Lithocarpus pachyphylla,* followed by *Magnolia campbelli* and *Rhododendron*, especially those with mossy trunks, *Betula utilis,* followed by *Ilex hookeri, Osmanthus* and *Sorbus cuspidata.*

Red panda was found mainly in the north-western part of Neora Valley NP, *i.e.* in five of the 25 compartments surveyed (20 per cent), mostly in Rashet 3 (four records) and Rachila 11 (three records) compared with Rhenock 4b (two records), Rashet 4 (one record) and Rachila 12 (one record). The highest altitude record was at 3,170 m in Jorepokhri, the lowest at 2,350 m in Chaudapheri. There were no sightings from the two adjoining blocks of Kalimpong Forest Division.

Red Pandas were found only in high, dense, moist temperate forest with thick undergrowth, where many trees are ancient and hollow, especially in the less accessible areas. However, Red Pandas were neither sighted nor reported in the Lauratious high forest. The park's other two habitats, the sub-tropical forests in the middle hills and the tropical forests in the lower hills, do not support Red Pandas.

Table 18.1: Location-Specific Sighting Records of Red Panda in Neora Valley NP and Adjacent Kalimpong Division

Date/Month/Year	Location	Source	Remarks
December 1999	Upper range	Questionnaire survey: forest staff	First rescue during Tiger census, sent to Darjeeling Zoo for treatment.
March 2000	Upper range	Questionnaire survey: forest staff	Second rescue, sent to Darjeeling Zoo, but died on the way.
25 November 2002 (morning)	Rashet 3	Prince, 2003	Panda bounding along the track, disappeared into the bamboo grove.
March 2004	Rashet 3	D. Ghose	Near Chaudapheri Red Panda camp.
April 2005	Rashet 3	D. Ghose	1st Mile Road on the Pankhasari ridge. Miscellaneous tree species.
April 2005	Rashet 3	Questionnaire survey: forest staff	Plantation area.
April 2006	Rahset 3	Questionnaire survey: forest staff	Plantation area.
May 2006	Rachila 11	D. Ghose	Upward trail from Rashet 3 to Rachila 11, then downward slope to Zero Point.
6 September 2006	Rashet 4	ATREE 2008	Panda eating bamboo leaves/shoots in mixed vegetation with 60 per cent bamboo undergrowth.
8 March 2007	Rashet 4	Questionnaire survey: forest staff	1st mile on Pankhasari ridge. Mixed vegetation.
2 April 2007	Ruka 4	Questionnaire survey: forest staff	Mostly *Y. maling* with scattered patches of *Rhododendron*. On the eastern slope of Thosum La *Lithocarpus* is found.
8 April 2007	Rachila 11	Questionnaire survey: forest staff	Broadleaved forest and bamboo thicket- *Rhododendron, Michelia, Alnus nipalensis, Q. lamellosa,* undergrowth of *Arundinaria racemosa, Eupatorium adenophorum, Maesa chisia, Aesculus,* also ferns and mosses.
18 May 2007 (within 15h30–16h55)	Alubari towards Hathidanda	ATREE 2008	Panda, sitting on an oak tree, photographed.
14 August 2007	Ruka 4	Questionnaire survey: forest staff	Beyond 100–200 m wide strip in the eastern boundary of the park.
5 November 2007	Rachila 13	Questionnaire survey: forest staff	Erstwhile forest village, evacuated and brought under plantation programme.
2 February 2008	Rashet 4	Questionnaire survey: forest staff	Above 2,200 m *Quercus, Rhododendron,* mixed with *Lithocarpus* and *Yushania* are found; *Y. maling* covers about 70 per cent of the area above 2,300 m. *Arundinaria* occurs elsewhere. A large area along the Pankhasari ridge is unproductive.

Contd...

Table 18.1–*Contd...*

Date/Month/Year	Location	Source	Remarks
10 March 2008	Rashet 4	Questionnaire survey: forest staff	Mixed vegetation.
17 April 2008	Rashet 3	Questionnaire survey: forest staff	Mixed plantation of native hardwoods.
11 May 2008	Pankhasari-1 (Kalimpong Division)	Questionnaire survey: forest staff report not available.	Mixed vegetation. Carcass of an adult male found. Post mortem
22 November 2008	Near Tiger Camp	J. Das	Sighted during Tiger census operation. *Lithocarpus* forest and scattered pure bamboo.
22 November 2008	Rachila 13	M. Roy	Oak forest and pure bamboo thickets. During Tiger census a Panda was sighted on a tree near Jorepokhri (3,128 m asl; on the way from Alubari), from 08h43 for 45 minutes.
12 December 2008	Rachila 2	Questionnaire survey: forest staff	*Lithocarpus* predominates, changing to *Rhododendron* and *Y. maling* above 2,800 m.
15 December 2008	Rhenock 4b	Questionnaire survey: forest staff	*Y. maling* covers about two-thirds of the area. Rhododendron is also found mixed with *Lithocarpus* and *Quercus*.
Early January 2009	Rashet 3	A British tourist	The Panda was sitting on a tree (Anon., 2009).
5 May 2009	Rachila 11	Study team	Miscellaneous tree species; some plantations.
7 May 2009	Rashet 3	Study team	Mixed plantation (*Rhododendron* and *Lithocarpus* area.
23 May 2009	Rashet 3	Study team	Eating acorns, sat on branch of an oak.
3 June 2009	Rachila 12	Study team	Sitting on a tree by Zero Point–Alubari trail.
9 June 2009	Rhenock 4b	Study team	Dwarf bamboo, *Rhododendron* mixed with oak and Buk-Oak.
12 June 2009	Rhenock 4b	Study team	Dwarf bamboo, *Rhododendron* mixed with oak and Buk-Oak.
24 July 2009	Rachila 11	Study team	Miscellaneous tree species; some plantations.
20 October 2009	Rashet 4	Study team	Miscellaneous tree species.
17 November 2009	Rashet 3	Study team	Mixed plantation area.
8 December 2009	Rachila 11	Study team	Miscellaneous tree species; some plantations.

Figure 18.4: Foraging Red Panda (Sourabh Basu).

In dense canopy and thick undergrowth (visibility not more than 5–10 m, often less than 2 m), direct Red Panda sighting was very limited. Sighting in the *Quercus–Lithocarpus* forest with undergrowth of *Yushania* (2,100–2,400 m) was also not frequent. Most sightings were recorded in the second storey of the high forests, generally occurring above 2,400 m, particularly in Rashet and Rachila blocks, where *Lithocarpus pachyphylla* is the predominant tree species. Red Panda was often sighted in the plantations of native hardwood species (comparatively younger trees) in Rashet 3 compartment.

Red Panda sightings were confined to four forest blocks of the park, spread over 37 km² (Table 18.2). These four forest blocks account for about 43 per cent of the total area of Neora Valley NP.

Even though there were no 2009 records, Red Panda was also reported during the questionnaire survey from outside the park in two adjoining blocks of Kalimpong Division, *i.e.* Pankhasari (48 km²), south of Rashet (Chaudapheri) block and Ruka (18 km²), east of Rachila block. Only three sightings of Red Panda in these two Reserve Forests, in 2007 and 2008, were reported by the questionnaires. The combined Red Panda habitat in Neora Valley NP and these adjoining blocks of Kalimpong Division is about 103 km².

There are no concrete evidences of seasonal migration of red panda in Neora Valley NP. Ghose *et al.* (2007) observed that "during spring the Red Pandas are everywhere [whether Singalila NP or Neora Valley NP or both, was not specified] but during the monsoon they are only found in small patches". This statement warrants further investigation.

Table 18.2: Blocks/Compartments Surveyed and Evidences of Red Panda in Upper Neora Valley NP

Block	Compartments Surveyed (Total)	Compartments with Evidence (Total)	Area (ha)	Latitude	Longitude	Altitude (m asl)
Rachila	1–17 (17)	2, 11, 12, 13, 14 (5)	1,759	27°05'–27°07' N	88°43'–88°45' E	1,400–3,150
Thosum	1–4 (4)	3 (1)	979	27°04'–27°05' N	88°45'–88°46' E	1,250–3,050
Rhenock	4b–5 (2)	4b (1)	691	27°07' N	88°43' E	1,200–3,000
Rashet	3–4 (2)	3, 4 (2)	299	27°05'–27°07' N	88°42'–88°44' E	1,450–2,650
4	25	9	3,728	27°04'–27°07' N	88°42'–88°46' E	1,200–3,150

The previous findings were more or less corroborated by Pradhan (2012). The sightings and evidences of red panda in upper Neora Valley areas were recorded between 2,230 m and 3,100 m such as Ake Mile (outside the protected area), Chaudpheri area (near Goth), Chaudpheri (Plantation area), Upper Chaudpheri, Alubari (on way to Mulkharga), Alubari (near Jorepokhari), Rechilla (on way to Rechilla), and Rechilla main, Path leading to Mulkharga, Hathi danda and on way to Tiger camp. Encounter rate of pellet group was highest in around Hathi danda, Jorepokhari and Rachilla, roughly included in compartments 5, 9, 6, 10 and 4b of Rhenok, and Compartments 14 and 13 of Rechilla, followed by areas in and around Chaudapheri with medium to high abundance and a low to medium abundance of red panda was in the areas in and above (ridge forest) of Alubari. No evidence of red panda was, however, found in the PHE source area, Jaributi, and areas below Alubari. 40 per cent of the red panda evidences were found at the lowest vegetation zone of 2,200 – 2,600 m. There was a significant dip to 13 per cent of evidence found in the vegetation zone 2,600 – 2,800 m, while evidences of red panda was highest with 47 per cent in the altitude zone above 2,800 m.

Quercus pachyphylla featured as the most important tree species used by the red pandas (63 per cent) followed by *Sorbus cuspidata* (16 per cent), *Rhododendron* species (10.5 per cent) and rest in other species. Some of the other important trees used were *Acer* spp, *Vitex heterophylla* and trees supporting the saprophytic tree locally known as 'Lahare tenga'.

Population Estimates

In Singalila, the recorded density was 1 mature individual per 1.67 km, when the population was estimated to no more than 26 (Bahuguna *et al.,* 1998) and subsequently a maximum of 47 individuals (Pradhan *et al.,* 2001). The last census estimate (2000) by the forest department was 78 individuals. It is difficult to estimate the relative red panda population size in this border area because the population is not confined to either the Nepalese side of the ridge or the Indian side at Singalila NP.

ATREE (2008) recorded sightings of two Red Pandas in Neora Valley NP, one each in 2006 and 2007, photographed one of them and collected their pellets from six transects. They estimated a population of 28–32 red pandas in the upper Neora Valley by extrapolating their probable density in the surveyed transects to the rest of the area.

On the basis of above records, total population of the wild red pandas in Darjeeling Himalayas is ca. 110.

Threats

The biotic and abiotic threats to the red pandas and habitats in the study area are:

(i) Fire

Fire has swept large areas of the parks in the past. Outbreaks of fire in the late 19th and early 20th century caused considerable damage to the forests resulting in extermination of *Abies densa* and some other trees. Such incidents were marked in Singalila than Neora Valley.

(ii) Cattle Grazing

Concessions were given to the Khasmahal tenants to allow grazing above 2,550 m. This has reduced the undergrowth and caused considerable damage to the flora. The earliest records of cattle, goats and sheep in these areas are by O'Malley (1907). It is known that grazing was restricted to areas above 2727 and 2879 m in Tonglu and Singalila Ranges for which grazing permits were issued (Anon., 1967). A minimum of 80-90 'Goths' were present in Singalila area. Each 'Goth' had a minimum average of 30-50 cattle. Establishment of a cattle station required clearing of vegetation, thus causing gaps in the canopy cover. The area was also bereft of bamboo, an important component of red panda habitat. Free range livestock do not allow forest regeneration and slowly turn once lush broad leaf forest into shrubland and grassland in the long run. After declaration of the Singalila area NP it was closed to cattle grazing and cattle stations known as 'Goths' between 1991 and 1993. But grazing is still widespread from northern Kalpokhri, Kaiyakatta, Gairibans up to Jaubari 3 km on the south. Free range livestock consume *e.g.* Malingo (*Yushania maling*) shoots causing its decline, but its regeneration is most important for red pandas. Malingo is also used for roofing, fencing, walls, baskets, bedding, to support crops, as combs, kitchen utensils and baby cribs. Less availability of this species is caused by overconsumption of its shoots and overharvesting. During the monsoon, Malingo and Nigalo shoots are extensively exploited. Household consumption was assessed to be 10.96 kg per family (Pradhan *et al.*, 2000). A household also uses about 1038 culms per year. Lack of regeneration of other potential food species of red panda also threatens survival of this species. Cattle grazing is not a threat in the upper Neora Valley. The lower valley is, however, affected by illegal grazing, but it is not a known habitat of red panda now-a-days.

(iii) Landslips

The area is prone to landslips which are facilitated by forest destruction, overgrazing. unterraced cultivation, poor drainage provision and excavation of pits or quarries. Landslides are more common in Singalila than Neora Valley.

(iv) Hunting

There were a few reports of hunters coming into Neora Valley NP from neighbouring areas (Ghose *et al.*, 2007). Live-trapping of red panda for supply to the zoos was a lucrative business in Darjeeling. There has been large scale capture from present Singalila NP for this purpose, particularly during 1960s after creation of Darjeeling Zoo. At least 300 individuals were captured from this area (Bahuguna *et al.*, 1998). This illegal trade nexus continued till the late 1970s (Pradhan, 2012). A trader in Rimbick area had received as many as 16 red pandas in a day. Among the animals received, there would also be pregnant females and incidences of cubs being born in the cages were also reported. Trade was, however, reduced after promulgation of the Wildlife (Protection) Act 1972. In the 1980s, there was a sharp decrease in trade though some illegal hunting probably still occurs (Walker, 1990). Cases of red panda cubs are still being taken from their nests, to be kept as pets in some of these remote areas (Pradhan *et al.*, 2001).

(v) Predation

The greatest predatory threat to red panda is unleashed household dogs of the villages. Such incidents often take place only in Singalila NP.

(vi) Illegal NTFP Collections

The core red panda habitat in the upper hills of Neora Valley is least affected of all the park's habitats by anthropogenic threats because exploitation of forest resources and human movement are controlled there. Firewood, fodder and timber collections are localized in Singalila NP. Firewood and fodder collections often occur simultaneously and within close proximity of the fringe villages in India and Nepal as well. Fodder is cut selectively. Timber is being cut in small old growth pockets. Exploitation is marked in Kalpokhri, Kaiyakatta and Jaubari.

(vii) Conversion of Natural High Forests into Plantations

Starting in 1962, Kalimpong Division clear-felled high forest (except a few good quality patches in Neora Valley NP) for replacement by exotics like *Cryptomeria japonica, Cupressus cashmeriana, Pinus patula* and others, mixed with indigenous hardwood (broad-leaved) species. These stands remain in East Nar Block of lower Neora Valley NP. Up to 1992 (30 years), over 26 km² of suitable Red Panda habitat in the contiguous blocks of the park (Pankhasari and part of Rashet) was lost. In 1996, the Supreme Court of India has imposed a ban on exploitation and collection in Reserve Forests, National Parks and Wildlife Sanctuaries. Since then, this practice was discontinued. In Singalila also conversion of natural forests to commercial plantations started during the British regime.

(viii) Encroachments

Before notification of Neora Valley NP, many neighbouring villagers used to spend several dry-season months in deep forests, making temporary cattle sheds (*gothh*) to facilitate cattle grazing. Moreover, there were a number of age-old forest villages inside the park. After notification, all forest villages in the upper range were shifted into the fringe areas and the land was regenerated through plantation of native species. For example, Rachilachawk was evacuated in 1996 and the village land was placed under a plantation programme with native species. During the ground survey, Red Panda was sighted in this area. But shifting of villages from the impact area of Singalila NP has not yet been done and the threat still remains.

(ix) Expansion of Tourism Facilities

Tourism was a burgeoning threat to red panda in Singalila NP with the rise of almost 250 per cent during last 10 years, while, it was also found that tourism to Upper Neora Valley areas was minimal with an average visitation of the Park with just 84 visitors/month (Pradhan, 2012). Due to difficult terrain and lack of communication facilities, only adventurous trekkers usually visit Neora Valley. Beginning in 2008, more facilities are being opened up at Choudapheri (Red Panda camp). Construction of an all-weather road beyond Choudapheri, increased vehicular traffic and tourists in this prime Red Panda habitat may lead to environmental pollution and cause disturbance to red pandas.

(x) Lack of Coordination among the Operating Agencies

There are many external agencies such as police, para-military force (SSB), tourism, Public Welfare department, Darjeeling improvement fund department, etc. operating in Sighalila NP. But coordination among those agencies and the custodian forest department for protection, conservation and sustainable development of the landscape is lacking, thereby affecting red panda and its habitat. But such use for non-forestry purpose is not marked in Neora Valley NP, except that of the Public Health Engineering (PHE) department.

Due to dense, contiguous, virgin forests and low anthropogenic pressures, status of red panda is comparatively much better in Neora Valley than those of Singalila. Hence, these two protected areas need different specific treatment (Pradhan, 2012). While Neora needs a more regulated and watchful vigilance and intervention for the preservation of its present state of intactness and conservation, Singalila needs a more aggressive protection and conservation strategy, which would range from top down urgent transboundary conservation strategies and policy implementation involving India (Sikkim and West Bengal), Nepal and Bhutan, to bottom up community involvement and management, along with coordination amongst the different sectoral agencies operating there.

Ex situ Conservation

There are about 80 zoos in the world with <300 red pandas as a part of the conservation effort in captivity and during the last two decades more than 300 red pandas were born in those zoos (Jha, 2012). The master plan for breeding of red pandas was initiated in 1990s by Dr. Angela Glaston, International Studbook Keeper for the red pandas. Accordingly, a conservation breeding programme "Project Red Panda" was first initiated in PNHZP (founded in 1958), as a part of the Global Red Panda Management Programme. PNHZP is ideally situated in a virgin patch of forest (Northern Montane/East Himalayan Wet Temperate Forest) on the Birch Hill (2150 m) spread over 67.56 acres, close to the natural range of red panda at Singalila. No other captive breeding facility is available in the vicinity of the habitat of the red panda.

Jha (2011) mentioned five objectives of this programme:

1. To establish a viable captive population of red pandas in the Indian subcontinent held in zoological collections based within the natural range of the species.

2. To distribute surplus stock to other subsidiary conservation breeding centres in suitable locations in Eastern Himalaya.

3. To provide surplus individuals for restocking the dwindling population of the species in Singalila and Neora Valley National Parks and to reintroduce red pandas into Senchal Wildlife Sanctuary.

4. To provide scientists and naturalists with the opportunity to study various aspects of the biology and behaviour of this rare species.

5. To stimulate public awareness of the plight of this endangered species.

PNHZP has been successful in conservation breeding of Red Pandas during the last two decades with the founder genomes from wild stock of four (1: 3) Singalila NP, *i.e.* one male (Basant, transferred on 31 December 1991) and three females (Anita, transferred on 1 December 1991; Chanda, transferred on 31 December 1991; and Divya, transferred on 1 December 1991). Under this programme, PNHZP received six red pandas from various foreign zoos (Bahuguna 1993). One more male 'Oscar' (d.o.b. 29 June 1992) was brought from Rotterdam Zoo on 1 April, 1993 to augment the existing stock. 'Hari' (male, d.o.b- 30 June 1993, Rotterdam), Gora (male, d.o.b.- 25 June 1993, Koln) and Indira (female, d.o.b.- 26 June 1993, Madrid) arrived in Darjeeling on 10 November 1994 to induce new blood. 'Omin' (original name Tsambo) (male, d.o.b. 17 July 1994, Rotterdam) and 'Vicky' (female, d.o.b. 26 June 1994, Antwerp) were further added on 25 December 1996.

The first successful breeding occurred on 20 June 1994, when two cubs 'Ekta' and 'Friend' were born (Bahuguna, 1995) to 'Basant' and 'Anita'. By 2002, the red panda population had plummeted and Darjeeling zoo had around 22 red pandas when it was decided to use a few selected red pandas to restock the natural population in the wild (Singalila NP) (Figure 18.5). Pre-release Monitoring of the resident red pandas and habitat in Singalila NP was done in November-December 2002. Compartment 3 of the Gairibans Beat was selected on the following criteria:

1. It is a prime red panda habitat;
2. It is well protected by the forest department; and

Figure 18.5: Back to the Wild (Padmaja Naidu Himalayan Zoological Park, Darjeeling).

3. It has a high density of red pandas as indicated by direct and indirect signs (Sharma *et al.,* 2005).

DNA based analysis of two selected females (Mini, born on 17 January 1998 and Sweetie, born on 25 June 1997; in both cases the sire and dam were Gora and Ekta) was performed with LaCONES at the Centre for Cellular and Molecular Biology, Hyderabad for Taxonomic Status analysis and Genetic Variability studies. An Intermediary release facility (4,850 m^2) for soft release of the animals was created near Gairibans Beat Office (around 2,800 m) under South Singalila Range, comprising of a southern and northern ridge with normal forest, bamboos and mature trees. Corrugated iron sheets, more than 2.5 m high, were erected surrounding the facility as a barrier and shrubs or trees near it were removed to check against predation on and escape of the captive animals. Their food was modified before at PNHZP so as to make them more dependent on a natural diet. There was another small enclosure within the soft release centre, which housed the animals during the initial one month, and slowly they were allowed to go out. By the time of the release, the animals were completely dependent on the natural food available in the enclosure, which included bamboos and wild fruits and berries.

Reintroduction

Reintroduction of captive-bred red pandas into their native habitat (Singalila NP) is the first of its kind in South Asia, a major milestone in wildlife conservancy. In the first phase, Mini and Sweetie were shifted to the intermediary release facility at Gairibans (2,626 m) on 15 August 2003 for acclimatization and being monitored. The release site at Gairibans was preferred on the basis of first ecological study during 1993-1996 (Table 18.3, vide Pradhan, 1999) and a pre-release survey and because it is a known habitat of the resident red pandas ranging from 2,550 m to 3,000 m with wet temperate evergreen forest, temperate evergreen deciduous forest, pure bamboo patches, pure rhododendron patches and plantations. The density of red panda in this site is higher than any other parts, which was one of the criteria so that the released females could find their wild mates within a short period, whereas the upper reaches of the area were disturbed owing to its proximity to a trekking route up to Sandakphu.

Before actual release, radio-collars (Telenoics, USA) of appropriate size and weight were fitted on the animals for post-release monitoring for at least 12 months (or till batteries of the collars were functional) by the forest department staff on alternate days using the non-triangulation location technique known as "Homing in on the Animal Method". Through radio-collaring and regular monitoring, births, death, movement and, to some extent, behaviour of the released female red pandas were recorded (Jha, 2012).

Information based on direct sightings reflect the habitat usage only at a particular point of time and a possible source of error could be the infrequent or chance sightings (Sharma *et al.,* 2005). Previous habitat studies on the red panda in Singalila NP had ended up discussing whether a particular area is an optimum red panda habitat or whether the area is colonised by the species at all. Some animals do not occupy all their potential range even though they are able to disperse into the unoccupied area, an individual behaviour that limits the distribution of the species. Alternately, it was also discussed that the density of the red panda population in the area is so low that the sub-adults do not disperse but get accommodated in areas where they are born.

Table 18.3: Location-Specific Sighting Records of Red Panda in Singalila NP

Month	Gairibans		Kaiyakatta-Kalipokhri		Sandakphu	
	Sighting	Number	Sighting	Number	Sighting	Number
January	2	2	1	1	0	0
February	0	0	2	2	0	0
March	0	0	1	1	3	3
April	0	0	3	3	1	1
May	0	0	1	1	1	1
June	1	1	0	0	1	1
July	2	2	1	1	0	0
August	2	2	0	0	1	1
September	0	0	0	0	0	0
October	1	1	2	4	3	4
November	0	0	1	1	0	0
December	1	1	1	3	0	0
Total	9	9	13	17	10	11

This hypothesis appears to be true in case of the post-release of the radio-collared captive-bred females in the wild. The two females behaved differently after release on 14 November 2003. While Mini remained close to her release site in the Middle area at ca. 2800 m between PWD Road and Nepal border most of the time (80 per cent) and sometimes explored the adjacent Pulkhola and Plantation areas, Sweetie was more agile than Mini. Mini first started interacting with the wild ones as one male appeared there on 18 November just after 4-5 days of her release. Her movement over a long distance from the released site to a quite unknown area made tracking-cum-monitoring difficult initially. However, subsequently, there were 13 encounters between Mini and a wild male- four times in December (4th, 13th, 21st and 31st), three times in January (1st, 16th and 23rd) and six times in February (9th, 11th, 13th, 20th, 23rd and 25th). But she entered the Nepal forest and it created problems for the trackers in ensuring its safety across the border. Despite all positive signs of adaptation and survival in the wild, she could not be traced since March and most probably was predated by a clouded leopard (*Neofelis nebulosa*) before she could reproduce after successful mating. The remains of her body- skull, tail, and a paw, were found together with the radio-collar on 15 March 2004.

Sweetie remained close to her release site for six days and then travelled about 2 km northwards during November 2003. In December, she settled at an area (26 EMV and Madan Vir) near M.R. Road (an oak forest habitat), about 1-1.5 km away from Gairibans, up to January 2004. Though a stray wild red panda was seen on this road on 4 December, pair bond was not perhaps established between them. Since M.R. Road area supports a very low density of wild red pandas, *i.e.* unoccupied or uncolonised habitat, as was observed previously by Pradhan *et al.* (2001) and the occasional reports of red panda sightings in this area were those of transient animals (Sharma *et al.,* 2005). Sweetie moved further in February towards the high density red panda habitat at Kaiyakatta (2,800-2,900 m), 4.2 km away from the release site, located in between Gairibans and Kalpokhri and preferred to stay in this better habitat from February to June.

She was seen with a wild male on twice in February (17th and 26th), once in March (11-12th), when mating was observed, and April (3rd). Time spent by her at Kaiyakatta area was 61 per cent in February, 95 per cent in March, 50 per cent in April, 54 per cent in May and 45 per cent in June (Sharma *et al.,* 2005). Within 7-14 July, she gave birth to a cub (sex could not be determined) in the hollow of her nest (Oak tree). But the cub was found missing after a month (predation anticipated).

In the second phase, two more female captive-bred red pandas- Neelam (born on 18 June 2001 to Gora-Ekta) and Dolma (born on 28 June 2001 to Omin-Kalita)- were taken to the soft release facility at Gairibans in November 2003, just a week before release of Mini and Sweetie. On a stormy night, Dolma escaped from the soft release facility climbing a broken tree on the fencing. Though she could not be radio-collared, she was later observed in the wild in good condition. Neelam was released in the wild in August 2004. They were being monitored, but no breeding with the wild males was observed by the monitoring team since then.

Recommendations

While periodic surveys were carried out in northern Bengal including red panda in Singalila NP in 2002, such a survey was not undertaken in Neora Valley NP. Periodic census of this species in both Singalila and Neora NPs is essential for ascertaining the status of this species and formulating management strategies.

The park authorities should also maintain a red panda database (time, locality where found including vegetation, physical characteristics of the animal, disease and treatment, post mortem, etc.) based on day-to-day monitoring for further analysis.

Due to ban on felling in the protected areas by the Supreme Court of India, felling of exotic plantations within Neora Valley NP to allow natural forest regeneration is not practicable. The red panda habitat should be extended by including two contiguous reserve forests (Ruka and Pankhasari blocks) of Kalimpong Division for better management and conservation of red panda.

The local Forest Protection Committees and Eco-development Committees should also be motivated to actively participate in the joint protected area management to reduce illegal resource exploitation.

The forest staff in the park should be provided with modern equipment and training to combat the poachers, timber mafias and illegal traders.

Conservation of this flagship species will also sustain the rich biodiversity in the study area as a whole.

Acknowledgments

I am specifically grateful to all the forest staff for helping the ground survey, all the respondents to questionnaire survey and also to Indranil Mitra for providing maps; Sourabh Basu and Padmaja Naidu Himalayan Zoological Park, Darjeeling, for photographs.

References

Agrawal, V.C., Das, P.K. Chakraborty, S., Ghose, R.K., Mandal, A.K., Chakraborty, T.K., Poddar, A.K., Lal, J.P., Bhattacharya, T.P., and Ghosh, M.K., 1992. Mammalia, pp. 27–169. In: Director, Zoological Survey of India (ed.). *State Fauna Series 3. Fauna of West Bengal,* Part 1. Zoological Survey of India, Kolkata.

Anon., 1967. *Tenth Working Plan for Darjeeling Division, Northern Circle 1967-68 to 1976-77, Vol. 1.* Government of West Bengal.

Anon., 2008. Red Panda survey in Neora Valley National Park. *Red Panda,* **1(2)**: 1.

Anon., 2009. A walk through Lava and Lolegaon. *India Digest* (High Commission of India, London) 281: 8.

Anon., 2010. *Protected areas of West Bengal.* Wildlife Wing, Government of West Bengal, Kolkata, West Bengal, India.

[ATREE] Ashoka Trust For Research in Ecology and the Environment, 2008. *Status survey of Red Panda (Ailurus fulgens) in the Neora Valley National Park, Darjeeling, West Bengal.* ATREE Eastern Hima-laya Programme, Darjeeling, West Bengal, India (draft interim report submitted to Forest Department, Government of West Bengal).

Bahuauna, N.C., 1993. Darjeeling Zoo receives red panda for captive breeding programme. *Zoo's Print,* August: 7-9.

Bahuauna, N.C., 1995. Red Panda birth in Darjeeling Zoo. *Zoo's Print,* April: 19-22.

Bahuguna, N.C., Dhaundyal, S., Vyas, P., and Sin-ghal, N., 1998. The Red Panda in Singalila National Park and adjoining forest: a status report. *Small Carnivore Conservation,* **19**: 11-12.

Bahuguna, N.C., and Mallick, J.K., 2010. *Handbook of the mammals of South Asia.* Natraj Publishers, Dehradun, India.

Biswas, B., and Ghose, R.K., 1982. Progress report 1 on 'Pilot Survey' of the World Wildlife Fund-India/Zoological Survey of India collaborative project on the '*Status Survey of the Lesser Cats in Eastern India'* (Project No. I.U.C.N. 1357- India). Zoological Survey of India, Calcutta, 52 pp. (unpublished).

Biswas G.G., Das, D., and Mukhopadhyay, A., 1999. Richness of mamma-lian species in the higher elevations of Neora Valley National Park. *Zoo's Print Journal,* **14(4)**: 10–12.

Chaudhuri, A.B., and Sarkar, D.D., 2003. *Megadiversity Conservation: Flora, Fauna and Medicinal Plants of India's Hot Spots.* Daya Publishing House, New Delhi, 300 pp.

Choudhury, A.U., 2001. An overview of the status and conservation of the Red Panda *Ailurus fulgens* in India, with reference to its global status. *Oryx,* **35**: 250–259.

Chowdhury, K. 1983. *First biological expedition to Neora Valley.* WWF–India, Eastern Region, Kolkata, West Bengal, India (unpublished re-port).

Ghose, D., Molur, S., and Leus, K., 2007. *Report of the Red Panda pre-PHVA meeting, 17– 19 Feb, 2007.* WWF-India, Deorali, Gangtok, Sikkim, India.

Jha, A.K. (2011). Release and Reintroduction of Captive Bred Red Pandas into the Singalila National Park, Darjeeling, India, pp. 435-446. In: Glatston, A.R. (ed.). *Red Panda, the Biology and Conservation of the First Panda.* Academic Press, London, UK.

Jha, A.K., 2012. *Ex-situ* and *In-situ* Linkages: Conservation of Red Panda. *Ex-situ Updates,* **1(2)**: 8-11.

Mallick, J.K., 2010a. Neora Valley - a new short-listed world heritage site. *Tigerpaper,* **37(3)**: 12–16.

Mallick, J.K., 2010b. Status of Red Panda *Ailurus fulgens* in Neora Valley National Park, Darjeeling District, West Bengal, India. *Small Carnivore Conservation,* 43: 30–36.

Mallick, J.K., 2012. Mammals of Kalimpong Hills, Darjeeling District, West Bengal, India. *Journal of Threatened Taxa,* **4(12)**: 3103–3136.

Mukhopadhyay, A., 1996. *Biodiversity of Animal Life of Neora Valley National Park: Mammals and Other Vertebrates (Part II).* Department of Zoology, University of North Bengal, Darjeeling, West Bengal. (unpublished).

O'Malley, L.S.S., 1907. *Bengal district gazetteer: Darjeeling.* Gyan Publishing House, New Delhi, India.

Pradhan, S., 1998. Studies on Some Aspects of the Ecology of the Red Panda, *Ailurus fulgens* (Cuvier, 1825) in the Singalila National Park, Darjeeling, India. *Ph.D. Thesis,* North Bengal University, India.

Pradhan, S., 1999. Observations of the Red Panda (*Ailurus fulgens*) in the Singalila Park, Darjeel-ing, India. *Small Carnivore Conservation,* **21**: 6-8.

Pradhan, S., 2006. Conserving the enigmatic Red Panda. *West Bengal,* **48(10)**: 31–33.

Pradhan, S., 2012. *Final Report: Develop first draft of Conservation strategy for red panda in Darjeeling Himalayas.* The Rufford Small Grants Project.

Pradhan, S., Bahuguna, N.C., and Saha, G.K., 1994. Observations on some behavioural aspects of red panda (*Ailurus fulgens*) in captivity at the Padmaja Naidu Himalayan Zoological Park, Darjeeling. *Journal of Bengal Natural History Society,* New Series **13(1)**: 63-67.

Pradhan, S, Saha, G.K., and Khan, J.A., 2000. Food habits of the red panda, *Ailurus fulgens* in the Singhalila National Park, Darjeeling, India. *Journal of the Bombay Natural History Society,* **98**: 224–230.

Pradhan, S., Saha, G.K., and Khan, J.A., 2001. Ecology of the Red Panda (*Ailurus fulgens*) in the Singhalila National Park, Darjeeling, India. *Biological Conservation,* **98**: 11–18.

Prince, M., 2003. *Birding trip report North Bengal, India; November 2002*. <http://www.kolkatabirds.com/mikereport.pdf>. Downloaded on 3 January 2009.

Saha, S.S., and Singhal, N., 1996. *Mammalian specimens of Bengal Natural History Museum, Darjeeling*. Department of Forest (Wildlife Wing), Government of West Bengal, Darjeeling, India.

Saha, S.S., Chattopadhyay, S., Mukherjee, R.P., and Alfred, J.R.B., 1992. Wildlife and its conservation in West Bengal, pp. 419–443. In: Director, Zoological Survey of India (ed.). State Fauna Series 3. *Fauna of West Bengal*, Part 1. Zoological Survey of India, Calcutta.

Sclater, W.L., 1891. *Catalogue of Mammalia in the Indian Museum. Vol. II.* Calcutta.

Sharma, B.R., 1990. *First management plan for the Neora Valley National Park in Darjeeling District, West Bengal, for the period 1990–1991 to 2000–2001*. Directorate of Forests, Government of West Bengal, Darjeeling, India.

Sharma, B.R., Gupta, B.K., and Chakrabarty, B., 2005. Conservation breeding and release of red pandas in West Bengal, India. pp. 157-162. In: Menon, V., Ashraf, N.V.K., Panda, P., and Mainkar, M (Eds.). *Back to the Wild: Studies in Wildlife Rehabilitation*. Conservation Reference Series 2. Wildlife Trust of India, New Delhi.

Singhal, N., 1999. *Neora Valley National Park at a Glance*. Jalpaiguri: Divisional Forest Officer, Wildlife Division-II.

Singhal, N., and Mukhopadhyay, A., 1998. *Management plan of Neora Valley National Park, West Bengal for the period 1998–99 to 2007–08*. Wildlife Circle, Government of West Bengal, Darjeeling, India.

Srivastav, A., Nigam. P., Chakraborty, D., and Nayak, A.K., 2009. *National Studbook of Red Panda (Ailurus fulgens)*. Wildlife Institute of India, Dehradun and Central Zoo Authority, New Delhi.

Tikader, B.K., 1983. *Threatened animals of India*. Zoological Survey of India, Calcutta, 307 pp.

UNESCO World Heritage Centre 2009. *Neora Valley National Park*. <http://whc.unesco.org/en/tentativelists/5447>. Downloaded on 29 July 2009.

Williams, B.H., 2004. The status of the red panda in Jamuna and Mabu villages of Eastern Nepal. *M. Sc. Thesis*. Department of Environmental Studies, San José State University.

Wright, A., 2004. History of WWF-India: Eastern India Region. *Panda* October: 13–14.

2015, **Animal Diversity, Natural History and Conservation, Vol. 5**   *Pages 305–312*
Editors: **V.K. Gupta and Anil K. Verma**
Published by: **DAYA PUBLISHING HOUSE, NEW DELHI**

Chapter 19

Habitat Characteristics and Foraging Ecology of Common Babbler, *Turdoides caudatus caudatus* (Dumont)

Ashima Anthal* and D.N. Sahi

P.G. Department of Zoology, University of Jammu,
Jammu – 180 006, J&K, India

ABSTRACT

The present study on habitat characteristics and foraging ecology of Common Babbler, *Turdoides caudatus caudatus* (Dumont) was carried out during June, 2010 to December, 2011 at district Jammu. It is a common resident bird in the dry open and scrubby habitats of the study area. The study revealed that these babblers have omnivorous feeding habit with insects forming the predominant portion of their diet. Their plant diet included berries, seeds and grains. Digging/probing into soil was found to be the most common strategy applied by these babblers for foraging during present study. Moreover, the interactions of Common Babblers with their feeding associates were also recorded.

Keywords: *Common Babbler, Turdoides caudatus caudatus, Omnivorous, Habitat, Diet, Foraging ecology.*

Introduction

Habitat selection by a species is the result of its intrinsic characteristics and the inherent features of the habitat, supplemented by environmental constrains like

---

* *Corresponding author.* E-mail: ashima.anthal@gmail.com

topography, climate and geology. Habitat characteristics such as vegetation height and prey abundance could affect foraging behaviour (Robinson and Holmes, 1982; Holmes and Schultz, 1988). Moreover, the knowledge of the ways in which birds exploit resources within their habitat not only increase understanding of how they use their environment but also the essential requirements for their survival. The Common Babbler (*Turdoides caudatus caudatus*) is the smallest Indian member of the genus Turdoides. It is one of the commonest birds of northern India but in south it is less common (Whistler, 1949). Throughout most of its range in India it is sympatric with two other species, the Jungle Babbler (*Turdoides striatus*) and Large Grey Babbler (*Turdoides malcolmi*) (Gaston, 1978a). It keep in flocks of six to twenty individuals spending its time on ground, hopping about rapidly with a bouncing gait, more commonly and scuttling like a rat under the sparse vegetation and hedges (Ali and Ripley, 1983). It is generally associated with areas of low or moderate rainfall and occurs up to about 2000m in the foothills of the Himalayas (Ali and Ripley, 1971). It inhabits dry open scrubland, semi-desert, thorn scrub, sandy floodplains and rocky hills and has four distinctive subspecies:

1. *salvadorii* (Iraq and Southwest Iran).
2. *huttoni* (Southeast Iran, Southern Afghanistan and Western Pakistan).
3. *eclipes* (Northwest Pakistan grading into nominate subspecies in northern Pakistan and in Himachal Pradesh and Punjab in northwest India).
4. *caudatus* (plains of India from Punjab east to Calcutta and from foot of Himalayas to southern India, Rameswaram island and Laccadives) (Cramp and Simmons, 1993).

Studies regarding Common Babblers have been undertaken by various workers (Rana, 1970; Gaston, 1978; Saini, 1982; Ali and Ripley, 1983; Hosseini Moosavi *et al.,* 2011) in several places in India but no such study has been done in and around Jammu. This study was thus carried out to investigate the characteristics of habitat used by Common Babblers and their foraging habits in light of their diminishing habitat due to increasing urbanisation in the study area.

Material and Methods

Study Area

The study was conducted in district Jammu that sprawls on both the banks of picturesque river Tawi, a tributary of river Chenab. Nestled between the Himalayan foothills, the Shivalik ranges and the Indo-Gangetic plains, the Jammu constitutes the south western district of the Jammu and Kashmir state. Bounded by district Udhampur in the north and northeast, Samba district in the east and southeast, Sialkot district of Rawalpindi (Pakistan) in west and Rajouri district and parts of district Bhimber (POK) in the northwest, it covers a geographical area of 3097 sq.km. Geographically, it lies between 32°27′ and 33°30′ North latitudes and 74°19′ and 75°20′ East longitudes. Altitudinally, it extends from 275 to 410 m above the msl. Being situated in the subtropical part, the climatic conditions in and around the study area are sub-humid to arid and characterized by four well marked seasons in a year.

Methodology

Present study on habitat characteristics and foraging ecology of Common Babblers (*Turdoides caudatus caudatus*) was made on 20 groups ranging in size from 3 to 13 individuals by adopting systematic field procedures and techniques. The data was collected during a continuous period of 19 months from June, 2010 to December, 2011. Data was collected from the field by direct observation method and by using binocular (Bushnell 7 X 50 U.S.A. made) whenever found necessary. The observations were made early in the morning or late in the evening when babblers are actively feeding with least disturbance to the birds. Besides, several erratic visits of shorter durations were also conducted during different hours of the day. For estimation of group size of babblers Visual census, Line transect method and Point transect method were used. Individual groups of Common Babblers were followed for periods varying from 2 to 6 hours. While following groups it was usually possible to keep some birds in view at all times, but rarely possible to see all members of the group together. During each scan, observations on the foraging habitat, type of feeding method employed, feeding session, size of the flock, type of diet and association with other bird species were recorded. Seasonal changes in the feeding habits of babblers were studied too. Photographs were taken with the aid of Canon EOS camera fitted with 300 mm zoom lens and digital cameras.

Results and Discussion

Population and Habitat

The Common Babbler, *Turdoides caudatus caudatus* is a common resident bird in the dry open and scrubby habitats of the study area. It is a small, slim bird with an earthy brown plumage and a long cross barred tail. A total of 20 groups of Common Babblers were found during the course of investigation. Overall, the size of the 20 groups varied from 3 to 13 individuals with a mean group size of 7.5 ± 0.5. However, Gaston (1978a) recorded the group size to range from 2 to 15 individuals. The members of a particular group were observed to live socially. Advantages likely to accrue to group territorial species are improved detection of predators, combined defence against them (mobbing) and increased foraging efficiency (Gaston, 1978b). They shared a common foraging area, defended a common territory, also roosted and nested together. Throughout the study period, the area of the territory maintained by each group was discerned to remain more or less the same. There was very little overlap between groups and no instances of territorial clashes or group mixing were recorded. These babblers were dependent upon low vegetation cover for escaping predators so they preferred to feed under or near bushes and only on few occasions ventured out in open areas that were dominated by short grass. Such areas were mainly covered with thorny bushes of *Zizyphus jujuba, Carissa opaca, Lantana camera* with clumps of *Saccharum munja* and few stunted trees. These babblers were perceived to be shy and took refuge in the bushes or grass even at minute disturbance. They flew low from bush to bush with a few rapid beats of the wings alternating with a sailing motion. Gaston (1978a) reported that they did not normally forage more than two metres from cover.

Foraging Habits

In Common Babblers, feeding was mainly confined to ground. These babblers were found to be omnivorous feeding on both animal and plant matter as reported in several other studies (Rana, 1970; Ali and Ripley, 1971; Saini, 1982; Narang, 1986; Kumar, 2006). The members of the groups moved together while criss-crossing different parts of their home range. The movement from one part of the home range to another, a characteristic of many territorial species, was not only for optimization of foraging but also for patrolling their territories to keep away conspecifics (Charnov *et al.*, 1976). During breeding season, babblers were often seen foraging singly or in small parties. Among animal matter, the main food item consumed by Common Babblers included insects belonging to 6 orders. These included grasshoppers, beetles, ants, moths, butterflies, cockroaches, termites, crickets, caterpillars and spiders (Table 19.1). In order to find these insects, they turned over dead leaves, small stones and sticks. For moving larger objects like stones they inserted their bills beneath them and then pushed forward and upward. Small sized insects were engulfed as a whole whereas, larger insects were first held with feet and then torn into pieces by pecking it several times. Hopping was their major method of movement during feeding. They also explored the clumps of grasses, dug around their roots, along the edges of the walls and probed holes on the ground in search of the prey. They were occasionally seen feeding on bushes and small trees. The plant matter consumed by them mainly comprised of grass seeds, grains and berries (Table 19.1). Grains of rice and wheat were consumed from the ground and never from the standing crops. Weed seeds were consumed too but in smaller quantities. During winter months, they were found feeding on grass seeds and on berries of *Zizyphus jujuba* (Ber) fallen on the ground.

Table 19.1: Different Food Items Consumed by Common Babblers
(*Turdoides caudatus*) during Study Period

Food Item	Order/Family	Name of Species
Insects	Orthoptera	Grasshoppers, Crickets
	Coleoptera	Beetles
	Lepidoptera	Moths, Butterflies, Caterpillars
	Isoptera	Termites
	Hymenoptera	Ants
	Blattaria	Cockroaches
Arachnids	Araneida	Spiders
Grains	Graminae	Rice (*Oryza sativa*)
	Graminae	Wheat (*Triticum aestivum*)
Seeds	Poaceae	*Saccharum munja*
Fruit	Rhamnaceae	*Zizyphus jujuba*

It was observed that during feeding one member of the group acts as 'sentinel' and scan the surrounding area for predators and alerts the foraging conspecifics against any threat by giving alarm calls. The sentinels of Common Babblers used mainly shrubs, small trees and seldom walls for this activity. In the absence of any disturbance, the sentinel watch silently and often preens itself. However, if a predator was seen, the sentinel becomes active and spread its wings and tail and flicks them fast. If the predator still persists, the sentinel hovers over it and calls loudly and repeatedly. The other group members in response move into the cover and join sentinel in mobbing it. Foraging was resumed after the predator moves away. The sentinel placed itself close to the group and changed its location when the group moved to another feeding site but it was sometimes also noticed to lead the group in movement. The sentinel duty was frequently rotated among the members of the group except juveniles. The change-over between sentinels were usually quiet involving one bird coming down and the other taking its place but didn't include any aggression, forced changes or physical contact as reported by Zahavi and Zahavi (1997). The poor flight ability as well as ground feeding in this species which make them more prone to predators necessitates such anti-vigilance system. Moreover, they are mainly insectivorous and spend a considerable time searching for food so feeding without a sentinel can lead to reduction in their feeding time. Thus, in the presence of a sentinel the group member's gain an increase in foraging success (Hollen *et al.,* 2008).

The various strategies employed by these babblers for foraging were classified into following categories (based on Andrle and Andrle, 1976):

1. *Hopping and Gliding:* This was the most common method used close to the ground to catch grasshoppers and crickets.
2. *Lifting of dead leaves:* In this method, the birds flicked dead leaves on the ground with their bill so as to reveal the prey residing beneath them.
3. *Probing:* The birds inserted their bills into curled-up leaves, gaps in the bark and holes present on the tree trunks as well as on ground with the purpose of collecting insects.
4. *Peering:* The birds twisted their heads to one side and peered under leaves in search of caterpillars.

Among these, digging/probing into soil (38.3 per cent) was noticed to be the most common method used for feeding during present study (Table 19.2). During summer and rainy months (breeding season of babblers), they consumed large quantities of animal matter. Orthopterans formed the major portion of diet followed by lepidopterans. This was probably to meet the high energy demands of their breeding activity. Plant matter was consumed in lesser quantities. However, in winter (November to March) the consumption of animal matter decreased. This may be due to less availability of insects in winter (Toor and Saini, 1986). Furthermore, they also supplemented their diet with items gleaned from the kitchen premises like chapattis, bread and vegetables throughout the year.

Feeding pattern in these babblers not only revealed seasonal fluctuations but also showed variations during different times of the day. The intensity of food-intake gets accelerated from the awakening and then gradually slows down. During hot summer days, they start feeding in early hours (0500 to 0530 hours) but during cold days, it starts as late as 0730 hours whereas during monsoon season they feed at irregular times depending upon the down pour. After feeding actively in morning hours they rest and preen sporadically with peak during mid-day in summers whereas in winters, feeding was continued throughout the day with relatively little rest during mid-day. The intensity of food intake again increases around 1530 hours in the summer and 1430 hours in winter. They spend on an average 37.5 per cent of the day time for feeding in summer (average day length-14 hours), 35.4 per cent in monsoon (average day length-13 hours) and 28.4 per cent in winter (average day length-11 hours).

Table 19.2: Frequency of different Foraging Methods of Common Babbler (*Turdoides caudatus*)

Methods	Hopping and Gliding	Peering	Probing into Curled-up Leaves on Trees	Digging and Probing into Holes on Fround	Flicking of Dead Leaves on Ground	Total
Frequency	28	5	9	49	37	128
Percentage	21.8	3.9	7.0	38.3	28.9	

Table 19.3: List of Birds Sharing Feeding Sites with Common Babbler (*Turdoides caudatus*)

Sl.No.	Name of the Species	Order	Family
1.	House Sparrow (*Passer domesticus*)	Passeriformes	Passeridae
2.	Common Myna (*Acridotheres tristis*)	Passeriformes	Sturnidae
3.	Yellow-Eyed Babbler (*Chrysomma sinense*)	Charadiformes	Timaliidae
4.	Black Drongo (*Dicrurus adsimilis*)	Passeriformes	Dicruridae
5.	Red-Vented Bulbul (*Pycnonotus cafer*)	Passeriformes	Pycnonotidae
6.	Hoopoe (*Upupa epops*)	Coraciiformes	Upupidae
7.	Little Brown Dove (*Streptopelia senegalensis*)	Columbiformes	Columbidae
8.	Ring-Necked Dove (*Streptopelia capicola*)	Columbiformes	Columbidae
9.	Indian Robin (*Saxicoloides fulicatus*)	Passeriformes	Muscicapidae
10.	Rufous Backed Shrike (*Lanius schach erythronotus*)	Passeriformes	Lanidae
11.	Jungle Babbler (*Turdoides striatus*)	Passeriformes	Timaliidae

Feeding Association with other Species

The Common Babblers due to their apparent shy nature did not often get mixed with other bird species however on some occasions they were seen to feed in association with House Sparrows, Common Myna, Jungle Babblers, Yellow-Eyed Babbler, Red-Vented Bulbul, Black Drongo, Hoopoe, Rufous Backed Shrike, Indian Robin, Little Brown Dove, Ring-Necked Dove (Table 19.3). They were also seen foraging in association with Five-striped Palm Squirrel. These feeding associates moved along with the babblers or joined the foraging flocks at times. Occasionally, clashes were recorded between babblers and other species over the same food item. In spite of these occasional conflicts between babblers and other species, it was detected that the attendant species get benefitted from the vigilance provided by their sentinel system. The other species responded to their alarm calls and joined them in mobbing predators. When two species have predators in common, they may benefit from responding to each other's alarm call as this allows them to reduce their investment in vigilance behaviour or increase the probability of escaping predators (Cheney and Seyfarth, 1990; Griffin *et al.,* 2005). Thus, it appears that there exists a sort of communal relationship between babblers and their feeding associates.

Conclusion

The present study reveals that Common Babblers successfully exploit both plant and animal resources available in the study area. But, compared to plant foods, the animal food mainly insects are consumed in larger proportions. These birds therefore help to keep an efficient check on various harmful insects like grasshoppers, beetles, termites, caterpillars etc., which are injurious to the agricultural crops and stored grains. So, in conclusion we can say that the economic status of Common Babblers in relation to mankind is useful.

Acknowledgements

The authors are highly acknowledged to the Department of Zoology, University of Jammu for providing the necessary facilities to carry out the study.

References

Ali, S., and Ripley, S.D., 1971. *"Handbook of the Birds of India and Pakistan"*. Vol. 6. Oxford University Press, London.

Ali, S., and Ripley, S.D., 1983. *"Hand Book of the Birds of India and Pakistan"*. Compact Edn. Oxford Univ. Press, Delhi. p. 737.

Andrle, R.F., and Andrle, P.R., 1976. The Whistling Warbler of St. Vincent, West Indies. *Condor.*, **78**: 236-243.

Charnov, E.L., Orians, G.H., and Hyatt, K., 1976. The ecological implications of resource depression. *Amer. Natur.*, **110**: 247-259.

Cheney, D.L., and Seyfarth, R.M., 1990. How monkeys see the world: inside the mind of another species. Chicago, Il: University of Chicago Press.

Cramp, S., and Simmons, K.E.L., 1993. Handbook of the birds of the Europe, the Middle East and North Africa. The Birds of the Western Palaearctic. Vol. V. Flycatchers to shrikes. Oxford: Oxford University Press.

Gaston, A.J., 1978a. Ecology of Common Babbler (*Turdoides caudatus*). *Ibis*, **120(4)**: 415-432.

Gaston, A.J., 1978b. The evolution of group territorial behaviour and cooperative breeding. *The American Naturalist*, **112(988)**: 1091-1100.

Griffin, A.S., Savani, R.S., Hausmanis, K., and Lefebvre, L., 2005. Mixed species aggregations in birds: zenaida doves (*Zenaida aurita*), respond to the alarm calls of carib grackles. *Anim. Behav.*, **70**: 505-515.

Hollen, L.I., Bell, M.B.V., and Radford, A.N., 2008. Cooperative sentinel calling? Foragers gain increased biomass intake. *Current Biol.,* **18(8):** 576-579.

Holmes, R.T., and Schultz, J.C., 1988. Food availability for forest birds: effects of prey distribution and abundance on birds foraging. *Can. J. Zool.,* **66:** 720-728.

Hosseini Moosavi, S.M., Behrouzi-Rad, B., and Amini-Nasab, S.M., 2011. Reproductive Biology and Breeding Success of the Common Babbler, *Turdoides caudatus* in Khuzestan Province, Southwestern Iran. *Podoces,* **6(1):** 72-79.

Kumar, S., 2006. Diversity of avian and mammalian fauna of district Kathua, J&K. Ph.D. Thesis. University of Jammu, Jammu.

Narang, M.L., 1986. Contribution to the food of Common Babbler, *Turdoides caudatus* (Dumont). *Indian J. Forest,* **9:** 140-145.

Rana, B.D., 1970. Some observations on food of the Jungle babbler *Turdoides striatus* and the Common Babbler (*Turdoides caudatus*) in the Rajasthan desert, India. *Pavo, Indian J. Ornithol.,* **8:** 35-44.

Robinson, S.K., and Holmes, R.T., 1982. Foraging behaviour of forest birds: the relationship among search tactics, diet and habitat structure. *Ecology,* **63:**1918-1931.

Saini, M.S., 1982. Feeding ecology of the babblers of Punjab with special reference to the Large Grey Babbler *Turdoides malcolmi* (Sykes). *M.Sc. Thesis.* Punjab Agricultural University, Ludhiana.

Toor, H.S., and Saini, M.S., 1986. Feeding ecology of the Large Grey Babbler, *Turdoides malcolmi. Proc. Indian Acad. Sci.* (Anim. Sci.), **95:** 429-436.

Whistler, H., 1949. "*Popular handbook of Indian birds*". Gurney and Jackson, London. p. 40-43.

Zahavi, A., and Zahavi, A., 1997. The handicap principle: a missing piece of Darwin's puzzle. *Oxford: Oxford University Press.*

2015, Animal Diversity, Natural History and Conservation, Vol. 5   Pages 313–333
Editors: V.K. Gupta and Anil K. Verma
Published by: DAYA PUBLISHING HOUSE, NEW DELHI

Chapter 20

On the Behavioural Ecology of Sloth Bear (*Melursus ursinus* Shaw 1791) in Mudumalai Wildlife Sanctuary, Western Ghats, India

N. Baskaran[1,2]*, S. Venkatesh[1], S.K. Srivasthava[3]
and Ajay A. Desai[1]

[1]Bombay Natural History Society,
Hornbill House, S.B. Singh Road, Mumbai – 400 023, M.S., India
[2]Department of Zoology and Wildlife Biology,
A.V. C. College (Autonomous), Mannampandal,
Mayiladuthurai – 609 305, Tamil Nadu, India
[3]Chief Conservator of Forests,
Tamil Nadu Forest Department, T.N., India

ABSTRACT

Sloth bear (*Melursus ursinus*), a vulnerable species, is widely distributed in Indian sub-continent–India, Nepal, Bhutan, Sri Lanka and perhaps in Bangladesh. Despite its wide distribution, the species is still lacking detailed data on its ecology and behaviour. We studied the activity pattern, diet composition, den characteristics and use pattern, home range, breeding season and social organization of sloth bear in Mudumalai Wildlife Sanctuary, Southern India. Activity pattern assessed based on 112 direct sightings of bears revealed that moving and feeding, that accounted for 91 per cent of sightings, were greater in the morning and evening hours, while resting primarily seen during the mid-day period, indicating that

* *Corresponding author.* E-mail: nagarajan.baskaran@gmail.com

bears are not entirely nocturnal and they tend to extend their activities to mornings and evenings. Diet composition assessed based on analysis of 474 scats revealed that both plant matter (48 per cent of the diet, in the form of fruits from 11 species) and animal matter (52 per cent, mostly ants and termites) equally contributed to the diet of sloth bear. Fruit availability positively influenced the fruit use by bears. Since availability of fruits tree species and their fruits varied among habitats and seasons, diet composition also varied significantly among the habitats and seasons. Den characteristics and their use pattern observed from evaluation and monitoring of 19 dens, indicated that dens types varied from small overhanging caused by the soil erosion of stream beds to deep excavated structures. The den use was more during June to August, possibly to take shelter against heavy rains from southwest monsoon. Home range size estimated based on 40 re-sightings of two individually known groups worked out to a mean rage of 12.3 ± 6.85 km^2 with an adult female along with two cubs ranged over large area (15 km^2) than the other group of two sub-adult bears (5.4 km^2). Breeding observed based on association of adult male and female as well as birth of cubs revealed that mating is likely to occur during late summer or early monsoon and cubs are born mostly during November-December. Litter size estimated based on 55 re-sightings of five different mother-young groups and two different independent sub-adult groups yielded a mean litter size of 1.7. Social organization in terms of mean group size, mother-young relationship and dispersal of young ones observed based on the 112 sightings of bears. Mean group size estimated was 1.8 bears, with solitary sightings were the most common and maximum group size recorded was three (adult female with two cubs), suggesting that sloth bears are solitary animals and adults are likely to be seen together only during the mating season. Mother-young association observed based on direct observation showed that cubs stay at the dens until three months and during this period mothers mostly remain with cubs. The cubs stay with mother at least a year, after which they disperse away from their mother gradually over a period of one year. Cubs dispersed from mothers were observed to live together up to one year time. The long association between mother and young would leads to mother normally breeding once in two to three years period. The study based on the findings suggests measures for the conservation of sloth bears in this region.

Keywords: *Melursus ursinus*, Activity, Feeding, Den use, Breeding, Social-organization.

Introduction

The Sloth bear (*Melursus ursinus*) is widely distributed on the Indian sub-continent– India, Nepal, Bhutan, and Sri Lanka and perhaps in Bangladesh–ranges from the base of the Himalayas to the southern tip of the Western Ghats including the island of Sri Lanka (Prater, 1965). The species was once very common and widely distributed throughout these ranges. Presently its population is declining in many parts of its range due to deterioration and loss of its habitat (Johnsingh, 1986). In addition, poaching for gall bladder, capture of cubs from dens some time killing their mother have resulted in sloth bears vanishing from large parts of their former range. Population surveys have indicated that their numbers are low and possibly declining (Krishnan, 1972: Singh, 1973: Spillet, 1967). IUCN lists the species in vulnerable category (IUCN Red List, 2014), Indian Wildlife Act (1972) in Schedule I and CITES in Appendix II (Servheen, 1991).

Despite its wide distribution, very little is known about the behavioural ecology of this species. Most of the earlier information available on the species is either anecdotal in nature or natural history observations (Prater, 1965; Schalle,r 1967). Only recently, there have been some studies on the behavioural ecology (Laurie and Seidensticker, 1977), on home range (Joshi *et al.*, 1995,), diet composition (Joshi *et al.*, 1997), population density (Garshelis *et al.*, 1999) in Chitwan National Park, Nepal, on diet composition and seed dispersal in southern India (Baskaran, 1990; Baskaran *et al.*, 1991; Gokula, 1991; Desai *et al.*, 1997; Sreekumar and Balakrishnan, 2002; Baskaran and Desai, 2010), on habitat use pattern (Aktar *et al.*, 2004) and some ecological aspects using radio-telemetry in Panna National Park (Yoganand *et al.*, 1999), Central India on home range and habitat use in Sri Lanka (Ratnayeke *et al.*, 2007). Nevertheless, site or regional specific information on population size, reproduction, social organization, mother-young relations and ranging that are essential for management are still lacking.

In India, there have been a few attempts of species-specific conservation programmes like Project Tiger, Project Elephant on a long-term basis across the country with specialized funding. Similarly, there have a few species-specific management actions to establish a new breeding population of Indian or Greater One-Horned Rhinoceros (*Rhinoceros unicornis*) in Dudhwa National Park, the attempt to find a second home for the Lion (*Panthera leo persicus*) and some captive breeding programs that have been successful. However, for a vast majority of the species listed in the Schedule I of the Indian Wildlife Act (1972), there are no explicit management plans that cater to the specific needs of these species. Largely the lack of site-specific ecological studies on these species has seriously hampered development of species-specific and site-specific management plans. This study presents data on (i) Activity pattern, (ii) diet composition in different season and habitat (iii) den site characteristics and use pattern (v) home range size, (v) breeding season and litter size and (iv) social organization based on group size, mother-young relationship and size of sloth bears in Mudumalai Wildlife Sanctuary, Tamil Nadu, India.

Methods

Study Area

The study was conducted in Mudumalai Wildlife Sanctuary that located at the tri-junction of Tamil Nadu, Karnataka and Kerala states (Figure 20.1) lies between 11° 32' and 11° 45' north latitude and 76° 20' and 76° 45' east longitude. It is bounded to the north by Bandipur National Park and to the west and northwest by Wayanad Wildlife Sanctuary and to the south and east by Nilgiri North Forest Division form the boundary and part of Nilgiri Biosphere Reserve. It has an undulating terrain with an average elevation of 900 to 1000 m and is drained by the Moyar river and its tributaries. The Moyar River and a few bigger streams that drain into it are perennial. There are several large man-made water holes that act as water source during the dry season. The study area has a long wet season and a short dry season. It receives

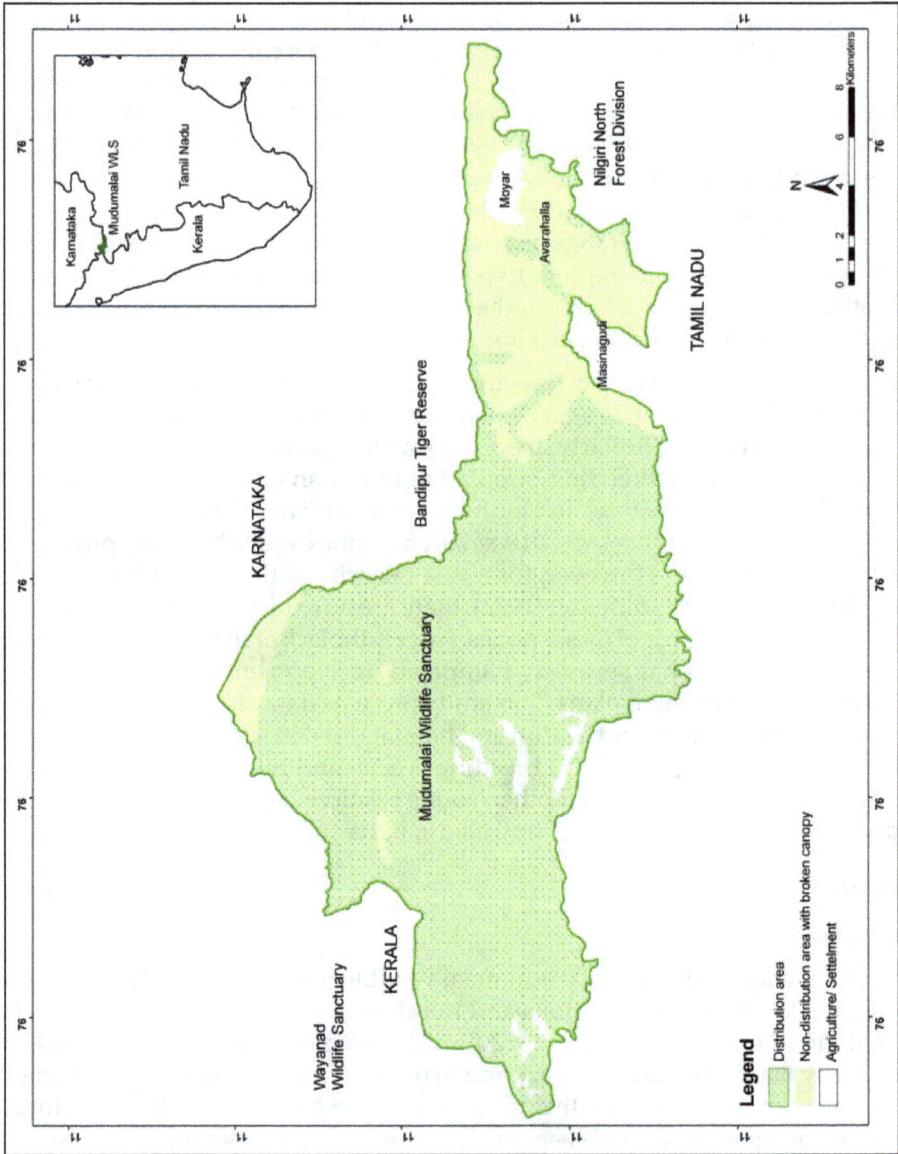

Figure 20.1: Map Showing the Study Area Mudumalai Wildlife Sanctuary with an Insert of its Location in Relation to Three Southern Indian States.

rainfall from southwest and northeast monsoon. The southwest monsoon starts by May and ends by August while the northeast monsoon starts by September and ends by December. The rainfall has a marked east-west gradient, with the eastern areas getting the least amount of rain (600 to 1000 mm) and the western regions the heaviest rains (1000 to 2000 mm). Temperature ranges from 8° C in December to 35° C in April (Baskaran, 1998). The vegetation follows a similar gradient as the rainfall with tropical dry thorn forests to the east of the sanctuary followed by the tropical dry deciduous short grass and dry deciduous tall grass forests in the middle, and the moist deciduous forests to the west. There are also a few patches of semi-evergreen and evergreen forests to the west (Champion and Seth, 1968). Intensive study on den site characteristics and its use pattern, population estimate techniques and diet composition were carried out between September 1996 and February 1997. Records on bear sightings and behavioural observations, home range size and social organization based on group sizes and, mother-young relationship and breeding season and litter size of sloth bear were gathered in Mudumalai Wildlife Sanctuary while the first author was working continuously on various other species in the park since 1990.

Activity Pattern and Social Organization

Though sloth bears like other myrmecophagus mammals are largely nocturnal in their habits, they are not strictly nocturnal and tend to be crepuscular (active in the early mornings and late evenings). But their largely nocturnal habits limit the data that can be gathered through direct observations. Thus, we used *ad-lib-tum* sampling method, and recorded the date, time, activity, group size and, age and sex (if possible based on association) on every sightings of sloth bear to understand their activity, social organization and breeding. Generally it is difficult to sex bears Laurie and Seidensticker (1977) and thus classified them into three different age classes *viz.* adult, sub-adult and cub based on body size and group composition.

Diet Composition

The nocturnal foraging habits of sloth bears did not permit direct observations on feeding behaviour and thus the study used scat analysis method as suggested by earlier studies on bears (Baskaran, 1990; Hamer and Herrero, 1987; Landers *et al.,* 1979; Laurie and Seidensticker, 1977; Mace and Jonkel, 1986; Maehr and Brady, 1984; Schaller, 1967, 1969). Dietary composition and its seasonal variation was assessed using scat analysis method (Baskaran, 1990; Baskaran *et al.,* 1997; Laurie and Seidensticker, 1977). The bear scats were collected along selected game roads from different vegetation types at fortnightly intervals. From each scat, undigested food items were segregated separately using sieves of various sizes in tape water and measured the volume of each food item to the nearest 1 m following water displacement method. Plant matter was segregated up to species level while animal matter was broadly classified into ants, termites and beetles. Bears feed on honeycombs, ingesting the entire honeycomb including the wax, larvae and eggs of the honeybees. The only evidence of such feeding is the presence of wax in the scat. Whenever wax was observed in the scat, it was classed as feeding on honeycombs. This method may not give the actual quantum of honey consumed, but it does give us the frequency of consumption and an approximate indication of its contribution to the diet of sloth bears. Those food items that could not be identified were categorized into unidentified plant and animal matter. The results are presented in percentage contribution of various diet items to the diet bears, by converting the volume different food items found in each scat into percentage and estimated the mean from them for individual food items in different habitats and seasons.

Assessment of Food Availability

Fruits and insects (ants and termites) form a major part of the diet of the bears (Baskaran, 1990; Laurie and Seidensticker, 1977). We estimated density and diversity of fruit tree species and quantified the fruit availability through systematic monitoring in different habitats to assess habitat quality. To estimate the fruit tree density, all the tree species (>20cm gbh) were enumerated using belt transects (20 x 100 m) laid perpendicular to game roads used for scat collection in all the vegetation types. The fruit tree diversity was calculated using Shannon-Wiener diversity index (H'). The phenology was monitored only for major fruit tree species whose fruits form a considerable part of the bears diet (Baskaran *et al.,* 1997). A total of five species, with ten individuals from each species in each habitat type was monitored on monthly interval and assessed the fruit quantity using visual percentage rating method.

Den Characteristics and Use Pattern

To study den characteristics and its use by bears, extensive ground survey was undertaken especially around dry streams, hillocks etc to locate and study the dens use by sloth bears. Data on location, substrate type, cover and approach paths to the den were gathered to assess the den characteristics. The identified dens were visited at fortnightly intervals to collect the data on den use and locate breeding females so as to gather information on their reproductive behaviour. Artificial dens constructed by the forest department were also monitored on regular basis to their use pattern by bears.

Home Range

During the study period, it was possible to monitor movement of a mother with two-well grown cubs (one of which had a slight limp which facilitated identification) and two sub adult bears (siblings which were still together constantly), on 25 and 15 occasions over a period 14 months. Thus we estimated home range for these two groups using the minimum convex polygon method (Dalke, 1938; Jennrich and Turner, 1969).

Breeding and Mother-Young Relationship

Data on the breeding season were gathered largely based on the birth of cubs. The duration of den use by the cubs, duration of the association of cubs with the mother and association between independent sub-adults (siblings) were also monitored to the extent possible.

Results and Discussion

Activity

Sloth bears were seen on 112 occasions, sighted mostly (95 per cent) during day light period between 06:00 and 18:00 h morning (Figure 20.2). Of the 112 observations, foraging was accounted for 38 per cent times and moving 53 per cent of the time. Resting was observed on 7 per cent of the occasions and threat display on 2 per cent of the occasions towards the observer. Both moving and feeding activities were greater in the mornings and evenings than in the middle of the day. Resting bears were seen primarily during the mid-day period. These observations certainly suggest that sloth bears are not entirely nocturnal in the study area and that they tend to extend their activities to mornings and evenings also. These observations go in support of fining by Laurie and Seidensticker (1977) who also suggest that sloth bears in Royal Chitwan National Park in Nepal are not entirely nocturnal.

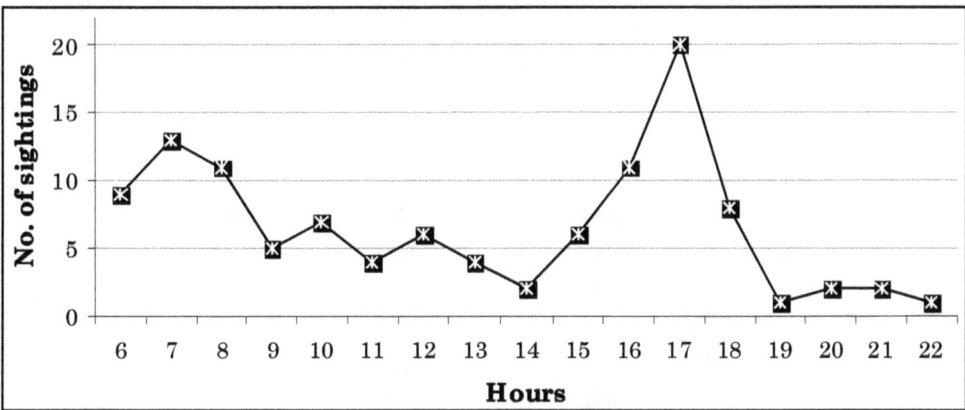

Figure 20.2: Direct Sighting of Sloth Bear at different Hours of Day in Mudumalai Wildlife Sanctuary.

Diet Resource Availability and Diet Composition

Food Resources

Among the four habitats, fruit tree diversity was highest in dry deciduous tall grass followed by short grass habitat (Table 20.1). Similarly, individual tree species varied in their densities among four habitat types for example *Cassia fistula* density was 0.82 tree/ha. in the dry deciduous short grass habitat and highest in 23 individuals/ha. in the dry deciduous tall grass habitat. Similarly, *Zyziphus mauritiana* was the most common fruit bearing trees edible to sloth bear in dry thorn forest and this species was not recorded in the other habitats.

Fruit Availability

Fruits species edible to sloth bear varied remarkably among habitats and seasons (Table 20.2). The dry thorn forests had the highest fruit abundance followed by dry deciduous tall grass habitat. Similarly, among the seasons, fruits edible to sloth bear

Table 20.1: Density, Relative Proportion and Diversity (H') of Fruit Species Edible to Sloth Bears Available in different Habitats of Mudumalai Wildlife Sanctuary

Habitat (Area Sampled in ha.)	Fruit Species	Density/ha.	Relative Proportion	(H')
Dry deciduous short grass (12.2)	Cordia domestica	1.15	10.8	1.737
	Cassia fistula	0.82	7.75	
	Cordia oblica	1.8	17	
	Diospyros montana	0.08	0.77	
	Diospyros melanoxylon	1.23	11.6	
	Ficus species	0.49	4.65	
	Grewia tiliifolia	4.18	39.5	
	Syzygium cuminii	0.82	7.75	
Dry deciduous tall grass (6)	Cordia domestica	0.83	2.31	1.127
	Cassia fistula	22.82	63.41	
	Diospyros montana	0.67	1.86	
	Ficus bengalensis	0.5	1.39	
	Grewia tiliifolia	6.5	18.05	
	Syzygium cuminii	4.17	11.59	
	Zyziphus oenoplia	0.5	1.39	
Moist deciduous (4.8)	Cordia domestica	8.67	27.08	1.434
	Cassia fistula	5	15.63	
	Diospyros montana	0.67	2.09	
	Ficus religiosa	0.5	1.56	
	Grewia tiliifolia	13.83	43.2	
	Olea dioeca	1.67	5.22	
	Syzygium cuminii	1.67	5.22	
Dry thorn forest (7.4)	Cordia domestica	2.33	6.56	1.476
	Cassia fistula	4.5	12.68	
	Cordia oblica	1.17	3.3	
	Diospyros montana	7	19.72	
	Grewia tiliifolia	0.33	0.93	
	Syzygium cuminii	0.5	1.41	
	Zyziphus mauritiana	17.83	50.24	
	Zyziphus oenoplia	1.83	5.16	

was highest during dry season followed by first wet season. The second wet season was largely devoid of fruiting with the exception of the early fruiting of *Z. mauritiana*, which had 4.31 per cent fruits in dry thorn habitat. *Syzygium cuminii* that suppose to fruit during first wet season, largely failed during present study. On the other hand, *C. fistula* was the major food species, as its fruits were available to bears in all the vegetation types, relatively for longer period in dry as well as during early first wet season. However, its availability during dry season varied from 3.8 per cent in the moist deciduous forests to 26 per cent in the dry deciduous tall grass forest. Overall the study shows that fruit availability varied temporally among seasons and spatially among habitats. Fruiting rate of a given species also varied among habitats indicating spatial variation fruiting success. Further, comparison of present results with that of earlier available for the same area (Baskaran, 1990), reveal a significant an inter-annual variation in fruiting success, especially in the case of *S. cuminii.*

Table 20.2: Per cent Availability of Fruits Edible to Sloth Bear Recorded in different Season among Habitats

Season	Species	Moist Deciduous	Dry Deciduous Short Grass	Dry Deciduous Tall Grass	Dry Thorn
Dry	*Cassia fistula*	3.8	26	14.6	17.9
	Cordia domestica	0	0.25	0	0
	Grewia tiliifolia	0	0	0	*
	Syzygium cuminii	0	0	0	0
	Zyziphus mauritiana	*	*	*	12.4
First wet	*Cassia fistula*	1	0.25	0.5	0
	Cordia domestica	0	–	9.44	0
	Grewia tiliifolia	4	2.5	0.75	*
	Syzygium cuminii	0	0	0	0
	Zyziphus mauritiana	*	*	*	0
Second wet	*Cassia fistula*	0	0	0.05	0
	Cordia domestica	0	0	0	0
	Grewia tiliifolia	0	0	0	*
	Syzygium cuminii	0	0	0	0
	Zyziphus mauritiana	*	*	*	4.31

*: Not monitored for phenology as these species are rare or absent in the given habitat.

Diet Composition (Annual and Seasonal)

As sample sizes were different for different vegetation types and for different seasons, the volume of different food items recorded in each scat were converted into percentages and from the percentage data, mean contribution of various food item was arrived for each season and vegetation types. Based on the analyses of 474 scats, 11 different fruit species, four different insects including honey bees and bee wax were identified as food of sloth bear (Table 20.3). The contribution of different food items (plant and animal matter) to the seasonal and annual diet varied remarkably. In an earlier study (Baskaran 1990) in the same area, 18 fruits species recorded in the diet. In the present study fruits from Ficus spp. (2.83 per cent), *Grewia tiliifolia* (2.05 per cent) and *Zygiphus oenoplia* (2.75 per cent) contributed >1 per cent to the diet of the sloth bear, whereas these species during 1990 contributed <1 per cent (0.43 per cent, 0.8 per cent and 0.21 per cent respectively) to the diet (Baskaran, 1990). On the other hand, *S. cuminii* that failed to fruit during 1996 was recorded in the bears' diet in >1 per cent, while the same observed to form 45 per cent of the overall diet of bears. As fruiting success varied remarkably between years, individual fruit species contribution to the diet of bear changed significantly. Similarly, as most of the fruit species are seasonal, their use pattern too changed according to their availability different seasons (Table 20.3). In the dry season *C. fistula, Z. mauritiana* and *Ficus* spp. constituted the bulk of the diet, each species contributing 34.7 per cent, 30 per cent and 5.67 per cent respectively. In the first wet season *Cordia domestica, G. tiliifolia* and *Lantana camara* formed 22.6 per cent, 10.4 per cent and 4.84 per cent respectively of the diet. In the second wet season the only species contributing reasonably to the diet was *Z. oenoplia* (4.8 per cent).

Table 20.3: Annual and Seasonal Diet Composition of Sloth Bear Estimated from 474 Scats

Food Items	Seasonal Diet (Per cent Vol.)			Annual Diet (n = 474)
	Dry (n = 224)	*First Wet (n = 96)*	*Second Wet (n = 154)*	
Plant matter	74.1	39.5	10.4	47.9
Anogeisus latifolia	0	0.001	0	0
Cassia fistula	34.7	0.05	0.79	17.6
Cordia domestica	0	22.6	0	4.45
Dispyros melanoxylon	0.26	0	0	0.13
Ficus spp.	5.67	0	0	2.83
Grewia tiliifolia	0	10.4	0	2.05
Syzygium cuminii	0	0.86	0	0.16
Zyziphus mauritiana	30.0	0	2.29	15.7
Grewia hirsuta	0.22	0	0	0.11
Lantana camara	0.35	4.84	1.25	1.51
Zyziphus oenoplia	2.59	0	4.8	2.75
Unidentified	0.11	0.66	1.25	0.57
Animal matter	25.9	60.5	89.6	52.1
Ants	16.6	45	70.6	38.6
Termites	7.52	14.3	14	10.8
Beetles	0.04	0.1	0	0.04
Honey	1.72	0.46	2.09	1.58
Unidentified	0.01	0.58	2.77	0.96

Animal matter was composed mainly of ants and termites contributed to 49 per cent of the overall diet (Table 20.3). Ants were the major part of animal matter in the bears' diet constituting 17 per cent, 45 per cent and 71 per cent in the dry, first wet and second wet season respectively. Termites constituted 8 per cent of the dry season diet and, 14 per cent each of the first and second wet seasons diet respectively. Bears feeding on honey identified based on honeybee and bee wax constituted 2 per cent, 0.5 per cent and 2 per cent of the diet of dry, first wet and second wet season respectively. Although sloth bear scats contained marginally a higher percentage (52 per cent) of animal matter than animal matter on annual basis, but on the seasonal basis consumption of plant and animal matter varied significantly ($\chi^2 = 83.77$, $df = 2$, $P < .001$) between seasons.

Availability calculated based on availability of fruits edible to sloth bear from individual species in each habitat type seasonally was compared with utilization of those species arrived based on their volume in the scats from each habitat in different season. Comparison of the utilization with their availability showed a positive relationship (Figure 20.3). For example, plant material dominated (74 per cent) the diet during the dry season when most of the trees were in fruiting. But fruits contribution to the diet declined to 40 per cent during the first wet season, when only a few species were in fruiting and further down to only 10 per cent during the second wet season. Animal matter in the form ants and termites dominated the diet during both wet seasons (first wet: 61 per cent and second wet 90 per cent) and declined to 26 per cent of the diet during dry season. Although, sloth bears eat plant and animal matters almost equally (plant matter 52 per cent and animal matter 48 per cent) over the annual scale, there was a distinct shift from plant matter diet during dry season to animal matter diet with the progress of wet season.

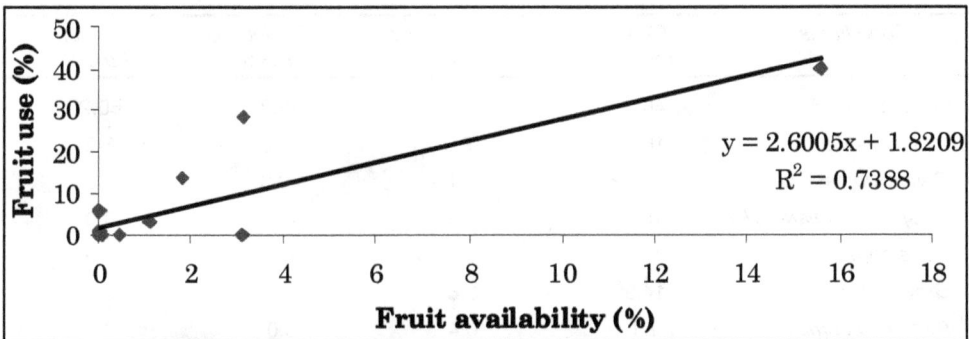

Figure 20.3: Relationship between Fruit Availability and their Use by Sloth Bear in different Habitats of Mudumalai Wildlife Sanctuary.

Seasonality of Food Sources

More seasonal a food item, the shorter would be its duration of availability in the environment and consequently lower its contribution to the annual diet. Most of the fruits are highly seasonal and contribute little, individually, to the annual diet of the sloth bears. But as most species are fruiting in the dry season and first wet season together they contribute substantially to the seasonal diet. Animal matter in contrast is the least seasonal and contributes the most to the annual diet. The non-seasonality of animal matter ensures a steady supply of food to bears in the absence of fruits. If we assume that the availability of ants and termites remain fairly constant through the year it would appear that fruits are the preferred food and when fruits are available, bears shift their diet towards plant matter. In time when both fruits and insect are available, it is likely that bears able to (i) gather the required quantum of fruits per unit time more easily than insects, due to the larger size and easier access of fruits. Although, insect food may offer bears a higher protein diet than fruits, being a large bodied animal, quantum of food required to quench the hungry at shortest time might influence the animal to go for fruits. However, their ability to switch to insect when fruits are absent or in short supply shows their adaptation.

Variation in Diet Composition among Habitats

As the fruit tree species diversity, and density, and fruiting phenology and fruiting success also varied among habitats, the diet composition was also expected to vary among the same. Among the four habitat types, the contribution of plant matter was in the order of dry thorn (60 per cent), moist deciduous (47 per cent), dry deciduous short grass (43 per cent) and dry deciduous tall grass (24 per cent) habitats (Table 20.4). The availability of fruits in different habitats also showed a similar order of habitats except for the moist deciduous habitat (that had the lowest fruit), but the same had second highest diversity of fruit trees. Number fruit species formed the diet of sloth bear was the highest in the dry deciduous short grass (9 species) followed by dry thorn and moist deciduous (6 species each) and lowest in dry deciduous tall

Table 20.4: Diet Composition of Sloth Bear in different Habitat Types

Food Items	Moist Deciduous	Dry Deciduous Tall Grass	Dry Deciduous Short Grass	Dry Thorn Forest
Plant matter	46.8	24.4	43.2	60.2
Cassia fistula	16.3	19.1	10.5	8.3
Cordia domestica	13.7	1.1	19.3	3.6
Diospyros melanoxylon	0.0	0.0	0.3	0.0
Ficus spp.	1.3	0.0	6.4	0.0
Grewia tiliifolia	14.3	3.8	0.1	0.0
Syzygium cuminii	1.1	0.4	0.0	0.0
Zyziphus mauritiana	0.0	0.0	4.6	24.7
Grewia hirsuta	0.0	0.0	0.2	0.0
Lantana camara	0.1	0.0	0.4	9.8
Zyziphus oenoplia	0.0	0.0	1.4	10.2
Unidentified	0.0	0.0	0.0	3.5
Animal matter	53.1	75.5	56.7	39.6
Ants	42.1	53.5	45.7	29.1
Termites	5.1	20.3	6.8	9.1
Beetles	0.1	0.1	0.0	0.0
Honey	5.9	1.4	2.0	0.0
Unidentified	0.0	0.3	2.2	1.4

grass habitats (4 species), which in turn fall in line with diversity of fruits species recorded in these habitats (Table 20.4).

Unlike the other myrmecophagus animals, sloth bears feed on fruits to a fairly high degree. Though fruits constitute 48 per cent of the diet they are not available throughout the year. In tropical habitats, where the sloth bear occupies on the Indian sub-continent, most trees are adapted to fruit in relation to the fairly distinct wet and dry seasons. In the study area a large number of plant species (whose fruits are edible to bears) come into fruit in the dry season and some in the first wet season. There was very low availability of fruits in the second wet season (see Table 20.2). Fruits therefore constitute a proportion of the sloth bears diet in relation to the fruiting phenology and fruiting success of individual tree species.

Den Characteristics

Sloth bears use dens for breeding as well for shelter. Out of 19 dens identified, most (73.6 per cent) were located in *nulla* beds (Table 20.5). Other than *nulla*, rocky areas both on hills and flat areas were used for dens. Most of the dens were either dug directly into the mud walls of *nulla* (36.8 per cent) or under roots of trees located along the *nulla* edges (26.6 per cent). The rest were under rocks either on hills, flat areas or in *nulla*. The vegetation cover around the den did not appear to influence the den site as all types of cover were recorded around the dens with 15.8 per cent of them were with dense vegetation cover and 42.1 per cent medium cover. A majority of the dens had a single entrance (78.9 per cent) and the rest had two entrances. Approximately

Table 20.5: Characteristics of Sloth Bear Den Assessed in Muduamlai Wildlife Sanctuary, Tamil Nadu (*n* = 19 dens)

Characteristics	Variables	Per cent of Dens
Den location	Rocky hill	15.7
	Flat rocky area	10.5
	Nulla (dry stream)	73.6
Substrate	Mud	36.8
	Under Tree Root	26.8
	Under Rock	36.8
Vegetation cover around den	Very dense	15.8
	Dense	42.1
	Semi-open	21.5
	Open	21.5
Number of entrance	One	78.9
	Two	21.1
Clearings in front of the den	Present	47.4
	Absent	52.6
Scat near den	Present	68.4
	Absent	31.6

half the dens had a single path leading to the den while the other half had two paths, only one had three paths. More than half the number of dens (53 per cent) had small clearings in front of the entrance indicating regular use and the tendency of bears to lie outside when the weather is good. Scats were found in 68 per cent of the dens. Dens types varied from small overhanging caused by erosion of the *nulla* wall due to heavy water flow in the wet season to deep structures that had been excavated. It is not clear whether bears excavate the dens entirely by themselves or expand existing dens dug by other species (*i.e.* porcupine) or natural cavities created by erosion or holes under rocks.

Den Use Pattern

Dens were used more during June, July and August (southwest monsoon), possibly to take shelter from the heavy rains (Figure 20.4). Den use was moderate between September–December the northeast monsoon period, but was the lowest in the dry season. It is likely that bears use the dens in the late wet season and dry season are mostly for breeding purpose by females to protect their newborn cubs. Other bears (adult male, sub-adult male and female) are likely to rest under whatever cover they find (*i.e.* dense bushes, bamboo thickets, tall grass, under fallen tree) during late wet and dry seasons.

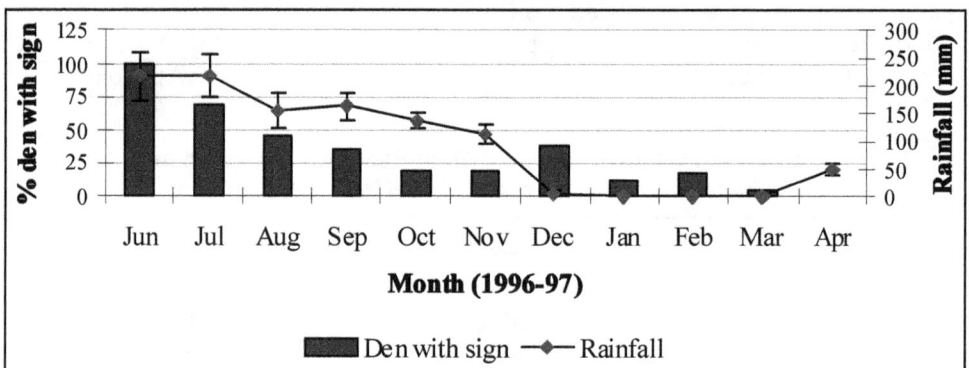

Figure 20.4: Percentage of Den Observed with Sign of Use by Sloth Bear in different Months in Mudumalai Wildlife Sanctuary, Tamil Nadu.

Artificial Den Use

Forest Department constructed four artificial dens in the study area. A systematic monitoring of the artificial dens revealed that only one of the nine artificial dens was used by bears. Since there are ample number of dens available under natural condition, and their use by bears itself is extremely low, as shown by our den use observations, creation of new dens may not be required. Bears were thus sighted directly on only five of the 162 den visits (little over 3 per cent use). Except for June when all the six dens located were found with fresh signs, in all the other months fresh signs were seen in 70 per cent or less cases. Lack of published data available on sloth bear den sites and its characteristic features or den use pattern to compare our observations pose greater uncertainty. A more detailed study using radio telemetry to evaluate den use by bears and their availability in different habitat types would give us a better understanding if management needs to make artificial dens, although the study does not indicate a deficiency in the number of dens.

Home Range

In course of our study it was possible to re-sight one group of three bears (a mother with two sub-adult cubs, with one of the cub was limping) on 25 different occasions. Another pair of sub-adults (one of which was half that of other), which were independent of their mother but were still together and this pair was re-sighted on 15 occasions. Home range for these two groups was estimated using the minimum convex polygon method (Table 20.6) and these home range sizes are likely to be the minimum size, as this study followed the re-sighting method, which is known to underestimate the range size (Baskaran 1998). The actual home range sizes are likely to be larger. Joshi *et al.* (1995) studying radio-collared sloth bears have estimated the mean home range size for females in Chitwan National Park as 9.4 km² with the maximum size recorded being 23.8 km². The home range estimated for the female with two sub-adult cubs approaches the maximum seen in the Chitwan study even though no radio telemetry studies were involved. It is therefore likely that home ranges in the present study area are equal to or larger than those in Chitwan. Home range sizes would be dependent on the resource distribution and abundance.

Table 20.6: Home Range of Sloth Bears in Mudumalai Wildlife Sanctuary

Bear Id.	No. of Locations (n)	Study Duration (months)	Home Range Size (km²)
Mother + two sub-adult cubs	25	14	19.1
Two sub-adults	15	11	5.54
Overall	40	–	12.3±6.85

Breeding Season

Sloth bear breeds during summer and after a gestation of seven months, cubs are born in December and January (Prater, 1965). During the present study only one occasion two adult bears (more likely an adult male and female) were seen together and this was late summer (May). Newborn cubs were recorded in a den in November indicating that mating could have taken place during early summer. In Royal Chitwan National Park, Nepal, cubs were mostly reported between Septembers and January (Laurie and Seidensticker,1977). Based on the information from Prater (1965), Laurie and Seidensticker (1977) and the present study, it appears that the breeding season of the sloth bear is largely similar for the entire sub-continent. Mating occurs during summer and cubs being born over an extended period from September to January.

Litter Size

Litter size was calculated based on a total of five different mother-young groups and two different, independent sub-adult groups (siblings that were yet separated from each other) recorded from 55 sightings of bears during the study period. Based on the number of young ones seen, the mean litter size estimated was 1.7. Though the norm is likely to be two cubs per litter, as five out of seven litters comprised two cubs, there were litters with one well-grown cub or sub-adult. It is not clear, in these two cases whether the original litter was one or two (with one cubs had died). Laurie and Seidensticker (1977) reported a similar litter size *i.e.* 1.6, but their data, similar to the present study, based entirely on observations of mother and cubs and not on the actual litter size at birth. Therefore, it is likely that the normal litter size in sloth bears is two with occasional births of a single cub.

Social Organization

Group Size

The most common group size was one and the largest group size recorded was three, the mean group size was 1.8 bears (n = 112). Single bears were seen on 38.4 per cent of the sightings and groups of two and three bears were seen on 37.5 per cent and 24.1 per cent of the sightings respectively. With the exception of one sighting of an adult male and female together all sightings of two and three bears were of mother and cubs or of independent sub-adults. These results suggest that sloth bears are solitary animals and adults are likely to be seen together only during the mating season. Other than that all other groups that comprised more than one individual are mother and dependent young. Sub-adult bears were also seen together for some time after their dispersal from their mother. The results of group composition are similar to that of from Royal Chitwan National Park in Nepal (Laurie and Sidensticker, 1977).

Mother-Young Relationship

During the present study we were able to record birth of one litter and to monitor the use of the den by the mother and cubs. During our study on den use by sloth bears, a bear was observed nearby a monitored den (identified in C9 den) on 24th September 1996, but when the same den was inspected again on 26 October 1996 there was no sign of use by bears. On 26th November when this den was re-inspected, two cubs were recorded in the den. Therefore it is likely that the cubs were born some time in November. The cubs remained in the den till 27th January 1997 and had left the den by 31 January when the den was inspected again and no cubs were recorded in the den. It was not possible for us to confirm if the cubs had become mobile enough to accompany the mother on her foraging trips or if the mother had moved them to another den. We feel it is most likely that the cubs were mobile when they left the den, as they were approximately three months old at the time when the den was found without the cubs. Laurie and Seidensticker (1977) reported that most of the young cubs accompanied their mothers during February-March period, when they were around two to four months old. In the present study if the cubs had born in early or late November they would have used the den for a maximum of three months and a minimum of two months. On the other hand, report from Sri Lanka also shows that cubs accompany their mother when they are as young as 3 weeks old (Norris, 1969). Nevertheless, the observations from Mudumalai and also in Chitwan (Laurie and Seidensticker, 1977) do not agree with that of Noris (1969). It appears that cubs are likely to accompany the mother on her foraging trips only after two to three months old.

During the period of cubs' presence in the den (26th November – 27th January), the den was visited on 13 occasions and the mother was seen at the den on only one occasion (7.7 per cent of the time) indicating that the mother does go out to forage when the cubs are young (the cubs were alone in the den on 12 occasions). Laurie and Seidensticker (1977) stated that the mother either feeds with movements around the dent or remain with their cubs feed without feeding. Nevertheless, our observations suggest that the mother does go out for feeing but is likely with restricted movement. Similar observations have been made earlier by Ajay Desai in the same study area (personal communication).

The duration of the association between the young and their mother is not known (Laurie and Seidensticker, 1977). In the present study we were able to monitor the association between one female and her cubs. This female was first seen with her two cubs on 1 August 1995, the cubs were likely to have born between November 1994 and January 1995. The mother and cubs group was monitored prior to the present study and we continued to monitor this group through the present study. The cubs were seen continuously with the mother throughout their second year. They were first seen independent of their mother in January 1997 but this was likely to be a break of short duration as they were seen foraging with the mother in May 1997. These observations suggest that the young remain with their mother for at least two years after which they gradually disperse away from their mother over a period of six moths or one year time.

In the course of our study, we were also able to monitor the association between two sub-adult cubs, which were independent of their mother (they were never seen with the mother) from 8 May 1996 to 6 April 1997. They were seen together on 15 occasions over a period of one year. It is likely that they were not entirely independent of their mother through this period and were possibly joining up with her occasionally especially during the earlier part of the separation. The process of separation between the mother and cubs, and between the cubs appears to be a long one, stretching over a year after the initial process of separation begins when the cubs are one year old. The cubs therefore remain or continue to associate with the mother at least until they are two years old.

The long association between the cubs and mother would result in the mother normally breeding only once in two to three years. A female with sub-adult cubs was reported associating with a male bear, the cubs were moving 30+ meters from the pair, (Ajay Desai, pers. comm.). It appears that the cubs continue to associate (possibly on a reduced scale) with the mother even after she mates and possibly continue to associate with her even during her subsequent pregnancy. Sloth bears are likely to have low reproductive rate considering the long association of the cubs with the mother and also the small litter size. Grizzly bears (*Ursus arctos*) with an average litter size of 1.6-2.5 (Craighead *et al.,* 1974; Herrero,1972) and black bears (*U. americanus*) with an average litter size of 1.6-2.6 (Jonkel and Cowan, 1971) appear to have larger litter size than the sloth bears.

Management Recommendations

The recommendations are made with reference to sloth bears and not specifically with reference to the study area. Mudumalai Sanctuary offers an opportunity to study the ecology of the sloth bear in a natural (as possible) habitat. Therefore some suggestions can be made that are useful to other areas also.

All along the roads in the sanctuary, all vegetation is cleared for a width of 10 meters on either side. This is done to facilitate better viewing of wildlife (view lines) and also to act as a firebreak. While clearing these view lines/fire breaks along the roads, saplings of fruit species like *Cassia fistula,* have been chopped annually. These saplings should be allowed to grow, as they are important food resources for bears.

In other areas with sloth bears, especially in degraded forest it is essential to improve the habitat quality by taking up planting of tree species whose fruits serve as food for sloth bears. Fruit tree species should be selected based on the sloth bear diet in the local area. If any fruit tree species are very rare or have become locally extinct due to human exploitation and do not occur in the diet of bears locally, but have been recorded elsewhere in the diet, then these species too can be planted.

Ants and termites form the bulk of the sloth bears diet and they are primarily ground dwelling species. Forest fires have their greatest impact on the ground flora and fauna. The fires not only burn out the dry grass and leaf litter which forms a food source for termites they eliminate these sources of cover for ants and other insects. The microhabitat and climate at the ground level, which harbors insects important to the sloth bears diet, is drastically altered by fires. Fire management therefore requires much greater attention.

Minor forest produce collection (honey/fruits) would adversely affect the sloth bears. This should be controlled or eliminated in habitats with sloth bears.

Natural den sites in the study area appear to be adequate (based on the low percentage of use of existing dens) and creation of man made dens may not be necessary for their use. If such structures are built, bears will use them but not continuously. But in areas where there are no adequate den sites, artificial structures may be useful as they are likely to be used if constructed properly.

The study of a nocturnal species through a short-term study and by indirect means has severe limitations. We still know very little about its ranging behaviour and spacing mechanisms and also of its population size and dynamics. These can only be studied through long-term studies using radio telemetry and/or using marking techniques. Thus more detailed studies are especially needed for the Eastern and Western Ghats of southern India that severely lack such details.

The image of sloth bears in the minds of most people is that of a performing bear (more or less something to be laughed at) or that of a dangerous animal out to maul and kill. There is a need for creating greater awareness among people about sloth bears, and their uniqueness and the role they play in nature (seed dispersers) and the need for their conservation.

Acknowledgments

We thank Mr. G. Kumaravelu (IFS) former Conservator of Forests (WL) Western Region for his encouragement in this project and Tamil Nadu Forest Department for financial support. We would like to express our gratitude to Dr. K. S. Devdass, former Range Officer Mudumalai Range for extending all logistic supports to the field work. We would like to specially acknowledge Late Mr. J.C. Daniel former Honorary Secretary, and Dr. A.R. Rahmani, Director Bombay Natural History Society and Mr. A. Udhayan (IFS), former Wildlife Warden, Mudumalai Wildlife Sanctuary for their critical suggestions to this work. We are grateful to our field assistant Mr. V. Maran for assistance in the field.

References

Akhtar, N., Bargali, H.S. and Chauhan, N. P. S., 2004. Sloth bear habitat use in disturbed and unprotected areas of Madhya Pradesh, India. *Ursus,* **15**:203-211.

Bargali, H. S., Akhtar, N., and Chauhan, N.P.S., 2004. Feeding ecology of sloth bears in a disturbed area in central India. *Ursus,* **15**:212–217.

Baskaran, N., 1990. An ecological investigation on the dietary composition and habitat utilization of Sloth bear (*Melursus ursinus*) at Mudumalai wildlife sanctuary, Tamil Nadu (South India). *M. Phil., Thesis,* A.V.C. College Mannambandal, Tamil Nadu.

Baskaran, N., 1998. Ranging and resource utilization by Asian Elephant (*Elephas maximus* Linnaeus) in Nilgiri Biosphere Reserve, South India. *Ph.D. Thesis.* Bharathidasan University, Tiruchirapalli, India, 138+iii.

Baskaran, N., Sivanagesan, N. and Krishnamoorthy, J., 1997. Food habits of the sloth bear at Mudumalai Wildlife Sanctuary, Tamil Nadu, South India. *Journal of the Bombay Natural History Society,* **94**:1-9.

Baskaran, N., and Desai, A.A., 2010. Does indigestible food remains in the scats of sloth bears (*Melursus ursinus*) represent actual contribution of various food items? *Journal of Threatened Taxa,* **2(13)**: 1387–1389.

Cowan, I. McT., 1972. The status and conservation of bears (Ursidae) of the World-1970. IUCN Publs. N.S. No. 23: 343- 367.

Craighead, J.J., Varney, J.R. and Craighead, F.C. Jnr., 1974. A population analysis of the Yellowstone Grizzly bears. *Univ., Mont. For. Conser. exp. Stat. Bull.,* **40** : 1-20.

Dalke, P.D., 1938. The cottontail rabbits of Connecticut. *State of Connecticut, Geological and Nat. Hist. Surv. Bull.,* **65** : 1-97.

Daniel, J.C., Desai, A.A., Sivaganesan, N. and Kumar, Ramesh, 1987. The study of some endangered species of wildlife and their habitats: The Asian Elephant. Report October 1987. Bombay Natural History Society.

Desai, A.A., Baskaran, N. and Venkatesh, S., 1997. Behavioural ecology of the sloth bear in Mudumalai Wildlife Sanctuary and National Park, Tamil Nadu. Report. Bombay Natural History Society, Bombay, India, and Tamil Nadu Forest Department, Chennai, India.

Garshelis, D.L., Joshi, A.R. and Smith, J.L.D., 1999. Estimating density and relative abundance of sloth bears. *Ursus, 11*: 87-98.

Gokula, V., 1991. Some aspects on the feeding habits of the Sloth bear (*Melursus ursinus*) at Mundanthurai wildlife sanctuary, Tamil Nadu (South India). *M.Sc., Thesis*, A.V.C. College, Mannambandal, Tamil Nadu.

Hamer, D., and Herrero, S., 1987. Grizzly bear food and habitat in the Front ranges of Banff National park, Alberta. *Int. Conf. Bear Res., and Manage.*, **7** : 199-213.

Herrero, S., 1972. Aspects of evolution and adaptation in American Black bears (*Ursus americanus*) and Brown and Grizzly bears (Ursus arctos) of North America. *Int. Conf. Bear Res. and Manage.*, **2** : 221-230.

IUCN, 2013. IUCN Red List of Threatened Species. Version 2013.2. <www.iucnredlist.org>. Downloaded on 11 April 2014.

Jaffeson, R.E., 1975. *Melursus ursinus* : Survival status and conditions. Washington, D.C. The author.

Jennrich, R.I., and Turner, F.B., 1969. Measurements of non- circular home range. *J. Theor. Biol.*, **22** : 227-237.

Johnsingh, A.J.T., 1986. Diversity and conservation of carnivorous mammals in India. *Porc. Indian Acad. Sci.*, 73- 89.

Jonkel, C.J., and Cowan, I.McT., 1971. The Black bears in the Spruce-fir forest. *Wildl. Monogr.*, 27.

Joshi, A.R., Garshelis, D.L., and Smith, J.L.D., 1995. Home ranges of Sloth bears in Nepal : Implication for conservation. *J. Wildl. Manage.*, **59 (2)** : 204-214.

Joshi, A.R., Garshelis, D.L., and Smith, J.L.D., 1997. Seasonal and habitat - related diets of Sloth bear in Nepal. *J. Mammol.*, **78 (2)** : 584-597.

Krishnan, M., 1972. An ecological survey of the larger mammals of Peninsular India. *J. Bombay Nat. His. Soc.*, **69 (1)**: 27-54.

Laurie, A., and Seidensticker, J., 1977. Behavioural ecology of the Sloth bear (*Melursus ursinus*). *J. Zool., Lond.*, **182** : 187-204.

Mace, R.D., and Jonkel, C.J., 1986. Local food habits of the Grizzly bear in Montana. *Int. Conf. Bear Res., and Manage.*, **6** : 105-110.

Maehr, D.S., and Brady, J.R., 1984. Food habits of Florida black bears. *J. Wildl. Manage.*, **48 (1)**: 230-235.

Norris, T., 1969. Ceylon sloth bear. *Animals*, **12** : 300-303.

Prater, S.H., 1965. *The Book of Indian Mammals*. Bombay Nat. Hist. Soc., 3rd edn., Oxford Univ. Press, Bombay.

Ratnayeke, S., van Manen, F.T., and Padmalal, U.K.G.K., 2007. Home ranges and habitat use of the sloth bear (*Melursus ursinus inornatus*) at Wasgomuwa National Park, Sri Lanka. *Wildlife Biology*, **13 (3)**: 272-284.

Schaller, G.B., 1967. *The Deer and Tiger: A Study of Wildlife in India*. University of Chicago Pres, London.

Schaller, G.B., 1969. Food habits of the Himalayan Black bear (*Selenarctos thibetanus*) in the Dachigam Sanctuary, Kashmir. *J. Bombay Nat. His. Soc.*, **66 (1)** : 156-159.

Servheen, C., 1991. Trade controls for North American Black bears. News letter of the Species Survival Commission 51.

Singh, A., 1973. *Tiger Haven*. London, Macmillan.

Spillett, J.J., 1996. A report on wildlife surveys in North India and Southern Nepal. *J. Bomb. Nat. Hist. Soc.*, **63 (3)** : 492-516.

Sreekumar, P.G., and Balakrishnan, M., 2002. Seed dispersal by the sloth bear (*Melursus ursinus*) in India. *Biotropica*, **34**:474-477.

Yoganand, K., Johnsingh, A.J.T., and Rice, C.G., 1999. Annual technical report (Oct. 1998 to Sep. 1999) of the project "Evaluating Panna National Park with special reference to the ecology of sloth bear". Wildlife Institute of India, Dehradun, India. Unpublished report.

2015, Animal Diversity, Natural History and Conservation, Vol. 5   Pages *335–366*
Editors: V.K. Gupta and Anil K. Verma
Published by: DAYA PUBLISHING HOUSE, NEW DELHI

Chapter 21

Ecological Crisis vis-à-vis Intraspecific Conflict: A Case Study with Rhinos in Jaldapara and Gorumara National Parks, West Bengal, India

Jayanta Kumar Mallick (Retd.)*
Wildlife Wing (Headquarters), Forest Directorate,
Government of West Bengal, W.B., India

ABSTRACT

In West Bengal, population of the Great Indian one-horned rhinoceros (*Rhinoceros unicornis*) in the floodplains of Jaldapara (216.51 km²) and Gorumara (79.45 km²) National Parks in Jalpaiguri district, the hunting and poaching hotspots up to early 1980s, has gradually been recovered from the bottom of depression (14+8= 22) to the carrying capacity threshold (186+43= 229) during the last two and half decades, consequent upon strict protection and successful reproduction. The burgeoning population has now posed a heavily skewed sex-ratio (excessive number of adult males than that of reproductive females). This has resulted in frequent infighting- as well as forced mating-casualties and dispersal of the victims to the safe range-edge or even far beyond to establish new territories. The probable solutions of this crucial management problem are round-the-clock monitoring, seclusion of the identified more aggressive males, reintroduction of more wild females from the neighbouring rangelands, development of new grasslands and translocation of the excess population to those habitats in the region in phases.

Keywords: *Riparian grassland, Population boom, Skewed sex-ratio, Infighting-casualties.*

* E-mail: jayantamallick2007@rediffmail.com

Introduction

The Great Indian One-horned Rhinoceros, *Rhinoceros unicornis* Linnaeus, 1758, is now restricted to almost exclusively within and around protected areas in India (Assam, Uttar Pradesh and West Bengal) and Nepal (extinct from Bangladesh and Bhutan) and its population has turned around from approximately 200 individuals in the late 19th century to a range of 3300-3350 at the end of August 2013. The recovery of the species is one of the greatest conservation success stories in both the countries. Despite some serious threats in past few decades, the park managers have been able to reduce the poaching pressure successfully and the rhino population increased considerably. *R. unicornis* is the only Asian large mammal species in recent history whose IUCN Red List status has been down-listed from 'Endangered' to 'Vulnerable' in 2008. But due to the most impressive increase of rhino populations and other associated mega herbivores- *Elephas maximus* (elephant) and *Bos gaurus* (gaur) in the protected areas (rhino lands), followed by excessive cattle grazing, the carrying capacity of rhino has been exceeded there, which leads to an ecological crisis and increased risk of rhino-human conflict as the rhinos frequently move out of the parks to meet their biological needs.

The re-introduction programmes of the excess park population have now begun and the species is starting to repopulate the former habitats where not so long ago they had become extinct from. But the threat of poaching is still present, so also the dwindling habitats. Most of the rhino habitat is suffering from tremendous biotic pressure. As a result, intraspecific competition and conflict is often encountered resulting in increasing mortality or injury of the rhinos, particularly the bulls. Moreover, many rhinos are now frequently straying from the safety of the parks to the fringe areas as well as far-flung (ca. 50 km) habitats and some of them have been staying there either seasonally or even permanently.

Objective of the Study

Though a number of studies have been undertaken in the popular rhino lands of Assam (India) and Nepal, very little attempt has been made to project the third important conservation site in northern West Bengal. The present paper discusses the nature and impact of heavily skewed sex-ratio (male-biased) as a result of the burgeoning rhino population in Jaldapara and Gorumara, reaching the carrying capacity threshold and probable management solutions to this problem.

Study Areas

In erstwhile Bengal, *R. unicornis* was distributed extensively in both the Upper Gangetic Plains and Lower Gangetic Plains (Biographical Zones 7A and 7B) during the nineteenth century, but presently the species is restricted to only two fragmented (approximate linear distance of 45 km) protected areas (National Parks) in the former zone- Jaldapara spread over 216.51 km² (25°58´-27°45´N and 89°08´-89°55´E) under Wildlife Division-III and Gorumara covering 79.45 km² (26°43´–26°47´N and 89°47´–88°52´E) under Wildlife Division-II (Bahuguna and Mallick, 2004). Jaldapara is the second largest natural home and *in situ* conservatory of this Schedule-I species in India after Kaziranga of Assam (Ghosh and Das, 2007). Moreover, the Ministry of

Environment and Forests, Govt. of India, has declared Gorumara as one of the best-managed National Parks in the country for the year 2009 and for the first time in the history of Greater One-horned Rhino census, advanced genetic tools have been used in Gorumara and the same programme will also be undertaken in Jaldapara.

Jaldapara is demarcated by Bhutan and Totopara village on the north, Falakata-Cooch Behar Road on the south, Reserve Forests, 14 tea gardens and 32 revenue villages on the east and west. Gorumara is bounded by Batabari-Nagrakata Road on the north, reserve forests of Jalpaiguri Forest Division on the south, the River Jaldhaka and tea gardens on the east and National Highway 31 on the west. More than 50 human settlements are located within its zone of influence.

The altitude of these rhino lands varies from 25 m to 275 m asl. Here the highest temperature recorded in the summer is 37°C with high humidity (75-95 per cent) and the lowest in the winter- 10°C. The average annual rainfall is about 380 cm, mostly occurring between mid-May to mid-October. December is driest and July wettest.

The Torsa-Malangi floodplains have formed the trouser-shaped rhino-habitat in Jaldapara. It consists of twelve forest blocks (46 compartments)- Bania (Compartments 1-4, 8b): 16.48 km², Barodabri (1b, 2, 6b, 7b): 6.09 km², Chilapata (1-2, 3b, 4b): 18.54 km², Dalsingpara (1-4): 14.78 km², Hasimara (1-4): 16.43 km², Jaldapara (1-5): 34.51 km², Jaigaon (1-2): 17.56 km², Malangi (1-3): 12.42 km², Mendabari (3-6): 10.22 km², Salkumar (1-4): 5.03 km², Titi (1-4 part): 38.19 km² and Torsa (1-3): 26.26 km² (Figure 21.1).

The combined Jaldhaka-Murti-Indong-Diana floodplains have formed Gorumara, which consists of 11 forest blocks (25 compartments)- Barahati (1–3): 9.31 km², Bhokolmardi (1–3): 8.35 km², Central (1): 2.48 km², Dhupjhora (1a, b, c): 4.24 km², Gorumara (1–2): 6.59 km², Jaldhaka (1b): 2.62 km², Kakurjhora (1–2): 5.65 km², Medlajhora (1–3): 8.50 km², Selkapara (1–2): 7.67 km², South Indong (1–3): 12.73 km² and Tondu (1; 2a, b; 3; 4a, b): 11.31 km² (Figure 21.2).

Next to Kaziranga (Assam, India) and Royal Chitwan (Nepal), Jaldapara is the third prime conservation area of this endemic megaharbivore of the floodplains (Vigne and Martin, 2012), a genetic isolate from Nepal and Assam for centuries (Groves, 1993, 2003; Ali *et al.*, 1999). Jaldapara (Sanctuary) has been upgraded to National Park in May 2012 for better management, whereas Ministry of Environment and Forests, Government of India, declared Gorumara as the best-managed National Park in the country for the year 2009.

Materials and Methods

A database was prepared by collating the published and unpublished records (census, monitoring, birth/casualty, rescue, chemical immobilization, treatment and release) of *R. unicornis* in the study area. Ground surveys were undertaken separately for Jaldapara and Gorumara during 1st January-31st December 2012 with the help of frontline forest department staff. A questionnaire survey was also conducted among the stakeholders including forest and fringe villagers, members of local Eco-development/Forest Protection Committees living in and around these two protected areas.

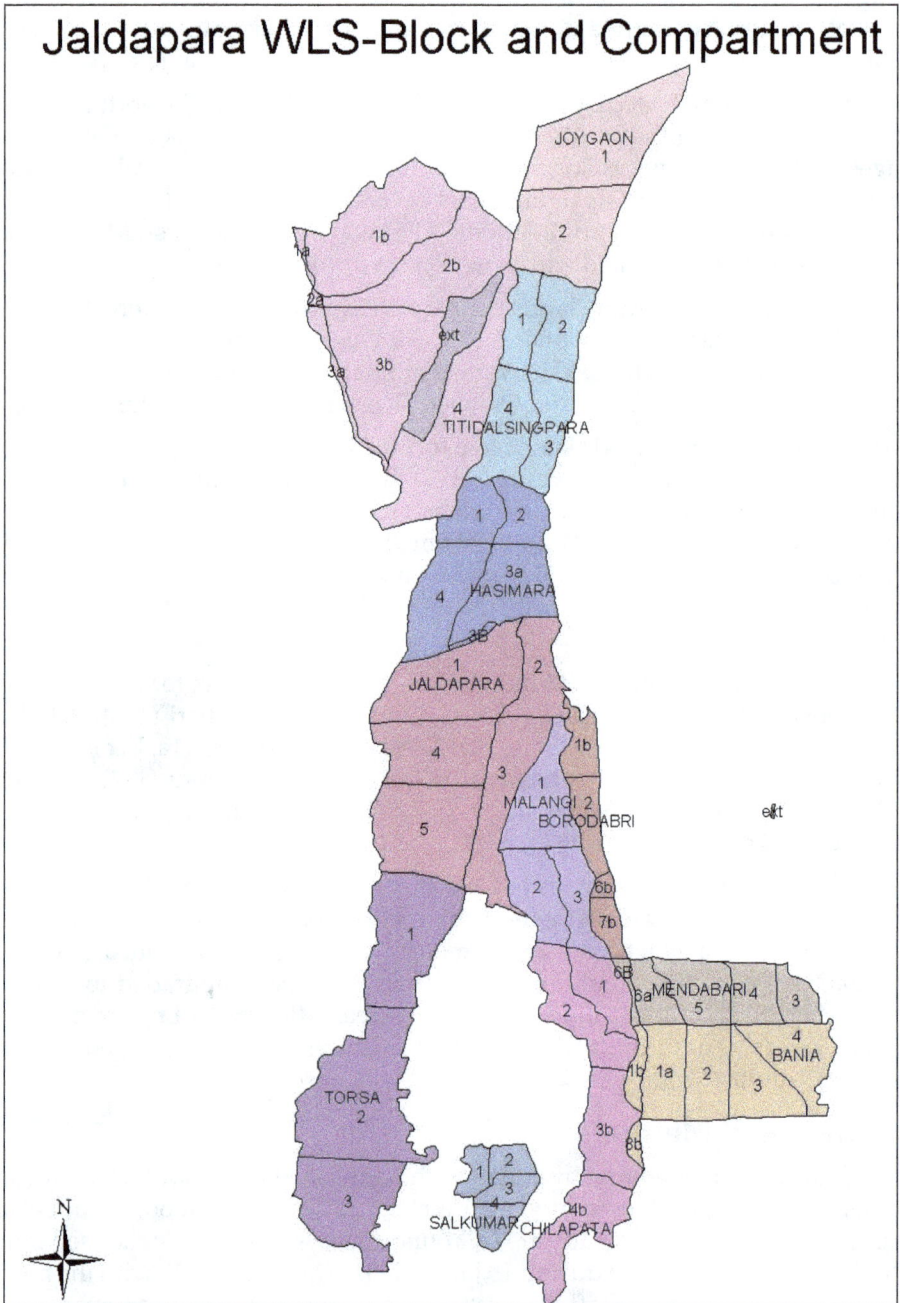

Figure 21.1: Jaldapara National Park.

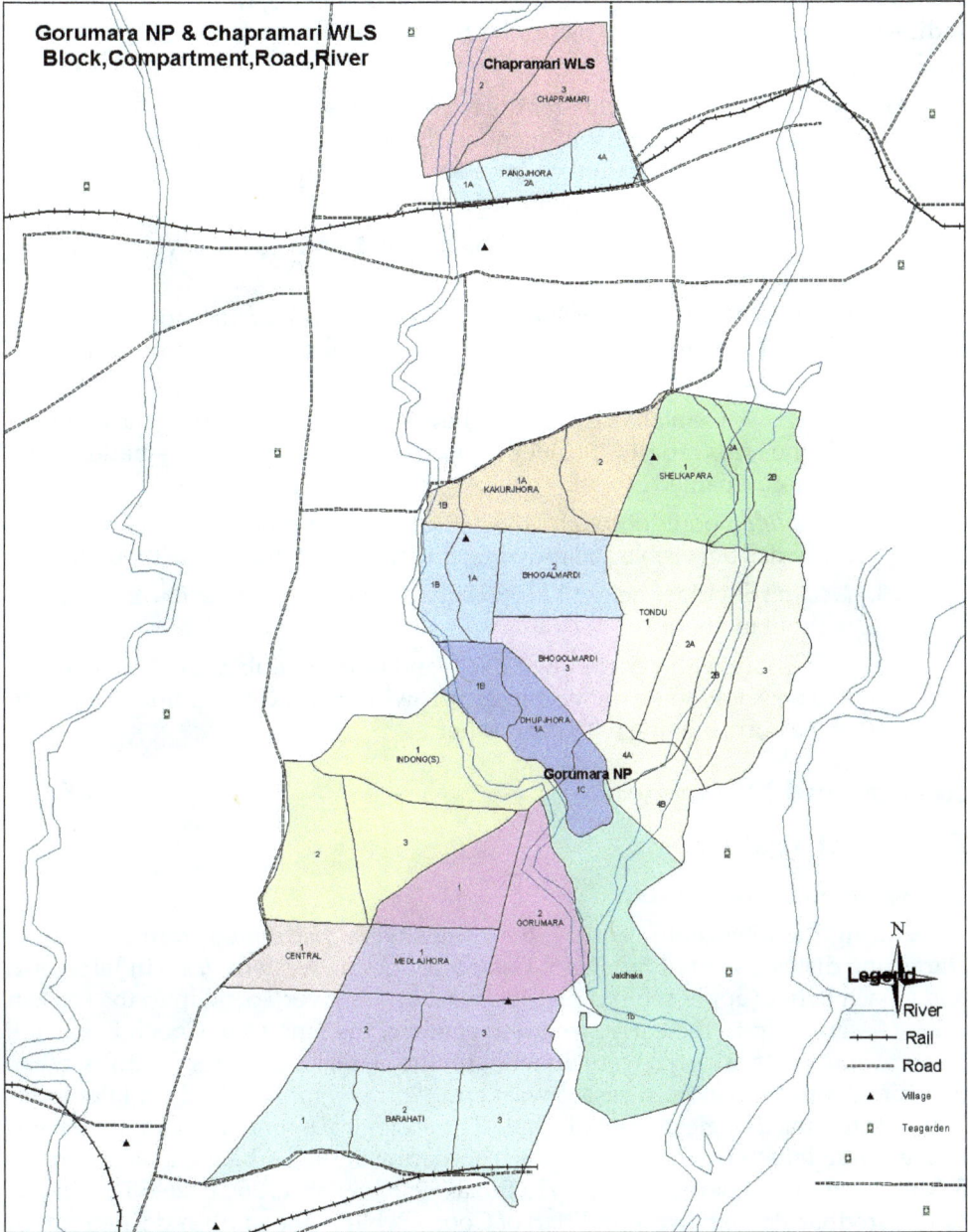

Figure 21.2: Gorumara National Park.

The rhinos were recorded in the thick and open grasslands, savannah woodlands, saltlicks and wallowing pools by direct sighting in the morning (06-10 h), late afternoon and evening (16-21 h) from elephant back, vehicle, watch tower and also indirect evidences like 'middens' or high community dung heaps, tracks and wallowing signs.

An attempt was made to classify the rhinos as to sex and age class (*i.e.* adults, subadults and calves including the juveniles). If there was any doubt as to the sex of an adult animal, it was classified as 'non-sexed'. The age/sex of the rhinos was determined in the field on the following basis:

1. *Adult*: Shoulder height above 160 cm and horn base above 18 cm with prominent neck folds;
   (a) *Male*: Penis either visible from the side or the rear (urination indicating the location); deeply folded skin around the neck; large horn with wider base;
   (b) *Female*: Genitalia visible from the rear (urination indicating the location); the folds around the neck and the horn comparatively smaller; often accompanied by calf.
2. *Sub-adult*: Shoulder height 135-160 cm and horn base 8-18 cm and neck fold small but visible and growing, the individuals moving without mother;
3. *Juvenile*: Shoulder height 120-135 cm and horn base up to 8 cm and moving without mother;
4. *Calf*: Shoulder height below 120 cm and horn base absent or just growing and neck fold small but visible and growing, practically without formation of the horn; moving with the mother.

Results and Discussion

Ecological History

(a) Fragmention of Habitat

During the nineteenth century, the rhinos were distributed from the *terai* of Darjeeling district, west of the River Tista, through the western *duars* in Jalpaiguri and Cooch Behar (upper region) districts, east of the River Torsa up to the eastern *duars* (Assam). Up to the early twentieth century, the jungles of Cooch Behar Raj (eastern and western *duars*) were linked with the Darjeeling *terai* or foothills, when the rhinos could travel from east to west and *vice versa* for hundreds of kilometres without crossing a single man-made path. The ecological boundary of Jaldapara was then extended up to Titi Block in the north bordering Bhutan, Chilapata Block (Bania-5, Mendabari-1 and Barodabri-8) up to Buxa (89°20'-89°55' E, 26°30'-26°55' N) in the east, Garodhat-Patlakhawa-Pundibari of Cooch Behar in the south and Khairbari in the west. The rhinos of Bholka Range (Buxa) moved freely between Assam and West Bengal across the Sankosh. The populations of Jaldapara and Gorumara are isolated by physical barriers and ecological boundaries. There is no record of rhino-movement in between Jaldapara and Gorumara. The ecological boundary of Gorumara was extended up to the River Tista in the west and surrounding forests of Jalpaiguri Forest Division in the north, east and south.

As a sequel to fragmentation of the habitat, no biological corridor is left linking the protected areas in northern Bengal- Buxa-Jaldapara, Jaldapara-Gorumara, Gorumara-Chapramari and Gorumara-Mahananda (Figure 21.3). So, the surviving subpopulations in Jaldapara and Gorumara have become isolated, which has restricted the potential gene-flow.

(b) Loss of Habitat

It is estimated that the extent of tall riparian grassland lost in three sectors of northern Bengal between 1900 and 1990 was 450-580 km² on account of-

1. Extension of tea gardens (200-250 km²),
2. Extension of agriculture (60-80 km²),
3. Encroachments or establishment of leasehold *Khasmahals* (20-30 km²),
4. Annual flood-erosions (20-30 km²), and
5. Forestry practices (conversion of grassland by commercial plantation): 150-190 km² [30-40 km² in the Sankosh-Rydak sector (89°44'-89°47' E, 26°34'-26°40' N) of Buxa and Garodhat, 70-90 km² in the Torsa-Hasimara sector of present Jaldapara including Patlakhawa of Cooch Behar and 50-60 km² in Jaldhaka-Diana sector of present Gorumara] (Bist, 1994).

(c) Population Decline

Whereas rhino populations in the floodplains of Ganges river system, *i.e.* Malda, Murshidabad and 24-Parganas (the Sundarbans) districts of southern Bengal were wiped out during the late nineteenth century (Mukherjee, 1963; Agrawal *et al.*, 1992; Das, 2008), at least 240 rhinos [120+ in Sakosh-Rydak, 100+ in Torsa and 20+ in Jaldhaka-Diana sectors] existed in northern Bengal during 1890s (Bist, 1994). Consequent upon large-scale hunting and poaching as well as extensive habitat loss, their range has been reduced to less than one-eighth (Banerjee, 1966) followed by extinction of most of the subpopulations in this region during the last two centuries (Maharaja of Cooch Behar, 1908; Menon, 1996; Martin, 1996a, 1996b, 1999; Pandit and Yadav, 1996; Bahuguna and Mallick, 2010). The rhino population of Jaldapara-Gorumara was shot down to 22 (14+eight respectively) in 1985 (Bhutia, 2008).

Even during three enumerations in Jaldapara between 1964 and 1966, out of a population range of 50-60 as a sequel to the sharp decline in 1966, Spillet (1966) observed, "36 per cent of the sexed adults were males and 52 per cent of the adult females were accompanied by young. In comparison, 54 per cent of the adults observed during the enumeration were males and only 31 per cent of the adult females were accompanied by young. In either case, the relatively high percentage of young would probably indicate that the population is thriving". But after thirty years, Wildlife Institute of India (1997) recorded a negative growth rate of rhinos in Jaldapara and Gorumara up to 1996. The annual growth rate of Jaldapara population was -0.0222 (95 per cent CL±0.039) per year between 1930 and 1996, whereas the figure was 0.0079 (95 per cent CL±0.012) for Gorumara between 1968 and 1996. The sex ratio for

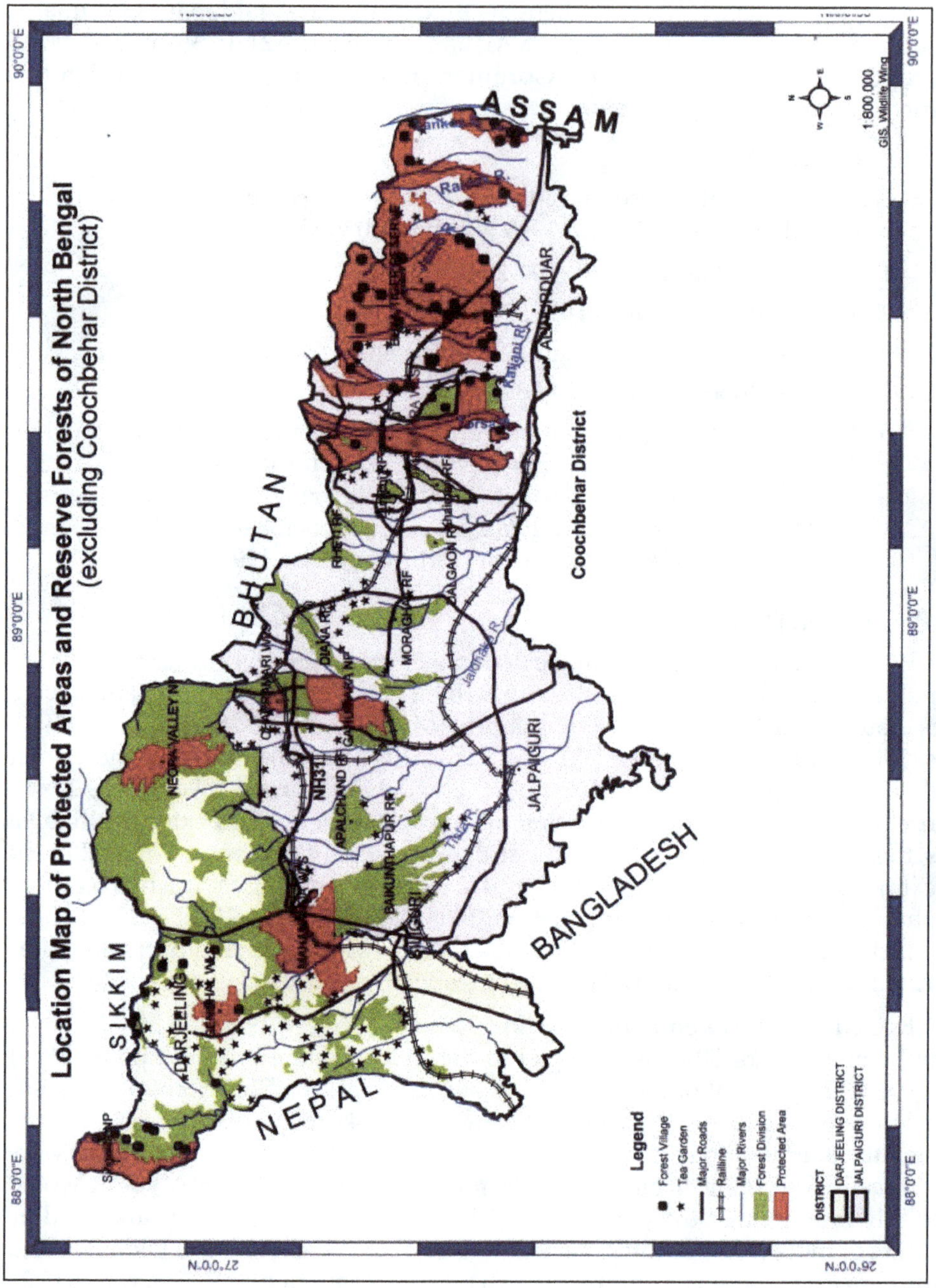

Figure 21.3: Protected Areas in North Bengal.

adult male to adult female in Jaldapara was 0.694:1.000 (95 per cent CL±0.133) during 1975-1996, while in Gorumara, it was 0.395:1.000 (95 per cent CL±0.077) during 1968-1996. The ideal ratio is considered to be 0.25:1.00 to 0.30:1.00. The calf-adult female ratio for Jaldapara was 0.509:1.000 (95 per cent CL±0.078) and in Gorumara it was 0.349:1.000 (95 per cent CL±0.111).

(d) Population Revival (Boom)

During early 1990s, the carrying capacity of Jaldapara and Gorumara was assessed to be 100 and 20 respectively (Molur et al., 1995). But during the last two and half decades, the rhino population had reached an all-time record of 229 (Jaldapara and surroundings: 186, an increase of 48.8 per cent between 2009 and 2013, *i.e.* 9.76 per cent per year + Gorumara and surroundings: 43, a rise of 22.85 per cent between 2010-2012, *i.e.* 7.61 per cent per year), as per direct counts by the park authorites (Tables 21.1 and 21.2). Aaranyak, Assam, also conducted a field survey in Gorumara for dung collection in April 2011 by using elephants. Laboratory work was carried out in the Wildlife Genetics Laboratory of Aaranyak at Guwahati. The unique "genetic profiles" of individual rhinos, popularly known as DNA fingerprints were determined from dung DNA extracts using a set of polymorphic microsatellite markers. In this process, candidate markers were tested on rhino samples of known individual identity, in order to determine the minimum number of such markers needed for identifying individuals from unknown samples. Through this process, the total number of individuals present (43) in the collected dung samples was determined (Barthakur *et al.*, 2012; Talukdar, 2012).

Table 21.1: Rhino population in Jaldapara

Year	AM	AF	Nonsexed SAM	SAF	Nonsexed	MC	FC	Nonsexed	Total
1975	7	7	4	–	–	–	5			23
1978	5	7	4	–	–	–	3			19
1980	5	7	6	–	–	–	4			22
1988	9	11	–	–	–	–	4			24
1989	9	13	–	–	–	–	5			27
1992	8	12	–	–	–	–	13			33
1993	4	7	2	–	–	–	11			26
1996	9	18	2	–	–	–	8			42
2002	17	11	2	6	2	2	16	4	14	74
2004	32	28	3	5	3	2	1	1	21	96
2006	35	33	2	4	3	5	3	3	20	108
2009	42	33	1	14	6	1	0	2	26	125
2013	62	55	4	8	6	9	42			186

AM: Adult Male; AF: Adult Female; SAM: Sub Adult Male; SAF: Sub Adult Female; MC: Male Calf; FC: Female Calf.

Table 21.2: Rhino Population in Gorumara

Year	AM	AF	Unsexed	SAM	SAF	Unsexed	Calf	Total
1978	1	3	3				1	8
1989	4	7	–				1	12
1993	4	7	–				4	15
1996	2	6	–				7	15
2002	6	11	–	–	–	–	5	22
2004	9	11	–	–	–	–	5	25
2005	13	9	2	–	–	–	3	27
2006	12	9	–	2	–	–	8	31
2009	17	12	–	1	–	–	5	35
2013	14	11	–	7	3	1	–	43

AM: Adult Male; AF: Adult Female; SAM: Sub Adult Male; SAF: Sub Adult Female.

The conservation success of rhinos in the study area could be achieved due to exemplary conservation efforts like intensive (24 h) patrolling (05-09 h on elephant back, 09-15 h on foot, 15-19 h on elephant back, 19-22 h by vehicle and 22-05 h by vehicle or on foot/elephant back), intelligence gathering and eco-development activities in and around these protected areas (Martin, 2006; Martin and Vigne, 2012), for which there was only one case of poaching during the twenty first century.

Habitat Use Patterns

Habitat utilisation pattern of rhinos (Figures 21.4–21.6) is dependent upon food, grass cover and water (Sarma, 2012). The rhinos used to exploit all the forest seral stages in Jaldapara and Gorumara to meet its year-round food and cover requirements (Rawat, 2005). The habitat types and extent in Jaldapara and Gorumara are shown in Tables 21.3 and 21.4.

(a) Jaldapara

The rhinos appear to be confined to the moist habitats supporting the semi-evergreen to evergreen forests, almost always in association with the alluvial plains and tall grassland. This largest tract of tropical grasslands (savannah) in the state consists of *Saccharum sponaneum, S. arundinaceum, Phragmites karka, Arundo donax, Narenga porphyrocoma, Themada villosa,* etc, dotted with associations of Khair-Sissoo (*Acacia catechu-Dalbergia sissoo*) and Simul-Siris (*Bombax ceiba-Albizia procera*) woodlands. Most of these grasslands have been lost to the woodland encroachment. Moreover, due to the use of mostly one particular species *i.e., Saccharum narenga* (Dhadda), which grows vigorously and, naturally, it suppresses the growth of other species of rhino fodder in the vegetation (Ghosh and Das, 2007).

The natural vegetation is in a state of flux due to recurring floods and succession (Bhattacharya, 2012). In Jaldapara, the River Torsa shifted its course in 1968 from west to east over a width of 20 km, creating *Mara* (dead), *Buri* (old) and *Char* (deserted)

Table 21.3: Habitat Types and Extent in Jaldapara

Habitat Type	Forest Block and Compartments	Area in km²	Percentage Cover
Dry mixed	Titi 1-4; Jaldapara 3-5; Torsa 1-3; Chilapata 4B; Hasimara 4	46.17	21.33
Wet mixed (Semi-evergreen)	Jaldapara 2; Hasimara 4; Mendabari 6; Bania 3, 4; Barodabri 1-2; Chilapata 2, 3B	12.41	05.73
Mixed Sal	Barodabri 1-2, 6B, 7B; Mendabari 6; Bania 1; Titi 1-3; Salkumar 1, 3	17.61	08.13
Grassland	Hasimara 3; Jaldapara 3-5; Malangi 1-3; Chilapata 1, 3B; Bania 8B; Barodabri 7B; Torsa 2	30.55	14.11
Grassland and Khair-Sissoo succession	Hasimara 1-4; Dalsingpara 1-4; Joigaon 1-2; Titi 4; Jaldapara 1-5; Malangi 1-2; Chilapata-3B	42.90	19.81
Grassland with Simul-Siris succession	Dalsingpara 1-4; Joigaon 1-2; Titi 1-4; Torsa-1	22.59	10.43
Bamboo brakes	Titi 1; Jaldapara 3; Bania 1	1.23	00.56
Plantations	Mendabari 3-6; Bania 1-4; Titi 2-4; Joigaon 1; Hasimara 2; Dalsingpara 3	26.40	12.20
Sandy riverbed, degraded patch, forest villages	Barodabri 6(a), 7b; Hasimara 2; Malangi 2-3; Chilapata 2	16.65	07.70
Total		216.51	100.00

Table 21.4: Habitat Types and Extent in Gorumara

Habitat Type	Forest Block and Compartments	Area in km²	Percentage Cover
Riverine forest	Tondu 1, 2a, 2b, 3, 4a, 4b; Selkapara 1-2	18.99	23.91
Sal forest	Gorumara 1-2; South Indong 1-3; Bhogolmardi 1-3	27.67	34.82
Wet and dry mixed forest	Barahati 1-3; Central 1; Medlajhora 1-3; Kakurjhora 1-2	25.93	32.63
Savannah forest	Jaldhaka 1b; Dhupjhora 1a, 1b, 1c	6.86	08.64
Total		79.45	100.00

**Figure 21.4: Foraging Rhinos in Jaldapara.
(Photo Credit Sourav Basu).**

**Figure 21.5: Rhino Feeding on Grass in Jaldapara.
(Photo Credit Sourav Basu).**

Figure 21.6: Statue in Gorumara.

Torsa and the small Malangi merged with it during the previous floods. Most of the area of earlier course of the Torsa is covered by grassland. However, the emerging trees have invaded the grassland extensively. Weeds and climber infestation has also degraded the habitat.

Many habitat improvement programmes have been undertaken like overwood removal in colonized areas, plantation of indigenous tall and short fodder grasses, eradication of weeds and climbers, selective burning of the grassland for regeneration of nutritive fodder, cutting back of tree species, control of wild fire, etc. A number of concrete water recharging structure was constructed at Harindangar char (Jaldapara-5) for maintaining the water-level of the stream Chirakhawa (preferred rhino-bathing site), which have yielded excellent results.

Spillet (1966) observed that wherever domestic livestock grazing is evident there are few, if any, rhino present. The rhinos used the whole park area up to mid-twentieth century, but subsequently the use-area has been reduced to about 160 km² due to various anthropogenic disturbances (Ghosh, 1991). The core habitat was roughly estimated to be about 100 km² (46.29 per cent) (Syangden *et al.,* 2007). During the present study, 96.04 km² (45 per cent of the total area) were found to be most suitable habitat of the rhinos, of which the pure grasslands cover 30.55 km² (14.11 per cent), grasslands with *Acacia-Dalbergia* succession 42.90 km² (19.81 per cent) and grassland with *Bombax-Albizia* succession 22.59 km² (10.3 per cent), whereas the secondary habitat was estimated to be about 50 km² and the seasonal (monsoon) upland habitat to be approximately 10 km². The rhinos are not presently known to visit Titi (39.19 km²), Jaigaon (17.56 km²) and Dalsingpara (14.78 km²) blocks, north of National Highway 31A as well as Salkumar (5.02 km²) in between the two legs.

Bhattacharya (1982) described the home range and daily movement pattern of the rhino at Jaldapara and Gorumara. Banerjee (1993) also observed that the rhinos ramble in maximum numbers in most of the seasons within those forest blocks because their preferred food plants (quality as well as quantity) are available there. According to him, the grasslands (dominant species being *Saccharum narenga, S. arundinaceum* and *S. longisetosum* var. *hookeri*) in Chilapata, Torsa and Jaldapara are quite productive (coverage 65-75 per cent and biomass 19.50-22.05 ton/ha) as compared to other areas (Hollong, Bengdaki, Daidaighat, Borodabri, NWCC Line) of the park (coverage 31-41 per cent and biomass 12.5-15.08 ton/ha). Hence, the rhinos are mainly concentrated in the central and southern parts of the park (prime grassland habitat).

Das *et al.* (2003) also analysed nine vegetation types in Jaldapara: (i) Riverine, (ii) Wet Mixed Forest (iii) Sal Forest (iv) Semi-evergreen Forest (v) Low Grassland, (vi) Tall Grassland, (vii) Savannah, (viii) Open Herbland and (ix) Hydrophytic vegetation. But they did not survey all the vegetation types in Chilapata and Nilpara Ranges, as the rhinos rarely or never visit these places. Some places of Mendabari and Bania (Compartments 6 and 7) Beats of Chilapata Range are visited by Rhinos to graze on *Imperata cylindrica, Digitaria ciliaris, Oplismenus compositus, Axonopus compressus* and *Saccharum narenga* for a very brief period before returning to Jaldapara, but never stay at Bania areas for a considerable duration. They observed that two ranges - Jaldapara East (Jaldapara, Sissamara, Daidaighat and Malangi Beats) and West (Hollong, Kunjanagar, Bengdaki and TEC Beats) are high rhino concentration areas. The rhinos visit the CC Line Camp area, across the river, less frequently and they generally graze on *Oplismenus compositus* in the open herbland and on *Saccharum spontaneum* and *S. narenga* (restricted in distribution) in the riverine vegetation. The gross total of rhino-fodder and with specieswise yield, separately for monsoon and post-monsoon, in Jaldapara was also determined (Tables 21.5–21.7). The Beat-wise highest amount of fodder is produced (total of two seasons' production) in Sissamara Beat area (34953.408 qt) which is followed by Jaldapara Beat (29026.037qt), Harindanga (29008.826 qt), Moiradanga Beat (20544.492 qt) and Malangi Beat (17508.463 qt). Range-wise Jaldapara East Range is the highest producer (85730.18 qt) of rhino-fodder followed by Jaldapara West Range (79738.15 qt). The Siltorsa Beat of Jaldapara North Range also produce good amount of fodder (11466.455 qt) and is also a good area for the rhinos to graze. But the yield recorded from Bania and C.C. Line is quite low and it is not possible for the rhinos to live there continuously. While the total recorded yield of fodder during monsoon is 88378.78 qt, it is 89026.21 qt and the total yield of two seasons is 1,77,404.986 qt.

Approximately 30 km² of Jaldapara have good quality grassland (Nandi, 1991). The present study has revealed that the pure grasslands occur in small patches at Jaldapara 3-5, Malangi 1-3, Chilapata 1 and 3b, Bania 8b, Barodabri 7b and Torsa 2 compartments. The grasslands with *Acacia-Dalbergia* succession are found as small islands in the riverbeds and along the old course of Torsa (western leg). During the study, the rhinos were most frequently found in their prime habitat at Jaldapara 3-5, Torsa 1-2 and Chilapata 1-2. They were also sighted at Malangi 1-3 and Torsa 3. Occasionally, a few rhinos used to visit some parts of the Bania [29.04 km² (89°22' E 26°38' N)] and Mendabari [17.38 km² (89°23' E 26°38' N)] blocks. Maximum number of rhinos was spotted in the East range (97), followed by West Range (36) and North range (34) (89°15'-89°23' E, 26°32'-26°42' N). In the secondary habitat of the eastern part, 15 rhinos were sighted in Kodalbusty Range and two in Chilapata Range.

Jaldapara 2 (3.03 km²) and Hasimara 3a (5.24 km²), 3b (0.05 km²) and 4 (5.24 km²) have good grassland with *Dalbergia sissoo, Bombax ceiba* and *Acacia catechu*, but suffer from water scarcity, especially during the dry season. The rhinos visit these areas for grazing only. During the rainy season, their concentration increases here, but, during the summer, their number decreases.

A number of wallow pools, artificial glades and salt licks used by the rhinos were located at Jaldapara-3 and 5, Malangi-2 and Chilapata-2 and 3. Regular visibility of the rhinos was recorded from all these sites, particularly during the hot hours of the summer. The rhinos were mostly seen between 6-8.30 am and 5-8.30 pm.

(b) Gorumara

Out of four major forest types in Gorumara, the riverine grassland and savannah woodland with *Acacia-Albizia-Bombax* succession occupy about 30 per cent of the total plant cover (Mandal, 2007). In Gorumara too, the resident rhinos do not use the whole park area. In fact, the core habitat used by them is only 8 km² (10.06 per cent)(Bhutia, 2008). Ghosh (2012) mentioned two distinct ecotones for the rhinos-Riverain Rolling Flood Land Forests (RFF), Riverine Riparian Forests (RRF). This habitat offers the best grazing ground for the rhinos. During the present study, the rhinos were found to graze mostly in the riverine grasslands of Dhupjhora and Jaldhaka blocks (6.86 km²). The rhinos usually avoided the wet mixed forests, but, in the dry season (summer), when there is threat of fire in the grassland, the rhinos frequently used this type of forests at Barahati 1, 3; Central 1; Medlajhora 1; Dhupjhora 1b, 2 and Kakurjhora 2 as resting place and browsed on the lower canopy. The rhinos were mostly seen in the wallow pools and saltlicks located at Tondu 4, Old Khunia, Jaldhaka 1b, Medla doba, Medla 3, South Indong 1, Dhupjhora 1b and Barahati 3.

Abundance of rhino-dung in various habitats of Jaldapara and Gorumara, as recorded during the field survey, is tabulated in Table 21.8. It appears that the rhinos usually prefer the savannah and natural grasslands for grazing than the plantations and riverine forests. In the tropical moist deciduous forest, the dung-heaps were found in Gorumara only, but not in Jaldapara.

In Jaldapara, the rhinos and many other sympatric species such as wild elephant, gaur, hog deer, wild pig and spotted deer partially share the habitat, whereas the rhino habitat in Gorumara is mainly shared by the gaur, the seasonally ranging wild elephants and hog deer existing in low density. Whereas the deer are selective feeders, the rhinos and elephants are bulk feeders, feeding on a variety of grass species. In Jaldapara, particularly in the forest block of Jaldapara, Chilapata, Torsa and Malangi it is the abundance of dung sites of rhino, elephant, gaur, spotted deer and wild pig

Table 21.5: Gross-Total of Palatable Biomass Available in Jaldapara (Das *et al.,* 2003)

Ranges	Beats	Monsoon		Post-monsoon		Two Seasons' Totalin qt
		Yield/sqmin g	Total Yieldin qt	Yield/sqmin g	Total Yieldin qt	
Jaldapara East	Jaldapara	248.361	14134.71	261.655	14891.33	29026.037
	Sissamara	358.767	20080.19	265.736	14873.22	34953.408
	Dhai Dhai Ghat	85.331	1987.94	72.857	1620.489	3608.429
	Valuka River	–	–	666.5	633.841	633.841
	Malangi	258.849	8552.378	271.068	8956.085	17508.463
Range Total:			44755.21		40974.97	85730.18
Jaldapara West	Hollong: JP 5 (Harindanga)	148.841	11492.49	226.857	17516.34	29008.826
	Hollong: Torsa 1	44.8	393.656	35.3	310.18	703.836
	Moiradanga	453.402	12241.86	307.505	8302.633	20544.492
	Kunjanagar	288.938	5656.826	293.264	5741.531	11398.357
	Bengdaki	231.655	4816.112	337.051	7007.291	11823.403
	T.E.C.	296.241	2843.914	325.762	3415.319	6259.233
Range Total:			37444.85		42293.29	79738.15
Jaldapara North	Siltorsa	279.953	12351.948	277.761	12255.23	24607.178
	50 ft Camp	32.741	596.546	21.143	385.236	981.782
Range Total:			12948.494		12640.466	25588.96
Chilapata	Bania	72.034	89.682	239.896	298.67	388.352
Kodalbasti	C.C. Line Camp	80.93	20.501	242.2	61.353	81.854
GRAND TOTAL:			95258.737		96268.749	1,91,527.496

Table 6: Species-wise yield of fodder in different parts of Jaldapara Wildlife Sanctuary in monsoon vegetation (Das *et al.,* 2003)

Sl.No.	Fodder Species	Jaldapara East Range						Jaldapara West Range				Jaldapara North Range		Chiapata	Kodalbasti	Total Yield
		Jal.	Sissa.	DD Ghat	Mala.	Hollo.	Harind.	Moirad.	Kunja.	Bengd.	TEC	50 ft	Siltorsa	Bania	CC Line	
1.	Alpinia nigra	50.415	300.457	*	105.97	*	201.339	*	39.809	*	*	15.639	*	7.679	*	721.308
2.	Arundinella bengalensis	118.13	170.39	*	57.309	*	120.002	25.955	130.52	112.014	50.88	12.147	197.636	2.219	0.651	997.853
3.	Arundo donax	633.279	991.75	*	485.417	19.331	827.846	70.519	97.564	54.18	23.36	16.064	160.686	3.221	1.008	3384.225
4.	Axonopus compressus	1158.06	916.509	*	790.39	115.55	665.553	681.3	178.943	194.09	66.88	27.938	961.359	8.569	3.052	5768.187
5.	Cymbopogon jwarancusa	114.64	38.479	*	36.263	*	324.646	90	618.665	418.509	153	52.93	56.868	*	*	1903.96
6.	Imperata cylindrica	1695.3	2233.521	476.39	782.833	60.63	1234.53	1978.14	1537.852	891.563	448.6	59.185	2617.778	13.729	3.051	14033.15
7.	Mikania micrantha	1644.63	4100.184	939.57	960.346	91.385	1342.629	2533.47	403.959	66.213	71.68	35.924	1621.304	22.533	4.135	13837.969
8.	Saccharum arundinaceum	174.53	1151.201	*	36.263	*	*	117.058	*	*	*	*	*	*	*	1479.052
9.	Saccharum bengalense	78.918	519.631	*	*	*	226.813	75.339	*	107.831	*	*	36.572	*	*	1045.104
10.	Saccharum longisetosum var. hookeri	*	*	*	*	*	*	29.613	*	*	*	*	*	*	*	29.613
11.	Saccharum longisetosum var. longisetosum	*	1232.294	*	19.485	*	*	233.419	291.06	*	*	*	*	*	*	1776.258
12.	Saccharum narenga	2753.17	4615.046	*	1763.355	*	1772.068	2669.95	1100.023	1304.65	877.8	129.971	3638.679	13.782	4.298	20642.753
13.	Saccharum spontaneum	3428.37	1366.177	*	2037.084	*	2375.645	16.113	*	216.783	380.8	107.013	818.699	9.235	3.384	10759.302
14.	Saccharum spontaneum (Ikra)	400.433	485.285	*	264.093	*	*	632.323	*	32.571	*	*	*	*	*	1814.705
15.	Themeda arundinacea	166.562	1289.536	*	377.346	*	*	1543.36	117.468	218.61	81.6	*	1451.387	*	*	5245.864
16.	Thysanolena latifolia	65.411	*	67.214	59.325	15.377	49.867	*	*	*	*	*	*	4.439	*	261.633
17.	Typha angustifolia	430.947	111.622	*	75.058	*	1065.861	*	*	*	*	70.3	168.082	*	*	1921.87

Table 21.7: Species-wise Yield of Fodder in different Parts of Jaldapara Wildlife Sanctuary in Post-monsoon Vegetation (Das *et al.*, 2003)

Sl.No. Fodder Species	Jaldapara East Range					Jaldapara West Range					Jaldapara North Range		Chiapata	Kodalbasti	Total Yield
	Jal.	Sissa.	DD Ghat	Mala.	Hollo.	Harind	Moirad.	Kunja.	Bengd.	TEC	50 ft	Siltorsa	Bania	CC Line	
1. Alpinia nigra	391.101	34.345	*	32.352	*	526.83	173.452	30.066	*	*	19.344	*	67.048	*	1274.538
2. Arundinella bengalensis	28.93	*	*	110.776	*	38.607	89.659	17.48	80.85	3.52	2.429	48.24	*	3.479	423.97
3. Arundo donax	765.771	909.767	*	179.38	19.331	117.601	225.552	28.604	261.03	82.4	16.034	220.65	*	1.452	2827.572
4. Axonopus compressus	1162.631	1484.223	*	357.566	82.158	347.063	412.728	183.181	439.593	81.68	24.962	1211.892	*	1.469	5789.146
5. Cymbopogon jwarancusa	78.254	*	*	45.43	*	2582.28	442.353	801.363	*	2.88	29.547	890.099	*	*	4872.206
6. Imperata cylindrica	1552.207	2748.89	347.928	775.706	52.283	3186.917	1410.015	1487.229	840.147	403.04	23.261	2622.467	4.564	2.364	15457.018
7. Mikania micrantha	2535.057	2465.987	853.14	3154.448	87.431	518.911	373.419	345.628	1011.78	868.24	28.575	1022.05	97.292	27.055	13389.013
8. Saccharum arundinaceum	*	805.205	*	*	*	*	110.328	*	*	*	*	*	*	*	915.533
9. Saccharum bengalense	94.853	27.985	*	125.185	*	1435.37	*	76.214	53.13	19.52	*	540.947	*	*	2373.204
10. Saccharum longisetosum var. hookeri	*	*	*	*	*	*	50.741	*	*	*	*	*	*	*	50.741
11. Saccharum longisetosum var. longisetosum	*	1943.686	*	*	*	*	482.835	*	*	*	*	*	*	*	2426.521
12. Saccharum narenga	2110.724	1722.349	*	1263.918	*	5060.223	1888.361	1490.28	1647.03	1052.5	106.892	3279.791	111.09	14.777	19747.915
13. Saccharum spontaneum (Kasia)	2532.415	488.466	*	1633.599	*	51.673	1083.259	*	132.825	63.04	56.27	657.312	7.003	7.481	6713.343
14. Saccharum spontaneum (Ikra)	189.707	961.666	*	110.133	*	*	266.276	174.804	102.795	182.4	*	*	*	*	1987.781
15. Themeda arundinacea	658.045	300.203	*	460.22	*	*	423.621	*	542.85	177.76	*	259.827	*	*	2822.526
16. Thysanolena latifolia	*	*	73.081	*	7.908	*	9.31	*	*	*	*	*	*	*	90.299
17. Typha angustifolia	675.83	122.116	*	392.35	*	2177.209	408.631	*	762.3	*	38.414	112.648	*	*	4689.498

were much higher as compared to other species like sambar and barking deer which favour the woodland and the savannah woodland areas. It was found that the plantation of grasses were mainly preferred by the rhino, elephant, gaur, wild pig, less by hog deer and spotted deer. Moreover, because of its linear shape with two trouser leg like projections and size, Jaldapara is facing severe problem of livestock grazing and human pressure. Thousands of cattle graze inside the forests every day. Contradictorily, Gorumara does not suffer from such acute anthropogenic threats. Due to inadequate availability of fodder during the summer months, it is a period of stress for the resident herbivores. This raises high possibility of the pachyderms straying in the neighbouring areas and causes a serious management problem.

Table 21.8: Abundance of Rhino-Dung in Various Habitats of Jaldapara and Gorumara during Field Survey

Habitat Type	Jaldapara	Gorumara
Natural grassland	104	19
Plantation	61	17
Tropical moist deciduous forest	–	15
Riverine forest	14	10
Savannah forest	261	10
Total	440	71

Skewed Sex-Ratio

Age composition generally reflects the status of rhinos in terms of its reproductive potential (Spillet, 1966). A high percentage of young animals generally indicates that a population is growing or thriving, whereas a relatively small proportion of young indicates a low producing or senile population. Sex ratios likewise are an indication of reproductive potential. Most mammals, particularly ungulates, are promiscuous in their mating and a single adult male is generally sufficient to cover three or more females. Therefore, within reasonable limits, a predominantly female population has a higher reproductive potential than one with more males. With reliable age composition and sex ratio data, it is possible to calculate the average annual rate of net increase or loss, as well as determine the population status.

The average sex-ratio in Jaldapara and Gorumara was 1:2 during early 1990s (Bist, 1994). But, the ideal or balanced sex-ratio (male:female) of the breeding rhinos is 1:3. Now, there are 61 adult males, 55 adult females, three unsexed, 23 sub-adults and 42 calves in Jaldapara. The figures for Gorumara were 21 adult males, 14 adult females, 7 calves and 1 unsexed including the dispersed rhinos in Chapramari (88°49'-88°51' E, 26°53'-26°54' N) in Wildlife II and Swaraswatipur (west) in Baikunthapur Divisions. So, the population figures for Jaldapara and Gorumara are heavily skewed or male-biased. Further, sex identity of all the individuals in Gorumara was determined using genetic markers developed by Wildlife Genetics Laboratory of Aaranyak and Centre for Conservation and Research of Endangered Wildlife, Cincinnati Zoo and Botanical Garden, USA. Dung provides the source of DNA sample of every individual and also helps identify its sex. This is achieved through selecting and using a panel of highly polymorphic microsatellite markers for individual identification and sex chromosome linked markers for sex identification. However, the conventional method of head counting by the forest department almost matches with the figure of genetic study, *i.e.* a skewed male to female sex-ratio of 4:1. This study exemplifies how genetic analysis of dung can successfully be used for identification of individual rhinos and their sex, which can be used for long term monitoring of the natural populations. But the animal's age vis-à-vis reproductive capability could not be ascertained from this study. The most serious observation is that the genetic diversity in the Gorumara rhino population is low. Aaranyak is working further in generating genetic information from the rhino population in Jaldapara and to understand spatial distribution of genetic diversity to assist genetic management of the species in future.

It is well known that maternal age influences the natal sex ratio. For the wild dams of *R. unicornis*, the age categories at birth of calf are less than or equal to 12 years, between 12 and 19 years and greater than 19 years. Since the species is polygynous and the average body condition varies among them, it appears that the females in good condition tend to produce male offspring, whereas females in poor condition tend to produce female offspring. The sex ratio of male calves to female calves is expected to be low in young and aging breeding females and high in females of prime breeding age. As they reach the end of their reproductive life, their body condition diminishes. The young or nulliparous dams are in superior body condition compared with middle-aged or multiparous females. Thus, the younger dams would be more likely to produce male offspring. On the other hand, better adult condition affects the reproductive success of the male more than the female. The male in better condition exclude other males from breeding and sire many offspring himself, whereas a male in poor condition is not likely to sire any offspring because he is driven away from the females by the better conditioned males. This tendency is explained below.

Agonistic Interactions and Intraspecific Fighting

In ethology, the term 'agonistic behaviour', coined by Scott and Fredericson in 1951 (Barrows, 2001), in practice, a broader term than 'aggressive behaviour', is any social behaviour related to fighting between contestants who are competing over limited resources including food, shelter and mates. Although agonistic behaviour varies among different species, agonistic interaction includes not only actual aggression between two animals to gain the targeted resource in spite of the high cost involved, such as severe injury or fatality due to the inflicted wounds, but also two more functionally and physiologically interrelated, much safer, ritualistic behaviours, normally occurring in sequence (McGlone, 1986; Manning, 1998).

1. Prelude warming up-cum-threat displays of strength before the final round of battle takes place so that the opponent moves back, and

2. Ultimate submission and/or retreat.

Intraspecific fighting of the resident rhinos is the major mortality factor affecting all sex and age groups in the study area. The adult male rhinos are usually solitary in nature and form temporary associations with the females during sexual encounters.

Since home ranges of the Indian Rhinos show much overlap between different individuals of each sex, the concept of exclusive territory is absent in this species. For example, the breeding bulls have loosely defined territories, which are well defended but can overlap with those of both "weak" males, and with known, neighbouring "strong" males. As a result, the types and outcome of interactions between different sex and age classes are varied because their aggressive or submissive interactions depend on the probability of winning, *i.e.* adult male over adult female or subadult male and the prime-aged bull over the old-aged.

The dominant adult male tries to establish its supremacy by showing off its strength with other adult males for mating with females in oestrus. Hence, fights between males are quite common and whenever two or more adult or subadult rhinos approach each other, either at wallows or foraging grounds, there is every possibility of reaction by at least one of them. Most interactions are short and one individual normally retreats or moves aside. But if both of them are stubborn, one of them starts chasing the opponent over a short distance at the primary stage, failing which a prolonged chase is made by the stronger rhino till the intruder crosses the safe limit. The most prolonged interactions may also take place between an adult male and an adult female and sometimes result in the death of one opponent (Dinerstein and Price, 1991).

Laurie (1978, 1982) also studied the aggressive behaviour in Royal Chitwan of Nepal. High levels of fighting mortality in translocated populations have also been recorded in Chitawan, which accounted for 28.6 per cent of the deaths due to causes other than poaching (Laurie, 1978). The various components of agonistic interactions between the rhinos are sudden turn, prolonged stare, curling back the lips to show the lower tusks, advancing slowly towards the antagonist, charging with head down, horn-to-horn stare, horn clash + lunge, submissive lying, tail curling, immediate fight after initial displays and prolonged chase. In Chitawan, 37 per cent of agonistic interactions escalated to some kind of horn-to-horn confrontation, resulting, in extreme cases, in horn clashes and lunges with the lower tusks to the neck, flanks, and rump of the opponent (Laurie, 1982).

In southern African populations, such incidents accounted for 41 per cent of natural deaths since 1989 (Brett, 1998). Intolerance between the breeding bulls is most frequent in the study area. Ghosh (1991) described in brief about male dominance relationship in Jaldapara. Hazarika and Saikia (2010) also described the non-breeding agonistic behaviour of the male rhinos in Rajib Gandhi Orang National Park in Assam. They have categorised five subtypes of performing threat and threat displays against competitor to chase other intruder from his territory or to defend from unwanted competitor in its own territory. These are:

1. Snorting: A vocal threat by producing a "*khaawk…-khaawk…*" sound at regular interval;
2. Physical threat display or aggression of the dominant individual by expression (erect head, ear pinna, making a mild sound) for pretending to attack the other individual approaching or being approached.
3. Aggressive chasing to displace the weak Rhino or adult Rhino or sub-

adult Rhino by the dominant rhino for a very longer distance.

4. Attacking the opponent and having a horn-to-horn fight and also to inflict deep wounds by using the "tushes", the outer pair of long lower incisors (up to 8 cm of the males and shorter in case of the females), from backside, when the weaker of the opponent fled when charged.

5. Escaping behaviour: Generally, the weak animals does not take part in fights or attack the dominant male, but runs away with a galloping motion.

Recently, Tripathy (2013) highlighted this feature in Dudhwa in the *terai* of Uttar Pradesh.

The rhino males are generally territorial and they mark this area by defecation and/or urine scent markings and compete for estrous females. In fact, it is a predominant feature of rhino sociology. Mostly, the bulls do not associate with one another. Territorial bulls trespassing into the territory of an adjacent bull normally take avoidance action and serious fights are usually averted. Encounters may take the form of short charges with much dust raising or, at closer quarters, horn clashing. Subtle displays may involve pulling the ears back as a sign to the others to keep off; advancing steps often accompanied by a snarl are used as a threat; charges; prodding with the horn or staring at each other, horn against horn, as intimidatory gestures. Horn against horn clashing is a more intense ritual attack, which may develop into horn-wrestling and finally jabbing with the horn. Where a territorial bull is accompanied by a female in oestrus, however, serious fighting may ensue. Wounding may be caused by the horn or by heavy shoulder battering and may lead to internal injury. A deposed territorial bull may be allowed to remain in the territory providing he clearly demonstrates his submissiveness. Subordinate bulls respond to territorial bulls with snorting, snarling or shrieking, but seldom actually engage in fighting, although they have been killed in such encounters. Under the circumstances, when a young bull does try to establish himself in a territory, he either has to do so in an unoccupied area, or fight another bull to win some turf. Such 'upstarts' have little hope of winning a territory off a prime-aged bull (ca 17-30 years old), but can drive out or kill older bulls, who are on the decline physically. Old bulls, if not killed, will move out to a quiet part of their former range or live a fringe existence until they die. But a marked tension always looms large as they are short tempered and become hostile, if encountered. There is a lot of 'scrapping' going on all the time between the dominant and the subordinate rhinos and there are a great number of rhino with scars, torn ears and gashes, particularly the bulls.

Infighting among the adult males is a common site in the study area during the mating season. Avari (1957) doubted if Jaldapara can support a larger population because of the cantankerous nature of the bull and the numerous fights which develop as a result cause quite a number of deaths. These fights usually start over the favours of a female during the breeding season. In some infights, the horns are split or broken off (Ghosh, 1991). Till the first decade of twenty first century, such infighting-casualties were low (1-4 per annum) in the study area. Spillet (1966) described such an encounter in Jaldapara: "On April 2 at 05.45, just after crossing the River Torsa north-west of the Jaldapara Forest Rest House, we heard the roar, followed by two honking snorts, of rhino somewhere along the river north of us. We started towards the noise, but within a short time encountered an adult rhino. We observed it until we were able to determine its sex and ensure that there were no other rhino present in the tall grass nearby. After leaving the solitary male we sighted a large male at 06.00 coming directly towards us across an open island of sand in the river bed. On the distant bank were two more rhino. The male approaching us had a large bleeding gash, about 24 inches (60 cm) long, which extended across the top of his neck and down onto the left shoulder. The courting male was noticeably smaller than the ousted one and had a much shorter horn".

During the post-independence period, there were very few records of infighting-casualty up to early 1990s and all the reported cases (n= 7) occurred in Jaldapara (Bist, 1994). Thereafter, when the male-female sex-ratio was almost 1: 2, a few recurring incidents of male-male aggression leading to death of the old bulls and subordinate males in both Jaldapara and Gorumara were reported by Yadav (2000a, 2000b). An old bull was severely injured by a dominant male at Jaldapara-5 in December 1992 and again at Torsa-2 in the last week of January 1993, even if the former moved almost 10 km away from the first site. Similar incidents of repeated attacks on an older bull of Jaldapara took place in February 1993, April and September 1994, leading to its death in December.

Yadav (2000a, 2000b) also observed: "Antagonistic behaviour of Rhinos was typical before and during the fight. The dominant male was observed to keep both its ears erect and forward and occasionally move its head upwards before approaching the old male Rhino from a distance of nearly 100 m. As the dominant male approached nearer the old male, it perfumed squirt-urination and rubbed the horn on a medium girth *Dalbergia síssoo* tree. This display was followed by dragging of the hind legs one by one making marks on the earth. Soon after, the dominant male ran towards the older male in full speed with lowered head and making typical sound. Older male Rhino was chased for a distance and was charged by the dominant male by pointed horn in the rump's flanks. During the fight dominant male also opened the mouth wide and displayed tusks. The older male Rhino once turned suddenly to face the other but again ran after a brief pause. Dominant male Rhino chased the opponent for more than 300 m with accompanying typical loud vocalízations. The weak old male Rhino ran in full speed with curled tail. In another incident of clash similar behaviour was observed with additional head to head confrontation. The dominant male Rhino attacked with the horn and the tusks on the head of the weak male Rhino. Attacks were also made from side to side with aim at the head region. However, due to the movement injuries were also received on the neck and side of front folds. Both the male Rhinos frequently performed squirt-urination during the ñght. At the end of every fight the dominant male Rhino totally overpowered the o1d weak Rhino."

Repeated chase and aggressive fighting between one fully grown adult male Rhino and a sub-adult Rhino (5-7 years old) was observed in Gorumara (Yadav, 2000a, 2000b). This particular sub-adult male Rhino was very timid and used to flee immediately on the approach of stronger bull. This sub-adult male Rhino was repeatedly chased by the stronger bull and forced to move out of the prime Rhino habitat. The sub-adult male changed its range use after confrontation with stronger adult male and shifted to central Diana block after crossing the river Jaldaka in the last week of August 1995. Central Diana block is fragmented patch of forest surrounded by tea garden and human habitation and was not considered safe from protection point of view. It was decided to drive this sub-adult Rhino towards Gorumara using departmental Elephants. However, every time it went up to the River Jaldhaka it again ran back to a small *Dalbergia sissoo* plantation patch on the bank of the river Jaldhaka outside forest area. Since all attempts to drive back the Rhino failed, a team of staff along with one departmental Elephant were stationed in that area to ensure safety of the animal, with the hope that the Rhino may, on its own, enter Gorumara. However, the Rhino was reluctant to cross the river and confined its territory across Jaldhaka river near zero-bundh. The attempts for chemical capture failed in January, 1996 and the operation was ultimately abandoned apprehending that shock and trauma caused by chemically induced capture may have fatal result. In the night of 29.2.1996 the sub-adult Rhino was attacked by the dominant bull Rhino which crossed the river Jaldhaka and fiercely charged the weak rhino. The staff camping nearby burst crackers and drove away the adult male Rhino using departmental Elephant. The fight lasted for around half an hour. During the night, the sub-adult Rhino recieved severe injuries in the right thighs, under portion of the belly and on the left part of the face. Blood was also oozing from these injuries. It was evident from the injuries that the sub-adult male Rhino was completely overpowered by the adult bull and felled to the ground. The dominant bull used its horn ñercely to attack and injure the weak Rhino. Again after a week the dominant bull crossed the river Jaldhaka and entered the patch forest where sub adult male Rhino had taken shelter. It chased the sub-adult male Rhino for around 4 km towards Gommara National Park and forced it to cross the Jaldhaka river. The chase continued upto the Garati Camp where adult bull overpowered the weaker one and again injured it in the right and left thighs. The injured Rhino took shelter in a secluded area of Medla-3 Compartment where Rhinos generally do not visit. The movement of the Rhino was tracked and it was observed that wounds were quite severe. It was decided to keep the injured Rhino within the electric fence to give proper medication. The sub-adult Rhino was guided with the help of departmental Elephant towards the electric fence area. The electric fence was switched off and part of it dismantled when it approached near the fence. When the injured Rhino entered that area the electric fence was made operational. Firstly for some days long acting antibiotics mixed with turpentine oil were sprayed on the wounds from a distance on Elephant back. The movement of the injured sub-adult Rhino was regularly monitored and it was observed grazing, wallowing and drinking water. After three days the injured Rhino was seen in the resting position when staff went to spray medicine. It was able to move very slowly and again it went near a tree and took resting position. The veterinary doctors administered intramuscular antibiotic and antipyritic injections including a life saving drug. The condition of the Rhino deteriorated fast and it was unable to move and died on 12th March 1996. According to the veterinary doctors the injuries were quite severe and Rhino immovable. It also became anaemic due to heavy blood loss.

Male undergoes a period of heat as does the females and these periods must coincide before mating takes place. Males come into heat when they are in rut. In females, the period of estrous cycle is about 46-48 days. In the study area, the rhinos prefer to breed between the months of March and September. But the urge for mating of the males and females do not synchronize all the times. When the estrous female refuses the approaching male, she was observed to attack the male (Hazarika and Saikia, 2010). The bulls become equally aggressive during the pre-mating phase when fights were also observed in the study area between an aggressive bull and a cow not in oestrus or accompanied by a calf. In fact, a female reluctant to mate is not tolerated by the approaching male. In the extreme case, the male rhinos often attack and kill the breeding females to ensure that the other male contender does not get possession over it. Such incidents also took place when the breeding females refused to respond to an approaching male and go to another one. In Gorumara, some females (particularly the immature, pregnant and lactating females) died due to struggle during forced mounting, when they often got stuck into the mud (Martin and Vigne, 2012). Old female rhinos incapable to mate are also prone to sustaining injuries. This seems to indicate that they are not being tolerated. Calves generally remain with their mothers for three to four years and are quite protected by mothers. Often the calves were also killed opportunistically to force the reluctant mother to breeding again. In Jaldapara, a lactating male calf was also brutally killed (torn into pieces) by an adult male in September 2011, when the mother did not allow it to mate. Male infanticide was also observed in the study area. After weaning from mothers, the juvenile males are attacked frequently.

The male-male and male-female encounters, resultant injuries and casualties have now become a crucial management issue in Jaldapara and Gorumara. If the injured animals are detected in the forests during regular monitoring, they are captured through chemical immobilization, treated and released after recovery. But sometimes the animals, already under excessive stress, died in spite of taking all possible measured to save them.

Since the sex-ratio has now become heavily imbalanced, 12 rhinos died of suspected infighting in 2012 and another one in January 2013. In the past three years, four males and one female have died in Gorumara because of infighting and two males lost their fight with a stronger one and strayed out of Gorumara permanently. A forty years old male was killed by a young bull in an infight over a female at Siltorsa riverbed of Jaldapara in September 2012. Three male rhinos died in the park within one month. Such incidents took place mostly at Jaldapara 1, 3, 5, Malangi1 1-3, Torsa 1-3, Chilapata 1-2, Barodabri 1, 6, Bania 1, 3, 4, and Torsa riverbed in Jaldapara and Dhupjhora and zero bundh (River Jaldhaka) in Gorumara. On 9th January 2013, a 40-year old male rhino was killed by a young one in Jaldapara. The carcass was found on the Holong River banks in the west range near Sibnathpur village.

Dispersal

Temporary or seasonal local migration for food and shelter is a common phenomenon among the adult rhinos, particularly the males. In the past, the rhinos confined to the grassland habitat between the Rivers Sankosh and Rydak moved freely between Assam and West Bengal. Similarly, the rhinos of Jaldapara and Chilapata travelled to the neighbouring Bhutri forests on the east and also to Cooch

Behar on the south. The Patlakhawa subpopulation also ranged further up to Pundibari near Cooch Behar. The rhinos were distributed across the forest of upper Tondu, lower Tondu and in the floodplains of the Rivers Diana and Jaldhaka on the north-east or Hiljhora forest on the north-west of Gorumara. They further moved up to Indo-Bhutan border on the north and Mynaguri (now non-forest civil area) on the south. In March 1989, one female rhino strayed up to the Bangladesh border and was later brought back. In 1992, another female strayed out to Mahananda Sanctuary by crossing the River Tista and died due to exertion of such longest journey.

Sometimes some of the comparatively weaker individuals (young or old) migrate from the core habitat to distant areas permanently, which may be associated with the competition for reproductive resources. After the deadly infighting between two males over a breeding female, the defeated or chased out rhino was often reported to stray out of the prime rhino habitat to the range-edge or new habitat far away, which is an indicator of the skewed male-female ratio in Jaldapara and Gorumara. Whereas there are contiguous reserve forests in the eastern side of Jaldapara, some older or subordinate male rhinos, after being chased out of the prime habitat by the dominant male, used to move towards the secondary habitat at their range-edge and settle in Bania-Mendabari-Chilapata or southern Patlakhawa forests of Cooch Behar. But they were not reported to go to the disturbed and degraded forests in the north beyond Madarihat or towards the west. The rhinos of Gorumara are known to travel about 8.5 km north up to Chapramari, a sanctuary in between the Jaldhaka and Murti floodplains spread over 9.15 km^2, which has limited area under grass. They also used to stray into the forests of Central Diana or often towards Sipchu-Khumani forests of Jalpaiguri Division.

In the study area, tendency of the rhinos to stray out of the protected areas due to increased biotic and abiotic pressure was observed more during the fall and winter than other seasons. The lists highlighting straying incidents (Table 21.9) and chemically immobilised rhinos (Table 21.10) in the study area will reveal that in most of the cases, bulls were involved and the majority strayed from Jaldapara. Traditionally these animals are tried to retrieve back to the respective protected area with the help of manpower and *kunkis* (trained captive elephants). But driving through the fragmented thickly populated disturbed areas is very difficult and risky too. Besides, such efforts consume lot of resources as well as time. When such an effort fails, chemical restraining and translocation of the disoriented animal becomes the only viable option left to save the life of the individual.

The Apalchand forest (around 80 km^2) of Baikunthapur Division, west of Gorumara at a distance of about 50 km, is close to the River Tista, where a suitable grassland has been developed. In June 1992, a highly vagile female strayed out into the forests of Apalchand range and moved as far west across the River Teesta into Mahananda (88°19'-88°31' E, 26°46'-26°55' N), following the route through Apalchand (88°38' E, 26°47' N) and Salugarah (88°27' E, 26°45' N) Reserve Forests. The rhino was, however, poached. On 15 September 2008, an adult male had strayed out of the core area of Gorumara and went marauding through five villages in Mynaguri block. After one week, a ten-year old female and a calf had strayed out from Gorumara. All of them were driven back to the forests. On 12 November 2009, a seven-year old adult

Table 21.9: Straying of Rhino in Jaldapara and Gorumara

Date	Loation	Sex	Extent of Damage	Management action	Remarks
15.11.2009	Senpara, Baradighi from Gorumara	Adult male	None	Driven back to forest	–
22.11.2009	Mongpong from Gorumara	Adult male	None	Driven back to forest	–
25.11.2009	Near Tista river beside Saraswatipur, Baikunthapur from Gorumara	Adult male	None	Driven back to forest	–
28.11.2009	Mahananda Wildlife Sanctuary from Gorumara via Baikunthapur	Adult male	None	Driven back to forest	–
02.12.2009	Mongpong from Gorumara	Adult male	None	Driven back to forest	–
11.12.2009	Kathambari, Phuljhora from Gorumara	Adult male	None	Driven back to forest	–
03.01.2011	Moiradanga, Jaldapara	Adult male	None	Tranquillised and released back to forest	–
19.01.2011	Uttar Majabari near Kranti, Baikunthapur from Gorumara	Adult male	None	Driven back to forest	–
20.01.2011	Gazoldoba, 10 no., Baikunthapur from Gorumara	Adult male	None	Driven back to forest	–
23.01.2011	Mongpong from Gorumara	Adult male	None	Driven back to forest	–
28.02.2011	Battala, Belacoba, from Baikunthapur	Adult male	Human injury	Driven back to forest	Prasanta Yadav, DR/Fr. injured

Table 21.10: Chemical Immobilisation of Injured Rhinos in Jaldapara and Gorumara

Sl.No.	Date	Location	Sex/Age	Drug Applied	Remarks
1.	23.10.2003	Torsa-1, Jaldapara	Adult male	Immobilon- 2.5 ml	Treated
2.	12.05.2004	Chilapata-2, Sissamara, Jaldapara	Adult female	Immobilon- 1.7 ml	Inventory of injury
3.	12.05.2004	Chilapata-2, Sissamara, Jaldapara	Calf	Immobilon- 1.7 ml	Inventory of injury
4.	25.05.2005	Dhupjhora, Gorumara	Calf	Immobilon- 0.3 ml	Treated
5.	30.09.2005	Malangi-1, Jaldapara	Adult male	HBM- 2.5 ml	Treated, died
6.	24.11.2005	Torsa-2, Jaldapara	Adult male	Immobilon- 4.7 ml	Treated
7.	10.12.2005	Barodabri-6, Kodalbasti, Jaldapara	Adult male	Immobilon- 2.5 ml	Treated
8.	10.01.2006	Bania-1, Chilapata, Jaldapara	Adult male	Immobilon- 7.6 ml	Treated (Dart missed)
9.	14.01.2006	Bania-3, Chilapata, Jaldapara	Adult male	Immobilon- 2.5 ml	Treated
10.	16.02.2006	Bania-4, Chilapata, Jaldapara	Adult male	Immobilon- 1.3 ml	Treated
11.	19.08.2006	Torsa riverbed, Jaldapara	Adult male	Immobilon- 2.5 ml	Treated
12.	28.09.2006	Barodabri-1, Kodalbasti, Jaldapara	Adult male	Immobilon- 3.0 ml	Treated
13.	29.09.2006	Barodabri-6, Kodalbasti, Jaldapara	Adult male	Immobilon- 3.0 ml	Treated
14.	24.02.2007	Jaldapara-5	Adult male	Immobilon- 2.5 ml	Treated, died
15.	18.03.2007	Malangi-3, Jaldapara	Adult male	Immobilon- 2.6 ml	Treated
16.	02.12.2008	Dhupjhora- 1b, Gorumara	Male, 7 years	Immobilon- 1.25 ml	Treated
17.	14.01.2009	Dhupjhora- 1b, Gorumara	Male, 7 years	Immobilon- 1.00 ml	Treated
18.	03.01.2011	Moiradanga, Jaldapara	Adult male	Xylazine- 2.25 ml	Treated
19.	20.09.2011	Jaldapara-3	Adult male	Immobilon- 1.90 ml, Revivon-1.90 ml	Treated
20.	30.11.2011	Jaldapara-2b	Adult male	Immobilon- 1.50 ml, Revivon-2.00 ml	Treated
21.	09.12.2011	Jaldapara-3	Adult male	Immobilon- 1.70 ml, Revivon-1.70 ml	Treated
22.	05.01.2012	Dhupjhora-1b, Gorumara	Adult male	Immobilon-2.00 ml, Revivon-2.50 ml	Treated

male left Gorumara after losing a fight with its rival, travelled to enter Apalchand forests, then reached Bodaganj under Belakoba range by crossing the Tista and finally entered the Eighth Mile forests of Mahananda. The reluctant rhino could not be driven back from Baikunthapur to Gorumara. In December 2010, another male rhino from Gorumara settled in the Apalchand forest. Ever since the grasslands in Gajoldoba area along the Tista serve as the third natural habitat for the dispersed rhinos. These rhinos occasionally visit the neighbouring forest areas of Saraswatipur and Targhera. Since these are not protected areas, regular patrolling on elephant back is required to monitor their movement.

Conclusions

The probable solutions to counter the problem of high infighting-casualties vis-à-vis skewed sex-ratio of the rhinos in Jaldapara and Gorumara are round-the-clock monitoring, seclusion of the identified more aggressive males in a separate large enclosure for promotion of eco-tourism, reintroduction of more wild females from the neighbouring rangelands, development of new grasslands and translocation of the excess population to other suitable habitats like Apalchand (Baikunthapur) in Jalpaiguri and Rasamati (Patlakahawa) forests in Cooch Behar in phases.

Acknowledgements

I am grateful to all the field staff of Jaldapara and Gorumara for conducting the ground survey and the respondents to the questionnaire survey. I also express thanks to Indranil Mitra for providing the maps of study area and also Sourabh Basu for the photographs.

References

Agrawal, V.C., Das, P.K., Chakraborty, S., Ghose, R.K., Mandal, A.K., Chakraborty, T.K., Poddar, A.K., Lal, J.P., Bhattacharya, T.P., and Ghosh, M.K., 1992. Mammalia. In *Fauna of West Bengal, Part I,* State Fauna Series 3 (Director, Zoological Survey of India, ed.), Zoological Survey of India, Calcutta, India, pp. 113-114.

Ali, S., Azfer, M.A., Basamboo, A., Mathur, P.K., Malik, P.K., Mathur, V.B., Raha, A.K., and Ansari, S., 1999. Characterization of a species-specific repetitive DNA from a highly endangered wild animal, *Rhinoceros unicornis,* and assessment of genetic polymorphism by microsatellite associated sequence amplification (MASA). *Gene,* **228(1-2)**: 33-42.

Avari, E.D., 1957. The Jaldapara Game Sanctuary, West Bengal. *Jour. Bengal Nat. Hist. Soc.,* **29 (3)**: 65-68.

Bahauguna N.C., and Mallick J.K., 2004. Ungulates of West Bengal and its adjoining areas including Sikkim, Bhutan and Bangladesh. In: *Ungulates of India* (Sankar K., Goyal S.P. eds.), ENVIS Bulletin: Wildlife and Protected Areas, **7(1)**: 219-238.

Bahauguna N.C., and Mallick J.K., 2010. *Handbook of the mammals of South Asia.* Natraj Publishers, Dehra Dun.

Banerjee, A.K., 1966. Rhinoceros- a historical treatment. In: *West Bengal Forests, Centenary Commemoration Volume.* Forest Directorate, Government of West Bengal, Calcutta, pp. 241–243.

Banerjee, L.K., 1993. *Plant resources of Jaldapara Rhino Sanctuary*. Botanical Survey of India, Calcutta.

Barrows, E., 2001. *Animal Behavior Desk Reference*. Florida: CRC Press LLC.

Bhattacharya, A., and Pal, B.C., 1982. Daily activity cycle of Great Indian one horned rhinoceros at Gorumara and Jaldapara WLS, West Bengal, India. *J. Bengal Nat. Hist. Soc.,* **1(2)**: 53-58.

Bhattacharya, M., 2012. Unwanted ground vegetation? *Current Science,* **102(5)**: 660.

Bhutia, P.T., 2008. Status of rhino and horn stockpiles in Jaldapara Wildlife Sanctuary and Gorumara National Park. In: *Report on the regional meeting for India and Nepal, IUCN/SSC Asian Rhino Species Group (AsRSG)* (Syangden B. *et al.*, eds.); March 5–7, 2007. Kaziranga National Park, Assam, India. pp. i–ii, 1–28.

Bist, S.S., 1994. Population history of Great Indian Rhino in North Bengal. *Zoo Print,* **9(3–4)**: 42-51.

Borthakur, S., Das, P.K., and Talukdar B.K., 2012. Genetic census of Greater One-horned Rhino in Gorumara National Park, West Bengal, India. *The Rhino Print* (Newsletter of the Asian Rhino Project), Winter: 13.

Brahmachary, R.L., Mallik, B., and Rakshit, B., 1971. An attempt to determine the food habits of the Indian Rhinoceros. *Jour. Bombay Nat. Hist. Soc.,* **67 (3)**:558-560.

Bret, R.A., 1998. Mortality factors and breeding performance of translocated black rhinos in Kenya: 1984-1995. *Pachyderm,* **26**: 69-82.

Das, A.P., Ghosh, C., and Bhowmick, D., 2003. *Project Report on estimation of Palatable Biomass in Jaldapara Wildlife Sanctuary with special reference to Rhinoceros unicornis L.* Department of Botany, University of North Bengal, India.

Das C., 2008. Further discovery of a plaque representing *Rhinoceros unicornis* from Dum Dum excavation, West Bengal: An observation. In: *Recent Researches in Indian Art and Iconography* (Sahai B., ed), Kaveri Books, Delhi, India, pp. 93–95.

Dinerstein, E., and Price, L., 1991. Demography and habitat use by greater one-horned rhinoceros in Nepal. *J. Wildl. Manage.* **55**: 401–411.

Ghosh, C., and Das, A.P., 2007. Rhino-Fodders in Jaldapara Wildlife Sanctuary in Duars of West Bengal, India. *Our Nature,* **5**: 14-20.

Ghosh D.K., 1991. Studies on the ecostatus of the Indian Rhinoceros (*Rhinoceros unicornis*) with special reference to the altered habitat due to the human interference in Jaldapara Wildlife Sanctuary, West Bengal. *Ph.D. Disseratation,* Ranchi University, Bihar, India.

Ghosh, S.B., 2012. Biodiversity and Wild Fodder of Gorumara National Park in West Bengal, India: Fodder Plants and Habitat of Gorumara National Park. *Journal of Environment and Ecology,* **3(1)**: 18-35.

Groves C.P., 1993. Testing rhinoceros subspecies by multivariate analysis. In *Proceedings of an International Conference: Rhinoceros Biology and Conservation* (Ryder O.A., ed), International Rhino Conference, May 9-11, 1991, Zoological Society of San Diego, California, USA, pp. 92-100.

Groves C.P., 2003. Taxonomy of ungulates of the Indian subcontinent. *Journal of the Bombay Natural History Society,* **100(2/3)**: 341-362.

Gupta, A.C., 1958. Gorumara Game Sanctuary. *Journal of Bengal Natural History Society,* **24(4)**: 133-139.

Hazarika, B.C., and Saikia, P.K., 2010. A study on the behaviour of great Indian one-horned rhinoceros (*Rhinoceros unicornis*) in the Rajiv Gandhi National Park, Assam, India. *NeBio.,* **1 (2)**: 62-74.

Laurie, W.A., 1978. The ecology and behavior of the greater one-horned rhinoceros. *Ph.D. Dissertation,* University of Cambridge, Cambridge, 449 pp.

Laurie, W. A., 1982. Behavioural ecology of the Greater One-Horned Rhinoceros (*Rhinoceros unicornis*). *Journal of Zoological Society of London,* **196**: 307-341.

Maharaja of Cooch Behar (Nripendra Narayan Bhupa), 1908. *Thirty-seven years of big game shooting in Cooch Behar, the Duars and Assam: A rough diary.* The Times Press, Bombay, 461 pp.

Mandal S., 2007. Wild fauna of Gorumara National Park, Jalpaiguri, West Bengal. *Intas Polivet,* **8(1)**: 257-261.

Manning, A., 1998. *An Introduction to Animal Behavior.* Cambridge University Press.

Martin E.B., 1996a. Smuggling routes for West Bengal's rhino horn and recent successes in curbing poaching. *Pachyderm,* **21**: 28-34.

Martin E.B., 1996b. The importance of park budgets, intelligence networks and competent management for successful conservation of the Greater One-horned Rhinoceros. *Pachyderm,* **22**: 10-17.

Martin E.B., 1999. West Bengal – Committed to rhino conservation yet a major entrepot for endangered wildlife products. *Pachyderm.,* **27**: 105-112.

Martin E.B., 2006. Policies that work for rhino conservation in West Bengal. *Pachyderm,* 41: 74-84.

Martin E., and Vigne, L., 2012. Successful rhino conservation continues in West Bengal, India. *Pachyderm,* **51**: 27-37.

Menon, V., 1996. *Under siege: poaching and protection of Greater One-horned Rhinoceroses in India.* TRAFFIC International, Cambridge, UK, 55 pp.

Molur S., Sukumar, R., Seal, U., and Walker, S. (eds.), 1995 *Report: Population and Habitat Viability Assessment (P.H.V.A) Workshop: Great Indian One-horned Rhinoceros: Jaldapara, 1993.* CBSG, Coimbatore, India, 85 pp.

Mukherjee, A.K., 1963. The extinct, rare and threatened game of the Himalayas and the Siwalik ranges. *Journal of the Bengal Natural History Society,* **32(1)**: 36-67.

Mukherjee, S., and Sengupta, S, 1998. Rhino poaching in Jaldapara Wildlife Sanctuary, North Bengal, India. *Tiger Paper,* **25 (1)**: 20-21.

Nandi, M., 1991. Is Terai grassland in West Bengal safe for Rhinos. *West Bengal,* 33(17): 518-521.

Pandit, P.K., and Yadav, V.K., 1996. Management of Jaldapara Wildlife Sanctuary, West Bengal, India. *Tigerpaper,* **23(4)**: 1-5.

Rawat, G.S., 2005. Vegetation dynamics and management of Rhinoceros habitat in *Duars* of West Bengal: An ecological review. *National Academy Science Letters,* **28(5 and 6)**: 179-186.

Sarma, P.K., 2012. Assessment of habitat utilization pattern of rhinos (*Rhinoceros unicornis*) in Orang National Park, Assam, India. *Pachyderm,* **51**: 38-44.

Spillet, J.J., 1966. The Jaldapara Wildlife Sanctuary, West Bengal. *Journal of Bombay Natural History Society,* **63(3)**: 534-556.

Syangden, B., Sectionov, S.E., Williams, A.C., Van Strien, N.,and Talukdar, B.K. (eds.), 2007. *Report on the Regional Meeting for India and Nepal IUCN/SSC Asian Rhino Specialist Group (AsRSG),* March 5-7, 2007, Kaziranga National Park, Assam, India, 28 pp.

Talukdar, B.K., 2012. Asian Rhino Specialist Group report. *Pachyderm,* **51**: 22-26.

Tripathy, A.K., 2013. Social and Reproductive Behaviour of Great Indian One-horned Rhino, *Rhinoceros unicornis* in Dudhwa National Park, U.P., India. *International Journal of Pharmacy and Life Sciences,* **4(11)**: 3016-3021.

Vigne, L., and Martin, E.B., 2012. A rhino success story in West Bengal, India. *Oryx,* **46(3)**: 327-328.

Wildlife Institute of India, 1997. *Study on the management of rhinoceros in West Bengal: Final Report.* West Bengal World Bank aided forestry project. Dehra Dun, India. pp. 201.

Yadav, V.K., 2000a. Male-male aggression in *Rhinoceros unicornis* – case study from North Bengal, India. *Indian Forester,* **126(10)**: 1030-1034.

Yadav, V.K., 2000b. Male-male aggression in *Rhinoceros unicornis*- An observation from North Bengal, India. *Zoos' Print Journal,* **15(9)**: 328-330.

2015, Animal Diversity, Natural History and Conservation, Vol. 5   *Pages 367–378*
Editors: V.K. Gupta and Anil K. Verma
Published by: DAYA PUBLISHING HOUSE, NEW DELHI

Chapter 22

State of the Asian Elephant (*Elephas maximus* Linnaeus 1758) and its Habitat in Nilgiri Biosphere Reserve, Southern India

Nagarajan Baskaran*[1, 2] and Raman Sukumar[2, 3]
[1]Department of Zoology, A.V.C. College (Autonomous),
Mannampandal, Mayiladuthurai – 600 001, Tamil Nadu, India
[2]Asian Nature Conservation Foundation, c/o Centre for Ecological Sciences,
Indian Institute of Science, Bangalore – 560 012, Karnataka, India
[3]Centre for Ecological Sciences, Indian Institute of Science,
Bangalore – 560 012, Karnataka, India

ABSTRACT

Nilgiri Biosphere Reserve in southern India perhaps encompasses the most diverse single landscape for elephants anywhere in Asia. The strong rainfall gradient from east (low) to west (high) resulting from a complex topography has resulted in a mosaic of vegetation types: from lowland tropical dry thorn forests through mid-elevation tropical dry and moist deciduous forests, high elevation tropical semi-evergreen and evergreen forests to montane shola–grasslands. This heterogeneous landscape also supports a wide assemblage of large mammals including gaur, sambar, chital and common langur at high density that, in turn, supports substantial populations of large carnivores such as the tiger, leopard and dhole. With one of the highest densities (max. of 2.4/km^2) of elephants in Asia, the NBR harbours a substantial part (>70 per cent) of the large population of >8000 elephants seen in the broader Nilgiri–Mysore-Wyanad Elephant Landscape.

---

* *Corresponding author.* E-mail: nagarajan.baskaran@gmail.com

With over 6000 elephants that show signs of an increasing trend in some areas, the number of elephants *per se* is not an issue in the NBR, with the exception of perhaps local overabundance, dispersal and conflict with agriculture. The long-term conservation of the elephant has to address issues relating to further fragmentation of habitat, proliferation of exotic invasive plants, revival in poaching of bulls for ivory, and escalation in human–elephant conflicts resulting in public antagonism toward the species. The goals of management should thus be to (1) consolidate habitats and protect corridors by declaration of Ecologically Sensitive Areas under the country's Environmental Protection Act (1986) to regulate development and prevent fragmentation, (2) take measures to manage invasive alien species such as *Lantana camara* and *Eupatorium odoratum* that may negatively impact native species, (3) take steps through integrated land-use planning at the landscape level to reduce elephant-human conflicts, and (3) build up a genetically viable elephant population by protecting the tusked males from ivory poaching.

Keywords: *Asian elephant, Population, Nilgiri Biosphere Reserve.*

Introduction

Asian elephant (*Elephas maximus*) an endangered species today, about 6000 years ago enjoyed a much wider geographic distribution and higher number than it does today. Its range was then extended from Mesopotamia in the west across the Indian subcontinent to Southeast Asia and China, as far north at least as the Yangtze-King (Santiapillai and Sukumar, 2006). Today there are about 30000–50000 elephants distributed discontinuously across 13 range states (Hedges, 2006) and IUCN lists this species as endangered (IUCN Red List, 2013), while CITES places it on Appendix–I. The range countries population varies from perhaps as low as 100 in Vietnam to well over 25,000 elephants India (Santipillai and Sukumar, 2006; Project Elephant, 2007). Within India, largest populations of Asian elephant are regionally seen in southern India (Baskaran, 2013; Baskaran *et al.,* 2013). Nilgiri Biosphere Reserve, located in southern India, perhaps harbours the highest density of elephants anywhere in Asia. Additionally, the Reserve being contiguous to its adjoining habitats both in Western and Eastern Ghats, supports the largest elephant population in Asia (Daniel *et al.,* 1995). The Reserve encompasses a wide range of habitats ranging from semi-evergreen to tropical dry thorn forests and shows distinct seasonality–dry versus two wet seasons, making it an ideal habitat for wide ranging mega herbivores like elephants. This paper reviews the present status of elephants and their habitats in Nilgiri Biosphere Reserve, Southern India.

Study Area

The Nilgiri Biosphere Reserve (NBR) lies between 76°0' E and 77°15' E longitude and 12°15' N and 10°45' N latitude. It covers an area of 5520 km² and is situated at the junction of three southern states *viz.* Tamil Nadu, Karnataka and Kerala (Figure 22.1). The strong rainfall gradient from east (low) to west (high) resulting from a complex topography has resulted in a mosaic of vegetation types: from lowland tropical dry thorn forests through mid-elevation tropical dry and moist deciduous forests, high elevation tropical semi-evergreen and evergreen forests to montane shola–

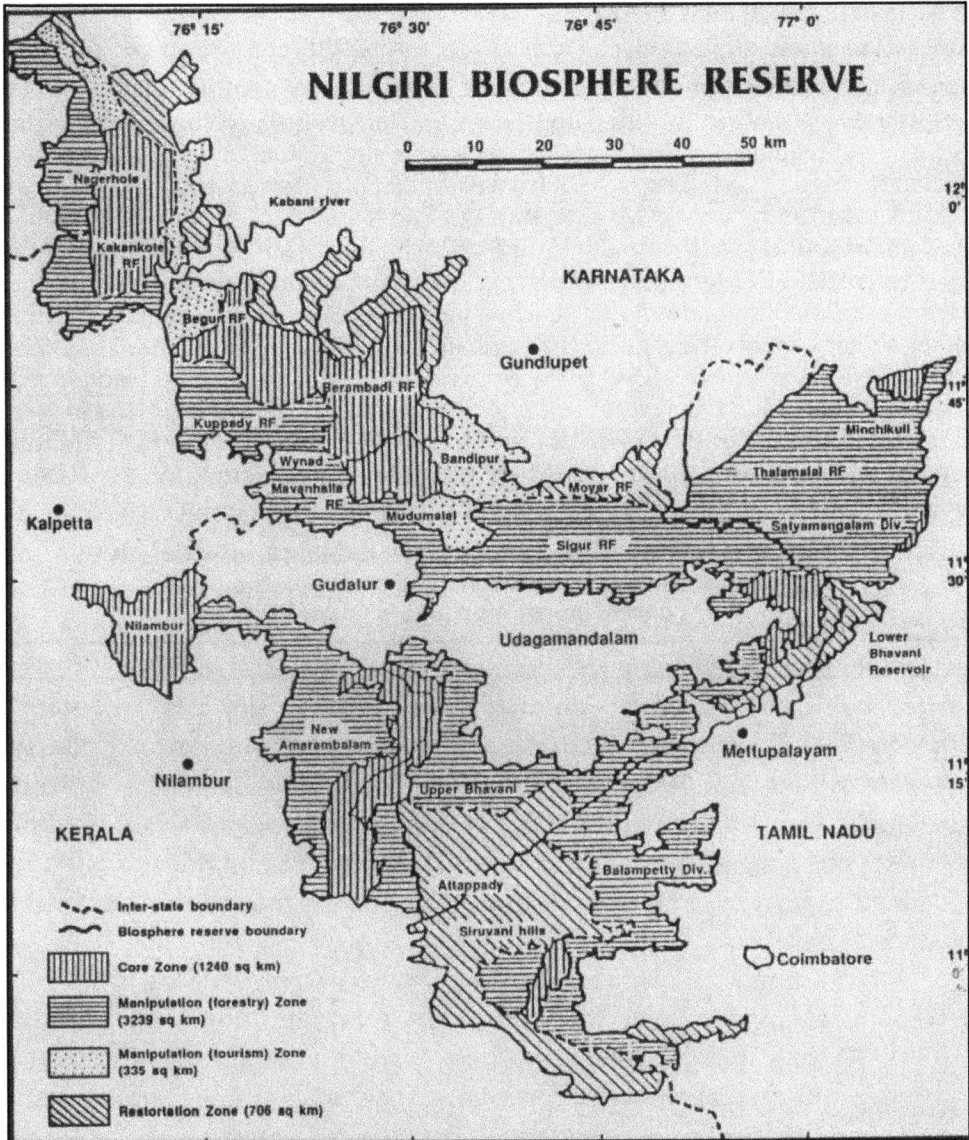

Figure 21.1: Map of Nilgiri Biosphere Reserve Showing Various Forest Divisions and Administrative Zones.

grasslands. This heterogeneous landscape also supports a wide assemblage of large mammals including gaur, sambar, chital and common langur at high density that, in turn, supports substantial populations of large carnivores such as the tiger, leopard and dhole. The NBR being the home of highest densities (max. of 2.4/km^2) of elephants in Asia, harbours a substantial part (>70 per cent) of the large population of more than 8000 elephants seen in the broader Nilgiri–Mysore-Wyanad Elephant Landscape.

Status of Elephant Habitat

The Nilgiri Biosphere Reserve comprises of 12 Forest Divisions (see Table 22.1), spread across three southern states of Tamil Nadu, Karnataka and Kerala and is an important elephant range with highest density of elephants anywhere in Asia. The reserve is spread across 5520 km² (shown as 6516 km² in the Table 22.1 due to the reason that exact area falling in NBR from Sathiyamanagalam and that Mannarghat are unknown) and consist of Mysore, Wyanad, and Nilgiri plateaus up to the Palghat Gap on the southern side, and the Sigur Plateau on the eastern side to Nilambur on the western side. The forest division such as Nagarahole, Bandipur, and Mudumalai Tiger Reserves, Wayanad Wildlife Sanctuary are the most important areas within the Reserve that support the highest (>2 elephants/km²). The strong rainfall gradient from east (low) to west (high) resulting from a complex topography has resulted in a mosaic of vegetation types: from lowland tropical dry thorn forests (on the eastern side) through mid-elevation tropical dry and moist deciduous forests (in the middle region), high elevation tropical semi-evergreen and evergreen forests (on the western side) to montane shola–grasslands (along the crest lines of the Ghats).

Table 22.1: Elephant Habitats Extent and Population Size in Various Forest Divisions of Nilgiri Biosphere Reserve, Southern India (*only part of the division area is under NBR)

Forest Division	Area (km²)	2010	2007	2005	2002	Mean
Nagarahole	643	643	600	804	1170	804
Bandipur	906	2175	1005	1217	1975	1593
Wayanad WLS	344	713	604	825	521	666
Mudumalai TR	321	995	578	294	703	643
Sathiyamangalam*	1455	393	845	700	873	703
Nilgiri North	854	427	598	92	557	419
Nilgiri South	321	0	14	0	–	5
Mukurthi NP	78	–	–	–	–	–
Nilambur South	100	62	11	50	18	35
Silent Valley	90	1	26	12	131	42
Coimbatore	738	295	262	119	178	214
Mannarghat*	665	472	46	200	58	194
Total	**6516**	**6176**	**4589**	**4312**	**6184**	**5315**

In fact central and eastern part of the Mysore-Nilgiri plateaus with extensive tracts of tropical deciduous forests that support a higher biomass of grass cover and dry thorn forests and with relatively less human interference in the deeper forests are potentially the most outstanding elephant habitats anywhere in Asia. This heterogeneous landscape also supports a wide assemblage of large mammals including gaur (*Bos gaurus*), sambar (*Cervus unicolar*), chital (*Axis axis*) and common langur (*Seminopithacus entellus*) in high densities (Karanth and Sunquist 1992; Varman and Sukumar, 1995) that support a substantial part of the large carnivorous population in this part of the Ghats.

Elephant Population

The Reserve is well known for harbouring the single largest population of Asian elephants anywhere in Asia. The census estimates reveal that the forested tracts in these parts of Landscape harbour about host 5000 elephants (Census 2002–05–07–10, Table 22.1). In addition, the area being contiguous with the Eastern Ghats on eastern side and Malnad Plateau on the northern side, the actual population exceeds 8000 elephants and which is undoubtedly largest single e actual population.

Conservation Problem

Among various conservation problems exist in the reserve, habitat loss, fragmentation and degradation are the most important factors threatening elephants. For example the western parts of the Wayanad plateau have witnessed a severe habitat fragmentation by developmental activities with most of the low-lying swamps (locally known as 'vayal'), a preferred microhabitat of elephants during the dry season, taken away for human habitation and cultivation. Besides, large tracts of natural forests have also been converted into monoculture plantations of teak, rubber, and pepper for commercial purposes, which has added to the fragmentation of natural habitats and increased human-elephant conflict. Similarly, the construction of a series of hydroelectric projects (Pykara), especially on the eastern side of Mudumalai, brought with them a large influx of human population and infrastructure development, which has created many bottlenecks threatening the habitat contiguity of Mudumalai Tiger Reserve with the Sigur Plateau that in turn connects the Eastern Ghats. Likewise, there were proposals for infrastructure developmental plans for: (i) creation of a highway from Kozhikode to Coimbatore via Masinagudi, Anaikatti and Sigur by widening the existing road from Vazhaithottam to Sigur and linking it to Bhavanisagar to bypass the existing Ghats section highway that goes via Nilgiris, and (ii) extending the Mysore–Chamarajanagar railway line to Coimbatore via Bhavanisagar–Sathiyamangalam cutting across the Moyar Valley, the connecting link between the Western and Eastern Ghats. Although these proposals have been shelved as of now, they may get revived in the future with pressure by political influence, as has happened several times earlier (Kozhikode–Coimbatore highway, for example). If any of these projects get implemented, it could decimate a large number of wild animals directly through road kills (Boominathan *et al.,* 2008; Baskaran and Boominathan, 2010) as well as fragmenting the link between the Western and Eastern Ghats (Baskaran, 1998). Additionally, such development would also encourage people to encroach onto the forested revenue patches available in the Sigur Plateau and further aggravate the habitat fragmentation, biotic pressure, and their resultant human-wildlife conflict in the region.

Elephant Corridor

Many of the elephant clans and bulls range in Mudumalai and Bandipur Tiger Reserve also range into Sigur Reserve Forest using the corridors exist on the eastern side of Mudumalai and western side of Nilgiri north Forest Division. Although the viability of the elephant population here is not in question, the threat of habitat loss and its resultant fragmentation makes it imperative to secure or strengthen vital corridors (especially in the Sigur plateau see Figure 22.2 and Table 22.2) to maintain landscape integrity for the free movement of elephants. Most important corridors that need immediate attention include *Singara-Masinagudi* that has both unoccupied revenue lands as well as Private land belongs to Singara estate (Sl.No. 1 and 2 in Figure 22.2 and Table 22.2).

The importance of this corridor in terms intensity of usage and management implications of elephant ranging behaviour (fidelity shown by elephant clans and bulls to the corridor) have been discussed in detail by various studies (Desai, 1991; Baskaran *et al.,* 1995; Baskaran, 1998). This corridor land falls under Private and Revenue control is presently forested but may perhaps be converted into non-forestry purpose at any time considering the fact that a large portion of land to south of this corridor that was part of Singara estate, is already converted into non-forestry purpose within the last 10 years. Therefore, this corridor needs immediate management attention. Similarly corridors (Sl.Nos. 3, 6, 8, 9 and 10 in Figure 22.2 and Table 22.2) also require management action of transferring lands from Revenue to Forest Department. In the past, there have been few attempts by Tamil Nadu Forest Department to Revenue Department requesting to transfer the unoccupied revenue lands to Forest Department to strengthen the corridors and reduce the human-elephant conflicts. However, this task has not been completed still and thus it needs consistent follow-up from Forest Department as well as committed NGO.

The recent up-gradation of Sathiyamangalam territorial Forest Division into a Tiger Reserve that forms part of the Moyar valley would considerably strengthen habitat viability in this region. As far as elephants of Nilgiri Biosphere Reserve are concerned, the habitats in the Sigur Plateau and Moyar Valley region are acting as an important corridor that connects the Western Ghats elephant population with Eastern Ghats population, which are with unequal sex ratios due to different levels of poaching pressures. Genetic study has shown interestingly that the entire elephant population here (Nilgiri%Eastern Ghats) constitutes a single matriline or mitochondrial haplotype (Vidya *et al.,* 2005). Ivory poaching has gradually distorted the sex ratio; for instance, in the Nilgiris, the adult male to female ratio has skewed from about 1:5 during the early 1980s to about 1:25-30 by about 2000 (Baskaran and Ajay, 2000; Arivazhagan and Sukumar, 2005). Since then the ratio has stabilized with the decline of poaching. Given the ranging behaviour of elephants (seasonal range and corridor fidelity) and ivory poaching, the contiguity between the Western and Eastern Ghats via Sigur Plateau and Moyar Valley is considered vital.

Similarly, the elephants from in Nilgiri North and Sathiyamangalam Forest Divisions move to the western part of Coimbatore Forest Division through Kallar corridor located nearby Mettupalayam (Baskaran, 1998). The bottleneck is created by the existence of Private land (Areca nut plantation) up to the foothills on the southern side and the steep escarpment on northern side in forest areas (Figure 22.3). The Coimbatore–Ooty highway, with high-level traffic round the day along with the Mettupalayam–Ooty railway line posing continuous disturbances to the movements of elephants in this corridor. This corridor too needs procurement of small patch of

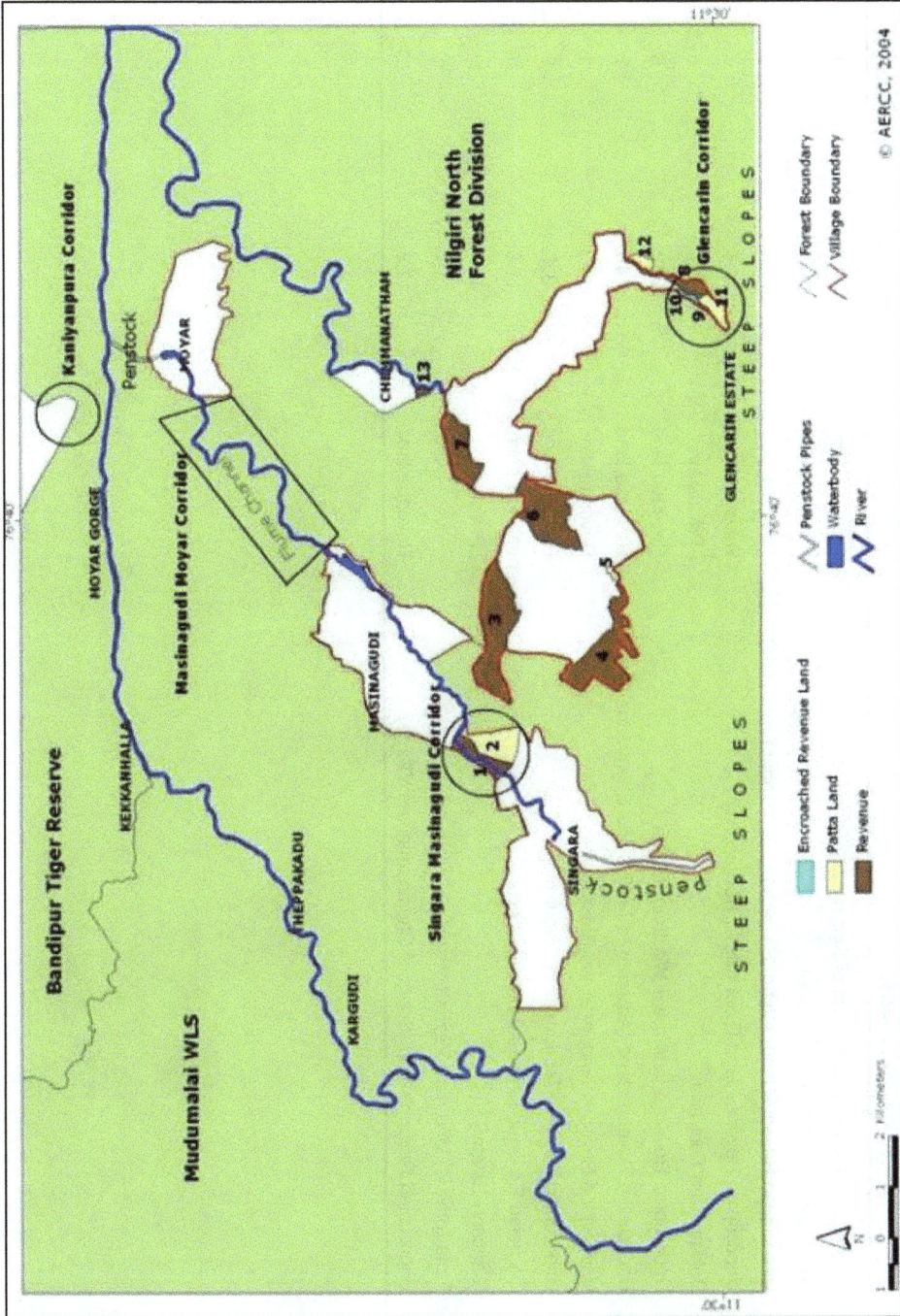

Figure 22.2: Map Showing the Corridors in the Nilgiri Region of Nilgiri Biosphere Reserve.

Table 22.2: Elephants Corridors and their Land Status Identified in the Nilgiri Biosphere Reserve, Southern India

Sl.No.	Location	Status	Area (Acres)	Map Index Reference
1.	Unoccupied Revenue land West of Singara estate	Revenue land	80.6	1
2.	Singara estate (Singara road side)	Private land	124.4	2
3.	Unoccupied Revenue land, east of Singara estate (Singara road side)	Revenue land	316.5	3
4.	Unoccupied Revenue land (Foot hills of Bokkapuram)	Revenue land	294	4
5.	Bokkapuram Foot hills	Private land	9.9	5
6.	Unoccupied Revenue land, near Westbury Estate	Revenue land	262	6
7.	Unoccupied Revenue land (Adjacent to MWLS and Chemmanatham)	Revenue land	148	7
8.	Unoccupied Revenue land, near the Sigur river	Revenue land	58.2	8
9.	Unoccupied Revenue land, near the Sigur river	Revenue land	3.4	9
10.	Occupied Revenue land near the Sigur river	Revenue land encroached	11.8	10
11.	Glencarin Estate	Private land	46.4	11
12.	Patta (private) land near Vazhathottam	Private Land	21.7	12
13.	Unoccupied Revenue land (South of Chemmanatham Village)	Revenue land	12.8	13

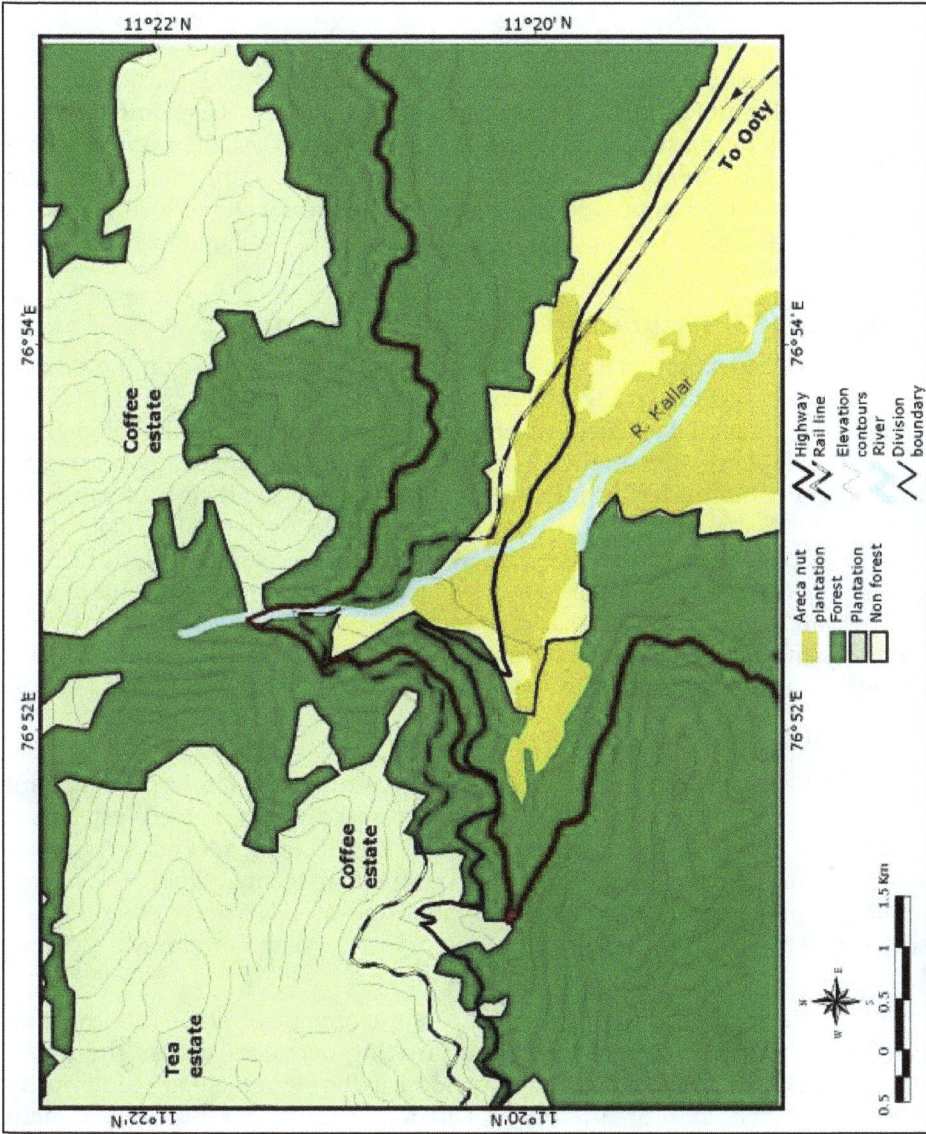

Figure 22.3: Map Showing the Kallar Corridor Located at Mettupalayam Region of Nilgiri Biosphere Reserve.

Private lands and construction of flyovers for the highway across the elephant crossing points (first and second hairpin bends).

Ivory Poaching and Skewed Sex Ratio

Illegal ivory poaching during 1984–86 selectively removed a large number of adult males in the population, especially on the southern side of the population in Sigur and the southern side of Mysore and Wynad plateaus (Daniel *et al.,* 1987), leading to a ratio significantly skewed towards females at the adult level (Daniel *et al.,* 1987, Baskaran and Desai 2000; Arivazhagan and Sukumar, 2005; Baskaran *et al.,* 2010). Though the number of adult male poaching incidents has come down in recent years, with decreasing adult males in the population poachers are targeting even sub-adults and juvenile males resulting in a remarkable difference in sex ratio even among the younger population segments (Baskaran and Desai, 2000; Baskaran *et al.,* 2010) indicating that ivory poaching continues with a similar intensity as in the past.

Anthropogenic Pressure and Habitat Degradation

The Mysore and Nilgiri plateaus in the Reserve starting from Nagarahole south to Bandipur and Mudumalai Tiger Reserves, with a moderate rainfall and free of much human habitation deep inside the forest, have retained their habitat integrity. However, the eastern fringes of the Mysore and Nilgiri (Sigur) plateaus experience relatively higher anthropogenic pressure, in the form of overgrazing by scrub cattle and firewood collection by people that severely degrade the natural habitats along the fringes on the eastern side (Silori and Mishra, 2001; Baskaran *et al.,* 2012). The degradation of natural habitats, besides depleting the food sources available to wild herbivores, alters the animal (Baskaran *et al.,* 2009; Baskaran *et al.,* 2011) and plant communities and devastates the biodiversity of the region. Thus measures are needed to regulate such anthropogenic pressure.

Conclusions and Recommendations

The elephants of Nilgiri Biosphere Reserve are part of the larger elephant populations that range further into the adjoining forest divisions in the Western and Eastern Ghats of Tamil Nadu, Karnataka and Kerala (and, recently, Andhra). Therefore, there is a need for integrated landscape management approach for co-ordinating elephant conservation with neighboring states. We make the following broad recommendations for the conservation of the elephant populations of Tamil Nadu.

1. Create long-term assets for the free movement of elephants through strengthening, augmenting and creating corridors across the major landscapes. In this respect, the corridors in the Sigur plateau need priority attention as these are in danger of attrition because of rapid development in the region. The Kallar corridor in the southern side also needs attention.

2. Ensure that new developmental proposals such as highways and railway lines are not allowed through critical elephant habitats.

3. Restore the natural vegetation in regions where invasive plants such as *Lantana* and *Chromolaena* have suppressed native plants on a large scale.

4. Manage elephant-human conflicts through community participation and low cost barriers where possible.

5. Ensure that poaching for ivory, witch has fortunately been on the downtrend in recent years, is not allowed to escalate once again not just in the Protected Areas but also in the territorial divisions.

Acknowledgments

We thank the State Forest Department Tamil Nadu, Karnataka and Kerala for their consistent support over the years in our research on the elephant populations of the state. We thank Ministry of Environment and Forests, Govt. of India, Whitley Fund for Nature, U.K. and US Fish and Wildlife Service for financial support for the field surveys in the southern Indian states.

References

Arivazhagan, C., and Sukumar, R., 2005. Comparative demography of Asian elephant (*Elephas maximus*) populations in India. Technical Report No. 106. Centre for Ecological Sciences, Indian Institute of Science, Bangalore – 560 012, September 2005.

Baskaran, N., 1998. Ranging and resource utilization by Asian elephant (*Elephas maximus* Linn.) in Nilgiri Biosphere Reserve, South India. Ph.D., Thesis, Bharathidasan Univ. Thiruchirapally.

Baskaran, N., 2013. An overview of Asian elephant (*Elephas maximus*) in Western Ghats and its implication for Western Ghats Ecology. *Journal of Threatened Taxa.*, **5 (14)**: 4854–4870.

Baskaran, N., and Desai, A.A., 2000. Population demography of Asian elephant (*Elephas maximus*) in Mudumalai Wildlife Sanctuary, Southern India. Final Report 1999-2000. Bombay Natural History Society, Bombay.

Baskaran, N., Udhayan, A., and Desai, A.A., 2009. Population distribution and conservation of four-horned antelope (*Tetracerus quadricornis*) in the tropical forests of southern India. *Journal of Scientific Transactions in Environment and Technovation*, **2(3)**: 139–144.

Baskaran, N., and Boominathan, D., 2010. Road kills of animals by highway traffic in the tropical forests of Mudumalai Tiger Reserve, Southern India. *Journal of Threatened Taxa.*, **2(3)**: 753–759.

Baskaran, N., Udhayan, A. and Desai, A.A., 2010. Status of Asian elephant (*Elephas maximus*) in Mudumalai Wildlife Sanctuary, Southern India. *Gajah,* **32**: 6–13.

Baskaran, N., Balasubramanian, M., Swaminathan, S. and Desai, A.A., 1995. Home range of elephants and its implications for the management of Nilgiri Biosphere Reserve, India. In: *A week with elephants. Proceedings of the International Seminar on the conservation of Asian elephant.* Mudumalai Wildlife Sanctuary, June 1993. (Eds. J. C. Daniel and H. S. Datye). *Bombay Natural History Society, Bombay*, 296–313.

Baskaran, N., Anbarasan, U. and Agoramoorthy, G., 2012. India's biodiversity hotspot under anthropogenic pressure: A case study of Nilgiri Biosphere Reserve. *Journal for Nature Conservation,* **20**: 56–61.

Baskaran, N., Kannan, V., Thiyagesan, K. and Desai, A.A., 2011. Behavioural ecology of four-horned antelope (*Tetracerus quadricornis* de Blainville, 1816) in the tropical forests of southern India. *Mammalian Biology,* **76**: 741–747.

Baskaran, N., Kannan, G., Anbarasan, U., Thapa, A. and Sukumar, R., 2013. A landscape-level assessment of Asian elephant habitat, its population and elephant–human conflict in the Anamalai hill ranges of southern Western Ghats, India. *Mammal. Biol.,* **78**: 470–481.

Boominathan, D., Asokan, S., Desai, A.A. and Baskaran, N., 2008. Impact of highway traffic on vertebrate fauna of Mudumalai Tiger Reserve, southern India. *Convergence Journal,* **10 (1–4)**: 52–63.

Daniel, J. C., Desai, A.A., Sivaganesan, N., Datye, H.S., Rameshkumar, S., Baskaran, N., Balasubramanian, M. and Swaminathan, S., 1995. Ecology of the Asian elephant. Final Report 1987-1994. Bombay Natural History Society, Bombay.

Daniel, J.C., Desai, A.A. Sivaganesan, N., and Rameshkumar, S., 1987. The study of some endangered species of wildlife and their habitats – The Asian elephant. Report October 1985 – September 1987. Bombay Natural History Society, Bombay.

Desai, A.A., 1991. The home range of elephants and its implications for the management of the Mudumalai Wildlife Sanctuary, Tamil Nadu. *J. Bombay nat. Hist. Soc.,* **88**: 145-156.

Hedges, S., 2006. Asian elephants in captivity: Status needs and values; in Elephant Range States Meeting. January 24-26, 2006, Kuala Lumpur, Malaysia, IUCN/ SSC Report.

IUCN, 2013. Red List of Threatened species. IUCN, Gland, Switzerland and Cambridge, UK. [htto://www.iucnredlist.org, accessed 25th December 2013].

Karanth, K.U., and Sunquist, S.E., 1992. Population structure density and biomass of large herbivores in the tropical forests of Nagarahole, India. *J. Tropical Ecology,* **8**: 21–35.

Santiapillai, C., and Sukumar, R., 2006. An overview of the status of the Asian elephant. *Gajah,* **25**: 3-8.

Silori, C.S., and Mishra, B.K., 2001. Assessment of livestock pressures in and around the elephant corridors Mudumalai Wildlife Sanctuary, southern India. *Biodiversity and Conservation,* **10**: 2181–2195.

Varman, K.S., and Sukumar, R., 1995. The line transects method for estimating densities of large mammals in tropical deciduous forest: An evaluation of models and field experiments. *Journal of Bioscience,* 20(2): 273–287.

Vidya, T.N.C., Fernando, P., Melnick, D.J., and Sukumar, R., 2005. Population differentiation within and among Asian elephant (*Elephas maximus*) populations in southern India. *Heredity,* **94**: 71-80.

2015, Animal Diversity, Natural History and Conservation, Vol. 5   *Pages 379–398*
Editors: V.K. Gupta and Anil K. Verma
Published by: DAYA PUBLISHING HOUSE, NEW DELHI

Chapter 23

Perceptions on the Incidence of the Invasive Giant African Snail, *Achatina fulica* (Bowdich) in Horticultural Areas of Bangalore Region

M. Jayashankar*

Division of Entomology and Nematology,
Indian Institute of Horticultural Research, Hessaraghatta Lake Post,
Bangalore – 560 089, Karnataka, India

ABSTRACT

The present research work is aimed in providing up to date information on the distribution of the giant African land snail, *Achatina fulica* (Bowdich) and other malacofauna in Bangalore region through field surveys, opinion polls from farmers and agri-horticulturists.

Keywords*: Achatina fulica, Bangalore, Invasive, Survey.*

Introduction

'Biosecurity' according to Penman (1998) is defined as effective management of risks by a system of coordinated pre-border, border, management and sector responses aimed at preventing the establishment and spread of organisms that may have adverse effects on the economy, environment and people's health. Many alien species of

---

* *Corresponding author.* E-mail: jay_zoology@rediffmail.com

terrestrial molluscs are established in India, data about which is sparse (Raut and Ghose, 1984; Kaur, 2003). The potential costs of these alien species on agriculture or the environment are largely unknown in the Indian context, but potentially immense as reported in other parts of the world. Not only are they reported to devour native plants, modifying the environment, but they probably also outcompete native gastropods (Leslie *et al.,* 2008). Reports on such alien species from the Bangalore region are scarce, also their impact on local environment is lacking. Hence, a questionnaire cum detection survey on occurrence of such malacofauna was undertaken. Questionnaire surveys are particularly useful in trace-forward and trace-back investigations planting material are suspected as the source of the pest. Face-to-face interviews or questionnaires distributed to farmers and property owners are a useful way of determining whether hosts are present on a property. Information gathered through pest surveillance may be used to:

☆ Aid the early detection of new pests

☆ Conduct pest risk analyses to justify regulating a particular pest and to require precautionary phytosanitary measures

☆ Establish and maintain pest-free areas

☆ Compile host lists and distribution records

The tropical giant African land snail, *Achatina fulica* Bowdich is a major crop pest across the globe (Mead, 1961, 1979; Raut and Barker, 2002; Silvana, 2007) (Figure 23.1) and one of the most extensively studied snails because of its economic, ecological and medical importance (Mead, 1979; Raut and Ghose, 1984; Srivastava, 1992; Kliks

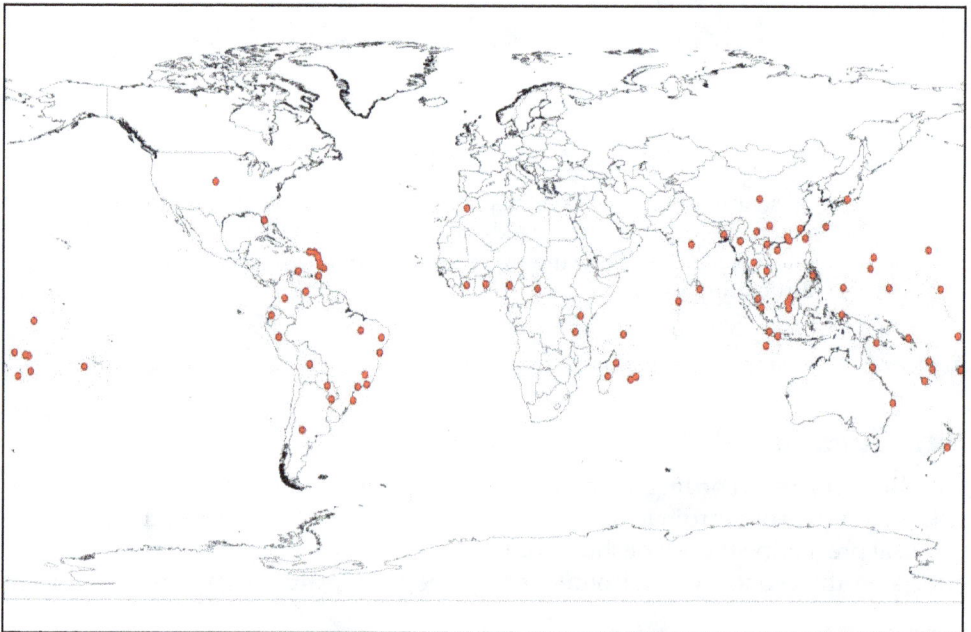

Figure 21.1: Current Distribution of *A. fulica* (Modified from CABI, 2014).

and Palumbo, 1992; Raut and Barker, 2002). The International Union for Conservation of Nature and Natural resources (IUCN) has listed *A. fulica* as one of the world's 100 most invasive species (Lowe *et al.,* 2000). It is a classic example of an introduced species and its success as an introduced species is attributed to several factors *viz.,* high reproductive capacity, voracious feeding habit, inadequate quarantine arrangements and human aided dispersal. In India, it was believed to have been introduced in Chouringhie gardens of Calcutta in 1847 by the British Conchologist William Henry Benson and from there on spread to many states of the country in course of time (Mead, 1961). According to Naggs (1997), although Benson had got the snail to India from Mauritius, he had handed them to a friend before leaving India and it was the friend who released them in his garden. This perhaps has lead to the large scale invasion of the snail in different parts of the country. Recent genetic studies indicate that all *A. fulica* now occurring throughout South Asia, Southeast Asia and the Pacific region are derived from one haplotype (Fontinilla *et al.,* 2007), the source being a single pair of specimens released in Chouringhie in 1847.

The snails has been thriving in some areas of India for a period of 100-150 years, with no clear evidence of abatement in its pest status (Raut and Ghose, 1984), which convinced Gude (1914) consider it as native to India (in the fauna of British India), though introduced. In spite of this long thriving history, reports on the agri-horticultural pest status of the snail are minimal. So far, the control measures of all types have failed to eradicate this pest, hence an integrated approach to contain the snail is recommended as an ultimate remedy (Srivastava, 1992). The snail has been reported for the first time causing damage to ornamental plants and vegetables in Bangalore during kharif (Veeresh *et al.,* 1979). Prerequisite to answering the question on pest management is the need for a thorough knowledge of the distribution, level of abundance of the species in the region. The research work is thus aimed in providing up to date information on the distribution of *A. fulica* and other malacofauna in Bangalore region through field surveys, opinion polls from farmers and agri-horticulturists.

Materials and Methods

Topographic Information of the Bangalore Region

Bangalore region comes under the eastern dry agro climatic zone of Karnataka with 35°C maximum and 14°C minimum temperature; 679.1 to 888.9mm of rainfall and situated at an elevation of 800-1500m. Soil type is red loamy, clay lateritic. The climate of Bangalore is that of a tropic savannah type, with summer season (March to May), rainy (June to September and November and December) and winter (December to January). The annual rainfall of Bangalore is 859mm spread over 6 months. It receives rains during both South-west and North-east monsoons. Bangalore Rural District has 951 inhabited villages with a total population of 8, 50,968 (2001 census); geographical area of 2, 29,519 (hectares) with 1, 29,522 (hectares) of total area sown with 84,023 hectares of food crops and 45,499 hectares of non food crops. Bangalore Urban district has 558 inhabited villages with a total population (2001 census) of 65, 37,124 geographical area of 2, 17,410 hectares with 60,814 hectares total area sown, total food grains in 3, 28,487 hectares, oil seeds in 1,364 hectares, fruits 3,845 hectares and vegetables in 2,077 hectares (Source: Office of the District Statistical Office, Bangalore Rural and Urban Districts, Records of the year 2008-2009).

Crop Carieties Vultivated in the Region

Bangalore Urban has peripheral agriculture and horticulture areas, while Bangalore Rural has larger areas under intensive farming along with horticulture areas. Vegetable crops include Pumpkin, Cucumber, Little finger, Gherkins, Snake gourd, Bitter gourd, Ridge gourd, Watermelon, Muskmelon, Potato, Tomato, Beans, Onion, Green Chillies, Tapioca, Cabbage, Knol-khol, Cauliflower, Peas, Lady's finger. Leafy vegetables include Menthe, Palak, Curry leaves, Radish, Beetroot, Carrot, Capsicum, Drumstick etc. Fruit crops grown include Mango, Banana, Lemon, Orange, Guava, Sapota, Pineapple, Pomegranate, Papaya, Grapes, Jack etc., Spice crops like Pepper, Cardamom, Tamarind, Ginger, Garlic and Coriander. Plantation crops include Coconut, Arecanut, Betel vine and Cashew. Commercial flowers grown in the region include Aster, Marigold, Jasmine, Chrysanthemum, Rose, Anthurium. Medicinal and aromatic plants including Periwinkle, Dhavana, Lemongrass etc., are cultivated in the region.

Detection Survey of *A. fulica* in Bangalore Region

The study was divided into two phases. Phase I was a pre-survey exercise, while the second involved a field exercise with data collection from permanent sample plots (villages) spread across eight taluks of Bangalore region. The Phase I conducted during 2008, involved interaction and interviews (formally and informally) with individual farmers, the other respondents include agriculture scientists, officials of Farm Owner's Association and Nurseries, Horticultural Department officials. The formal interview involved the use of structured interview questionnaires (enclosed below). The informal interviews entailed more of courtesy calls on opinion leaders and socialization chats with individuals and households. More villages, in addition to the ones laid during the pre-survey exercise were laid at the onset of the Phase II study. Information from a total of 114 villages (57 each from Bangalore and rural districts) were used in the study conducted during 2009-10. Continuous questionnaire–Opinionnaire survey and field observations were conducted during the monsoon (July-September) and the period just after the monsoon (October-January), before the dry spell began (February/May) in the study region, involving discussions with local people in the areas visited. The interviewees were provided with informational photographs, shells and contact numbers to communicate on the status of the snails.

The signs of snail presence were evaluated observing evidences including juvenile and adult snails, eggs, empty snail shells, defoliation, extensive rasping, slime trails and ribbon-like feces. Questionnaire was prepared consulting earlier surveys (Asamoah, 1999, USDA–APHIS, 2005). Twenty four private nurseries in Kumbalagodu, Bangalore South taluk were also part of the survey, results of which are presented separately. Since snails are nocturnal in habit being active during early morning, overcast and rainy days, surveys were carried out accordingly during such occasions, as surveys at night are logistically difficult. Visual search method was employed to get the data on pest snails. Information from 150 farmers was recorded. The Malacofauna detected during the detection survey were photographed using Cannon Digital Camera, 8 mega pixcels.

In the field, exact location of each sample collection site was recorded with a handheld Garmin e-trex, 12 Channel global positioning system (GPS) device to determine coordinates for each study point and of district and taluk map (Land records of 2001) of Karnataka State Remote Sensing Applications Centre (KSRAC), Government of Karnataka. All points were within the administrative boundary of the locations. Arc GIS Software (8.1 version) was used for the data analysis and projection. The live specimens were first drowned in water for a day, then kept in 10 per cent formaldehyde, for an hour and finally preserved in 70 per cent alcohol. The identification is mainly based on the Fauna of British India series of mollusca (Blanford and Godwin-Austen, 1908; Gude, 1914, 1921). Identification and confirmation assistance was sought from Dr. N. A. Aravind, Malacologist, Ashoka Trust for Research in Ecology and Environment (ATREE), Bangalore and Dr. Basudev Tripathy, Malacolgy Division, Zoological Survey of India (ZSI), Kolkata.

Data on village survey parameters from 114 villages were analyzed using Chi-squared tests of homogeneity for significance of differences among samples. The significance level was set up at 5 per cent level of significance. Chi-square test for distribution of Malacofauna taluk wise: To test the null hypothesis that there is no evidence of difference in terms of proportions between the Species, i.e., $H_o: p_{1j} = p_{2j} = p_{3j} = p_{4j} = p_{5j} = p_j, j=1, 2,$ (p- denotes the proportion/ probability) the Chi-square test was applied.

QUESTIONNAIRE–OPINIONNAIRE
FIELD SURVEY–GIANT AFRICAN SNAIL DETECTION

1.	Interview only	2. Interview and field survey	3. GAS detected	4. Survey date
5.	Survey area	6. Educational Qualification	7. Experience	8. Age

9. Resident/landowner's Name:

10. Address

☆ Hobli

☆ Taluk

☆ Contact numbers

11. Property type Business name/Farm/Nursery/Residence/Agriculture

12. Property Size: Acres

13. Longitude-Latitude: GPS

I. Introduction History

14. Suggested method of introduction : market goods/intentional/others/don't know

15. Estimated time since introduction recent-(Recent-5yrs) (6-10) (>10) (don't know)

16. Do you facilitate or check spread how?

17. Describe vegetation and crops

☆ Perennial

☆ Biennial

☆ Annual

☆ Seasonal

☆ Ornamental

☆ Cover or fence

18. Cropping patterns: Mixed/Rotation

19. Describe habitat and irrigation (greenery/water sources)

20. Substrate found

☆ on or in

☆ average number/day in rainy

☆ average number/day in summer

21. Hiding places

Cool moist area   Bricks   Crevices   Walls   Plant detritus   Canopy

22. Plants feeding on:

☆ During field survey

☆ Frequently noticed on

23. Specially preferred (Why) plant
    species/vulnerable age of plants

24. Plants (Why?) avoided by snails/act
    as repellent

25. Other behavioral observations

Egg laying:

Number/Depth/Hatchability

Mating:

Duration/Time of the day:

II.   Signs of snail presence: Defoliation/Slime trails/Ribbon like feces

III.   Survival success

Food Source   Lack of control and predators

Rainfall and humidity (weather)   Changes in the population over time

IV.   Nuisance factor

☆ For households ( )   ☆ Not possible to walk a path way without crushing ( )

☆ Defacement on plants ( )   ☆ Dead and decomposing snails ( )

☆ Slime trials on walls ( )   ☆ Low market value due to bite mark/fecal matter ( )
floor and soil

V. Economic loss (annual expenses/gain)

Loss>gain ( )   Loss<gain ( )

VI. Health awareness

Threat to human health and plant ( )   Food Source to human and domestic animals ( )

VII. Awareness

a. Unaware/not heard off/absent in the village

b. Aware of presence elsewhere and in same village but unaware of its threat

c. Aware of Snails presence but nuisance only

d. Aware of Snails presence and threat but no action

e. Aware of Snails presence threat and action taken poor result

f. Aware of Snails presence threat and action taken successful

g. Snail beneficial and revered

VIII. Management

1. Cultural (Sanitation in crop fields/Crop rotation/Detractor plants)

a. Sanitation in crop fields   b. Crop rotation   c. Detractor plants

2. Mechanical:

a. Duration of hand collection   b. How do you dispose?

b. Labour number and expense (if any)   d. Barriers-mechanical wire/bunds

3. Chemical:

Quantity/acre: Name: Impact:

4. IPM followed how?

5. Natural enemies/predators:[quantity fed/mode of feeding/when observed
Birds/Reptiles/Mammals (Bandicoot/Mongoose)/others (flat worms/insects like beetles.

IX. Other malacofauna/related cohabitants

Sl.No.	Other Malacofauna/ Related Cohabitants	+/-	Threat/Not (Mention crops and kind)	Since and How	Management (+/- and what)
1.	*Crytozona bistrialis*				
2.	*Macrochlamys indica*				
3.	*Laevicaulis alte alte*				

Sl.No.	Other Malacofauna/ Related Cohabitants	+/-	Threat/Not (Mention crops and kind)	Since and How	Management (+/- and what)
4.	*Mariella dussumeri*				
5.	Millipede				
6.	Insect pests				
7.	Others				

The following general norms of a questionnaire (Ramli Mohamed and Khor Yoke Lim, 1988) surveys were followed during the present survey.

1. Conducting the conversation in an friendly atmosphere by being friendly, courteous, conversational, and unbiased.

2. If the respondent had any difficulty in understanding the questions the questions were suitably adjusted slightly to facilitate your respondent's understanding.

3. The questions were asked in the same order as in the questionnaire to avoid any confusion to the respondent.

4. Specific information's were carefully recorded without interpretation of a respondent's answers.

5. Caution was taken on not suggest a possible reply.

6. If the response given indicated that the respondent misunderstood the question, the question was repeated to clarify.

Results

Demographic Data of Farmers

Results of opinions gathered from agri-horticulturists are reported for eight taluks of Bangalore region comprising Bangalore Urban and Bangalore Rural districts of Karnataka state. Of the 150 farmers interviewed, ~20.14 per cent of the interviewees had received formal education up to the primary level and/or secondary level and 9.72 per cent had received tertiary education (PUC/Diploma/Degree). The rest had no formal education. Age wise percent of adults interviewed was >20 years (2.78), 21-40(34.03), 41-60(33.33) and >60(9.72) while the data of 20 per cent of the remaining was not available. In terms of possession of agricultural land (acreage) by farmers interviewed, it was, <1(5.9 per cent), 1-5 (80 per cent), 5.1-6(7.6 per cent) and >10 acres (5.9 per cent). The Global Positioning System (GPS) coordinates of villages of the detection survey are tabulated and represented in Map 23.1 (Geo-tagging), indicating the occurrence status (present or absent) of giant African land snails and other malacofauna (OMF) in the study area.

Malacofaunal Diversity and Distribution

Surveys during 2008-2010 have uncovered remarkable records on distribution of terrestrial snails and slugs in the region (Figures 23.2–23.7). Ten different molluscs

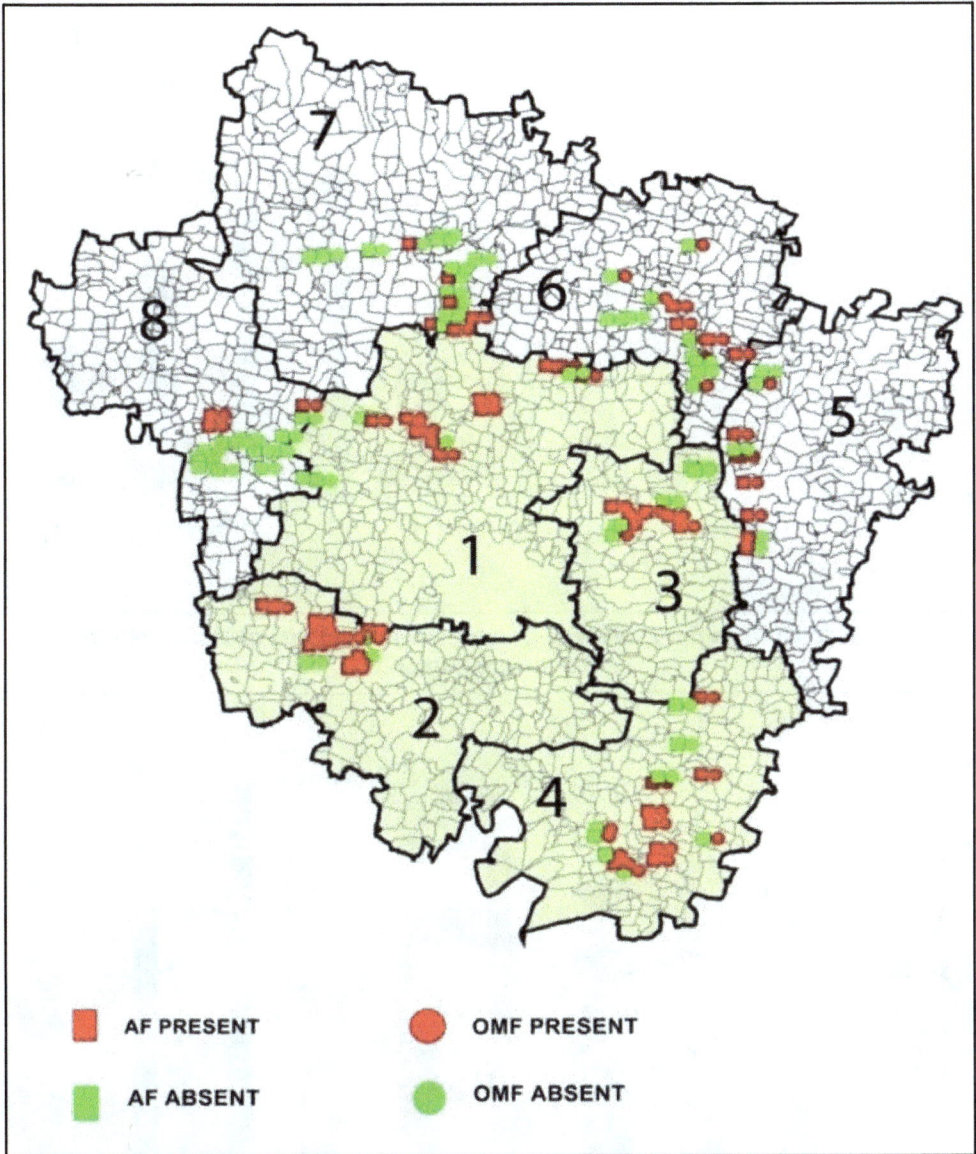

1: Bangalore North; 2: Bangalore South; 3: Bangalore East; 4: Anekal;
5: Hoskote; 6: Devanahalli; 7: Doddaballapura; 8: Nelamangala.

Map 23.1: Occurrence of *Achatina fulica* in 114 Villages of Bangalore Region.

Study Sites: BN: Bangalore North; BS: Bangalore South; BE: Bangalore East; AN: Anekal; HK: Hoskote; DV: Devanahalli; DB: Doddaballapur and NM: Nelamangala; BR: Bangalore region.

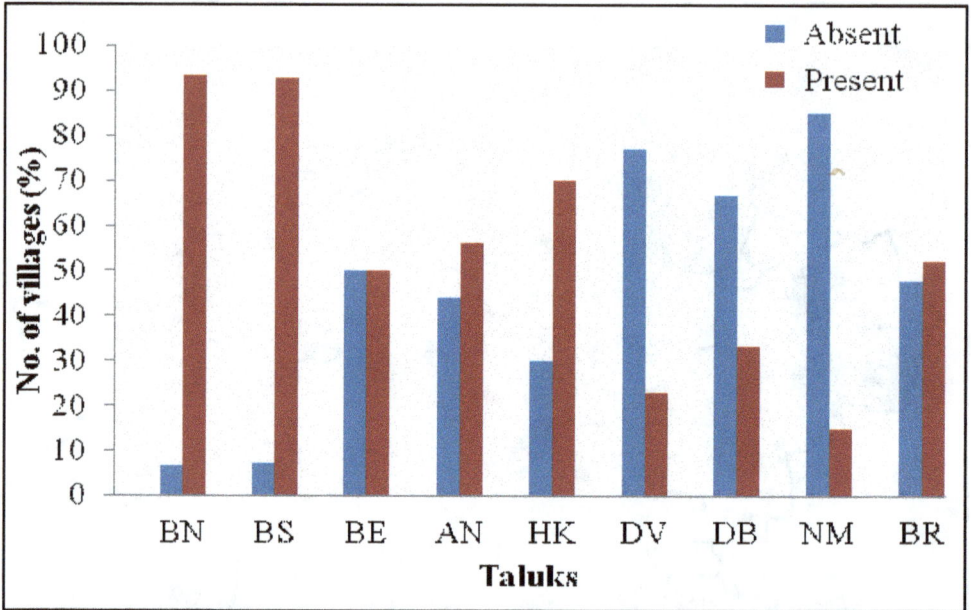

Figure 23.2: Multiple Bar Plots with Occurrence Percentage of *Achatina fulica* in Bangalore Region.

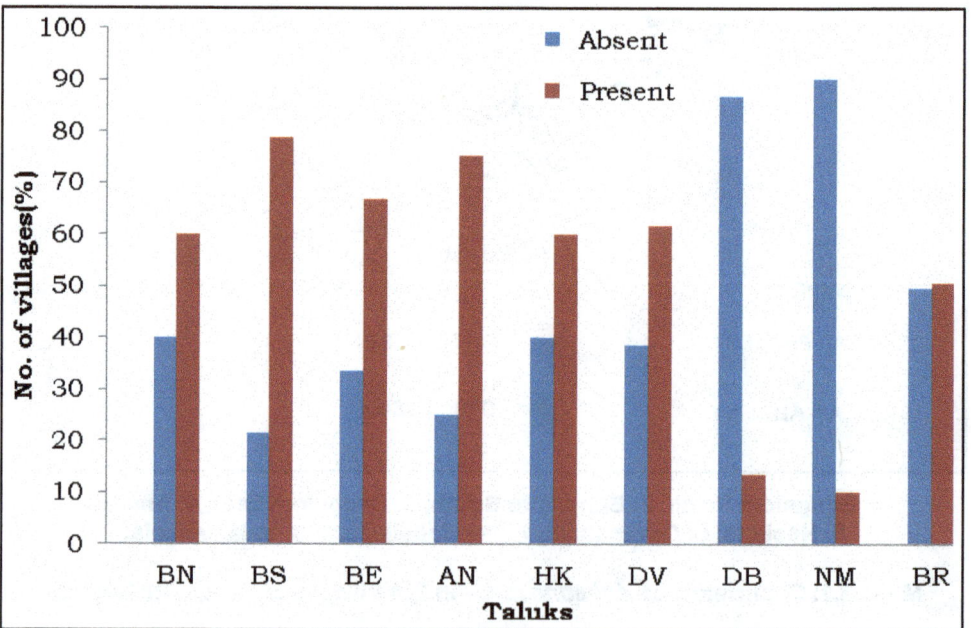

Figure 23.3: Multiple Bar Plots with Occurrence Oercentage of *Laevicaulis alte* in Bangalore Region.

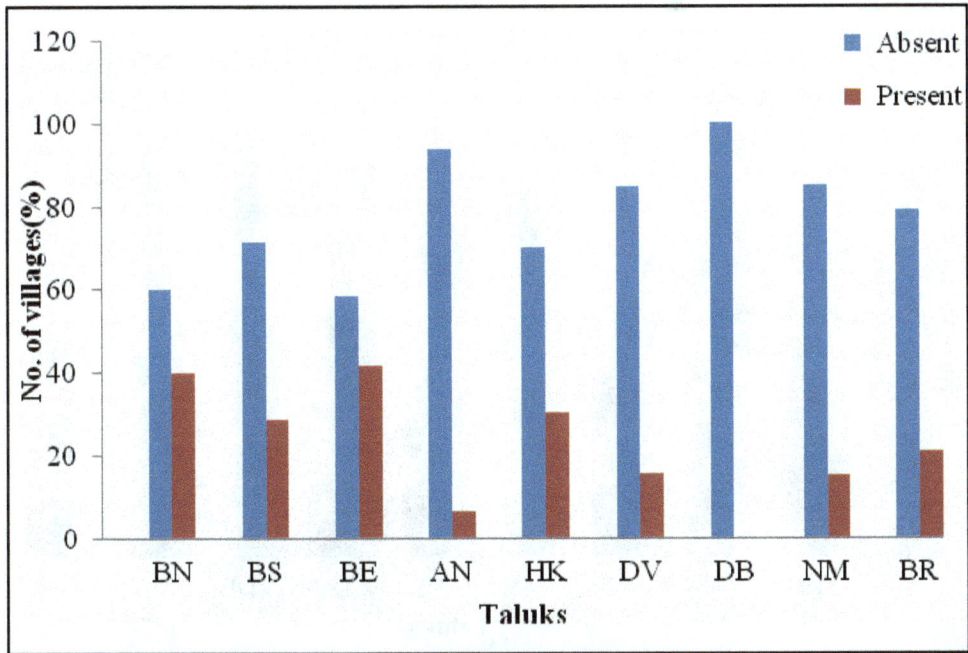

Figure 23.4: Multiple Bar Plots with Occurrence Percentage of *Mariella dussumeri* in Bangalore Region.

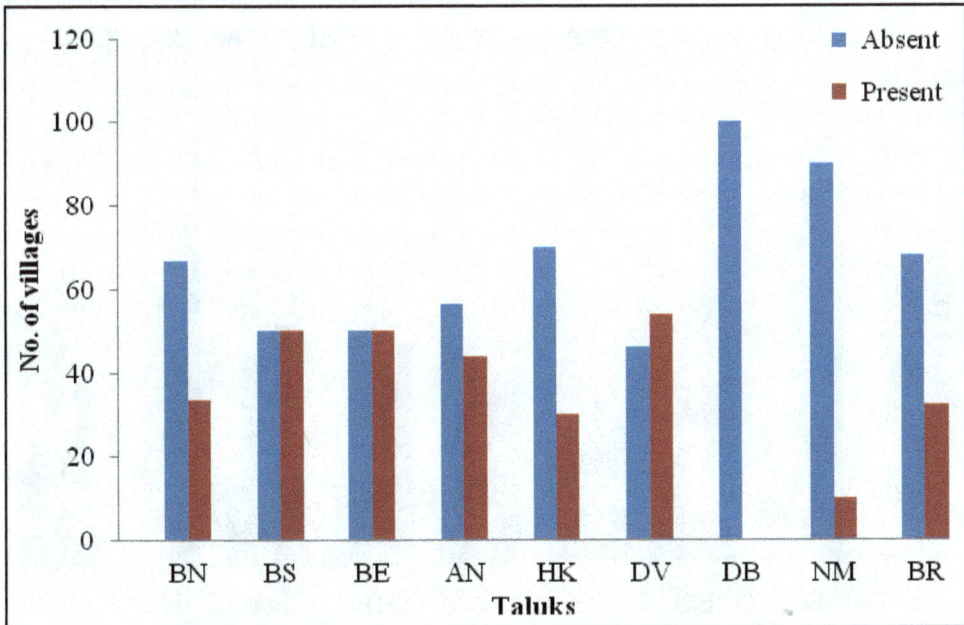

Figure 23.5: Multiple Bar Plots with Occurrence Percentage of *Cryptozona bistrialis* in Bangalore Region.

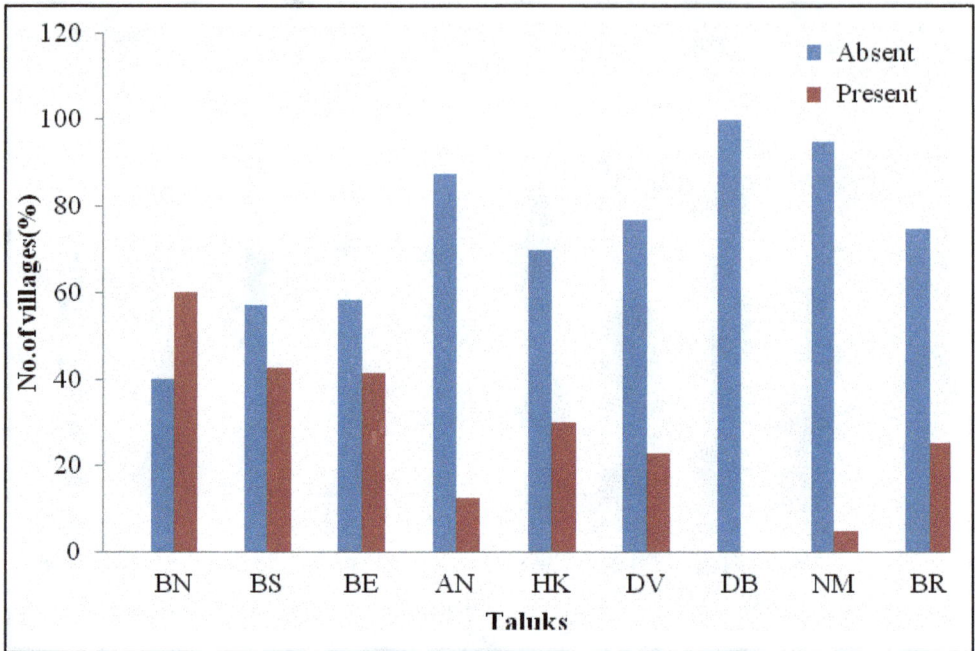

Figure 23.6: Multiple Bar Plots with Occurrence Percentage of *Macrochlamys Indica* in Bangalore Region.

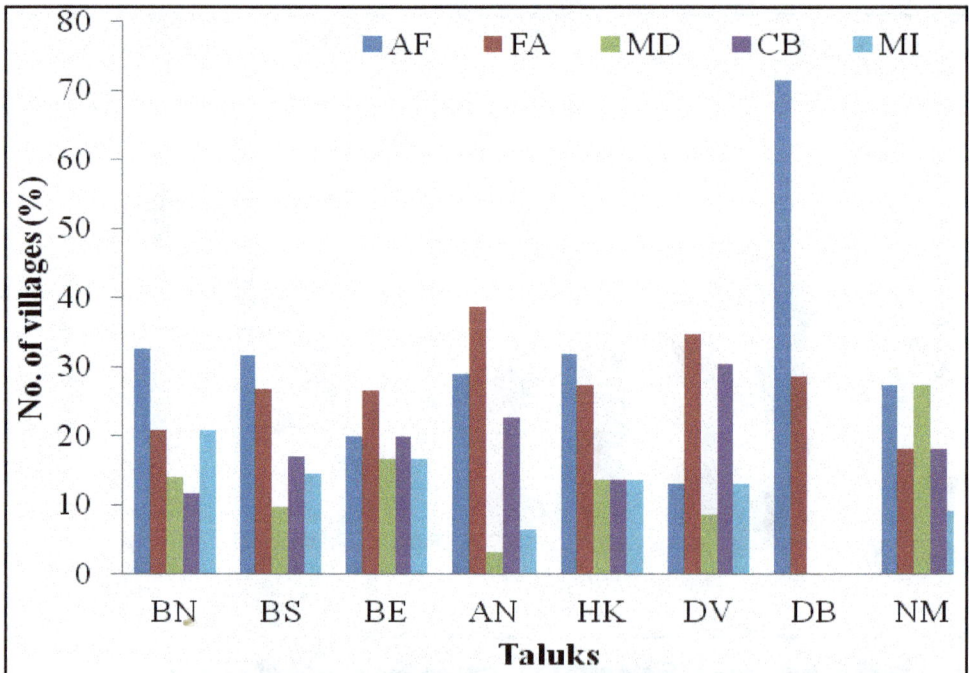

Figure 23.7: Multiple Bar Plots Comparing the Occurrence Percentage of Malacofauna between Taluks.

species were found to occur in the study sites (Table 23.1). According to the information gathered from respondents during the survey, different malacofauna species occur in the area and the most common among these is the GALS, *Achatina fulica*. Other terrestrial snails include *Cryptozona bistrialis* (Beck), *Cryptozona semirugata* (Beck), *Macrochlamys indica* Godwin-Austen *Rachis punctatus* (Anton), *Glessula bravis* (Pfeiffer) and the aquatic snail, *Lymnea* (*Pseudosuccinea*) *luteola* Lamarck. Three species of terrestrial slugs, *Mariella dussumeri* Gray, *Laevicaulis alte* (Ferrussac) and *Deroceras laeve* (Müller) were also recorded. The malacofauna represent six different families *viz.*, Ariophantidae (4), Achatinidae (1), Lymnaeidae (1), Subulinidae (1), Cerastidae (1), Veronicellidae (1) and Agriolimacidae (1).

Table 23.1: List of Malacofauna Recorded in Bangalore Region

Sl.No.	Family	Scientific Name
1.	Achatinidae	*Achatina fulica fulica* (Bowdich)**
2.	Ariophantidae	*Cryptozona bistrialis* (Beck)†
3.	Ariophantidae	*Cryptozona semirugata* (Beck)†
4.	Ariophantidae	*Macrochlamys indica* Godwin-Austen†
5.	Cerastidae	*Rachis punctatus* (Anton)†
6.	Subulinidae	*Glessula bravis* (Pfeiffer)†
7.	Lymnaeidae	*Lymnea* (*Pseudosuccinea*) *luteola* Lamarck*
8.	Ariophantidae	*Mariella dussumeri* Gray†
9	Veronicellidae	*Laevicaulis alte* (Ferrussac)**
10.	Agriolimacidae	*Deroceras laeve* (Müller)**

*: Fresh water snail; **: Introduced pest; †: Endemic pest.

Table 23.2: Chi-square Test for Occurrence Percentage of Malacofauna in Eight Taluks of Bangalore Region

Taluk	Chi-Square Co-efficient	df	p-Value
Bangalore North	13.40	4	0.00*
Bangalore South	16.13	4	0.00*
Bangalore East	2.00	4	0.73[NS]
Anekal	22.85	4	0.00*
Hoskote	6.16	4	0.18[NS]
Devanahalli	9.82	4	0.04
Doddaballapura	15.12	4	0.00*
Nelamangala	1.40	4	0.84[NS]

*: Significant; NS: Not Significant.

From Table 23.2, it is inferred that, since the p-value is less than 0.05, at 5 per cent level of significance it is concluded that there is strong evidence that, the status (presence or absence) of malacofauna in five taluks *viz.,* Bangalore North, Bangalore

South, Anekal, Devanahalli and Doddaballapur, is different between the different species. Whereas, the p-value more than 0.05, at 5 per cent level of significance indicates that, there is difference in the occurrence (presence or absence) of malacofauna in three taluks *viz.,* Bangalore East, Hoskote and Nelamangala, between different species.

Table 23.3: Chi-square Test for Distribution of Individual Mollusc Species between Taluks

Species	Chi-square Co-efficient	df	p-Value
Achatina fulica	37.521	7	.000
Laevicaulis alte	30.494	7	.000
Mariella dussumeri	13.943	7	.052
Cryptozona bistrialis	18.703	7	.009
Macrochlamys indica	24.107	7	.001

From Figure 23.7 and Table 23.3, it is inferred that, since the p-value is less than 0.05, at 5 per cent level of significance it is concluded that there is a strong evidence that, the occurrence of malacofauna between the five taluks *viz.,* Bangalore North, Bangalore South, Anekal, Devanahalli and Doddaballapur, is different for the four species *viz., Achatina fulica, Laevicaulis alte, Cryptozona bistrialis* and *Macrochlamys indica.* Whereas, the p-value more than 0.05, at 5 per cent level of significance indicates that, the occurrence of *Mariella dussumeri* between the five taluks *viz.,* Bangalore North, Bangalore South, Anekal, Devanahalli and Doddaballapur, is not significant.

Co-occurrence of Malacofauna

Co-occurrence of four different species of malacofauna *viz., Macrochlamys indica, Cryptozona bistrialis, Mariella dussumeri* (endemic snails) and *Laevicaulis alte* (exotic slug) in the study region is represented in a Bar graph. During the present survey, *A. fulica–M. dussumeri* combination co-occured in 27 per cent of villages followed by *A.fulica–C. bistrialis* 25 per cent, *A.fulica–L. alte* in 25 per cent and *A. fulica–M. indica* combination in 23 per cent of 114 villages in Bangalore region. Among the nurseries the order of malacofaunal combination was recorded as follows: *A.fulica–C. bistrialis* (83.3 per cent) followed by *A.fulica -M. dussumeri* (79.1 per cent), *A.fulica–M. indica* (75 per cent) and *A.fulica–L. alte* (53.3 per cent).

Local Dispersal

Human aided agency is reported to be the main means of dispersal for *A. fulica.* Collection and dispersal linked to agri-horticultural goods was often given as a mechanism for dispersal by respondents to our Questionnaire-Opinionnaire during interviews. But, its impact as pest seems more recent and still unaware of in many villages. From the opinions gathered it was found out that, 24.44 per cent of the respondents replied that *A. fulica* was recently introduced in the region, 2.63 per cent feel it was 6-10 years, 6.14 per cent feel more than 10 years and 65.79 per cent are not aware of the duration of the introduction of the snail. In the nurseries it was found that, 45.8 per cent opined it was 6-10 years, 41.6 per cent recent to 5 years and 8.3 per cent felt the snail was introduced more than 10 years ago. It was found that, 2.63 per cent of the respondents opine that *A. fulica* probably got introduced along with market goods, 3.51 per cent feel it was by intentional means as pet/harmless/beneficial animal, 3.51 per cent feel other means like migration would have facilitated the introduction and 90.35 per cent are not aware of the mode of introduction. In the nurseries, 66.6 per cent felt it was via market goods, 25 per cent were not aware and 4.1 per cent felt it was by other means. A similar trend was noticed in case of other four malacofaunal pests in the study sites.

The snails were observed in refuse disposal sites, home gardens, pump house doors, around old rough walled buildings and boundary walls. It was found that, 22 per cent of the respondents felt that cool and moist conditions in the field favored the snails and minimum (5 per cent) of opine dense canopy favored the snails while, 28 per cent were not aware of the factors favoring the snails (Figure 23.8). From the survey of nurseries, majority (47.1 per cent) of the respondents felt that cool and moist conditions in the field favored the snails, 17.6 per cent is due to plant detritus, 15.7 per cent crevices, 13.7 per cent walls and 5.9 per cent felt bricks.

With regard to indications for the snail presence, a majority opinion by 34 per cent of the respondents opined that slime trails were easy signs of snail detection in the fields and a minority 20 per cent opined defoliation as signs indicating snail presence, 21 per cent were not aware of any signs as modes of detection (Figure 23.9). In the survey of nurseries from 24 nurseries, 34.3 per cent of the respondents opined that slime trails were easy signs of snail detection in the nurseries indicating snail presence, 34.3 per cent opined Presence of ribbon like faeces as a detection sign and 31.4 per cent opined defoliation as signs indicating snail presence.

It was found that, a majority (33 per cent) of respondents replied that *A. fulica* in the region survived and spread due to better (favorable) rainfall and humidity weather conditions and minimum of 8 per cent felt it was it due to lack of control measures and predators and 30 per cent were not aware of the factors facilitating the survival success of the snail (Figure 23.10). In the survey of nurseries 35.8 per cent of the respondents opined that availability of constant food supply as the major factor favoring the snail survival success in the region, 35.8 per cent opined rainfall and humidity as factors and 28.4 per cent opined lack of control measures and predators as factors favoring the snail survival success.

It was found that, a majority of 49 per cent of the respondents replied that they were unaware of *A. fulica*/not heard off/absent in the village, and minimum of 3 per cent replied the Snails were beneficial as decomposers (Figure 23.11). From the opinions obtained from nurseries, 54.1 per cent of the respondents opined that, they were aware of snail's presence, threat and action taken but yielded poor results. 45.8 per cent of the respondents were aware of snail's presence and threat but have not taken serious action to control the pest population.

Impacts and Control Measures

High densities of GALS population were observed in Hoskote taluk of Bangalore Rural district. Many farmers complained that they could not grow vegetables because they were eaten by *A. fulica* at the seedling stage. The snails were reported to cause

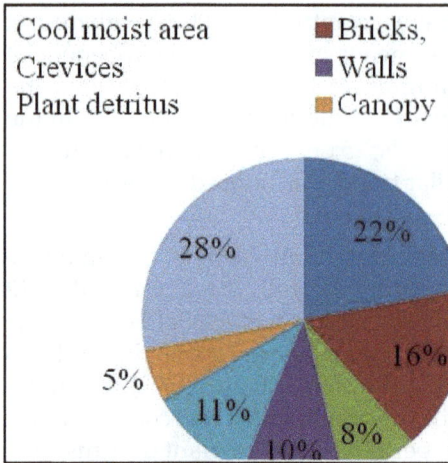

Figure 23.8: Pie-Chart of Opinions of Farmers on the Preferred Hide-Outs of the Snail in the Fields.

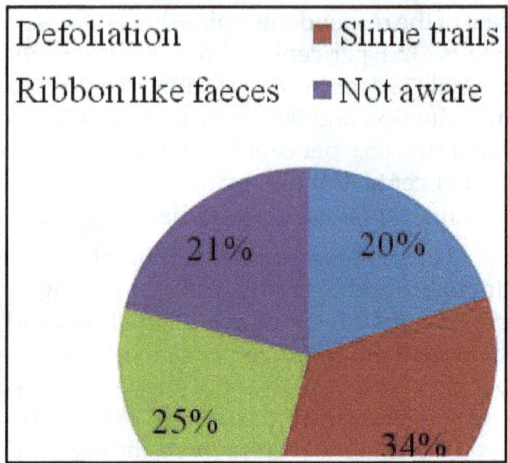

Figure 23.9: Pie-Chart of Opinions of Farmers on the Presence of Snail Signs as Means of Detection.

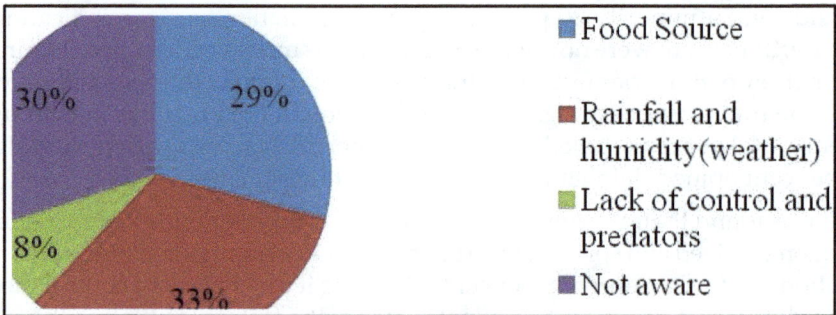

Figure 23.10: Pie-Chart of Opinions of Farmers on Snail Survival Success in the Region.

damage to vegetable crops such as cauliflower, potato, cabbage, pumpkin, cucumber, bottle gourd, white gourd, spinach, radish and tomato. Fruits such as banana, guava, papaya, and jack fruit are all considered to be vulnerable to *A. fulica's* infestation. One of the respondents in Adur village, Bangalore East taluk told that prior to the survey visit almost 1 acre of newly cultivated vegetable crop saplings were completely destroyed within just one night and had decided to quit agriculture. Similar opinions were sounded by farmers across the region. In view of Health Hazard and Nuisance, the snails produce numerous eggs, grow rapidly and are abundant on foot trails. Several respondents complained of the wounds made by the snail's sharp broken shells and the problem of navigating such paths. They leave sticky slime trails and excreta wherever they wander such as on, nursery ornamental plant parts, vegetable leaves and walls of houses; in addition to being unpleasant and unsightly these are potential sources of infection and attract entomofaunal scavengers. One common and widespread practice is to throw snails onto high way roads in the hope that they will be crushed by passing traffic. The density of rotting snails imparts an offensive

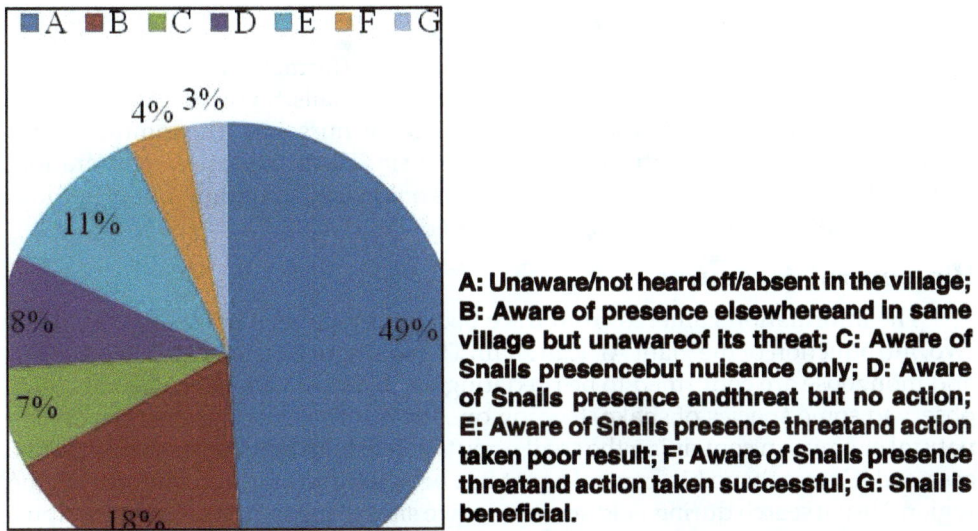

A: Unaware/not heard off/absent in the village; B: Aware of presence elsewhereand in same village but unawareof its threat; C: Aware of Snails presencebut nuisance only; D: Aware of Snails presence andthreat but no action; E: Aware of Snails presence threatand action taken poor result; F: Aware of Snails presence threatand action taken successful; G: Snail is beneficial.

Figure 23.11: Pie-Chart of Opinions of Farmers on Snail Presence Impacts and Control Initiatives.

odour in the stretch of the road as reported by passerby and villagers of the vicinity. None of the respondents were aware of the vector status of the snail and its potential as food source to human and domestic animals when reared under hygienic conditions.

Mechanical Control Measures

During the present survey it was found that 21 per cent (farmers) and 54.1 per cent (horticulturists) have adopted mechanical control measures. *A.fulica* has thrived and poses threat to crops even by routine monitoring, eradication of weeds and disposal of garbage/garden waste, on a regular base being practiced. They tend to climb on trees and stone pillars erected to support creeper crops concealing their presence and deny access to eradicate them. As the snail is regarded sacred and referred as Basavana hula in Kannada–The holy bullock incarnation, there is sense of stigma in eradicating it, more so some farmers believe it to be harmless and seem to facilitate crop growth. It was also noticed that snails were hand collected and burnt or buried in soil as such. Eradication operations like this would favor snail escape and thereby facilitate its spread, hence farmers were asked to refrain from such practices. Polythene covers used in nurseries has been successful in minimizing the snail populations build up. Laborers are paid for eradication work to hand collect and eliminate the snails as observed in the study area. A common method practiced, that was observed during the survey was to pack snails in gunny bags or plastic bags and throw them onto roads to be crushed by passing traffic. People living nearby irrigation canals or huge sewages throw their daily collections directly into flowing water. The locals were suggested to refrain from such practices as it would facilitate further spread.

Chemical Control Measures

Based on the respondents reply, 11 per cent (farmers) and 45.8 per cent (horticulturists) use chemicals specifically to control snails. Snail-kill (metaldehyde – active ingredient) is used as snail bait in most of the nurseries. The common salt is largely used and delivers satisfactory results causing death of snail by dehydration. Hand collected snails are dumped in large pits onto which large amount of Sodium Chloride is added for mass killing of the snails.

Bio-control Measures

The information gathered from trace back investigations point at a decade or two, introduction of the giant African snail. The southern Coucal, *Centropus sinensis* and mongoose are confirmed to be predating on *A. fulica* in addition; farmers have observed some species of snake, feeding on *A.fulica*. The extent of the role of such natural enemies in containing the snail population in high density areas needs to be assessed. Currently no bio-control agent has been used to eradicate the snail in the region. Visual search during field surveys in two sites (Bangalore University campus, Hebbal Forest Department nursery) both *A. fulica* and flatworms *Diversibipalium* sp. 1 and *Diversibipalium* sp. 2 were recorded and regarded as areas of coexistence at the time of the search. Around 68 per cent of the farmers interviewed have not initiated specific control measures.

Discussion

To identify pests associated with a host or a neighboring area, a pest list needs to be developed. Hence, the main purpose of pest surveillance is to generate information about the presence or absence of pests in a way that is internationally acceptable as reliable and sound. Such pest lists serve as an inventory of pests in an area, the list of pestiferous malacofauna is prepared for the present survey. The survey also revealed information on serious or emerging pest problems where there are currently few available and affordable control options and additional research is needed. A priority for survey managers will be to determine the likelihood that the pest may have been dispersed with commodities from planting stock from commercial nurseries. Also, such surveys would be useful in setting the research agenda, testing research hypotheses, designing extension strategies, evaluating the effectiveness of projects and development interventions.

Despite different management measures it has been observed that naturalized populations of *A. fulica* often eventually decline greatly following establishment. Such a decline is often preceded by an exponential increase phase and a stable phase (Mead, 1979, Raut and Barker, 2002). Simberloff and Gibbons (2004) suggest a do-nothing approach to management as a potential rationale for species in which spontaneous collapse has been repeatedly observed. In conclusion an integrated approach is emphasized by various workers to manage the outbreak of the snail menace in different agro-ecosystems of the country. Tackling the menace of *A.fulica* warrants community approaches as in case of other pests like rats, for its effective minimization or eradication.

Acknowledgements

The author is grateful to Dr.M.S.Reddy (Bangalore University) and Dr.Abraham Verghese (NBAII, Bangalore) for their valuable guidance and suggestions. The author is also thankful to Dr. Balakrishna, Dr.Francis and Mr.Rajashekar (KSRSAC) for extending laboratory space and facility for plotting the maps required.

References

Asamoah, S. A., 1999. Ecology and Status of the giant African snails in the Bia Biosphere Reserve in Ghana. Final Report (September, 1999): Ecological Studies on the Giant African Snails, Submitted to: The division of Ecological sciences, UNESCO (MAB) Young Scientist Research Award Scheme, *Paris Cedex 15 France.* 1-31.

Barker, G. M., 2001. Gastropods on Land: Phylogeny, Diversity and Adaptive Morphology. *In*: Barker, G.M. (ed.), The Biology of Terrestrial Molluscs. *CABI Publishing, U.K,* 1-146.

Blandford, W.T., and Godwin-Austen. H.H., 1908. The fauna of British India. Mollusca. Testacellidae and Zonitidae. London: *Taylor and Francis.*

CABI, 2014. Invasive Species Compendium. Wallingford, UK: CAB International. www.cabi.org/isc

Fontanilla, I.K.C., 2010. *Achatina (Lissachatina) fulica* Bowdich: its molecular phylogeny, genetic variation in global populations, and its possible role in the spread of the rat lungworm *Angiostrongylus cantonensis* (chen). Ph.D. Thesis, University of Nottingham.

Fontanilla, I.K.C., Hudelot, C. Naggs, F., and Wade, C.M., 2007. *Achatina fulica*: its molecular phylogeny and genetic variation in global populations. *Abstracts. World Congress of Malacology,* Antwerp, Belgium. 63.

Gude, G.K., 1914. The Fauna of British India. Mollusca II. Trochomorpidae and Janellidae. Taylor and Francis, London.

Gude, G.K., 1921. The fauna of British India. Mollusca III. Land operculates. Taylor and Francis, London.

Kaur, H., 2003. Life cycle of the slug, *Filicaulis (Eleutherocaulis) alte ferussac* (Mollusca: Veronicellidae) infesting plant nurseries and vegetable fields in Punjab. *Geobios,* **30 (4)**:225-228.

Kliks, M.M., and Palumbo, N.E., 1992. Eosinophilic meningitis beyond the Pacific Basin: the global dispersal of a peri domestic zoonosis caused by *Angiostrongylus cantonensis,* the nematode lungworm of rats. *Social Science and Medicine,* **34 (2)**: 199-212.

Leslie, A. R., Mark, F. D., and Robert O. H. JR., *2008.* Invasive species impact: asymmetric interactions between invasive and endemic freshwater snails. *J. North American Benthological Society,* **27(3)**: 509-520.

Lowe, S., Browne, S.M., Boudjrlas, and De Poorter, M., 2000. 100 of the world's worst invasive alien species: A selection from the global invasive species database.

The Invasive Species Specialists Group of the Species Survival Commission of the World Conservation Union. 12 pp. First published in Aliens 12, December 2000. Reprinted November 2004. Hollands Printing, Auckland, New Zealand.

Mead, A.R., 1961. The Giant African Snail. *The University of Chicago Press*, 257.

Mead, A.R., 1979. Economic malacology with particular reference to *Achatina fulica*. *In: Fretter, V., and Peake, J.* (eds.) *Pulmonates,* **2B.** *Academic Press, London,* 150 pp.

Naggs, F.,1997. William Benson and the early study of land snails in British India and Ceylon. *Archives of Natural History,* **24**: 37-88.

Penman, D.R., 1998. Managing a leaky border: towards a biosecurity research strategy. Wellington, New Zealand, Ministry of Research, Science and Technology, 61p.

Ramli Mohamed and Khor Yoke Lim., 1988. A Report of the Focus Group Interview on the Strategic Extension Campaign on Integrated Weed Management in the Muda Irrigation Scheme, Malaysia. Consultants' Report submitted to the Muda Agricultural Development Authority and FAO.

Raut, S.K., and Barker, G.M., 2002. *Achatina fulica* Bowdich and other Achatinidae as Pests in Tropical Agricultural. In: Barker G.M (eds.), *Mollusc as Crop pests.* CABI Publishing, Wallingford: 55-114.

Raut, S.K., and Ghose, K.C., 1984. Pestiferous Land Snails of India, *Z.S.I. Tech. Monog.* **11**: 151. Zoological Survey of India, Calcutta.

Schileyko, A.A., 1999. Treatise on Recent Terrestrial Pulmonate Molluscs, Part 4: Draparnaudiidae, Caryodidae, Macrocyclidae, Acavidae, Clavatoridae, Dorcasiidae, Sculptariidae, Corillidae, Plectopylidae, Megalobulimidae, Strophocheilidae, Cerionidae, Achatinidae, Subulinidae, Glessulidae, Micractaeonidae, Ferussaciidae. *Ruthenica,* Moscow, 129 pp.

Silvana, C.T., Fabio, A.F., Norma, C.S., Cowie, R.H., and Monica, A F., 2007. Rapid spread of an invasive snail in South America: The giant African snail, *Achatina fulica*, in Brazil. *Biol. Invasions,* **9**: 693-702.

Simberloff, D., and Gibbons, L., 20 4. Now you See them, Now you don't!–Population Crashes of Established Introduced Species. *Journal Biological Invasions,* **6**(2): 161-172.

Srivastava, P. D., 1992. Problem of land snail pests in Agriculture: A study of the giant African snail. Concept publishing company, New Delhi, P.234.

USDA–APHIS., 2005. New Pest Response Guidelines. Giant African Snails: Snail Pests in the Family Achatinidae. USDA – APHIS – PPQ – Emergency and Domestic Programs – Emergency Planning, Riverdale, Maryland. *http:// www.aphis.usda.gov/import_export/plants/manuals/index.shtml.*

Veeresh, G. K., Rajagopal, D., and Puttarudraiah, M., 1979. First record of African giant snail, *Achatina fulica* (Bowdich) (Mollusca: Gastropoda) as a serious pest of ornamental crops in Bangalore. *Current Research,* **8**: 202-204.

2015, Animal Diversity, Natural History and Conservation, Vol. 5   Pages *399–406*
Editors: V.K. Gupta and Anil K. Verma
Published by: DAYA PUBLISHING HOUSE, NEW DELHI

Chapter 24

A Comprehensive Review on the Eco-biology of the Endemic Fruit Fly *Phorticella striata* Sajjan and Krishnamurthy, 1975

D.S. Pranesh Sujaymeendra[1], S.N. Hegde[2],
B.P. Harini[3] and M. Jayashankar[4]*

[1] No. 307, Mars Enclave, Vidyaranyapura, Bengaluru – 560 097, Karnataka, India
[2] Department of Molecular Biology,
Yuvaraja College, Mysore – 570 005, Karnataka, India
[3] Department of Zoology, Bangalore University,
Bengaluru – 560 056, Karnataka, India
[4] Department of Entomology and Nematology, IIHR,
Bengaluru – 560 089, Karnataka, India

ABSTRACT

Recently there is an increase in scientific studies pertaining to the drosophilid *Phorticella striata* Sajjan and Krishnamurthy, 1975 from different parts of India, particularly from the southern states. *P. striata* is described from Karnataka, India. Realizing this emerging significance of the species the present review is presented incorporating all related aspects concerning the eco-biology of the species. However, further studies in future are needed to analyze the influence of ecosystem properties and species interactions to validate the existing range and potential spread of *P. striata* in India.

Keywords: *Phorticella striata, Karnataka, Endemic, Distribution, Eco-biology, Drosophillid.*

---

* *Corresponding author.* E-mail: jay81zoology@gmail.com

Introduction

The family Drosophilidae is a schizophoran dipteran group under the superfamily Ephydroidea. The family is subdivided into two subfamilies Steganinae and Drosophilinae. The Steganinae is a small, primitive and poorly understood subfamily while the Drosophilinae is one of the most studied and described subfamily, both by morphological characters as well as by molecular studies. The family Drosophilidae is known by 320 species under 30 genera from India, with 57 species under 8 genera of subfamily Steganinae and 263 species under 22 genera of subfamily Drosophilinae (zsi.gov.in/checklist/Drosophilidae.pdf). Some species are endemic to certain regions and few are cosmopolitan.

Drosophilids are considered as one of the best experimental model for both genetic and taxonomical studies involving theoretical issues or of the general relevance of the experimental. They have a long and distinguished history ever since T H Morgan in 1909 chose them as experimental models (Carpenter, 1905; Castle, 1906; Castle *et al.,* 1906, Lutz, 1911; Bizzo and Sene, 1982). *Drosophilids* are excellent as a model in the field of genetics, evolution and ecology. These insects are highly sensitive to environmental variations which lead to fluctuation in their population. They have provided important information in understanding their evolutionary process. They serve as excellent environmental indicators due to their cosmopolitan distribution, the sensitivity of the flies to environmental variables and the simplicity in their field collections (Parsons, 1991; Francisco Rodríguez-Trelles *et al.,* 1998; Foote and Carson, 2004).

Recent records show the evolve of the drosophilid *Phorticella striata* in different parts of India. *P. striata* is described from Karnataka, India. The species was named for the characteristic striations on the head, mesonetum and scutellum (*P. striata* flies are adorn with two prominent chalky white stripes, well marked by a dark hue on either side of the head and thorax and another pair arranged laterally on the thorax) (Figures 24.1 and 24.2). It has characteristic bigger size than most of the *Drosophila* species. Like *Drosophila* this species also has all the characteristics of a good laboratory tool to analyze genetic and evolutionary problems (Yenisetti and Hegde, 2003). Realizing this emerging significance of the species the present review is presented incorporating all related aspects concerning the eco-biology of the species.

Taxonomic Entitlement of the "Genus Phorticella"

Genus *Phorticella* and *Zaprionus* are closer with their characters and hence are difficult to differentiate morphologically. Both possess silvery stripes on their mesonota. Genus *Phorticella* consists of two subgenera, *Phorticella* with seven species, occurring in oriental and Australian regions and *Xenophorticella* with four known species of which three are reported from oriental region and fourth from Madagascar (Ashburner *et al.,* 2005). Although the binomial name, *Phorticella striata* Sajjan and Krishnamurthy, 1975 is widely used in scientific communications, reported synonyms include *viz., Drosophila chamundiensis* Sajjan and Krishnamurthy, 1972, *Drosophila bicolovittata* Singh, 1974, *Phorticella carinata* Takada, 1981 and *Phorticella flavipennis* (Duda, 1929) reported as the accepted name (http://www.diptera.org, http://www.catalogueoflife.org/annual-checklist/2013/details/database/id/101). The

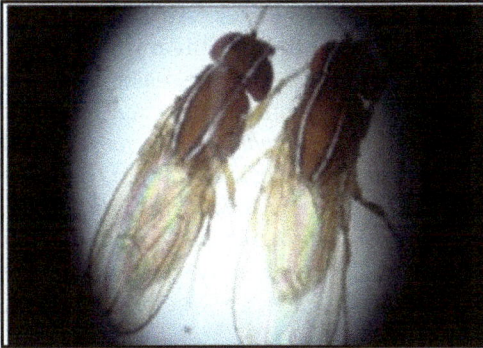

Figure 24.1: *Phorticella striata*
(Male and Female).

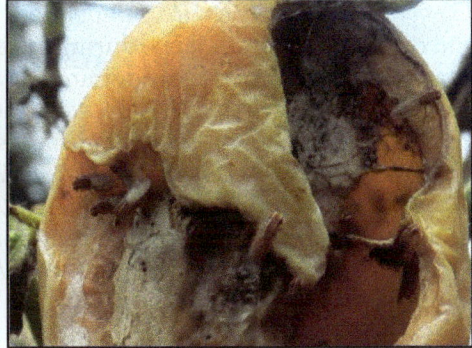

Figure 24.2: *P. striata* Congregating on
Rotting Tomato.

genera are reported by the scientific name, *P. flavipenis* from Western Ghats (http://bioinfo.lowtem.hokudai.ac.jp/).

Distribution

Different workers have employed both bottle trapping and net sweeping methods for fly collection. *P. striata* is a drosophillid insect discovered from Karnataka, India belonging to group Drosophillidae (Sajjan and Krishnamurthy, 1975). Ever since, scientific literature on its distribution is available from different parts of the country (Table 24.1).

Table 24.1: Distribution Records of *P. striata* from different Parts of India

Sl.No.	State	Locations	References
1.	Karnataka	Brindavan Gardens, Mysore	Yenisetti and Hegde, (2003)
		M.M. Hills, Chamarajanagara	Pranesh and Harini, (2012)
		Nandi Hills, Chikkaballapura	Pranesh (2013)
		Dharwad	Srinath and Shivanna, (2014)
		Bannerghatta National Park, Bengaluru Urban Distict	Dilip *et al.* (2014)
		Prominent market places in Bangalore North and South Taluks of Bengaluru Urban District.	Pranesh *et al.* (2014)
2.	Tamil Nadu	Palini Hills	Hegde *et al.* (1998)
3.	Gujarat	Ahmedabad	Yenisetti and Hegde, (2002)

Host Range and Seasonality

Hegde *et al.* (1998) recorded *P. striata* up to an altitude of 1450 m in Palani Hills of Tamil Nadu. The prominent vegetation in the moutain peaks included huge eucalyptus trees, *Acanthospermum hispidium, Grewia hirsuta, Hibiscus* species,

Euphorbia species. *P. striata* is dependent on mango fruiting season during summer and decreases its abundance during other seasons (Srinath and Shivanna, 2014). *P. striata* were found to be more dominating than other species at all the seasons at Male Madeshwara Hills in South Karnataka, the prominent flora included *Acacia concinna, Zizipus jujube, Vitex negundo, Phyllanthus* species, *Andrographis serpellifolia, Bridalia* species, *Tectona grandis, Tamarindus indica, Vitex negundo* besides mango orchards (Pranesh and Harini, 2012). Pranesh (2013) recorded *P. striata* in Nandi hills, Chikkaballapura district. The hill range had *Mangifera indica, Anacardium occidentale, Vitis ripari, Ficus religosa, Ficus bengalensis, Pongamia glabra*, and many shrubs including cactus as prominent vegetation.

Reproductive Success

Laboratory studies on body size and life history traits such as fecundity, dispersal ability and mating success in *Drosophila* have demonstrated strong linkage between them. Quick stationary displays are made by the *P. striata* male towards the rear of the female, at 90°, 135°, or 180° to her long axis. He then positions himself close to and facing the tip of the female's abdomen and the fluttering male grapples the female. If the female is non-receptive, she then decamps. In spite of this extent of rejection, the male usually chase and tries to grapple; then the female rocks to avoid the male. The female also depresses her abdomen. In many cases, though, the male mounts the female; intromission does not follow as the female is not receptive. In such a situation the female rocks terribly to dismount him. Males of this species attempt to mount and copulate with non-receptive females, but they invariably are unsuccessful. If she is receptive, the moment the male grapples she spreads her wings to accommodate the male. He then lunges onto her and copulates. The moment the male mounts, the female vibrates both the wings for a few seconds continuously, and then stays still, whereas the male vibrates his wings while grappling and stops soon after that. Later the male vibrates the wings intermittently for 1-2 seconds during the process of copulation. Copulating females may walk around, whereas the male stays still. While copulating he keeps his front pair of legs below her wings and holds her abdomen with his middle tarsi. At the end of copulation, the male dismounts the female and both involve in cleaning (Yenisetti *et al.,* 2001).

Yenisetti and Hegde (2002) demonstrated male and female remating in two mass culture stocks raised from flies collected from Mysore and Ahmadabad. Study revealed that in male *Phorticella* variation in remating frequency, mean matings between two populations was insignificant. Variation in mating durations for different matings in both the strains was also insignificant. However, significant interstrain mating duration variations for the first four matings were observed. Variation in female remating latency between the populations was found statistically insignificant. Insignificant variation was observed for the duration of copulation between the first and second matings in Mysore population. However, remating duration was significantly shorter in Ahmadabad population. Observed inter-strain variations can be attributed to geographic isolation which may lead to 'response variation' of the genetic system of a species to different physical environments. (Yenisetti and Hegde, 2003) proved that size-assortative mating exists in *P. striata* with the help of male and female choice experiments. In female preferential mating (female choice), males with long wings were observed to be successful in mating with both large and small females, indicating the success of large over small males in male rivalry, which may also be due to selection by females. In male preferential mating (male choice) both large and small males preferred mating with large females. The authors substantiate their findings as follows, it is evident that large males have higher reproductive success as they being highly vigorous can move faster, encounter females more quickly, and mate faster. Large females also have higher reproductive success, as they have more ovarioles and are able to produce a greater number of fertile eggs when mated with large males. Yenisetti *et al.* (2008) showed a direct relationship between wing length and mating success. Males mating in the first fifteen minutes (0-15 min) time interval had significantly longer wings. In females also there was significant difference in wing length that mate in different time intervals. Females mating in the first thirty minutes (0-30 min) bear significantly longer wings when compared with those that mate during 31-60 min. This showed large flies mate early in both the sexes and that sexual selection acting on body size.

Latha and Krishna (2014a) studied age of male effect on male remating and progeny production in *P. striata* has been premeditated. It was noticed that middle aged male had greater percentage of remating than those of young and old males. In all the three male age classes male mated with two females in 1 hour had copulated significantly for longer time, mated females laid greater number of eggs and more progeny than male mated with one female in 1 hour. Further, among male mated with two females in 1 hour, male mated with 1st female copulated longer, mated female laid greater number of eggs and progeny production than male mated with 2nd female in 1 hour. Latha and Krishna (2014b) noticed that in *P. striata* females of all age preferred older males more frequently than young or middle aged males. Copulation duration ranges from 01 min to 04 min. Older males showed significantly greater courtship acts, mated faster and copulated longer than young or middle aged males. Females mated with old males laid significantly greater number of eggs and progeny than females mated with either young or middle aged males. Further, female mated with older males lived significantly for a shorter time than when she mated with either young or middle aged males. Thus these studies confirmed that the females evolved towards their mating with older males than younger. In *Drosophila* this has an advantage because mating is resource independent in both the parents and there is no parental care.

Discussion

There are a number of factors that may influence the species richness of a community *viz.*, geographical, *e.g.*, latitude and longitude; environmental as an instrument with a greater variety of niches would be able to host a greater variety of species with climatic variability, age, and rigidity of the environment; and biological relationships of predation, competition, and population density etc. These factors may have important consequences on the number of species and its distribution, and the changes in natural environment produced by alteration of seasons would result in the change in relative frequency of different species. As a consequence it is at the level of morphological expression and function that developmental changes have an effect on organismal fitness, and therefore this level is critical for establishing the link between developmental and evolutionary processes.

Eco-distributional analyses based on available literature indicate the expansion of *P. striata* distribution range. The aforesaid factors indicate that reproductive success could have been a leading attribute enabling *P. striata* to extend its distribution range ever since its description six decades ago. However, most of the available literature is from Southern India although drosophilid diversity is vivid in the northern and eastern parts of the country. Differences in the prevailing climatic conditions may perhaps be responsible for this variation (Rajpurohit, 2013), such variations in species distribution is expected to change with the changing climatic conditions (Rajpurohit, 2008; Singh, 2012). Further studies are needed to analyze the influence of ecosystem properties and species interactions to validate the existing range and potential spread of *P. striata* in India.

References

Ashburner, M., Golic, K.G., and Scott, R.H., 2005. *Drosophila* A Laboratory Manual, Second Edition. Cold Spring Harbor Laboratory Press.

Bizzo, N., and Sene, F., 1982. Studies on the natural populations of *Drosophila* from Perulibe (SP) Brazil (Diptera: Drosophilidae). *Revista Brasileira de Biologia*, **42**:539–544.

Carpenter, F.W., 1905. The reactions of the pomace fly (*Drosophila ampelophila* Loew) to light, gravity and mechanical stimulation. *The American Naturalist,* **39**: 157-171.

Castle, W.E., 1906. Inbreeding, crossbreeding and sterility in *Drosophila. Science,* **23**: 153.

Castle, W.E., Carpenter, A.H., Clark, Mast, S.O., and Barrows, W.M., 1906. The effects of inbreeding, crossbreeding and selection upon the fertility and variability of *Drosophila. Proceedings of the American Academy of Arts and Sciences,* **41**: 731-786.

Dilip, A.S., Pranesh, D.S.S., Alexander, R., Avinash, K., Phalke, S., and Jayashankar, M., 2014. First report of Drosophilid diversity in an ecotone adjoining Bannerghatta National Park (Karnataka, India). *Drosophila Information Service,* **97** (Accepted)

Foote, D., and Carson, H.L., 2004. *Drosophila* as monitor of change in Hawaiian ecosystems. p. 368–372. In: Our Living Resources: A Report to the Nation on the Distribution, Abundance, and Health of U.S. Plants, Animals, and Ecosystems. Senior Science Editor and Project Director: Edward T. LaRoe. U.S. Department of Interior – National Biological Service, Washington, DC, 1995.

Francisco Rodríguez-Trelles., Miguel, A. R., and Samuel, M. S., 1998. Tracking the genetic effects of global warming: *Drosophila* and other model systems. *Conservation Ecology,* [online]. **2**(2): 2. URL: http://www.consecol.org/vol2/iss2/art2/

Hegde, S.N., Vasudev, V., Shakunthala, V., and Krishna, M.S., 1998. Drosophila fauna of Palni Hills: Tamilnadu, India. *Drosophila Information Service,* **81**:138-139.

Latha, M., and Krishna, M. S., 2014a. Age based female mate preference in *Phorticella striata. International Journal of Current Research,* **6** (1): 4705-4713.

Latha, M., and Krishna, M. S., 2014b. Male age effect on male remating and fitness in *Phorticella Striata. American Journal of Bioscience and Bioengineering,* **2**(1): 8-14. doi: 10.11648/j.bio.20140201.12

Lutz, F.E., 1911. Experiments with *Drosophila* ampelophila concerning evolution. *Carnegie Institution of Washington publication,* **143**: 1-40.

Mateus, R.P., Buschini, M.L.T., and Sene, F.M., 2006. The Drosophila community in xerophytic vegetations of the upper Parana- Paraguay River Basinet. *Brazilian Journal of Biology,* **66**(2B):719-729

Parsons, P.A., 1991. Biodiversity conservation under global climatic change: the insect *Drosophila* as a biological indicator? *Global Ecology and Biogeography,* **1**(3):77–83.

Pranesh, D.S.S., and Harini, B.P., 2012. Drosophilids of Male Mahadeshwar hills of Chamarajanagar District Karnataka State, India. *Drosophila Information Service,* **95**:62-65.

Pranesh, D.S.S., Arpana, H., Chaithra, R., Dilip, A.V., Rohini, G., and Jayashankar, M., 2014. A rapid diversity survey of drosophilids in selected market places in Bengaluru north and south taluks of Bengaluru urban district. National Seminar on Management of Uran Biodiversity: Issues, Challenges and Solutions, organized by Christ University during 1-2 September 2014.24pp

Pranesh, D.S.S., 2013. Ph.D Thesis, Study on *Drosophila* Biodiversity in Certain ecological Niches of South Karnataka (Unpubl.).

Rajpurohit, S., Parkash, R., and Ramniwas, S., 2008. Climatic changes and shifting species boundaries of Drosophilids in the Western Himalaya. *Acta Entomologica Sinica,* **51** (3): 328-35.

Rajpurohit, S., Nedved, O., and Gibbs, A.G., 2013. Meta-analysis of geographical clines in desiccation tolerance of Indian drosophilids. *Comparative Biochemistry and Physiology: Part A, Molecular and Integrative Physiology,* **164**(2):391-8. doi: 10.1016/j.cbpa.2012.11.013. Epub 2012 Nov 24.

Sajjan, S. N., and Krishnamurthy, N. B., 1975. Two new drosophilids from South India (Diptera: Drosophilidae). *Oriental Insects,* **9**:117-119.

Singh, S., 2012. Phenotypic plasticity mediates range retraction of *Drosophila nepalensis* in the Western Himalayas: Impact of global warming on stress resistance and adaptations. *SOAJ Entomological Studies,* **1**: 100-118

Srinath, B. S., and Shivanna, N., 2014. Seasonal variation in natural populations of *Drosophila* in Dharwad. *India Journal of Entomology and Zoology Studies,* **2** (4): 35-41

Yenisetti, S. C., Hegde, S.N., Krishna, M.S., and Venkateshwarlu, M., 2001. Courtship behaviour of *Phorticella striata* (Drosophilidae). *Drosophila Information Service*, **84**: 5-6.

Yenisetti, S. C., and Hegde, S. N., 2002. Remating in a Drosophilid: *Phorticella striata*. *Korean Journal of Genetics,* **24** (2): 113-118.

Yenisetti, S. C., and Hegde, S. N., 2003. Size-related mating and reproductive success in a drosophilid: *Phorticella striata. Zoological Studies*, **42**:203–210.

Yenisetti, S.C., Hegde, S.N., and Venkateshwarlu, M., 2008. Relationship between Mating Success and Wing Length in a Drosophilid: *Phorticella striata. Korean Journal of Genetics,* **30** (1): 1-9.

Previous Volumes

— Volume 1 —

2011, xv+480p., col. plts., figs., tabls., ind., 25 cm   Rs. 2200

ISBN 978-81-7035-752-0

— Volume 2 —

2013, xiii+490p., col. plts., figs., tabls., ind., 25 cm   Rs. 3500

ISBN 978-81-7035-831-2

— Volume 3 —

2013, xiii+460p., col. plts., figs., tabls., ind., 25 cm   Rs. 3500

ISBN 978-81-7035-830-5

— Volume 4 —

2015, xiii+441p., figs., tabls., ind., 25 cm   Rs. 2495

ISBN 978-93-5124-615-2

Index

www.ingramcontent.com/pod-product-compliance
Lightning Source LLC
Chambersburg PA
CBHW050506190326
41458CB00005B/1452